T0235866

Essential Oils

Mozaniel Santana de Oliveira

Editor

Essential Oils

Applications and Trends in Food Science
and Technology

 Springer

Editor
Mozaniel Santana de Oliveira
Museu Paraense Emílio Goeldi
Belém, Brazil

ISBN 978-3-030-99478-5   ISBN 978-3-030-99476-1   (eBook)
https://doi.org/10.1007/978-3-030-99476-1

This Springer imprint is published by the registered company Springer Nature Switzerland AG
The registered company address is: Gewerbestrasse 11, 6330 Cham, Switzerland

The researcher must be motivated by curiosity and doubts about the world and the phenomena of nature, in addition, they must dream and seek a better world, with peace, fraternity, and fairness for all.
The editor Prof. Dr. Mozaniel Santana de Oliveira, dedicate this academic and scientific book, to the people who were victims of this terrible Covid-19 pandemic. Also, I dedicate this book to my family, brothers, friends, my parents Mrs. Maria de Oliveira and Mr. Manoel de Oliveira, and my lovely wife Joyce Fontes.

Foreword

Over the last three decades, efforts have been made to investigate bioactive molecules produced by plants and microorganisms. Thus, several groups and new laboratories have been established in different parts of the world. Brazil has followed this trend, especially in universities and research centers throughout the country, where innovative projects have resulted in the development of theses and doctoral dissertations, proving that these studies, especially on plants, have the potential to be applied in various human activities, such as hygiene, cleaning, and medicine.

Over the recent years, efforts have been directed to utilize these chemical compounds in other activities, such as agriculture, food preservation, and as functional foods. Chemically diverse compounds, which can be apolar, low-polar, and polar substances, are often produced by plants. Among the apolar and low-polar compounds are the terpenoids. Many of them are components of essential oils, which have great potential for use in human nutrition and help maintain and improve the quality of life.

This work, which is entitled **Essential Oils Applications and Trends in Food Science and Technology,** summarizes in a didactic and thematic way much of the knowledge that has recently been gained on the importance of essential oils in their various applications, providing a clear and in-depth journey on the most important topics in this scientific field. At the same time, the quality of the topics covered and the professional contributions guarantee this work the relevance necessary to support the different segments of related fields in undergraduate and postgraduate studies.

Belém, Brazil

Antônio Pedro da Silva Souza Filho
Embrapa Amazônia Oriental

Acknowledgments

First of all, the editor would like to unreservedly thank all the coauthors who generously devoted part of their most priceless asset, time, to contribute to this project that has now become a reality. I am also very grateful to the research institutions that are linked to this book.

I would also like to acknowledge all the support provided by the Springer Nature team and thank the section heads and their colleagues on the editorial board. Special thanks to Daniel Falatko, Editor for Food Science, who believed in the potential of this project. I thank Prof Dr Eloisa Helena de Aguiar Andrade, for her supervision, and friendship at the Adolpho Ducke Laboratory - Museu Paraense Emílio Goeldi. I thank Prof Dr Antônio Pedro da Silva Souza Filho, and Prof Dr Ely Simone Cajueiro Gurgel, Embrapa Amazônia Oriental, and Museu Paraense Emílio Goeldi, respectively.

I thank *Museu Paraense Emílio Goeldi* for the opportunity to do science in the Amazon, which has been contributing to the regional scientific growth since 1866.

I thank *Programa de Capacitação Institucional* (PCI), linked and supervised by the *Ministério da Ciência, Tecnologia e Inovações* – MCTI, for the scholarship (process number 300983/2022-0).

About the Book

Plants are a primary source for the search and extraction of innovative chemical molecules that can meet the current demands of society in terms of their various uses. Today's values require organic-based products that are free from pollutants and chemical residues that endanger human health and the environment. In this context, plants represent promising sources of compounds that can replace those currently available on the market and allow obtaining products with a unique profile.

Plants produce countless chemically diverse substances with different properties and broad applications. Apolar, low-polar, and high-polar chemical molecules, such as polyphenols, are widely used. This wide range of possibilities also includes essential oils with a diverse composition of monoterpenes, diterpenes, triterpenes, and hydrocarbons. Certain plant species have a different profile based on sulfur compounds such as monosulfides, disulfides, and trisulfides.

Research in recent decades has focused on ways to harness the properties of essential oils for specific human activities such as cleaning, hygiene, nutrition, and, more recently, agricultural defense, with an emphasis on controlling pests and diseases that affect food production.

Given the current state of knowledge, it is extremely important to compile and publish a book that clearly focuses on the application of essential oils in a wide variety of fields and, more recently, in food science and technology.

In compiling this book, the main focus has been to develop an up-to-date reference work for its readers, based on recent scientific advances. It is well known that food science is constantly changing and advancing in the quest for a better quality of life for humans and other animals. We also realize that scientific knowledge renews itself over time and that advances are necessary for us to move forward in life.

Being aware that countries with a great biodiversity can provide new sources of chemically active molecules, which are reported mainly in the context of essential oil composition, we have tried to attract authors from different countries to write this work. Thus, 77 authors were invited and 18 chapters were written. To balance the diversity of scientific information, the book was divided into six parts: Part I – Essential Oils: General Concepts; Part II – Essential Oils: Food System Applications;

Part III – Essential Oils: Agricultural System Applications; Part IV – Essential Oils: Food Antiparasitics; Part V – Essential Oils: Food Applications in Degenerative Diseases; and Part VI – Essential Oils: In Silico Study.

Last but not least, we believe that this book will help and encourage many researchers, students, and teachers around the world to devote themselves to the study of the scientific topics covered in this work.

Belém, Brazil   Antônio Pedro da Silva Souza Filho
                                                                        Embrapa Amazônia Oriental

Contents

Contributors

Ana Lúcia Abreu-Silva Pós-graduação em Ciência Animal, Universidade Estadual do Maranhão, São Luís, MA, Brazil

Ahmed A. Almarie College of Agriculture – University of Anbar, Ramadi, Iraq

José Weverton Almeida-Bezerra Regional University of Cariri – URCA, Crato, CE, Brazil

Fernando Almeida-Souza Pós-graduação em Ciência Animal, Universidade Estadual do Maranhão, São Luís, MA, Brazil

Laboratório de Imunomodulação e Protozoologia, Instituto Oswaldo Cruz, Fiocruz, Rio de Janeiro, Brazil

November Rianto Aminu Chemistry Department Faculty of Science and Mathematics, Universitas Kristen Satya Wacana, Salatiga, Indonesia

Eloisa Helenade de Aguiar Andrade Museu Paraense Emílio Goeldi, Brazil

Coordenação de Botânica, Museu Paraense Emílio Goeldi, Belém, Pará, Brazil

Adolpho Ducke Laboratory, Botany Coordination, Museu Paraense Emílio Goeldi, Belém, Pará, Brazil

Sushila Arya Department of Chemistry, College of Basic Sciences and Humanities, G.B. Pant University of Agriculture and Technology, Pantnagar, Uttarakhand, India

Fernanda Wariss Figueiredo Bezerra Federal University of Pará, Belém, Pará, Brazil

José Jailson Lima Bezerra Federal University of Pernambuco – UFPE, Recife, PE, Brazil

Kátia S. Calabrese Laboratório de Imunomodulação e Protozoologia, Instituto Oswaldo Cruz, Fiocruz, Rio de Janeiro, Brazil

Renan Campos e Silva Programa de Pós-graduação em Química, Universidade Federal do Pará, Belém, Para, Brazil

Márcia Moraes Cascaes Universidade Federal do Pará, Programa de Pós-Graduação em Química, Belém, Pará, Brazil

Kauê Santana da Costa Instituto de Biodiversidade, Universidade Federal do Oeste do Pará, Santarém, Pará, Brazil

Jorddy Neves Cruz Museu Paraense Emílio Goeldi, Belém, Pará, Brazil

Adolpho Ducke Laboratory, Botany Coordination, Museu Paraense Emílio Goeldi, Belém, Pará, Brazil

Rafael Pereira da Cruz Regional University of Cariri – URCA, Crato, CE, Brazil

Jorddy Neves Cruz Museu Paraense Emílio Goeldi, Belém, Pará, Brazil

Adolpho Ducke Laboratory, Botany Coordination, Museu Paraense Emílio Goeldi, Belém, Pará, Brazil

Clêidio da Paz Cabral Pernambuco Department of Education and Sports, Recife, PE, Brazil

Tatiane A. da Penha-Silva Universidade Estadual do Maranhão, São Luís, MA, Brazil

Raimundo Junior da Rocha Batista Chemical Analysis Laboratory, Coordination of Earth Sciences and Ecology, Museu Paraense Emílio Goeldi, Belém, Pará, Brazil

Johnatan Wellisson da Silva Mendes Regional University of Cariri – URCA, Crato, CE, Brazil

Gabriel Messias da Silva Nascimento Regional University of Cariri – URCA, Crato, CE, Brazil

Francisco Sydney Henrique da Silva State University of Ceará – UECE, Fortaleza, CE, Brazil

Viviane Bezerra da Silva Federal University of Pernambuco – UFPE, Recife, PE, Brazil

Sandra Alves de Araújo Universidade Estadual do Maranhão, São Luís, MA, Brazil

Programa de Pós-graduação em Biotecnologia/RENORBIO, Universidade Federal do Maranhão, São Luís, MA, Brazil

Ingryd Nayara de Farias Ramos Center for Research in Oncology/Institute of Biological Sciences, Federal University of Pará, Belém, PA, Brazil

Celeste de Jesus Pereira Franco Adolpho Ducke Laboratory, Coordination of Botany Museu Paraense Emílio Goeldi, Belem, Para, Brazil

Roberta Dávila Pereira de Lima Federal University of Cariri – UFCA, Crato, CE, Brazil

Chrystiaine Helena Campos de Matos Programa de Pós Graduação em Agroquímica, Instituto Federal de Educação, Ciência e Tecnologia Goiano, Rio Verde, Goias, Brazil

Saulo Almeida de Menezes Federal University of Rio Grande do Sul – UFRGS, Porto Alegre, RS, Brazil

Wendel F. F. de Moreira Universidade Estadual do Maranhão, São Luís, MA, Brazil

Yang Deng College of Food Science and Engineering, Qingdao Agricultural University, Qingdao, China

Qingdao Special Food Research Institute, Qingdao, China

Marcelli Geisse de Oliveira Prata da Silva Center for Research in Oncology/ Institute of Biological Sciences, Federal University of Pará, Belém, PA, Brazil

Anderson de Santana Botelho Adolpho Ducke Laboratory, Coordination of Botany Museu Paraense Emílio Goeldi, Belem, Para, Brazil

Chemical Analysis Laboratory, Coordination of Earth Sciences and Ecology, Museu Paraense Emílio Goeldi, Belem, Para, Brazil

Priscilla Augusta de Sousa Fernandes Regional University of Cariri – URCA, Crato, CE, Brazil

Alicia Ludymilla Cardoso de Souza Federal University of São Carlos, São Carlos, São Paulo, Brazil

Programa de Pós-graduação em Química, Universidade Federal de São Carlos, São Carlos, São Paulo, Brazil

Jamile Maria Pereira Bastos Lira de Vasconcelos Pernambuco Department of Education and Sports, Recife, PE, Brazil

Bianca M. Dolianitis Laboratory of Agroindustrial Processes Engineering (LAPE), Federal University of Santa Maria (UFSM), Cachoeira do Sul, RS, Brazil

Lidiane Diniz do Nascimento Coordenação de Botânica, Museu Paraense Emílio Goeldi, Belém, Pará, Brazil

Marcio Pereira do Nascimento Regional University of Cariri – URCA, Crato, CE, Brazil

Allyson Francisco dos Santos Pernambuco Department of Education and Sports, Recife, PE, Brazil

Antonia Thassya Lucas dos Santos Regional University of Cariri – URCA, Crato, CE, Brazil

Maicon S. N. dos Santos Laboratory of Agroindustrial Processes Engineering (LAPE), Federal University of Santa Maria (UFSM), Cachoeira do Sul, RS, Brazil

Oberdan Oliveira Ferreira Adolpho Ducke Laboratory, Coordination of Botany Museu Paraense Emílio Goeldi, Brazil

Federal University of Pará, Brazil

Maria Eliana Vieira Figueroa Pernambuco Department of Education and Sports, Recife, PE, Brazil

Ailésio R. M. Filho Universidade Estadual do Maranhão, São Luís, MA, Brazil

Lorena Filip Department of Bromatology, Hygiene, Nutrition, Iuliu Hatieganu University of Medicine and Pharmacy, Cluj-Napoca, Romania

Victor Juno Alencar Fonseca Regional University of Cariri – URCA, Crato, CE, Brazil

Allana C. Guedes Pós-graduação em Ciência Animal, Universidade Estadual do Maranhão, São Luís, MA, Brazil

Simona Codruța Hegheș Department of Drug Analysis, Iuliu Hatieganu University of Medicine and Pharmacy, Cluj-Napoca, Romania

Himani Department of Chemistry, College of Basic Sciences and Humanities, G.B. Pant University of Agriculture and Technology, Pantnagar, Uttarakhand, India

Jian Ju College of Food Science and Engineering, Qingdao Agricultural University, Qingdao, China

André Salim Khayat Adolpho Ducke Laboratory, Botany Coordination, Museu Paraense Emilio Goeldi, Belém, PA, Brazil

Ravendra Kumar Department of Chemistry, College of Basic Sciences and Humanities, G.B. Pant University of Agriculture and Technology, Pantnagar, Uttarakhand, India

Chang Jian Li State Key Laboratory of Food Science and Technology, Jiangnan University, Wuxi, Jiangsu, China

Mi Li State Key Laboratory of Food Science and Technology, Jiangnan University, Wuxi, Jiangsu, China

Isadora F. B. Magalhães Pós-graduação em Ciência Animal, Universidade Estadual do Maranhão, São Luís, MA, Brazil

Sonu Kumar Mahawer Department of Chemistry, College of Basic Sciences and Humanities, G.B. Pant University of Agriculture and Technology, Pantnagar, Uttarakhand, India

Suraj Narayan Mali Department of Pharmaceutical Sciences and Technology, Institute of Chemical Technology, Mumbai, Maharashtra, India

Maria Flaviana Bezerra Morais-Braga Regional University of Cariri – URCA, Crato, CE, Brazil

Adenilde N. Mouchrek Universidade Federal do Maranhão, São Luís, Maranhão, Brazil

Mozaniel Santana de Oliveira Museu Paraense Emílio Goeldi, Belém, Pará, Brazil

Adolpho Ducke Laboratory, Botany Coordination, Museu Paraense Emílio Goeldi, Belém, Pará, Brazil

Carolina E. D. Oro Department of Food Engineering, Regional Integrated University of Upper Uruguai and Missions (URI), Erechim, RS, Brazil

Benedito Yago Machado Portela Federal University of Ceara – UFC, Fortaleza, CE, Brazil

Om Prakash Department of Chemistry, College of Basic Sciences and Humanities, G.B.Pant University of Agriculture and Technology, Pantnagar, Uttarakhand, India

Amit P. Pratap Department of Oils, Oleochemicals and Surfactants Technology, Institute of Chemical Technology, Mumbai, Maharashtra, India

Maria Ivaneide Rocha Regional University of Cariri – URCA, Crato, CE, Brazil

Felicidade Caroline Rodrigues Federal University of Pernambuco – UFPE, Recife, PE, Brazil

Vanessa M. Santana Pós-graduação em Ciência Animal, Universidade Estadual do Maranhão, São Luís, MA, Brazil

Adriane Gomes Silva Faculty of Pharmacy, Centro Universitário, FIBRA, Belém, Pará, Brazil

Hartati Soetjipto Chemistry Department Faculty of Science and Mathematics, Universitas Kristen Satya Wacana, Salatiga, Indonesia

Srushti Tambe Department of Pharmaceutical Sciences and Technology, Institute of Chemical Technology, Mumbai, Maharashtra, India

Amanda M. Teles Universidade Federal do Maranhão, São Luís, Maranhão, Brazil

Marcus V. Tres Laboratory of Agroindustrial Processes Engineering (LAPE), Federal University of Santa Maria (UFSM), Cachoeira do Sul, RS, Brazil

Valdicley Vale Faculty of Pharmacy, Federal University of Pará, Belém, Pará, Brazil

Lilian Cortez Sombra Vandesmet Catholic University Center of Quixadá, Quixadá, CE, Brazil

Cícero Jorge Verçosa Pernambuco Department of Education and Sports, Recife, PE, Brazil

Oliviu Voştinaru Department of Pharmacology, Physiology and Physiopathology, Iuliu Hatieganu University of Medicine and Pharmacy, Cluj-Napoca, Romania

João H. C. Wancura Department of Teaching, Research and Development, Sul-Rio-Grandense Federal Institute (IFSul), Charqueadas, RS, Brazil

Giovani L. Zabot Laboratory of Agroindustrial Processes Engineering (LAPE), Federal University of Santa Maria (UFSM), Cachoeira do Sul, RS, Brazil

Part I
Essential Oils, General Concepts

Chapter 1
Essential Oils and Their General Aspects, Extractions and Aroma Recovery

Alicia Ludymilla Cardoso de Souza, Renan Campos e Silva,
Fernanda Wariss Figueiredo Bezerra, Mozaniel Santana de Oliveira,
Jorddy Neves Cruz, and Eloisa Helenade de Aguiar Andrade

1.1 Introduction

Essential oils (EOs) are complex mixtures of various constituents such as phenyl-propanoids, esters, and homo-, mono-, sesqui-, di-, tri-, and tetraterpenes. Their therapeutic uses are related to the treatment of cancer, diabetes, and cardiovascular and neurological diseases, in addition to having anti-aging, antioxidant, and antimi-crobial effects (Saljoughian et al. 2018; Benny and Thomas 2019; Bezerra et al. 2020b). The mechanisms involved in the pharmacological action of essential oils are complex due to their extensive and varied composition. Thus, *in vivo, in situ,* and *in silico* studies have been carried out to clarify and confirm the traditional eth-nopharmacological uses and make them a viable alternative to current therapeutic drugs, which in their vast majority bring side effects to patients (da Costa et al. 2019; Leão et al. 2020; Araújo et al. 2020).

Essential oils have been used as a complementary therapy in the treatment of anxiety, pain, bipolar disorder, attention-deficit hyperactivity disorder, and depres-sion through oral administration, inhalation, applied in diffusers, baths, and other uses. Their effects are due to the possible action on modulating the GABAergic system and inhibiting Na + channels, resulting in the balance between neural excita-tion and inhibition, culminating in the proper functioning of the central nervous

A. L. C. de Souza
Federal University of São Carlos, Sao Paulo, Brazil

R. Campos e Silva
Programa de Pós-graduação em Química, Universidade Federal do Pará, Belém, Pará, Brazil

F. W. F. Bezerra
Federal University of Pará, Belem, Para, Brazil

M. S. de Oliveira · J. N. Cruz · E. H. de Aguiar Andrade (✉)
Museu Paraense Emílio Goeldi, Belem, Para, Brazil

system (Wang and Heinbockel 2018). In the study by Abuhamdah et al. (2015), the essential oil of *Aloysia citrodora* showed antioxidant activity and neuronal protection due to inhibition of nicotine binding. Anaya-Eugenio et al. (2016) intraperitoneally administered the essential oil of *Artemisia ludoviciana* in rats, and the authors found that this EO has antinociceptive effects that may have been partially mediated by the opioid system. Heldwein et al. (2012) administered *Lippia alba* essential oil on silver catfish and found that it had a central anesthetic effect related to the GABAergic system.

Regarding the upper and lower respiratory tract, essential oils can be used in the treatment of diseases such as laryngitis, epiglotitis, pharyngitis, abscesses, rhinitis, bronchitis, pneumonia, etc. The mechanism of action may be related to their high volatility that can easily reach the parts to be treated through inhalation or oral administration (Horváth and Ács 2015; Bezerra et al. 2020; Leigh-de Rapper and van Vuuren 2020; de Oliveira et al. 2021). Ács et al. (2018) performed *in vitro* tests with different EOs against respiratory tract pathogens, and they found that thyme essential oil was effective against *Streptococcus mutans* and *Moraxella catarrhalis*; cinnamon bark was efficient against *S. pneumoniae* and *Haemophilus* spp.; and clove EO had action on *S. pyogenes*. The authors also suggested their use in combination with reference antibiotics to determine the effective dose and possible side effects and toxicity.

Essential oils are also able to prevent and improve the clinical picture of atherosclerosis, vasorelaxation, heart failure, myocardial infarction, and hypotension. The pharmacological effect of EOs on cardiovascular diseases may be related to their structure. For instance, the location of OH groups on the benzene ring may influence its effectiveness; and monoterpene alcohols may be more effective than monoterpene hydrocarbons (Monzote et al. 2017; Yu et al. 2020; Kaur et al. 2021). Alves-Santos et al. (2016) investigated the cardiovascular effects of *Croton argyrophylloides* in normotensive rats. The authors reported that treatment with this EO was able to decrease blood pressure and the effect may be related to active vascular relaxation. Ribeiro-Filho et al. (2016) evaluated the antihypertensive effect of the monoterpene β-citronellol through intravenous injection in anesthetized rats, and the treatment induced biphasic cardiovascular effects and direct and endothelium-independent cardioinhibitory vasodilatation. According to the study, the effects of this essential oil are possibly related to its vasodilator effect and consequent hypotensive action.

EOs present constituents that may also be related to chemopreventive effects due to increased detoxification, antioxidant, antimutagenic, antiproliferative, and enzyme induction properties (Bhalla et al. 2013; Bayala et al. 2014). The essential oil of *Rosmarinus officinalis* showed cytotoxic activity against three human cancer tumor cell lines (SK-OV-3, HO-8910, and Bel-7402). The authors (Wang et al. 2012) attributed this activity to the synergistic action of the EO compounds: mainly α-pinene, β-pinene, and 1,8-cineole. In the work by Chen et al. (2013), the essential oil of *Curcuma zedoaria* showed cytotoxic effect *in vitro* and *in vivo* on non-small cell lung carcinoma and on cell apoptosis that plays an important role in the effectiveness of chemotherapy. The essential oil of *Pinus Roxburghii* was evaluated in the 3-(4,5-dimethylthiazol-2-yl)-2,5-diphenyltetrazolium bromide assay and showed induction of cytotoxicity in colon, leukemia, multiple myeloma, pancreatic, head and neck, and lung cancer cells, in addition to having inhibited cell proliferation and induced apoptosis, which was correlated with NF-κB suppression (Sajid et al. 2018).

EOs also can be used in the treatment of other diseases. *Coriandrum sativum* and clove essential oils showed neuroprotective effects on patients with Alzheimer's disease (Chen et al. 2013; Cioanca et al. 2013). The essential oils of *Croton matourensis, Syzygium cumini, Psidium guajava,* and *Melissa officinalis* showed potential anti-inflammatory activity (Bounihi et al. 2013; Bezerra et al. 2020b). Also, EOs of *Citrus bergamia, Coriandrum sativum, Pelargonium graveolens, Helichrysum italicum, Pogostemon cablin, Citrus aurantium, Santalum album, Nardostachys jatamans,* and *Cananga odorata* showed antiproliferative effects on neonatal human skin fibroblast cells (Han et al. 2017; Sihoglu Tepe and Ozaslan 2020). In fact, despite the strong indications, clinical studies with patients are still necessary in order to verify the real efficacy and toxicity of EOs.

1.2 Plant Organs Where EOs Are Found

Essential oils are secondary metabolites that play an important ecological role in plant defense, mediating the relationship between the plant and abiotic (light, temperature, oxygen, CO_2, ozone, etc.) and biotic factors (competing organisms, harmful pathogens, and beneficial animals). Biosynthesis can occur in different plant organs such as leaves, bark, flowers, buds, seeds, twigs, fruits, rhizomes, and roots (Najafabadi et al. 2017; Dhakad et al. 2018; Cascaes et al. 2021a). The synthesis, accumulation, and storage can occur in secretory glands, which are specialized histopathological structures that can be located: (i) on the surface of plants, thus having exogenous secretion; or (ii) inside organs, occurring endogenous secretion. Plants may also possess other secretory structures such as epidermal papillae, secretory bristles or glandular trichomes, schizogenous or secretory pockets, secretory channels, and intracellular secretory cells (El Asbahani et al. 2015; Lange 2015; Sharifi-Rad et al. 2017). Table 1.1 shows the chemical composition of some plant species and the organ from which their essential oil was obtained.

Table 1.1 Chemical composition of some essential oils according to the species and organs used for extraction

Species	Part	Composition	References
Syzygium cumini	Leaves	α-Pinene, β-pinene, and trans-caryophyllene	Mohamed et al. (2013)
Croton matourensis	Leaves	Larixol, manool oxide, linalool, E-caryophyllene, α-pinene, and α-phellandrene	Bezerra et al. (2020b)
Hornstedtia bella	Leaves	Germacrene D, viridiflorene, E-β-caryophyllene, and α-humulene	Donadu et al. (2020)
	Rhizomes	β-pinene, α-humulene, β-selinene, and epiglobulol	
	Whole plant	β-pinene, 1,8-cineole, and α-pinene	
Eucalyptus globulus	Fruits	Globulol, aromadendrene, and eucalyptol (1,8-cineole)	Said et al. (2016)
Salvia hydrangea	Flowers	Caryophyllene oxide, 1,8-cineole, and trans-caryophyllene	Ghavam et al. (2020)
Syzygium aromaticum	Bud	Eugenol, eugenyl acetate, and β-caryophyllene	Razafimamonjison et al. (2014)
	Leaves	Eugenol, eugenyl acetate, and β-caryophyllene	
	Stems	Eugenol and β-caryophyllene	
Citrus aurantium	Green leaves/ twigs	4-Terpineol, D-limonene, 4-carvomenthenol, and linalool	Okla et al. (2019)
	Small green branches	D-Limonene, dodecane, oleic acid, and trans-palmitoleic acid	
	Wooden branches	D-Limonene, dimethyl anthranilate, (−)-β-fenchol, and dodecane	
	Branch bark	D-Limonene, γ-terpinene, dodecane, and dimethyl anthranilate	

1.3 Chemical Composition of Essential Oils

1.3.1 Background

Essential oils (EOs), also called volatile oils, are odorous products obtained from plant materials that have already been identified by botanists. These complex mixtures, full of volatile compounds, are biosynthesized by plants in different parts and can be obtained by conventional techniques such as hydrodistillation, steam distillation, and cold pressing, as well as by innovative methods, which increase the efficiency of extraction, either concerning extraction time or to obtain higher-quality extracts (El Asbahani et al. 2015).

EOs can be obtained with variable yield (in quantity, quality, and composition). Climatic factors, soil composition, age of the plant material, stage of the vegetative cycle, and the time of the year when the material was collected can also influence the yield of EOs. In addition, one or more components can be found in high contents (from 20% to 70%) or trace amounts (Akthar et al. 2014).

As an example of the diversity of EOs, we can highlight 1,8-cineole, present in high quantity in the oil of *Eucalyptus globulus* (eucalyptus), being its main constituent; unlike *Coriandrum sativum* (coriander) that has linalool as the main component. Furthermore, the occurrence of chemotypes is common, since the same plant species may have different chemical characteristics, depending on where it was cultivated. An example is *Thymus vulgaris* (thyme), which has chemotypes related to its main compounds, such as thymol, carvacrol, terpineol, and linalool (Regnault-Roger et al. 2011).

Volatility is the main difference between EOs and fixed oils (lipid mixtures) obtained from seeds. In addition, EOs also carry pleasant and intense aromas that have led them to be sometimes called essences. These characteristics are due to their main constituents, terpenes and terpenoids, which originate from different biosynthetic pathways. Terpenes are part of the main group that constitutes EOs, and their chains are formed by the successive juxtaposition of several isoprene units (C_5), which originate other terpenes, such as monoterpenes (C_{10}), sesquiterpenes (C_{15}), and diterpenes (C_{20}). On the other hand, terpenoids, such as alcohols, esters, ketones, are terpenes that contain oxygen and that comprise a wide variety of organic functions (Baptista-Silva et al. 2020) (Fig. 1.1).

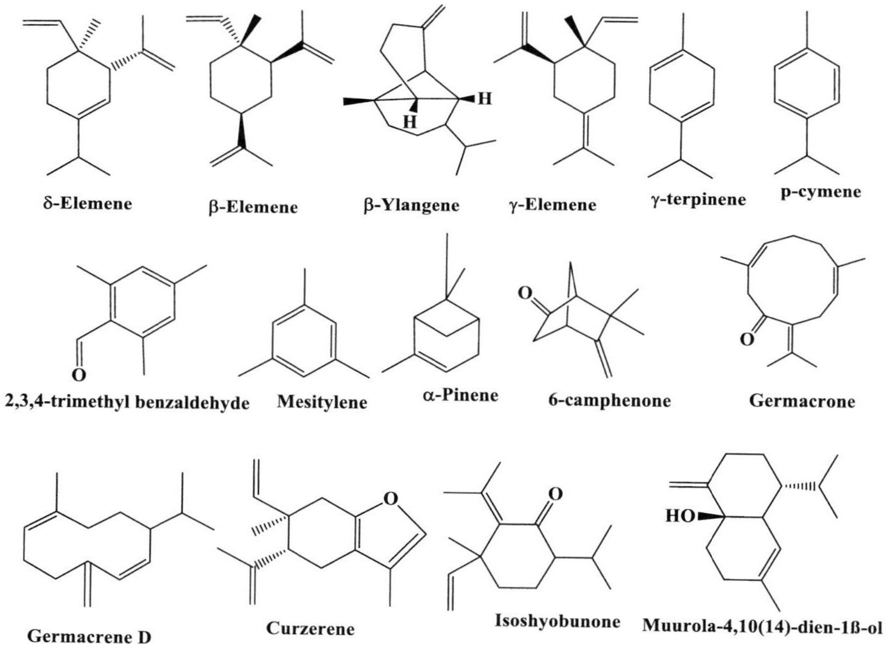

Fig. 1.1 Chemical structures found in essential oils. (Source: Adapted from Hyldgaard et al. 2012)

1.3.2   Terpenes

Terpenes, which have the general formula $(C_5H_8)n$, are derived from the mevalonic acid pathway from acetyl-CoA and are produced by specialized plant tissues. Isoprene (C5) is their basis and originates all other terpenes, being a subunit of these new molecules. Thus, according to the general formula, two isoprene units generate monoterpenes $(C_{10}H_{16})$; three units generate sesquiterpenes $(C_{15}H_{24})$; four isoprene units generate diterpenes $(C_{20}H_{32})$, and so on (Hyldgaard et al. 2012; Cascaes et al. 2021b).

Terpenes also present biological activities, such as anticonvulsant (De Almeida et al. 2011), anticancer (Bhalla et al. 2013), antifungal (Nazzaro et al. 2017), antibacterial (Guimarães et al. 2019), phytotoxic (with potential for biopesticide production) (Werrie et al. 2020), and several other properties and applications. This behavior guarantees a great advantage since EOs are environmentally friendly, safe when used with responsibility, natural, renewable, and biodegradable (Pandey et al. 2017).

1.3.3   Monoterpenes

Monoterpenes, as mentioned above, have in their structure two isoprene units $(C_{10}H_{16})$, and are present in 90% of the EOs. Important monoterpene hydrocarbons are limonene, p-cymene, α-pinene, and α-terpinene; and remarkable oxygenated monoterpenes are carvacrol, thymol, and camphor (Nazzaro et al. 2017). Carvacrol and thymol are some of the most active oxygenated monoterpenes ever identified, showing antimicrobial activity with MICs of 300 and 800 µg/mL, respectively (Hyldgaard et al. 2012). The *in vivo* and *in vitro* antitumor activities of 37 monoterpenes found in EOs have also been described (Sobral et al. 2014) and a recent study reports α-pinene as a compound that exhibits enantioselective biological activities, being promising for further research and consequent development of new drugs (Allenspach and Steuer 2021).

1.3.4   Sesquiterpenes

Sesquiterpenes are formed by three isoprene units $(C_{15}H_{24})$, being one of the most important terpenes (Ferreira et al. 2020). By the extension of the chain, the increase in the number of cyclizations is favored and thus, the formation of diverse structures is favored as well (Nazzaro et al. 2017). Pandey et al. (2017) reported that essential oils exhibit antifungal activity (acting by disintegration of fungal hyphae) due to the presence of monoterpenes and sesquiterpenes. There are also reports of two sesquiterpenes, valerena-4,7(11)-diene (VLD) and β-caryophyllene, that showed proven anxiolytic activity (Zhang and Yao 2019); other authors state that sesquiterpenes such as valeranone and β-eudesmol have anticonvulsant activity (De Almeida et al. 2011); and the anti-inflammatory activity of 12 sesquiterpenes, among them farnesol, was also studied (de Cássia Da Silveira e Sá et al. 2015).

1.3.5 Diterpenes

Diterpenes, formed by four isoprene units ($C_{20}H_{32}$), are called diterpenoids when they have oxygen in their structure. Compounds such as retinol, retinal, taxol, and phytol have diterpenes as the basis of their structure. According to several studies, diterpenes have the greatest antioxidant and cytotoxic capacities (Islam et al. 2016; Santana de Oliveira et al. 2021). An example is a diterpene ester, ingenol-3-angelate, isolated from the sap of *Euphorbia peplus*, which presents anticancer properties, being cytotoxic for different tumor cells (Greay and Hammer 2015). Also, diterpenes showed efficacy against ten neglected tropical diseases such as Chagas disease, chikungunya, echinococcosis, dengue, leishmaniasis, leprosy, lymphatic filariasis, malaria, schistosomiasis, and tuberculosis (de Alencar et al. 2017).

1.3.6 Alcohols

Alcohols in EOs may appear in their free form, combined with terpene chains or with esters. Thus, terpenes are called alcohols when they bind to hydroxyl and may assume the nomenclature of monoterpenols, sesquiterpenols, diterpenols, etc. for the varied structures (Hanif et al. 2019). Guimarães et al. (2019) reported that the presence of hydroxyl groups, in phenolic and alcoholic compounds, such as carvacrol, l-carveol, eugenol, trans-geraniol, and thymol, induced significant antimicrobial activity. Also, there are reports of the occurrence of synergism, i.e., interaction of antimicrobial compounds that present greater activity in combination than individually. In this case, linalool or menthol combined with eugenol is more effective, suggesting that a monoterpenoid phenol combined with a monoterpenoid alcohol presents greater antimicrobial activity (Hyldgaard et al. 2012).

1.3.7 Esters

Esters can also be extracted from EOs and as mentioned earlier, can be combined with alcohols, which may exhibit antimicrobial activity (Hanif et al. 2019). Esters can be found in plants such as lavender (*Lavandula angustifolia*) and, sometimes, they can occur as lactones (in the cyclic form derived from lactic acid): γ-lactones (five-membered rings), δ-lactones (six-membered rings), or as coumarins (Zuzarte and Salgueiro 2015). Studies focusing on *Lavandula angustifolia* oil have shown anxiolytic effect, being used for the production of drugs such as Silexan. Esters are also found in ylang-ylang (*Cananga odorata*) oils, relieving and reducing symptoms of stress and anxiety (Zhang and Yao 2019).

1.3.8 Ketones

As one of the constituents of plant EOs, ketones have different activities and can even be toxic. They may present expectorant and wound-healing properties (Hanif et al. 2019). Some examples of ketones that are expectorants and have intense aroma are camphor, carvone, fenchone, and mentone. With toxic effects, we can mention pulegone, which can cause changes in the central nervous system, as well as liver and kidney failure, lung toxicity, and finally lead to death. Thujone, which has two isomeric forms, α-thujone and β-thujone, has greater toxicity in its alpha form. α-Thujone is used in the production of absinthe (an alcoholic beverage prohibited in many countries) (Zuzarte and Salgueiro 2015; da Silva Júnior et al. 2021).

1.4   Extraction Methods of EOs

Essential oils (EOs) obtained from aromatic plants represent a diverse and unique source of natural products, which are widely used for bactericidal, fungicidal, antiviral, antiparasitic, insecticidal, medicinal, or cosmetic applications, especially in the pharmaceutical, health, cosmetic, food, and agricultural industries. Their use is boosted by the growing interest of consumers in natural substances (Reyes-Jurado et al. 2015).

   Before EOs can be used or analyzed, they must be extracted from the plant matrix, which might be its leaves, bark, peels, flowers, buds, seeds, and other parts (Tongnuanchan and Benjakul 2014). The main methods for extracting essential oils are hydrodistillation (HD), steam distillation, water-steam distillation, maceration, and empyreumatic distillation. Among these methods, HD has been the most common approach to extract essential oils from medicinal plants (Djouahri et al. 2013). Although these techniques have been used for many years, their application has some drawbacks, such as loss of some volatile compounds, low extraction efficiency, degradation of esters or unsaturated compounds by thermal or hydrolytic effects, and possible toxic residual solvents in extracts or EOs (Reyes-Jurado et al. 2015).

   New approaches, such as microwave-assisted extraction, supercritical fluid extraction, and ultrasound-assisted extraction, which emerged from the so-called green extraction techniques, have been applied to shorten extraction time, improve extraction yields, and reduce operating costs, optimizing production over traditional methods (Li et al. 2014; Reyes-Jurado et al. 2015). In the next sections, we will discuss in detail hydrodistillation and steam-distillation techniques, the most traditional methods, and supercritical fluid and microwave-assisted extraction, the green methods with more current use, addressing their main aspects.

1.4.1 Hydrodistillation

Hydrodistillation (HD) has been used since ancient times for the extraction of essential oils. Despite the intrinsic limitations of this technique, it remains the most common method applied both in the laboratory and on an industrial scale (Orio et al. 2012; Azmir et al. 2013; de Oliveira et al. 2020). The principle of extraction is based on azeotropic distillation. Indeed, at atmospheric pressure and during the extraction process, the oil molecules and water form a heterogeneous mixture that reaches its boiling temperature at a point near 100 °C. The EO/water mixture is then simultaneously distilled as if they were a single compound (El Asbahani et al. 2015; Rassem et al. 2016; Bezerra et al. 2020c). Hydrodistillation is a variant of steam distillation, indicated by the French Pharmacopoeia for the extraction of essential oils from dried plants. The distillation time depends on the plant material being processed. Prolonged distillation produces only a small amount of essential oil, but adds unwanted high-boiling point compounds and oxidation products (Rassem et al. 2016; Silva et al. 2019; Castro et al. 2021).

In hydrodistillation, the plant materials are packed in a distillation flask; and water is then added in sufficient quantity to boil. Alternatively, direct steam is injected into the plant sample. Hot water and steam act as the main influencing factors in the release of bioactive compounds from plant tissues. An indirect water cooling system condenses the mixture of water vapor and oil, which flows from the condenser to a separator, where they are separated (Azmir et al. 2013).

In short, the hydrodistillation system consists of a container, usually a volumetric flask, connected to a Clevenger-type apparatus coupled to a cooling system, with temperatures ranging from 10 to 15 °C. The solid-liquid mixture is heated, at atmospheric pressure, until it reaches the boiling temperature of water, allowing the odor molecules to evaporate together with the water, forming an azeotropic mixture. This combination is led to the condenser, where it liquefies and is collected at the end of the process. Due to its hydrophobic character, the oil does not mix with water and can be separated by decantation. After separation, the oil is completely dehydrated with anhydrous Na_2SO (Rassem et al. 2016).

1.4.2   Steam Distillation

Steam distillation has some characteristics that make it one of the most widely used methods for obtaining essential oils on an industrial scale, such as low cost, simplicity, and ease of design when compared to other advanced techniques (Muhammad et al. 2013).

There are two types of steam distillation: direct and indirect. In the indirect method, the plant material is soaked in water and heated to boiling. The resulting steam from the boiling water carries the volatile compounds with it. Then, cooling and condensation separate the oil from water. The disadvantage of this technique is the degradation of materials and unpleasant smell due to constant exposure to heat. On the other hand, in direct steam distillation, the most commonly used method for obtaining essential oils, no water is placed inside the distillation flask. Instead, steam is directed into the flask from an external source. The essential oils are released from the plant material when the steam bursts the sacs containing the oil molecules (Chemat and Boutekedjiret 2015; do Nascimento et al. 2020).

In the steam distillation process, water boils above 100 °C, at a pressure higher than atmospheric pressure, which facilitates the removal of the essential oil from the plant material, reducing the formation of artifacts (El Asbahani et al. 2015; Yadav et al. 2017).

1.4.3   Supercritical Fluid Extraction

Supercritical fluid extraction (SFE) has become the most widely used method for extracting and isolating EOs from aromatic plants. This technique provides fast and effective extraction, requires only moderate temperatures, eliminates cleaning steps, and avoids the use of harmful organic solvents (Yousefi et al. 2019; de Oliveira et al. 2019; de Carvalho et al. 2019). Due to these attributes, SFE is considered environmentally friendly in various fields such as natural material extraction (Ghasemi et al. 2011; Sodeifian et al. 2017).

Generally, the solvent used in supercritical extraction is CO_2 because this gas has ideal properties, such as low viscosity, high diffusivity, and density close to that of liquids. In addition to being non-toxic, non-aggressive, non-flammable, bacteriostatic, non-corrosive, non-explosive, CO_2 is chemically inert, available in high purity at a relatively low cost, and its polarity is similar to pentane, which makes it suitable for the extraction of lipophilic compounds (El Asbahani et al. 2015; Sodeifian et al. 2017; Yousefi et al. 2019). Furthermore, the extraction is performed at temperatures and pressures above the CO_2 critical point, 7.4 MPa and 31.1 °C, or close to this region (Sovová 2012); so a simple pressure relief is able to separate CO_2 from the extracted essential oils, leaving no solvent residue and providing a high-purity product (El Asbahani et al. 2015). The low viscosity and high diffusivity of the supercritical fluid increase the penetration power based on the high mass transfer rate of the solutes into the fluid, allowing efficient extraction of compounds (Sovilj et al. 2011; Silva et al. 2021).

The laboratory-scale SFE system basically consists of a carbon dioxide cylinder, cooling bath, high-pressure pump, oven, extraction vessel, flask, air compressor, flow meter, and flow control valves (Cruz et al. 2020).

The extraction process begins when the liquefied CO_2 contained in a cylinder enters a high-pressure pump. The liquid carbon dioxide is then compressed to a desired pressure by the pump and is also heated to a determined temperature. Optionally, a required volume of co-solvent can be added to increase its solvation properties. Then, supercritical CO_2 containing the extracted solutes flows through a depressurization valve at the extractor outlet. This results in the precipitation of solutes, which are collected in a separator, whereas CO_2 is easily released (Fornari et al. 2012; Ahangari et al. 2021).

Despite being an efficient extraction technique with high-purity products, the elevated cost of equipment installation and maintenance is still an obstacle to the development of supercritical fluid extraction, which makes the final product more expensive (El Asbahani et al. 2015).

1.4.4   Microwave-Assisted Extraction

Another extraction technique considered green and sustainable that has didactic, scientific, and commercial applications is microwave-assisted extraction. EO obtaining under microwave irradiation, without organic solvent or water, is an extraction method that can offer high reproducibility in shorter times, with simplified manipulation, reduced solvent consumption, lower energy input, and lower CO_2 emission (Cardoso-Ugarte et al. 2013; Kokolakis and Golfinopoulos 2013). It also provides high-value products and higher yields when compared to traditional extraction techniques (Karimi et al. 2020).

Microwaves are a form of non-ionizing electromagnetic radiation at frequencies ranging from 300 MHz to 30 GHz and wavelengths ranging from 1 cm to 1 m. However, the frequency commonly used in extractions is 2450 MHz, which corresponds to a wavelength of 12.2 cm. This energy is transmitted in the form of waves, which can penetrate biomaterials and interact with polar molecules in materials such as water to generate heat (Cardoso-Ugarte et al. 2013; El Asbahani et al. 2015). Although in most cases dried plant materials are used for extraction, plant cells still contain microscopic traces of moisture that serve as a target for microwave heating. These residual water molecules, when heated due to the microwave effect, evaporate and generate tremendous pressure on the cell wall due to the swelling of the plant cell (Mandal et al. 2007).

The heating of microwave-assisted hydrodistillation is based on its direct impact on polar materials/solvents and is ruled by two phenomena: ionic conduction and dipole rotation, which in most cases occur simultaneously (Rassem et al. 2016). Ionic conduction refers to the electrophoretic migration of ions influenced by the varying electric field. The resistance offered by the solution to ion migration generates friction, which ultimately heats the solution. Dipole rotation represents the realignment of the dipoles with the changing electric field. Heating is affected only at 2450 MHz frequency. The electrical component of the wave changes 4.9×10^4 times per second (Mandal et al. 2007).

Advances in microwave-assisted extraction have led to the development of various techniques such as compressed air microwave distillation (CAMD), vacuum microwave hydrodistillation (VMHD), microwave-assisted hydrodistillation (MWHD), solvent-free microwave extraction (SFME), microwave accelerated steam distillation (MASD), and microwave hydrodiffusion and gravity (MHG) (Cardoso-Ugarte et al. 2013; Reyes-Jurado et al. 2015).

1.5 Conclusion

The present review allowed us to evaluate several aspects of essential oils, such as their diversified chemical composition, which presents phenylpropanoids, homo-, mono-, sesqui-, di-, and tri-tetraterpenes, alcohols, esters and ketones, which can be obtained from organs present in vegetables such as leaves, bark, flowers, buds, seeds, twigs, fruits, rhizomes, and roots. The extraction process can be performed using conventional methods such as hydrodistillation and steam distillation or more innovative methods such as supercritical fluid extraction and microwave-assisted extraction. In this scenario, it can be concluded that essential oils have great potential to be used in place of synthetic inputs, because studies demonstrate their antimicrobial (bactericidal, fungicidal), antiviral, antiparasitic, insecticidal, antioxidant, anticancer, antitumor, neuroprotective, anti -inflammatory, among others, so that they can be used as inputs with less toxicity in pharmaceutical, cosmetic, food and agrochemical products.

Acknowledgments The author Dr Mozaniel Santana de Oliveira, thanks PCI-MCTIC/MPEG, as well as CNPq for the scholarship process number: 302050/2021-3.

Conflict of Interests   There are no conflicts of interest in this study.

References

Abuhamdah S, Abuhamdah R, Howes MJR, Al-Olimat S, Ennaceur A, Chazot PL (2015) Pharmacological and neuroprotective profile of an essential oil derived from leaves of Aloysia citrodora Palau. J Pharm Pharmacol 67:1306–1315. https://doi.org/10.1111/jphp.12424
Ács K, Balázs VL, Kocsis B, Bencsik T, Böszörményi A, Horváth G (2018) Antibacterial activity evaluation of selected essential oils in liquid and vapor phase on respiratory tract pathogens. BMC Complement Altern Med 18:1–9. https://doi.org/10.1186/s12906-018-2291-9

Ahangari H, King JW, Ehsani A, Yousefi M (2021) Supercritical fluid extraction of seed oils – a short review of current trends. Trends Food Sci Technol 111:249–260. https://doi.org/10.1016/j.tifs.2021.02.066

Akthar MS, Degaga B, Azam T (2014) Antimicrobial activity of essential oils extracted from medicinal plants against the pathogenic microorganisms: a review. Issues Bio Sci Pharma Res 2:1–7

Allenspach M, Steuer C (2021) α-Pinene: a never-ending story. Phytochemistry 190:112857. https://doi.org/10.1016/j.phytochem.2021.112857

Alves-Santos TR, de Siqueira RJB, Duarte GP, Lahlou S (2016) Cardiovascular effects of the essential oil of Croton argyrophylloides in normotensive rats: role of the autonomic nervous system. Evid Based Complement Alternat Med 2016:1–9. https://doi.org/10.1155/2016/4106502

Anaya-Eugenio GD, Rivero-Cruz I, Bye R, Linares E, Mata R (2016) Antinociceptive activity of the essential oil from Artemisia ludoviciana. J Ethnopharmacol 179:403–411. https://doi.org/10.1016/j.jep.2016.01.008

Araújo PHF, Ramos RS, da Cruz JN, Silva SG, Ferreira EFB, de Lima LR, Macêdo WJC, Espejo-Román JM, Campos JM, Santos CBR (2020) Identification of potential COX-2 inhibitors for the treatment of inflammatory diseases using molecular modeling approaches. Molecules 25:4183. https://doi.org/10.3390/molecules25184183

Azmir J, Zaidul ISM, Rahman MM, Sharif KM, Mohamed A, Sahena F, Jahurul MHA, Ghafoor K, Norulaini NAN, Omar AKM (2013) Techniques for extraction of bioactive compounds from plant materials: a review. J Food Eng 117:426–436. https://doi.org/10.1016/j.jfoodeng.2013.01.014

Baptista-Silva S, Borges S, Ramos OL, Pintado M, Sarmento B (2020) The progress of essential oils as potential therapeutic agents: a review. J Essent Oil Res 32:279–295. https://doi.org/10.1080/10412905.2020.1746698

Bayala B, Bassole IHN, Scifo R, Gnoula C, Morel L, Lobaccaro JMA, Simpore J (2014) Anticancer activity of essential oils and their chemical components – a review. Am J Cancer Res 4:591–607

Benny A, Thomas J (2019) Essential oils as treatment strategy for Alzheimer's disease: current and future perspectives. Planta Med 85:239–248. https://doi.org/10.1055/a-0758-0188

Bezerra FWF, do Nascimento Bezerra P, Cunha VMB, Salazar M de LAR, Barbosa JR, da Silva MP, de Oliveira MS, da Costa WA, Pinto RHH, da Cruz JN, de Carvalho Junior RN (2020) Supercritical Green Solvent for Amazonian Natural Resources. In: Inamuddin, Asiri AM (eds) Nanotechnology in the Life Sciences. Springer, Cham, Cham, pp 15–31

Bezerra FWF, Marielba de Los Angeles Rodriguez Salazar, Freitas LC, de Oliveira MS, dos Santos IRC, Dias MNC, Gomes-Leal W, de Aguiar Andrade EH, Ferreira GC, de Carvalho RN (2020b) Chemical composition, antioxidant activity, anti-inflammatory and neuroprotective effect of Croton matourensis Aubl. leaves extracts obtained by supercritical CO2. J Supercrit Fluids 165:104992. https://doi.org/10.1016/j.supflu.2020.104992

Bezerra FWF, de Oliveira MS, Bezerra PN, Cunha VMB, Silva MP, da Costa WA, Pinto RHH, Cordeiro RM, da Cruz JN, Chaves Neto AMJ, Carvalho Junior RN (2020c) Extraction of bioactive compounds. In: Green sustainable process for chemical and environmental engineering and science. Elsevier, pp 149–167

Bhalla Y, Gupta VK, Jaitak V (2013) Anticancer activity of essential oils: a review. J Sci Food Agric 93:3643–3653. https://doi.org/10.1002/jsfa.6267

Bounihi A, Hajjaj G, Alnamer R, Cherrah Y, Zellou A (2013) In vivo potential anti-inflammatory activity of Melissa officinalis L. essential oil. Adv Pharmacol Sci 2013:1–7. https://doi.org/10.1155/2013/101759

Cardoso-Ugarte GA, Juárez-Becerra GP, Sosa-Morales ME, López-Malo A (2013) Microwave-assisted extraction of essential oils from herbs. J Microw Power Electromagn Energy 47:63–72. https://doi.org/10.1080/08327823.2013.11689846

Cascaes MM, Dos O, Carneiro S, Diniz Do Nascimento L, Antônio Barbosa De Moraes Â, Santana De Oliveira M, Neves Cruz J, Skelding GM, Guilhon P, Helena De Aguiar Andrade E, Vico C, Cruz-Chamorro I (2021a) Essential oils from Annonaceae species from Brazil: a systematic review of their phytochemistry, and biological activities. Int J Mol Sci 22:12140. https://doi.org/10.3390/IJMS222212140

Cascaes MM, Silva SG, Cruz JN, Santana de Oliveira M, Oliveira J, de Moraes AAB, da Costa FAM, da Costa KS, Diniz do Nascimento L, Helena de Aguiar Andrade E (2021b) First report on the Annona exsucca DC. Essential oil and in silico identification of potential biological targets of its major compounds. Nat Prod Res. https://doi.org/10.1080/14786419.2021.1893724

Castro ALG, Cruz JN, Sodré DF, Correa-Barbosa J, Azonsivo R, de Oliveira MS, de Sousa Siqueira JE, da Rocha Galucio NC, de Oliveira BM, Burbano RMR, do Rosário Marinho AM, Percário S, Dolabela MF, Vale VV (2021) Evaluation of the genotoxicity and mutagenicity of isoeleutherin and eleutherin isolated from Eleutherine plicata herb. using bioassays and in silico approaches. Arab J Chem 14:103084. https://doi.org/10.1016/j.arabjc.2021.103084

Chemat F, Boutekedjiret C (2015) Extraction//steam distillation. In: Reference module in chemistry, molecular sciences and chemical engineering. Elsevier, Amsterdam, pp 1–12

Chen CC, Chen Y, Hsi YT, Chang CS, Huang LF, Ho CT, Der WT, Kao JY (2013) Chemical constituents and anticancer activity of Curcuma zedoaria Roscoe essential oil against non-small cell lung carcinoma cells in vitro and in vivo. J Agric Food Chem 61:11418–11427. https://doi.org/10.1021/jf4026184

Cioanca O, Hritcu L, Mihasan M, Hancianu M (2013) Cognitive-enhancing and antioxidant activities of inhaled coriander volatile oil in amyloid β(1-42) rat model of Alzheimer's disease. Physiol Behav 120:193–202. https://doi.org/10.1016/j.physbeh.2013.08.006

Cruz JN, da Silva AG, da Costa WA, Gurgel ESC, Campos WEO, Silva RC e, Oliveira MEC, da Silva Souza Filho AP, Pereira DS, Silva SG, de Aguiar Andrade EH, de Oliveira MS (2020) Volatile compounds, chemical composition and biological activities of Apis mellifera bee propolis. In: da Silva AG (ed) Essential oils – bioactive compounds, new perspectives and applications. IntechOpen, Rijeka, p 16

da Costa KS, Galúcio JM, da Costa CHS, Santana AR, dos Santos CV, do Nascimento LD, Lima AHLE, Neves Cruz J, Alves CN, Lameira J (2019) Exploring the potentiality of natural products from essential oils as inhibitors of odorant-binding proteins: a structure- and ligand-based virtual screening approach to find novel mosquito repellents. ACS Omega 4:22475–22486. https://doi.org/10.1021/acsomega.9b03157

da Silva Júnior OS, de Jesus Pereira Franco C, de Moraes AAB, Cruz JN, da Costa KS, do Nascimento LD, de Aguiar Andrade EH (2021) In silico analyses of toxicity of the major constituents of essential oils from two Ipomoea L. species. Toxicon 195:111–118. https://doi.org/10.1016/j.toxicon.2021.02.015

de Alencar MVOB, de Castro e Sousa JM, Rolim HML, das Graças Freire de Medeiros M, Cerqueira GS, de Castro Almeida FR, das Graças Lopes Citó AM, Ferreira PMP, Lopes JAD, de Carvalho Melo-Cavalcante AA, Islam MT (2017) Diterpenes as lead molecules against neglected tropical diseases. Phytother Res 31:175–201. https://doi.org/10.1002/ptr.5749

De Almeida RN, De Fátima Agra M, Maior FNS, De Sousa DP (2011) Essential oils and their constituents: Anticonvulsant activity. Molecules 16:2726–2742. https://doi.org/10.3390/molecules16032726

de Carvalho RN, de Oliveira MS, Silva SG, da Cruz JN, Ortiz E, da Costa WA, Bezerra FWF, Cunha VMB, Cordeiro RM, de Jesus Chaves Neto AM, de Aguiar Andrade EH, de Carvalho Junior RN (2019) Supercritical CO2 application in essential oil extraction. In: Inamuddin RM, Asiri AM (eds) Materials research foundations, 2nd edn. Materials Research Foundations, Millersville, pp 1–28

de Cássia Da Silveira e Sá R, Andrade LN, De Sousa DP (2015) Sesquiterpenes from essential oils and anti-inflammatory activity. Nat Prod Commun 10:1767–1774. https://doi.org/10.1177/1934578X1501001033

de Oliveira MS, da Cruz JN, Gomes Silva S, da Costa WA, de Sousa SHB, Bezerra FWF, Teixeira E, da Silva NJN, de Aguiar Andrade EH, de Jesus Chaves Neto AM, de Carvalho RN (2019) Phytochemical profile, antioxidant activity, inhibition of acetylcholinesterase and interaction mechanism of the major components of the Piper divaricatum essential oil obtained by supercritical CO2. J Supercrit Fluids 145:74–84. https://doi.org/10.1016/j.supflu.2018.12.003

de Oliveira MS, da Cruz JN, da Costa WA, Silva SG, da Paz Brito M, de Menezes SAF, de Jesus Chaves Neto AM, de Aguiar Andrade EH, de Carvalho Junior RN (2020) Chemical composition, antimicrobial properties of Siparuna guianensis essential oil and a molecular docking and dynamics molecular study of its major chemical constituent. Molecules 25:3852. https://doi.org/10.3390/molecules25173852

de Oliveira MS, Cruz JN, Ferreira OO, Pereira DS, Pereira NS, Oliveira MEC, Venturieri GC, Guilhon GMSP, da Silva Souza Filho AP, de Aguiar Andrade EH (2021) Chemical composition of volatile compounds in Apis mellifera propolis from the northeast region of Pará state, brazil. Molecules 26:3462. https://doi.org/10.3390/molecules26113462

Dhakad AK, Pandey VV, Beg S, Rawat JM, Singh A (2018) Biological, medicinal and toxicological significance of Eucalyptus leaf essential oil: a review. J Sci Food Agric 98:833–848. https://doi.org/10.1002/JSFA.8600

Djouahri A, Boudarene L, Meklati BY (2013) Effect of extraction method on chemical composition, antioxidant and anti-inflammatory activities of essential oil from the leaves of Algerian Tetraclinis articulata (Vahl) Masters. Ind Crops Prod 44:32–36. https://doi.org/10.1016/j.indcrop.2012.10.021

do Nascimento LD, de AAB M, da Costa KS, Galúcio JMP, Taube PS, CML C, Cruz JN, de Aguiar Andrade EH, de Faria LJG (2020) Bioactive natural compounds and antioxidant activity of essential oils from spice plants: new findings and potential applications. Biomolecules 10:1–37. https://doi.org/10.3390/biom10070988

Donadu MG, Le NT, Ho DV, Doan TQ, Le AT, Raal A, Usai M, Marchetti M, Sanna G, Madeddu S, Rappelli P, Diaz N, Molicotti P, Carta A, Piras S, Usai D, Nguyen HT, Cappuccinelli P, Zanetti S (2020) Phytochemical compositions and biological activities of essential oils from the leaves, rhizomes and whole plant of hornstedtia bella Škorničk. Antibiotics 9:1–16. https://doi.org/10.3390/antibiotics9060334

El Asbahani A, Miladi K, Badri W, Sala M, Addi EHA, Casabianca H, El Mousadik A, Hartmann D, Jilale A, Renaud FNR, Elaissari A (2015) Essential oils: from extraction to encapsulation. Int J Pharm 483:220–243. https://doi.org/10.1016/j.ijpharm.2014.12.069

Ferreira OO, Neves da Cruz J, de Jesus Pereira Franco C, Silva SG, da Costa WA, de Oliveira MS, de Aguiar Andrade EH (2020) First report on yield and chemical composition of essential oil extracted from Myrcia eximia DC (Myrtaceae) from the Brazilian Amazon. Molecules 25:783. https://doi.org/10.3390/molecules25040783

Fornari T, Vicente G, Vázquez E, García-Risco MR, Reglero G (2012) Isolation of essential oil from different plants and herbs by supercritical fluid extraction. J Chromatogr A 1250:34–48. https://doi.org/10.1016/j.chroma.2012.04.051

Ghasemi E, Raofie F, Najafi NM (2011) Application of response surface methodology and central composite design for the optimisation of supercritical fluid extraction of essential oils from Myrtus communis L. leaves. Food Chem 126:1449–1453. https://doi.org/10.1016/j.foodchem.2010.11.135

Ghavam M, Manca ML, Manconi M, Bacchetta G (2020) Chemical composition and antimicrobial activity of essential oils obtained from leaves and flowers of Salvia hydrangea DC. ex Benth. Sci Rep 10:1–10. https://doi.org/10.1038/s41598-020-73193-y

Greay SJ, Hammer KA (2015) Recent developments in the bioactivity of mono- and diterpenes: anticancer and antimicrobial activity. Phytochem Rev 14:1–6. https://doi.org/10.1007/s11101-011-9212-6

Guimarães AC, Meireles LM, Lemos MF, Guimarães MC, Endringer DC, Fronza M, Scherer R (2019) Antibacterial activity of terpenes and terpenoids present in essential oils. Molecules 24:2471

Han X, Beaumont C, Stevens N (2017) Chemical composition analysis and in vitro biological activities of ten essential oils in human skin cells. Biochim Open 5:1–7. https://doi.org/10.1016/j.biopen.2017.04.001

Hanif MA, Nisar S, Khan GS, Mushtaq Z, Zubair M (2019) Essential oils BT – essential oil research: trends in biosynthesis, analytics, industrial applications and biotechnological production. In: Malik S (ed) . Springer International Publishing, Cham, pp 3–17

Heldwein CG, Silva LL, Reckziegel P, Barros FMC, Bürger ME, Baldisserotto B, Mallmann CA, Schmidt D, Caron BO, Heinzmann BM (2012) Participation of the GABAergic system in the anesthetic effect of Lippia alba (Mill.) N.E. Brown essential oil. Braz J Med Biol Res 45:436–443. https://doi.org/10.1590/S0100-879X2012007500052

Horváth G, Ács K (2015) Essential oils in the treatment of respiratory tract diseases highlighting their role in bacterial infections and their anti-inflammatory action: a review. Flavour Fragr J 30:331–341. https://doi.org/10.1002/ffj.3252

Hyldgaard M, Mygind T, Meyer R (2012) Essential oils in food preservation: mode of action, synergies, and interactions with food matrix components. Front Microbiol 3:12

Islam MT, da Mata AMOF, de Aguiar RPS, Paz MFCJ, de Alencar MVOB, Ferreira PMP, de Carvalho Melo-Cavalcante AA (2016) Therapeutic potential of essential oils focusing on diterpenes. Phytother Res 30:1420–1444. https://doi.org/10.1002/ptr.5652

Karimi S, Sharifzadeh S, Abbasi H (2020) Sequential ultrasound-microwave assisted extraction as a green method to extract essential oil from Zataria multiflor. J Food Bioprocess Eng 3:101–109

Kaur H, Bhardwaj U, Kaur R (2021) Cymbopogon nardus essential oil: a comprehensive review on its chemistry and bioactivity. J Essent Oil Res 33:205–220. https://doi.org/10.1080/1041290 5.2021.1871976

Kokolakis AK, Golfinopoulos SK (2013) Microwave-assisted techniques (MATs); a quick way to extract a fragrance: a review. Nat Prod Commun 8:1493–1504. https://doi.org/10.117 7/1934578x1300801040

Lange BM (2015) The evolution of plant secretory structures and emergence of terpenoid chemical diversity. Annu Rev Plant Biol 66:139–159. https://doi.org/10.1146/annurev-arplant-043014-114639

Leão RP, Cruz JVJN, da Costa GV, Cruz JVJN, Ferreira EFB, Silva RC, de Lima LR, Borges RS, dos Santos GB, Santos CBR (2020) Identification of new rofecoxib-based cyclooxygenase-2 inhibitors: a bioinformatics approach. Pharmaceuticals 13:1–26. https://doi.org/10.3390/ph13090209

Leigh-de Rapper S, van Vuuren SF (2020) Odoriferous therapy: a review identifying essential oils against pathogens of the respiratory tract. Chem Biodivers 17. https://doi.org/10.1002/cbdv.202000062

Li Y, Fabiano-Tixier A-S, Chemat F (2014) Essential oils: from conventional to green extraction. In: Sharma SK (ed) SpringerBriefs in green chemistry for sustainability, 1st edn. Springer Nature, Switzerland, pp 9–20

Mandal V, Mohan Y, Hemalatha S (2007) Microwave assisted extraction – an innovative and promising extraction tool for medicinal plant research Vivekananda. Pharmacogn Rev 1:7–18

Mohamed AA, Ali SI, El-Baz FK (2013) Antioxidant and antibacterial activities of crude extracts and essential oils of Syzygium cumini leaves. PLoS One 8. https://doi.org/10.1371/journal.pone.0060269

Monzote L, Scull R, Cos P, Setzer WN (2017) Essential oil from Piper aduncum: chemical analysis, antimicrobial assessment, and literature review. Medicines 4:49. https://doi.org/10.3390/MEDICINES4030049

Muhammad Z, Yusoff ZM, Noor M, Nordin N, Kasuan N, Taib MN, Hezri M, Rahiman F, Haiyee ZA (2013) Steam distillation with induction heating system: analysis of kaffir lime oil compound and production yield at various temperatures (Penyulingan Wap Dengan Sistem Pemanasan Aruhan: Analisis Komposisi Minyak Limau Purut Dan Hasil Pengeluaran Pada Suhu Pelb). Malaysian J Anal Sci 17:340–347

Najafabadi AS, Naghavi MR, Farahmand H, Abbasi A, Yazdanfar N (2017) Chemical composition of the essential oil from oleo-gum-resin and different organs of Ferula gummosa. J Essent Oil-Bear Plants 20:282–288. https://doi.org/10.1080/0972060X.2016.1263582

Nazzaro F, Fratianni F, Coppola R, De Feo V (2017) Essential oils and antifungal activity. Pharmaceuticals 10:1–20. https://doi.org/10.3390/ph10040086

Okla MK, Alamri SA, Salem MZM, Ali HM, Behiry SI, Nasser RA, Alaraidh IA, Al-Ghtani SM, Soufan W (2019) Yield, phytochemical constituents, and antibacterial activity of essential oils from the leaves/twigs, branches, branch wood, and branch bark of sour orange (Citrus aurantium L.). Processes 7. https://doi.org/10.3390/pr7060363

Orio L, Cravotto G, Binello A, Pignata G, Nicola S, Chemat F (2012) Hydrodistillation and in situ microwave-generated hydrodistillation of fresh and dried mint leaves: a comparison study. J Sci Food Agric 92:3085–3090. https://doi.org/10.1002/jsfa.5730

Pandey AK, Kumar P, Singh P, Tripathi NN, Bajpai VK (2017) Essential oils: sources of antimicrobials and food preservatives. Front Microbiol 7:1–14. https://doi.org/10.3389/fmicb.2016.02161

Rassem HHA, Nour AH, Yunus RM (2016) Techniques for extraction of essential oils from plants: a review. Aust J Basic Appl Sci 10:117–127

Razafimamonjison G, Jahiel M, Duclos T, Ramanoelina P, Fawbush F, Danthu P (2014) Bud, leaf and stem essential oil composition of Syzygium aromaticum from Madagascar, Indonesia and Zanzibar. Int J Basic Appl Sci 3:224–233. https://doi.org/10.14419/ijbas.v3i3.2473

Regnault-Roger C, Vincent C, Arnason JT (2011) Essential oils in insect control: low-risk products in a high-stakes world. Annu Rev Entomol 57:405–424. https://doi.org/10.1146/annurev-ento-120710-100554

Reyes-Jurado F, Franco-Vega A, Ramírez-Corona N, Palou E, López-Malo A (2015) Essential oils: antimicrobial activities, extraction methods, and their modeling. Food Eng Rev 7:275–297. https://doi.org/10.1007/s12393-014-9099-2

Ribeiro-Filho HV, De Souza Silva CM, De Siqueira RJ, Lahlou S, Dos Santos AA, Magalhães PJC (2016) Biphasic cardiovascular and respiratory effects induced by β-citronellol. Eur J Pharmacol 775:96–105. https://doi.org/10.1016/j.ejphar.2016.02.025

Said ZB-OS, Haddadi-Guemghar H, Boulekbache-Makhlouf L, Rigou P, Remini H, Adjaoud A, Khoudja NK, Madani K (2016) Essential oils composition, antibacterial and antioxidant activities of hydrodistillated extract of Eucalyptus globulus fruits. Ind Crops Prod 89:167–175. https://doi.org/10.1016/j.indcrop.2016.05.018

Sajid A, Manzoor Q, Iqbal M, Tyagi AK, Sarfraz RA, Sajid A (2018) Pinus Roxburghii essential oil anticancer activity and chemical composition evaluation. EXCLI J 17:233–245. https://doi.org/10.17179/excli2016-670

Saljoughian S, Roohinejad S, Bekhit AEDA, Greiner R, Omidizadeh A, Nikmaram N, Mousavi Khaneghah A (2018) The effects of food essential oils on cardiovascular diseases: a review. Crit Rev Food Sci Nutr 58:1688–1705. https://doi.org/10.1080/10408398.2017.1279121

Santana de Oliveira M, Pereira da Silva VM, Cantão Freitas L, Gomes Silva S, Nevez Cruz J, de Aguiar Andrade EH (2021) Extraction yield, chemical composition, preliminary toxicity of Bignonia nocturna (Bignoniaceae) essential oil and in silico evaluation of the interaction. Chem Biodivers 18:e2000982. https://doi.org/10.1002/cbdv.202000982

Sharifi-Rad J, Sureda A, Tenore GC, Daglia M, Sharifi-Rad M, Valussi M, Tundis R, Sharifi-Rad M, Loizzo MR, Oluwaseun Ademiluyi A, Sharifi-Rad R, Ayatollahi SA, Iriti M (2017) Biological activities of essential oils: from plant chemoecology to traditional healing systems. Molecules 22:1–55. https://doi.org/10.3390/molecules22010070

Sihoglu Tepe A, Ozaslan M (2020) Anti-Alzheimer, anti-diabetic, skin-whitening, and antioxidant activities of the essential oil of Cinnamomum zeylanicum. Ind Crops Prod 145:112069. https://doi.org/10.1016/j.indcrop.2019.112069

Silva SG, da Costa RA, de Oliveira MS, da Cruz JN, Figueiredo PLB, do Socorro Barros Brasil D, Nascimento LD, de Jesus Chaves Neto AM, de Carvalho RN, de Aguiar Andrade EH (2019) Chemical profile of Lippia thymoides, evaluation of the acetylcholinesterase inhibitory activity of its essential oil, and molecular docking and molecular dynamics simulations. PLoS One 14:e0213393. https://doi.org/10.1371/journal.pone.0213393

Silva SG, de Oliveira MS, Cruz JN, da Costa WA, da Silva SHM, Barreto Maia AA, de Sousa RL, Carvalho Junior RN, de Aguiar Andrade EH (2021) Supercritical CO2 extraction to obtain Lippia thymoides Mart. & Schauer (Verbenaceae) essential oil rich in thymol and evaluation of its antimicrobial activity. J Supercrit Fluids 168:105064. https://doi.org/10.1016/j.supflu.2020.105064

Sobral MV, Xavier AL, Lima TC, de Sousa DP (2014) Antitumor activity of monoterpenes found in essential oils. Sci World J 2014:953451. https://doi.org/10.1155/2014/953451

Sodeifian G, Sajadian SA, Ardestani NS (2017) Experimental optimization and mathematical modeling of the supercritical fluid extraction of essential oil from Eryngium billardieri: application of simulated annealing (SA) algorithm. J Supercrit Fluids 127:146–157

Sovilj MN, Nikolovski BG, Spasojević MD (2011) Critical review of supercritical fluid extraction of selected spice plant materials. Maced J Chem Chem Eng 30:197–220. https://doi.org/10.20450/mjcce.2011.35

Sovová H (2012) Modeling the supercritical fluid extraction of essential oils from plant materials. J Chromatogr A 1250:27–33. https://doi.org/10.1016/j.chroma.2012.05.014

Tongnuanchan P, Benjakul S (2014) Essential oils: extraction, bioactivities, and their uses for food preservation. J Food Sci 79:1231–1249. https://doi.org/10.1111/1750-3841.12492

Wang ZJ, Heinbockel T (2018) Essential oils and their constituents targeting the GABAergic system and sodium channels as treatment of neurological diseases. Molecules 23:1–24. https://doi.org/10.3390/molecules23051061

Wang W, Li N, Luo M, Zu Y, Efferth T (2012) Antibacterial activity and anticancer activity of Rosmarinus officinalis L. essential oil compared to that of its main components. Molecules 17:2704–2713. https://doi.org/10.3390/molecules17032704

Werrie P-Y, Durenne B, Delaplace P, Fauconnier M-L (2020) Phytotoxicity of essential oils: opportunities and constraints for the development of biopesticides. A review. Foods 9:1291

Yadav AA, Chikate SS, Vilat RB, Suryawanshi MA, Student UG, Mumbai N, Mumbai N (2017) Review on steam distillation: a promising technology for extraction of essential oil. Int J Adv Eng Res Dev 4:667–671. https://doi.org/10.21090/ijaerd.33095

Yousefi M, Rahimi-Nasrabadi M, Pourmortazavi SM, Wysokowski M, Jesionowski T, Ehrlich H, Mirsadeghi S (2019) Supercritical fluid extraction of essential oils. Trends Anal Chem 118:182–193. https://doi.org/10.1016/j.trac.2019.05.038

Yu CY, Zhang JF, Wang T (2020) Star anise essential oil: chemical compounds, antifungal and antioxidant activities: a review. J Essent Oil Res 33:1–22. https://doi.org/10.1080/10412905.2020.1813213

Zhang N, Yao L (2019) Anxiolytic effect of essential oils and their constituents: a review. J Agric Food Chem 67:13790–13808. https://doi.org/10.1021/acs.jafc.9b00433

Zuzarte M, Salgueiro L (2015) Essential oils chemistry. In: de Sousa DP (ed) Bioactive essential oils and cancer, 1st edn. Springer International Publishing, Cham, pp 19–61

Part II
Essential Oil, Food System Applications

Chapter 2
Antibacterial Activity of Essential Oil in Food System

Jian Ju, Yang Deng, Chang Jian Li, and Mi Li

2.1 Introduction

In recent years, food safety has become an important global public health issue, in which food safety problems caused by foodborne pathogens have attracted more and more attention (Ju et al. 2017, 2018a). Common foodborne pathogens include *Escherichia coli*, *Staphylococcus aureus*, *Salmonella* and *Listeria monocytogenes* (Ju et al. 2018b; Mishra et al. 2011). These foodborne microorganisms are not only the main culprits of food corruption, but also pose a serious threat to human health. According to related reports, the mortality rate of patients caused by food poisoning caused by *Listeria monocytogenes* is as high as 30% (Abdollahzadeh et al. 2014; Ca Leja et al. 2016).

At present, most of the antimicrobial agents used in the food industry are still chemical synthetic preservatives because of their effectiveness and low price (Ju et al. 2019c, 2020a). However, long-term use of chemical synthetic preservatives may lead to microbial drug resistance and pose a potential threat to human health (Cizeikiene et al. 2013). With the continuous improvement of people's living standards and health awareness, consumers are more inclined to use natural preservatives. Therefore, it is imperative to seek safe, efficient and green natural preservatives (Ca Leja et al. 2016).

J. Ju (✉)
College of Food Science and Engineering, Qingdao Agricultural University, Qingdao, China

Y. Deng
College of Food Science and Engineering, Qingdao Agricultural University, and Qingdao Special Food Research Institute, Qingdao, China

C. J. Li · M. Li
State Key Laboratory of Food Science and Technology, Jiangnan University, Wuxi, China

© The Author(s), under exclusive license to Springer Nature Switzerland AG 2022
M. Santana de Oliveira (ed.), *Essential Oils*, https://doi.org/10.1007/978-3-030-99476-1_2

Essential oils have attracted more and more attention of food scientists because of their broad-spectrum antibacterial activity and significant antioxidant activity (Ju et al. 2019a, 2020c). Compared with chemical synthetic antimicrobial agents, essential oil has the characteristics of easy volatilization, biodegradability, low residue and low toxicity, so it is an ideal natural bacteriostatic agent (Ju et al. 2019a). Therefore, this paper systematically introduces the main active components and basic characteristics of essential oil, and analyzes the action mechanism of essential oil on bacteria and fungi in detail. Finally, the application cases of essential oil in food preservation are summarized. The main goal of this chapter is to provide a reference basis for making better use of the antibacterial activity of essential oils in the future, and to provide a scientific basis for ensuring food safety and related research.

2.2 The Main Active Components of Essential Oil

Essential oils, also known as volatile oils, are a general term of concentrated oily liquids (Silvestre et al. 2019). Essential oils are derived from different parts of plants, such as flowers, leaves, roots, pericarp and bark (Silvestre et al. 2020). In general, high concentrations of essential oils can be obtained by concentration, distillation, fermentation and solvent extraction. At present, steam distillation is the most commonly used method for the separation of essential oils in commercial applications (Chouhan et al. 2019). Essential oil is a complex mixture containing a variety of volatile components, usually composed of dozens or hundreds of compounds (Ju et al. 2018b). Generally can be divided into the following four categories: (1) Terpenoids. terpenes are a kind of natural pharmaceutical chemical components with a wide range of biological activities, which are composed of isoprene polymers and their derivatives. Monoterpenes, sesquiterpenes and their oxygen-containing derivatives are the main terpenoids in plant essential oils. These active components mainly include citral, linalool, menthol, menthone, borneol, citronellal, et al. (2) Aromatic compounds. Aromatic compounds are the second largest group of essential oils after terpenoids. For example, eugenol is the main component of clove essential oil, anisole is the main component of fennel essential oil, paeonol is the main component of moutan bark. These aromatic compounds usually have antibacterial, anti-inflammatory and antioxidant and other physiological activities. (3) Aliphatic compounds. There are also some small molecular aliphatic compounds in plant essential oil, such as n-heptane in turpentine, methyl ionone in houttuynia cordata essential oil and sunflower alkane in sweet-scented osmanthus essential oil. In addition, essential oils also contain small molecules of alcohols, aldehydes, acids and other compounds, such as isovaleraldehyde is often found in peppermint essential oil, lemon essential oil and citrus essential oil, isovaleric acid is often found in rosemary essential oil and valerian essential oil. (4) Compounds containing sulfur and nitrogen. Sulfur-containing and nitrogen-containing compounds are a kind of compounds with less content but have great influence on the quality of essential oil. Such as dimethyl sulfide in ginger essential oil, trisulfide in onion essential oil, methyl o-aminobenzoate in jasmine essential oil, isothiocyanate in black mustard essential oil, allicin in garlic essential oil, etc. The main active components of some commonly used essential oils and their inhibitory microorganisms are described in Table 2.1.

The chemical composition of essential oil samples extracted from the same plant may be highly variable, because different parts of the plant, growth time,

Table 2.1 The main active components of some commonly used essential oils and their inhibitory microorganisms

Plant name	Main components	Content	Mainly inhibited microorganisms	References
Thyme	Thymol	10–64%	*Listeria monocytogenes*	Karabagias et al. (2011)
Sichuan Pepper	Linalool	56.1%	*Aflatoxin*	
Cinnamon	Trans cinnamyl alcohol	68.4%	*Staphylococcus aureus*	Baratta et al. (2015)
Clove	Eugenol	7.5%	*Escherichia coli*	Naveed et al. (2013)
Oregano	Carvacrol	30%	*Listeria monocytogenes*	Ultee and Smid (2001)
Rosemary	1,8-cineole	46.6%	*Escherichia coli; Salmonella*	Mounia et al. (2006)
Peppermint	Menthol	29–48%	*Escherichia coli*	Azimychetabi et al. (2021)
Chrysanthemum	β-eugenol	19.83%	*Escherichia coli; Staphylococcus aureus*	Cui et al. (2018)
Lemon balm	Geranal	45.7%	*Staphylococcus aureus*	Baratta et al. (2015)
Origanum majorana (L.).	Terpene alcohol	20.8%	*Escherichia coli*	Baratta et al. (2015)
Fennel	Thyme quinone	37.6%	*Salmonella*	Sunita et al. (2014)
Eucalyptus	1,8-cineole	4.5–70.4%	*Staphylococcus aureus*	Elaissi et al. (2012)
Garlic	Diallyl disulfide	20–30%	*Bacillus cereus*	Jin et al. (2021)
Lavender	Linalyl acetate	27.6%	*Magnaporthe grisea; Botrytis cinerea*	Virgiliou et al. (2021), and Maietti et al. (2013)
Perilla frutescens (L.)	Perilla aldehyde	54.37%	*Staphylococcus aureus; Escherichia coli; Bacillus subtilis*	Zhang et al. (2018), and Zhong et al. (2020)
Camellia sinensis (L.)	Terpene-4-alcohol	40%	*Alternaria solani*	Hendges et al. (2021)
Ocimum basilicum (L.)	Linalool	41.3%	*Escherichia coli*	Amor et al. (2021)
Rose	Citronellol	34.0%	*Staphylococcus aureus; Escherichia coli*	Cebi et al. (2021), and Li et al. (2009)

geographical location, harvest time, extraction methods and storage conditions will
have an important impact on the chemical composition of essential oil (Ju et al.
2018b; Sharma et al. 2020). Therefore, the term chemical type has been used to
describe essential oils separated from specific plant species. For example, there are
six chemical types of thyme essential oils. The different chemical types mainly
depend on the content and type of the main active components in the essential oil.
Figure 2.1 shows the structural formula of some active components of essential oils
commonly used in food preservation.

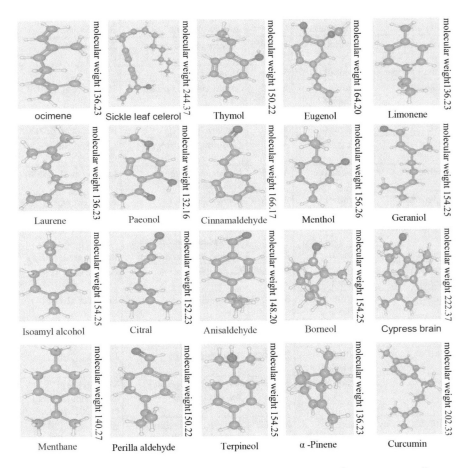

Fig. 2.1 Structural formula and molecular weight of some common active components of essential oils

2.3   Basic Characteristics of Essential Oils

The appearance of essential oil is mostly colorless or yellowish transparent liquid, which is composed of compounds with low molecular weight and low boiling point (Burt 2004). In general, essential oils have a strong aromatic smell. Essential oils are sensitive to light, heat and air, and are highly volatile at room temperature (Ni et al. 2021). The pH of essential oils is generally acidic or neutral. The specific gravity of essential oil is generally between 0.85–1.065, and the boiling point is between 70–300 °C. Essential oils are insoluble in water and soluble in less polar organic solvents such as petroleum ether and n-hexane. In addition, the essential oil also has a certain optical rotation and refractive index, usually its optical rotation is between 15–35, and refractive index is between 1.43–1.61 (Mandal et al. 2015).

2.4   Antibacterial Mechanism of Essential Oil

At present, many publications describe the relationship between the main active components and antibacterial activity of essential oils. Among them, the components with high antibacterial activity are mainly phenols, followed by oxygen-containing terpenes. Terpenes and other components including ketones and esters in essential oils, showed weaker antibacterial activity than phenols and terpenoids (Burt 2004; Ju et al. 2020e). Figure 2.2 shows a schematic diagram of the possible antibacterial mechanism of essential oils.

Because essential oil is a complex mixture, the effect and mechanism of essential oil on bacteria are different (Ju et al. 2019b). Current studies have shown that the

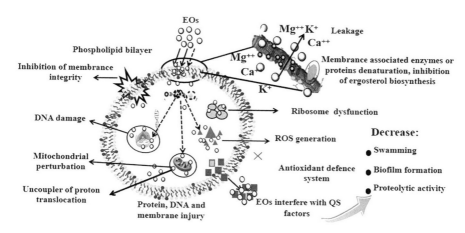

Fig. 2.2 Schematic diagram of the possible action mechanism of essential oil (Ju et al. 2019a)

main mechanisms of anti-bacterial effects of essential oils are: (1) destroying the integrity of bacterial cell wall membrane; (2) affecting membrane potential; (3) affecting the synthesis of bacterial protein and DNA; (4) interfering with the metabolic process of bacterial cells.

First of all, the essential oil exerts its bacteriostatic effect by changing the integrity of the cell wall membrane. Essential oil is a hydrophobic compound and can interact with the cell wall membrane of bacteria, which will result in changes in the permeability of the cell wall membrane, irreversible damage to the membrane structure, leakage of cell contents and eventually apoptosis (Lv et al. 2011; Ju et al. 2018b). For example, cinnamaldehyde can dissolve in the fatty acyl chain of the cell membrane, thus destroying the structure of the outer membrane, resulting in increased membrane permeability and cell death (Yin et al. 2020; Zhang et al. 2016). Red sand essential oil can affect the cytoskeleton structure of *Escherichia coli* cell membrane, thus affecting the integrity of the membrane and leading to the leakage of important biological macromolecules (Guo et al. 2016). In addition, thymol, carvol and cinnamaldehyde could affect the structure and proportion of cell membrane fatty acids in *Escherichia coli* O157:H7, *Staphylococcus aureus*, *Salmonella typhimurium* and *Pseudomonas fluorescens*, resulting in a decrease in the content of unsaturated fatty acids C18:2 trans and C18:3 cis and an increase in saturated fatty acids (especially C17: 0) (Marchese et al. 2016). In general, Gram-positive bacteria are more sensitive to essential oils than that of Gram-negative bacteria. This is mainly because the cell wall of Gram-positive bacteria is composed of peptidoglycan, which makes it easy for hydrophobic molecules to penetrate the cell wall and play a role in the cell wall and cytoplasm. In addition to peptidoglycan, the cell wall of Gram-negative bacteria has a phospholipid outer membrane connected by lipopolysaccharide. As a result, gram-negative bacteria are more resistant (Ju et al. 2018b).

Secondly, essential oil affects cell membrane potential. The change of membrane potential is an important index to evaluate the life activity of microorganisms (Ju et al. 2018b; Kong et al. 2019). Plant-derived natural products can cause the depolarization or hyperpolarization of cell membrane potential, and change the acid-base environment of intracellular fluid, thus affecting the growth of bacteria. The phenomena of depolarization and hyperpolarization are mainly due to the damage of the key functions of the cell membrane caused by the change of intracellular ion concentration (Duan et al. 2017). Among them, one of the mechanisms of pomegranate peel polyphenols on *Listeria monocytogenes* is that the hyperpolarization of cell membrane potential leads to the leakage of a large amount of intracellular potassium ions, which eventually leads to apoptosis (He et al. 2020). Similarly, amaranth extract can significantly increase the relative fluorescence value of *Staphylococcus aureus* cell membrane and cause depolarization (Li et al. 2014).

Thirdly, the effect on the synthesis of protein and intracellular genetic material. Protein, as the material basis of life, is the agent of bacteria to play a variety of physiological functions. Some active components in essential oils can affect the life activities of pathogens by acting on the proteins of pathogens (Muthaiyan et al. 2012). Similarly, related studies include *Kaempferia galanga Linn* essential oil that can reduce protein synthesis in *Escherichia coli*, *Salmonella typhimurium* and *Staphylococcus aureus* (Yang et al. 2018). Berberine can also inhibit the synthesis of *Streptococcus* protein (Peng et al. 2015).

Fourthly, the effect on energy metabolism. In cellular respiration, the electron transport chain on the cell membrane produces a transmembrane proton gradient, which is necessary for the synthesis of ATP. This process is catalyzed by a variety of enzymes with ATP enzyme activity, including ATP-dependent transporters and F1Fo-ATP enzyme complexes (Andrés and Fierro 2010). Essential oils can affect ATP synthesis by interfering with proton dynamics, changing the conformation of ATP and inhibiting the expression of ATP-related subunits (Turgis et al. 2009). For example, some studies have shown that trans-cinnamaldehyde can down-regulates the F1Fo-ATP enzyme of *Enterobacter sakazakii* and thus inhibiting the synthesis of ATP (Amalaranjou and Venkitanarayanan 2011). Vanillic acid can change the concentration of ATP in Carbapenem-resistant Enterobacter cloacae cells. Similarly, thymol can inhibit the activity of ATP synthetase in *Salmonella typhimurium*, affect the citric acid metabolic pathway and interfere with the tricarboxylic acid cycle (Pasqua et al. 2010; Barbosa et al. 2019).

2.5 Antifungal Mechanism of Essential Oil

Compared with bacteria, fungi have thicker cell walls. The inhibition mechanism of essential oil against fungi is generally the synergistic effect of multiple targets (Li et al. 2018; Cortés et al. 2019). At present, there are mainly four ways on the antifungal mechanism of essential oils. First of all, essential oils and their active components can change the morphology of microorganisms, such as the degradation of cell walls and the destruction of phospholipid bilayers (Vasconcelos et al. 2018; Ju et al. 2020d). Ju et al. (2020a, d) found that the combination of eugenol and citral could degrade the cell wall of *Penicillium roqueforti* and *Aspergillus niger* significantly. In addition, Qi et al. (2020) have also confirmed that cinnamaldehyde can damage the integrity of cell membrane and intracellular ultrastructure of *Aspergillus niger* (Qi et al. 2020). The second is to inhibit the synthesis of ergol. In previous reports, thymol was found to reduce the content of ergosterol in the cell membrane of *Fusarium graminearum* and confirmed that cyp51A and KES1 genes play an important role in regulating the synthesis of ergosterol (Diao et al. 2018). In addition, the antifungal effect of Polyene drugs is also due to their ability to reduce the content of ergosterol and change the permeability of cell membrane (Revie et al. 2018). The third is to inhibit tricarboxylic acid cycle pathway and energy metabolism. Similarly, Ju and his colleagues have shown that the combination of eugenol and citral can inhibit the activity of key enzymes in the tricarboxylic acid cycle

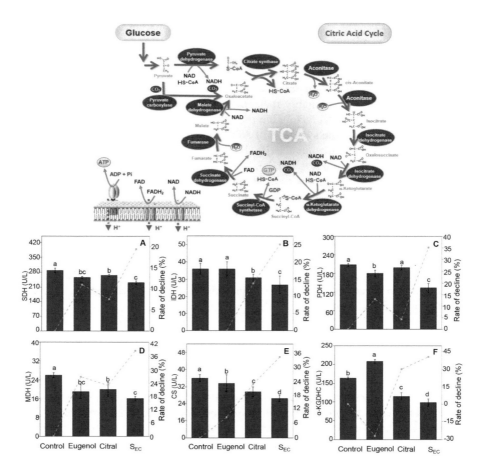

Fig. 2.3 The effects of S_{EC} on six key enzymes in TCA pathway (Ju et al. 2020a)

pathway of *Penicillium roqueforti* and cause a disorder in the normal energy metabolism of cells (Fig. 2.3) (Ju et al. 2020a). The fourth is to hinder the replication of genetic material and microbial reproduction. Essential oils can inactivate genetic material, causing DNA damage or genetic codon changes. According to related studies, it is reported that polyphenols can degrade plasmid DNA (Brudzynski et al. 2012). Similarly, Ju et al. (2020c) also confirmed that the mixture of eugenol and citral could lead to the degradation of DNA of *Penicillium roqueforti*. At present, the specific mechanism of DNA damage caused by essential oil is still unclear and needs the further study (Ju et al. 2020c) (Fig. 2.4). At present, the specific mechanism of DNA damage caused by essential oil is still unclear and needs to be further studied.

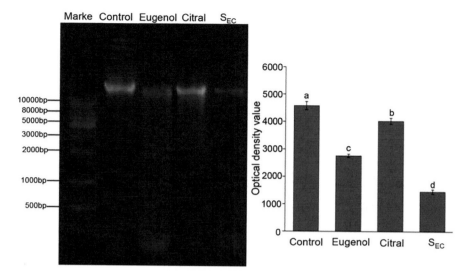

Fig. 2.4 The combination of eugenol and citral resulted in DNA degradation of *Penicillium roque-forti* (Ju et al. 2020c)

2.6   Application of Essential Oil in Food Fresh-Keeping

2.6.1   Application of Essential Oil in Fresh-Keeping of Fruits and Vegetables

Fruits and vegetables have high nutritional value, and proper consumption can supplement essential nutrients such as vitamins and minerals. However, fruits and vegetables are generally difficult to be preserved for a long time, especially in the process of transportation, they are more vulnerable to the invasion of pathogens, resulting in corruption. The use of preservatives can effectively inhibit the growth of spoilage microorganisms, reduce the decay rate of fruits and vegetables, and prolong their shelf life (Otoni et al. 2017; Jafarzadeh et al. 2021). However, due to the current safety and residue of chemical preservatives can not meet the high requirements of consumers for food safety. In recent years, economic, environmentally friendly and efficient natural preservatives have gradually become a research hotspot in the field of fruit and vegetable preservation. However, plant essential oils have attracted much attention in these natural preservatives (Tahir et al. 2019; Yousuf et al. 2021). The inhibition cases of essential oil on mold in fruits and vegetables are shown in Table 2.2.

The common treatment methods of essential oil in the preservation of fruits and vegetables are spraying, impregnation, fumigation, coating, emulsification, et al. Spraying is mainly carried out before the harvest of fruits and vegetables. Spraying essential oil can reduce the waste of essential oil and make more

Table 2.2 The inhibition cases of essential oil on mold in fruits and vegetables

Essential oil	Types of fruits and vegetables	Mold species	MIC	References
Thyme EO	Mango; citrus	*Alternaria alternata*; *Penicillium digitatum*	0.33 ~ 1.0 (μL/mL)	Combrinck et al. (2011)
Cinnamon EO	Citrus; Grape	*Rhizopus nigricans*; *Aspergillus flavus*; *Penicillium expansum*	1.0% ~ 2.0%	Xing et al. (2010)
Peppermint EO	Citrus	*Penicillium digitatum*	500 (μL/L)	Xing et al. (2010)
Anethum graveolens L. EO	Cherry; tomato	*Aspergillus flavus*; *Aspergillus oryzae*	2.0 (μL/mL)	Plooy et al. (2009)
Clove EO	Jujube; Citrus; Grape	*Botrytis cinerea*; *Alternaria alternata*	600 (μg/mL)	Guan and Li (2005)
Rosemary EO	Grape	*Aspergillus flavus*; *Spergillus niger*	0.25 ~ 1.0 (μL/mL)	Sousa et al. (2013)
Oregano EO	Plum	–	0.5, 1, 1.5 and 2% (w/w)	
Cambessedes EO	Mangabas	*Bacillus cereus*; *Serratia marcescens*	1.25% (w/w)	
Carvacrol	Vegetables	*Listeria monocytogenes*; *Aeromonas hydrophila*; *Pseudomonas fluorescens*	1.25 (μL/mL)	
Galangal	Mango	–	8% (w/w)	Zhou et al. (2020)
Cumin EO	Fish fillet	–	4 (μl/l)	Cai et al. (2015)
Pepper EO	–	*Aflatoxin*	0.6 (μl/ml)	Prakash et al. (2012)

efficient use of essential oil to keep fruits and vegetables fresh. For example, spraying citrus with *Citrus aurantiifolia* essential oil and thyme essential oil before harvest can significantly reduce the decay rate of the fruit and prolong the shelf life of the fruit compared with the untreated group (Badawy et al. 2011). The impregnation method is to soaking the fruits and vegetables in the essential oil emulsion with appropriate concentration gradient. After soaking, the fruits and vegetables are taken out and dried, and then stored in a fresh-keeping box. The main advantage of this method is that it is easy to operate, but it is also easy to cause secondary damage to fruits and vegetables. The related case study is that papaya was soaked in thyme essential oil and *Citrus aurantiifolia* essential oil solution to prolong the storage life of papaya (Bosquez-Molina et al. 2010). Fumigation treatment of essential oil refers to the use of essential oil as fumigant to evenly cover the surface of fruits and vegetables in the form of gas diffusion in a closed space, so as to effectively inhibit the growth and reproduction of pathogenic bacteria and rot causing bacteria. This method combined with modified atmosphere packaging may be a promising preservation technology. However, unfortunately, the relevant research has not been paid attention to. Microemulsification of essential oil refers to a thermodynamically stable system with a particle size of 1–100 nm prepared by mixing surfactant, cosurfactant, water phase and oil phase in a suitable proportion (Yao et al. 2021). The problems of volatile, unstable and short aging of essential oil can be solved by emulsifying. At the same time, liquid essential oil can be transformed into gel or solid state, which is conducive to its storage, transportation and use. At present, some gratifying results have been achieved in the research on the use of edible coating containing essential oil in the preservation of fruits and vegetables. For example, coating oranges with chitosan and tea tree essential oil can better maintain the color, hardness and solubility of oranges during shelf life (Cháfer et al. 2012). In addition, coating cherry and tomato fruits with chitosan and peppermint essential oil can also effectively reduce fruit decay during storage (Guerra et al. 2015).

2.6.2 Application of Essential Oil in Fresh-Keeping of Meat Products

Meat products are perishable foods, which need proper processing and treatment to prolong their shelf life. In addition, crushed meat products deteriorate more easily than fresh meat, because crushing increases the surface area and contact area of meat, exposing it to air and microorganisms. Under hygienic processing conditions, the microbial load is small, but processing and frequent treatment may bring harmful microorganisms. The growth and proliferation of these microorganisms lead to physical, chemical and sensory changes in meat products (Shah et al. 2014; Umaraw et al. 2020). Therefore, the film and coating containing essential oil has become a promising method for green preservation of meat products. Related research cases include the use of grape seed extract coated with chitosan to prolong the shelf life of chicken breast under refrigerated conditions (Hassanzadeh et al. 2017). Similarly, under the condition of cold storage, the combination of oregano essential oil and whey protein film applied to the preservation of chicken breast can effectively control the growth of spoilage microorganisms during storage. By using this active coating, the shelf life of chicken breast was extended from 6 to 13 days at 4 °C (Fernandes et al. 2018). In addition, the shelf life of chicken breast can be extended by 20 days at 4 °C when chitosan is used together with garlic essential oil (Bazargani-Gilani et al. 2015). Figure 2.5 shows the action mechanism of bioactive compounds in edible films and coatings (Umaraw et al. 2020).

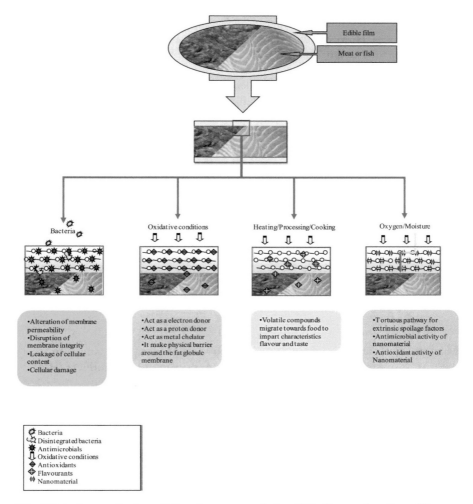

Fig. 2.5 Action mechanism of bioactive compounds in edible films and coatings (Umaraw et al. 2020)

2.6.3 Application of Essential Oil in Fresh-Keeping of Baked Goods

Baked food is one of the most important basic foods consumed by people all over the world, and it is also an important part of daily diet (Gavahian et al. 2018; Manzocco et al. 2020; Garcia and Copetti 2019). However, due to the particularity of baked products, they usually can not be directly placed after baking in the appropriate mold, and fungal spores are commonly found in the atmosphere.

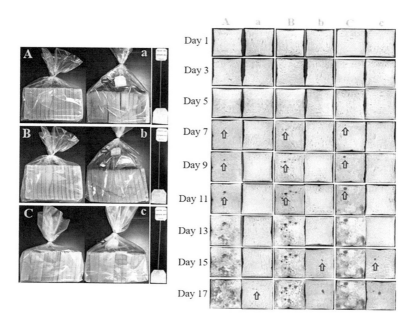

Fig. 2.6 The shelf life of bread and changes of microbial growth of bread during storage. A. B and C represent LDPE, PP and HDPE packaging, respectively. Breads in three different packages (LDPE, PP and HDPE) had a total shelf life of 13, 13, and 15 days, respectively, while control breads had a shelf life of 5 days (Ju et al. 2020b)

There are as many as 1000 fungal spores/m³ air found in some processing environments (Debonne et al. 2018). So baked products are usually contaminated by fungal spores in the air when exposed to air and the surface of processing or equipment. In addition, the baked products are rich in nutrients, which are especially suitable for the growth and reproduction of molds. According to relevant statistics, about 5% of bread products in Western Europe lose their edible value due to mold pollution every year, resulting in economic losses of more than 200 million US dollars (Els et al. 2018). This will not only cause significant economic losses, but also produce strongly pathogenic and carcinogenic toxins because some rotten fungi, which pose a serious threat to our health (Ju et al. 2018b).

At present, essential oils have attracted more and more researchers' attention in the bakery industry because of their remarkable antibacterial and antioxidant properties. In this regard, microcapsules containing essential oil can gradually release volatile compounds into packaging bags, and this method has been proved to have significant inhibitory effect on *Penicillum roqueforti* and *Aspergillus niger* (Fig. 2.6) (Ju et al. 2020b). In addition, this innovative system containing essential oil active ingredients has been proved to be effective in inhibiting the growth of yeast and mold on laboratory culture media and sliced bread (Conto et al. 2012; Anandharamakrishnan et al. 2015). Although this method will not affect the texture of bread, it will produce an unpleasant smell when the concentration of essential oil is high. As a substitute for modified atmosphere packaging, the active packaging prepared by the combination of essential oil and resin also has a good inhibitory effect on some common fungi on bread. Among them, some authors have shown that mustard essential oil has the strongest inhibitory effect on fungi in the gas phase. Through further study on allyl isothiocyanate, the main active ingredient in mustard essential oil, it was found that AITC had bactericidal effect on all tested fungi when its content was ≥3.5 mg/mL in gas phase. However, the fungi grown in hot dog bread were more sensitive to AITC than those in rye bread. The lowest inhibitory concentration of AITC to rye bread was 2.4 mg/mL and the lowest inhibitory concentration to hot dog bread was 1.8 mg/mL (López-Malo et al. 2007). In addition, the addition of 6% cinnamon essential oil as an active coating material into solid paraffin wax also had a significant inhibitory effect on mold growth in rye bread (RodrGuez et al. 2008). Similarly, microcapsules containing mustard essential oil have a better fresh-keeping effect than adding mustard essential oil directly to bread packaging material as an ingredient (Clemente et al. 2019).

2.7 Conclusion

This chapter first describes the main active components and basic characteristics of essential oil, then summarizes the antimicrobial mechanism of essential oil against bacteria and fungi in detail, and then systematically analyzes the application cases of essential oil in food preservation. At present, the studies on essential oils are mostly focused on their antibacterial and antioxidant properties in vitro, but there are few studies on the effects of essential oils on microorganisms in various food substrates. In particular, there are few reports about the chemical changes of essential oils in food and their effects on bacterial succession. On the other hand, the aromatic smell of essential oil is not suitable for application in some foods. While, the synergy between essential oil and other antibacterial agents could improve the antibacterial performance of essential oil and reduce the dosage of essential oil in food preservation. Therefore, it is necessary to study the synergistic effect of essential oil and other antibacterial agents.

Acknowledgments The work described in this article is supported by the China Scholarship Council.

Conflict of Interest   The authors declare no competing interests.

References

Abdollahzadeh E, Rezaei M, Hosseini H (2014) Antibacterial activity of plant essential oils and extracts: the role of thyme essential oil, nisin, and their combination to control Listeria monocytogenes inoculated in minced fish meat. Food Control 35(1):177–183

Amalaranjou MAR, Venkitanarayanan K (2011) Effect of trans-cinnamaldehyde on inhibition and inactivation of Cronobacter sakazakii biofilm on abiotic surfaces. J Food Prot 74(2):200–208

Amor G, Sabbah M, Caputo L, Idbella M, Feo VD, Porta R (2021) Basil essential oil: composition, antimicrobial properties, and microencapsulation to produce active chitosan films for food packaging. Foods 10(121):1–16

Anandharamakrishnan C, Ezhilarasi PN, Indrani D (2015) Microencapsulation of green tea polyphenols and its effect on incorporated bread quality. LWT-Food Sci Technol 64(1):289–296

Andrés MT, Fierro JF (2010) Antimicrobial mechanism of action of transferrins: selective inhibition of H+-ATPase. Antimicrob Agents Chemother 54:4335–4342

Azimychetabi Z, Nodehi MS, Moghadam TK, Motesharezadeh B (2021) Cadmium stress alters the essential oil composition and the expression of genes involved in their synthesis in peppermint (Mentha piperita L.). Ind Crops Prod 168:113602

Badawy F, Sallam M, Ibrahim AR, Asran MR (2011) Efficacy of some essential oils on controlling green mold of orange and their effects on postharvest quality parameters. Plant Pathol J 10(4):168–174

Baratta MT, Dorman H, Deans SG, Figueiredo AC, Ruberto G (2015) Antimicrobial and antioxidant properties of some commercial essential oils. Flavour Fragr J 13(4):235–244

Barbosa LN, Alves F, Andrade B, Albano M, Junior AF (2019) Proteomic analysis and antibacterial resistance mechanisms of Salmonella enteritidis submitted to the inhibitory effect of Origanum vulgare essential oil, thymol and carvacrol. J Proteome 214:103625

Bazargani-Gilani B, Aliakbarlu J, Tajik H (2015) Effect of pomegranate juice dipping and chitosan coating enriched with Zataria multiflora Boiss essential oil on the shelf-life of chicken meat during refrigerated storage. Innovative Food Sci Emerg Technol 29:280–287

Bosquez-Molina E, Jesús RD, Bautista-BaOs S, Verde-Calvo JR, Morales-López J (2010) Inhibitory effect of essential oils against Colletotrichum gloeosporioides and Rhizopus stolonifer in stored papaya fruit and their possible application in coatings. Postharvest Biol Technol 57(2):132–137

Brudzynski K, Abu Ba Ker K, Miotto D (2012) Unraveling a mechanism of honey antibacterial action: polyphenol/H_2O_2-induced oxidative effect on bacterial cell growth and on DNA degradation. Food Chem 133(2):329–336

Burt S (2004) Essential oils: their antibacterial properties and potential applications in foods—a review. Int J Food Microbiol 94(3):223–253

Ca Leja C, Barros L, Antonio AL, Carocho M, Oliveira MBPP, Ferreira I (2016) Fortification of yogurts with different antioxidant preservatives: a comparative study between natural and synthetic additives. Food Chem 210:262–268

Cai LY, Ailing CY, Lia C, Zhuo S, Liping L, Li J (2015) The effects of essential oil treatment on the biogenic amines inhibition and quality preservation of red drum (Sciaenops ocellatus) fillets. Food Control 56:1–8

Cebi N, Sagdic O, Arici M (2021) The famous Turkish rose essential oil: characterization and authenticity monitoring by FTIR, Raman and GC–MS techniques combined with chemometrics. Food Chem 354:129495

Cháfer M, Sánchez González L, González Martínez C (2012) Fungal decay and shelf life of oranges coated with chitosan and bergamot, thyme, and tea tree essential oils. J Food Sci 77(8):E182–E187

Chouhan K, Tandey R, Sen KK, Mehta R, Mandal V (2019) Critical analysis of microwave hydrodiffusion and gravity as a green tool for extraction of essential oils: time to replace traditional distillation. Trends Food Sci Technol 92:12–21

Cizeikiene D, Juodeikiene G, Paskevicius A, Bartkiene E (2013) Antimicrobial activity of lactic acid bacteria against pathogenic and spoilage microorganism isolated from food and their control in wheat bread. Food Control 31(2):539–545

Clemente I, Aznar M, Nern C (2019) Synergistic properties of mustard and cinnamon essential oils for the inactivation of foodborne moulds in vitro and on Spanish bread. Int J Food Microbiol 298:44–50

Combrinck S, Regnier T, Kamatou G (2011) In vitro activity of eighteen essential oils and some major components against common postharvest fungal pathogens of fruit. Ind Crop Prod 33(2):344–349

Conto LCD, Oliveira RSP, Martin LGP (2012) Effects of the addition of microencapsulated omega-3 and rosemary extract on the technological and sensory quality of white pan bread. LWT-Food Sci Technol 45(1):103–109

Cortés JCG, Curto MÁ, Carvalho VSD (2019) The fungal cell wall as a target for the development of new antifungal therapies. Biotechnol Adv 37(6):107352

Cui H, Mei B, Sun Y, Abdel-Shafi A, Lin L (2018) Antibacterial activity and mechanism of Chuzhou chrysanthemum essential oil. J Funct Foods 48:159–166

Debonne E, Maene P, Vermeulen A (2018) Validation of in-vitro antifungal activity of the fermentation quotient on bread spoilage moulds through growth/no-growth modelling and bread baking trials. LWT-Food Sci Technol 117:108636

Diao X, Hang Y, Liu C (2018) The fungicidal activity of Tebuconazole enantiomers against Fusarium graminearum and its selective effect on DON production under different conditions. J Agric Food Chem 66(14):3637–3643

Duan F, Xin G, Niu H, Huang W (2017) Chlorinated emodin as a natural antibacterial agent against drug-resistant bacteria through dual influence on bacterial cell membranes and DNA. Sci Rep 7(1):12721

Elaissi A, Rouis Z, Salem N, Mabrouk S, Salem YB, Salah K (2012) Chemical composition of 8 eucalyptus species' essential oils and the evaluation of their antibacterial, antifungal and antiviral activities. BMC Complement Altern Med 12(1):81–92

Els D, Filip VB, Simbarashe S (2018) The use of essential oils as natural antifungal preservatives in bread products. J Essent Oil Res:1–10

Fernandes RPP, Trindade MA, Lorenzo JM, de Melo MP (2018) Assessment of the stability of sheep sausages with the addition of different concentrations of Origanum vulgare extract during storage. Meat Sci 137:244–257

Garcia MV, Copetti M (2019) Alternative methods for mold spoilage control in bread and bakery products. Int Food Res J 26(3):737–749

Gavahian M, Chu YH, Rodriguez JML (2018) Essential oils as natural preservatives for bakery products: understanding the mechanisms of action, recent findings, and applications. Crit Rev Food Sci Nutr 60(2):310–321

Guan WQ, Li SF (2005) Inhibitory effect of clove essential oil on postharvest pathogens of fruits and vegetables. J Food Sci 12:227–230

Guerra ICD, De Oliveira PDL, Lima DSPA (2015) Coatings comprising chitosan and Mentha piperita L. or Mentha x villosa Huds essential oils to prevent common postharvest mold infections and maintain the quality of cherry tomato fruit. Int J Food Microbiol 214:168–178

Guo N, Zang YP, Cui Q, Gai QY, Jiao J, Wang W (2016) The preservative potential of Amomum tsaoko essential oil against E. coil, its antibacterial property and mode of action. Food Control 75(Complete):236–245

Hassanzadeh P, Tajik H, Rohani SMR, Moradi M, Hashemi M, Aliakbarlu J (2017) Effect of functional chitosan coating and gamma irradiation on the shelf-life of chicken meat during refrigerated storage. Radiat Phys Chem 141:103–109

He W, Shi C, Long X, Liu X, Zhao X (2020) Antimicrobial activity and mechanism of action of Perilla essential oil against Staphylococcus aureus. E3S Web Conf 145:01015

Hendges C, Stangarlin JR, Zamban VC, Mascaro M, Carmelo DB (2021) Antifungal activity and control of the early blight in tomato through tea tree essential oil. Crop Prot 148:105728

Jafarzadeh S, Nafchi AM, Salehabadi A, Oladzadabbasabadi N, Jafari SM (2021) Application of bio-nanocomposite films and edible coatings for extending the shelf life of fresh fruits and vegetables. Adv Colloid Interf Sci 291:102405

Jin Z, Li L, Yin ZA, Peipei A (2021) Diallyl disulfide, the antibacterial component of garlic essential oil, inhibits the toxicity of Bacillus cereus ATCC 14579 at sub-inhibitory concentrations. Food Control 126:108090

Ju J, Xu X, Xie Y, Guo Y, Cheng Y, Qian H (2017) Inhibitory effects of cinnamon and clove essential oils on mold growth on baked foods. Food Chem 240:850–855

Ju J, Yao W, Sun S, Guo Y, Cheng Y, Qian H (2018a) Assessment of the antibacterial activity and the main bacteriostatic components from bayberry fruit extract. Int J Food Prop 21(1):1043–1051

Ju J, Xie Y, Guo Y, Cheng Y, Qian H, Yao W (2018b) The inhibitory effect of plant essential oils on foodborne pathogenic bacteria in food. Crit Rev Food Sci Nutr 59(20):3281–3292

Ju J, Chen X, Xie Y, Yu H, Yao W (2019a) Application of essential oil as a sustained release preparation in food packaging. Trends Food Sci Technol 92:22–32

Ju J, Xie Y, Guo Y, Cheng Y, Qian H, Yao W (2019b) Application of edible coating with essential oil in food preservation. Crit Rev Food Sci Nutr 1–62

Ju J, Chen X, Xie Y, Yu H, Cheng Y, Qian H (2019c) Simple microencapsulation of plant essential oil in porous starch granules: adsorption kinetics and antibacterial activity evaluation. J Food Process Preserv 43:14156

Ju J, Xie Y, Guo Y, Cheng Y, Qian H, Yao W (2020a) Major components in Lilac and Litsea cubeba essential oils kill Penicillium roqueforti through mitochondrial apoptosis pathway. Ind Crop Prod 149:112349

Ju J, Xie Y, Guo Y, Cheng Y, Qian H, Yao W (2020b) A novel method to prolong bread shelf life: sachets containing essential oils components. LWT-Food Sci Technol 131:109744

Ju J, Xie Y, Guo Y, Cheng Y, Qian H, Yao W (2020c) Analysis of the synergistic antifungal mechanism of eugenol and citral – sciencedirect. LWT-Food Sci Technol 123:109128

Ju J, Xie Y, Yu H, Guo Y, Cheng Y, Zhang R (2020d) Synergistic inhibition effect of citral and eugenol against Aspergillus niger and their application in bread preservation. Food Chem 310:125974.1–125974.7

Ju J, Xie Y, Yu H, Guo Y, Yao W (2020e) Synergistic interactions of plant essential oils with antimicrobial agents: a new antimicrobial therapy. Crit Rev Food Sci Nutr 2:1–12

Karabagias I, Badeka A, Kontominas MG (2011) Shelf life extension of lamb meat using thyme or oregano essential oils and modified atmosphere packaging. Meat Sci 88(1):109–116

Kong J, Zhang Y, Ju J, Xie Y, Guo Y, Cheng Y (2019) Antifungal effects of thymol and salicylic acid on cell membrane and mitochondria of Rhizopus stolonifer and their application in post-harvest preservation of tomatoes. Food Chem 285(1):380–388

Li YJ, Liu XL, Liu X, Wu N, Peng X, Liang L (2009) Chemical constituents and antibacterial activity of rose essential oil. Plant Res 029(004):488–491

Li G, Xu Y, Wang X, Zhang B, Shi C, Zhang W (2014) Tannin-rich fraction from pomegranate rind damages membrane of Listeria monocytogenes. Foodborne Pathog Dis 11(4):313–319

Li JH, Zhao LQ, Wu AY (2018) Insights on the ultra-high antibacterial activity of positionally substituted 2′- O -hydroxypropyl trimethyl ammonium chloride chitosan: a joint interaction of -NH 2 and -N + (CH3)3 with bacterial cell wall. Colloids Surf B: Biointerfaces 9(173):429–436

López-Malo A, Barreto-Valdivieso J, Palou E (2007) Aspergillus flavus growth response to cinnamon extract and sodium benzoate mixtures. Food Control 18(11):1358–1362

Lv F, Liang H, Yuan Q, Li C (2011) In vitro antimicrobial effects and mechanism of action of selected plant essential oil combinations against four food-related microorganisms. Food Res Int 44(9):3057–3064

Maietti S, Rossi D, Guerrini A, Useli C, Romagnoli C, Poli F (2013) A multivariate analysis approach to the study of chemical and functional properties of chemo-diverse plant derivatives: lavender essential oils. Flavour Fragr J 28(3):144–154

Mandal S, Mandal M, Medicine ME, Zoology DO, Banga U, Physiology DO (2015) Coriander (Coriandrum sativum L.) essential oil: chemistry and biological activity. Asian Pac J Trop Biomed 5(6):421–428

Manzocco L, Romano G, Calligaris S (2020) Modeling the effect of the oxidation status of the ingredient oil on stability and shelf life of low-moisture bakery products: the case study of crackers. Foods 9(6):749–762

Marchese A, Orhan IE, Daglia M, Barbieri R, Lorenzo AD, Nabavi SF (2016) Antibacterial and antifungal activities of thymol: a brief review of the literature. Food Chem 210:402–414

Mishra BB, Gautam S, Sharma A (2011) Shelf life extension of sugarcane juice using preservatives and gamma radiation processing. J Food Sci 76(8):573–578

Mounia O, Stéphane C, Monique L (2006) Mechanism of action of Spanish oregano, Chinese cinnamon, and savory essential oils against cell membranes and walls of Escherichia coli o157:h7 and Listeria monocytogenes. J Food Prot 69(5):1046–1055

Muthaiyan A, Martin EM, Natesan S, Crandall PG, Wilkinson BJ, Ricke SC (2012) Antimicrobial effect and mode of action of terpeneless cold-pressed Valencia orange essential oil on methicillin-resistant Staphylococcus aureus. J Appl Microbiol 112(5):1020–1033

Naveed R, Hussain I, Tawab A, Tariq M, Iqbal M (2013) Antimicrobial activity of the bioactive components of essential oils from Pakistani spices against salmonella and other multi-drug resistant bacteria. BMC Complement Altern Med 13(1):265–265

Ni ZJ, Wang X, Shen Y, Thakur K, Wei ZJ (2021) Recent updates on the chemistry, bioactivities, mode of action, and industrial applications of plant essential oils. Trends Food Sci Technol 110(1):78–89

Otoni CG, Avena-Bustillos RJ, Azeredo HMC, Lorevice MV, Moura MR, Mattoso LHC (2017) Recent advances on edible films based on fruits and vegetables—a review. Compr Rev Food Sci Food Saf 16:1151–1169

Pasqua RD, Mamone G, Ferranti P, Ercolini D, Mauriello G (2010) Changes in the proteome of Salmonella enterica serovar Thompson as stress adaptation to sublethal concentrations of thymol. Proteomics 10:1040–1049

Peng L, Shuai K, Yin Z, Jia R, Bo J (2015) Antibacterial activity and mechanism of berberine against Streptococcus agalactiae. Int J Clin Exp Pathol 8(5):5217–5223

Plooy WD, Regnier T, Combrinck S (2009) Essential oil amended coatings as alternatives to synthetic fungicides in citrus postharvest management. Postharvest Biol Technol 53(3):117–122

Prakash B, Singh P, Mishra PK, Dubey NK (2012) Safety assessment of Zanthoxylum alatum Roxb. essential oil, its antifungal, antiaflatoxin, antioxidant activity and efficacy as antimicrobial in preservation of Piper nigrum L. fruits. Int J Food Microbiol 153(1–2):183–191

Qi S, Li J, Sun Y, Chen Q, Le T (2020) The antifungal effects of cinnamaldehyde against Aspergillus niger and its application in bread preservation. Food Chem 317:126405

Revie NM, Iyer KR, Robbins N (2018) Antifungal drug resistance: evolution, mechanisms and impact. Curr Opin Microbiol 45:70–76

RodrGuez A, NerN C, Batlle R (2008) New cinnamon-based active paper packaging against Rhizopusstolonifer food spoilage. J Agric Food Chem 56(15):6364–6369

Shah MA, Bosco SJD, Mir SA (2014) Plant extracts as natural antioxidants in meat and meat products. Meat Sci 98(1):21–33

Sharma S, Barkauskaite S, Jaiswal AK, Jaiswal S (2020) Essential oils as additives in active food packaging. Food Chem 343(8):128403

Silvestre WP, Livinalli NF, Baldasso C (2019) Pervaporation in the separation of essential oil components: a review. Trends Food Sci Technol 93:42–52

Silvestre WP, Baldasso C, Tessaro IC (2020) Potential of chitosan-based membranes for the separation of essential oil components by target-organophilic pervaporation. Carbohydr Polym 247:116676

Sousa LD, Andrade SD, Athayde A, Oliveira CD, Sales CD, Madruga MS (2013) Efficacy of Origanum vulgare L. and Rosmarinus officinalis L. essential oils in combination to control postharvest pathogenic Aspergilli and autochthonous mycoflora in Vitis labrusca L. (table grapes). Int J Food Microbiol 165(3):312–318

Sunita S, Das S, Shing G (2014) Composition, in vitro antioxidant and antimicrobial activities of essential oil and oleoresins obtained from black cumin seeds (Nigella sativa L.). Biomed Res Int 14(2):1–10

Tahir HE, Zou X, Mahunu GK, Arslan M, Li Z (2019) Recent developments in gum edible coating applications for fruits and vegetables preservation: a review. Carbohydr Polym 224:115141

Turgis M, Han J, Caillet S, Lacroix M (2009) Antimicrobial activity of mustard essential oil against Escherichia coli o157:h7 and Salmonella typhi. Food Control 20(12):1073–1079

Ultee A, Smid EJ (2001) Influence of carvacrol on growth and toxin production by Bacillus cereus. Int J Food Microbiol 64(3):373–378

Umaraw P, Munekata P, Verma AK, Barba FJ, Lorenzo JM (2020) Edible films/coating with tailored properties for active packaging of meat, fish and derived products. Trends Food Sci Technol 98:10–24

Vasconcelos NG, Croda J, Simionatto S (2018) Antibacterial mechanisms of cinnamon and its constituents: a review. Microb Pathog 120:189–203

Virgiliou C, Zisi C, Kontogiannopoulos KN, Nakas A, Iakovakis A, Varsamis V, Gika HG, Assimopoulou AN (2021) Headspace gas chromatography-mass spectrometry in the analysis of lavender's essential oil: optimization by response surface methodology. J Chromatogr B 1179:122862

Xing Y, Li X, Xu Q, Yun J, Lu Y (2010) Original article: antifungal activities of cinnamon oil against Rhizopus nigricans, Aspergillus flavus and Penicillium expansum in vitro and in vivo fruit test. Int J Food Sci Technol 45:1837–1842

Yang Y, Tian S, Wang F, Li Z, Li Y (2018) Chemical composition and antibacterial activity of Kaempferia galanga essential oil. Int J Agric Biol 20(2):457–462

Yao YN, Sun MJ, Li Z, Feng LL, Li LL (2021) Research progress of plant essential oil in fruit and vegetable preservation. J Light Ind Technol 37(5):17–21

Yin L, Chen J, Wang K, Geng Y, Ouyang P (2020) Study the antibacterial mechanism of cinnamaldehyde against drug-resistant Aeromonas hydrophila in vitro. Microb Pathog 145:104208

Yousuf B, Wu S, Siddiqui MW (2021) Incorporating essential oils or compounds derived thereof into edible coatings: effect on quality and shelf life of fresh/fresh-cut produce. Trends Food Sci Technol 108:245–257

Zhang YB, Llu XY, Wang YF (2016) Antibacterial activity and mechanism of cinnamon essential oil against Escherichia coli and Staphylococcus aureus. Food Control 59:282–289

Zhang ZJ, Li N, Li HZ, Li XJ, Cao JM, Zhang GP (2018) Preparation and characterization of biocomposite chitosan film containing Perilla frutescens (L.) Britt. essential oil. Ind Crops Prod 112:660–667

Zhong Y, Zheng Q, Hu P, Huang X, Zhang M (2020) Sedative and hypnotic effects of Perilla frutescens essential oil through GABAergic system pathway. J Ethnopharmacol 279(6):113627

Zhou W, He Y, Liu F, Liao L, Li J (2020) Carboxymethyl chitosan-pullulan edible films enriched with galangal essential oil: characterization and application in mango preservation. Carbohydr Polym 256(3):117579

Chapter 3
Activity of Essential Oils Against Food Spoilage Fungi

Anderson de Santana Botelho, Oberdan Oliveira Ferreira, Raimundo Junior da Rocha Batista, and Celeste de Jesus Pereira Franco

3.1 Introduction

Fungi are defined as heterotrophic, unicellular and multicellular organisms. The multicellulars are characterized by the formation of filamentous structures called "hyphae" that constitute the mycelium. Through the mycelium, during the reproductive phase, fungi form the spores that are responsible for the propagation of the species, which are present in different environments, some very small to the naked eye, but others easily observable, such as: molds, mildews, and mushrooms (Maia and de Carvalho Junior 2010).

Fungi are eukaryotic organisms that get their food from organic matter, and through it they get their nutrition by acting as parasites on living hosts. Thus, they can cause harmful effects to human health, such as the appearance of mycoses (da

A. de Santana Botelho (✉)
Adolpho Ducke Laboratory, Coordination of Botany Museu Paraense Emílio Goeldi, Belem, Para, Brazil

Chemical Analysis Laboratory, Coordination of Earth Sciences and Ecology, Museu Paraense Emílio Goeldi, Belem, Para, Brazil
e-mail: andersonbotelho@museu-goeldi.br

O. O. Ferreira
Adolpho Ducke Laboratory, Coordination of Botany Museu Paraense Emílio Goeldi, Belem, Para, Brazil

Federal University of Pará, Belem, Para, Brazil

R. J. da Rocha Batista
Chemical Analysis Laboratory, Coordination of Earth Sciences and Ecology, Museu Paraense Emílio Goeldi, Belem, Para, Brazil

C. de Jesus Pereira Franco
Adolpho Ducke Laboratory, Coordination of Botany Museu Paraense Emílio Goeldi, Belem, Para, Brazil

© The Author(s), under exclusive license to Springer Nature Switzerland AG 2022
M. Santana de Oliveira (ed.), *Essential Oils*, https://doi.org/10.1007/978-3-030-99476-1_3

43

Silva and do Nascimento Malta 2017) and infectious diseases (Dong et al. 2020) including intestinal disorders, vomiting, and diarrhea, caused by fungi during the deterioration of food products (Ye et al. 2013). The metabolism and growth of these microorganisms in food products has become a global problem (Villa and Veiga-Crespo 2014), as well as their strong resistance to commercial fungicides and the contamination from residues left by these chemical products (Hasheminejad et al. 2019).

This culminated in the search for new natural fungicides as alternatives, and among these alternatives, there are essential oils considered as natural antimicrobials that protect plants against microorganisms (Huang et al. 2010; Jing et al. 2014; Stević et al. 2014). The antifungal activity of essential oils can be mainly attributed to their hydrophobic nature, which causes loss of membrane integrity and leakage of cellular material from fungi (Dambolena et al. 2010). These mixtures of volatile compounds can be promising to preserve the quantity and quality of foods, because they tend to have low toxicity and do not cause environmental damage (Rana et al. 2011; Soliman et al. 2013). Given the above, the present work aimed to carry out a review on essential oils promising in combating the main genera of food contaminating fungi and their constituents.

3.2   Essential Oils and Food Spoilage Fungal Genera

3.2.1   *Aspergillus*

Aspergillus is an anamorph genus belonging to the phylum Ascomycota and comprises more than 250 species of saprophytic filamentous fungi (Rokas 2013). These species share a common structure of asexual spores called "aspergillum", which makes them morphologically very similar to each other; however, they present a high degree of phylogenetic and biological diversity, to the point of being classified into about 10 teleomorph genera (Geiser 2009). Such species can be differentiated using genome sequencing, which has proved to be a very useful tool in identification (Samson et al. 2014) and shown that some of them are as related to each other as fish are to humans (Galagan et al. 2005; Krijgsheld et al. 2013).

Fungi of the genus *Aspergillus* are among the most abundant on Earth and are not very selective regarding the conditions of abiotic growth, being able to grow in environments with temperatures from 6 to 55 °C and relatively low humidity, which allows these fungi to be present in several environments and feed on a wide variety of substrates (Krijgsheld et al. 2013). However, they are commonly found in spoiled products, such as grains, cereals, vegetables, fruits, juices, cakes, among other products, and are directly linked to the deterioration that make these foods unsuitable for consumption (Negeri et al. 2014; Snyder et al. 2019).

The main aggravating factor for foods contaminated by these fungi is the fact that some strains, such as *A. flavus* and *A. niger*, produce mycotoxins highly

Table 3.1 Essential oils active against *Aspergillus* spp.

Fungus spp	Active essential oil			References
	Plant	Part	Major compounds	
A. alternata	*Mentha spicata*	Aerial parts	Dextro-carvone and limonene	Kedia et al. (2014)
A. carbonarius	*Eremanthus erythropappus*	Trunks	Alpha-bisabolol	Brandão et al. (2020)
A. flavus	*Curcuma longa*	Roots	ar-tumerone, tumerone, beta-sesquiphellandrene and curcumene.	Hu et al. (2017)
A. flavus	*Cinnamomum jensenianum*	Bark	1,8-cineole and alpha-terpineol	Tian et al. (2012)
A. flavus	*Thymus vulgaris*	–	Thymol, p-Cymene and γ-Terpinene	Oliveira et al. (2020)
A. flavus	*Thapsia villosa*	Aerial parts	Limonene and Methyleugenol	Pinto et al. (2017)
A. flavus	*Piper betle*	Leaves	Chavibetol, linalool, beta-cubene, chavicol and caryophyllene	Basak and Guha (2017)
A. flavus	*Eugenia caryophyllata*	Flower buds	Eugenol and Caryophyllene	Kujur et al. (2021)
A. flavus	*Ageratum conyzoides*	Aerial parts	Precocene II and Precocene I	Nogueira et al. (2010)
A. flavus	*Eremanthus erythropappus*	Trunks	Alpha-bisabolol	Brandão et al. (2020)
A. flavus	*Salvia officinalis*	Aerial parts	1,8-cineole and camphor	Abu-Darwish et al. (2013)
A. flavus	*Cymbopogon citratus*	Bunches	E-citral and Z-citral	Sonker et al. (2014)
A. flavus	*Ocimum sanctum*	Leaves	Eugenol and β-Caryophyllene	Kumar et al. (2010)
A. flavus	*Zingiber officinale*	Rhizomes	α-zingiberene and geranial	Nerilo et al. (2016)
A. flavus	Thymus x viciosoi	Aerial parts	Carvacrol, thymol and p-cymene	Vale-Silva et al. (2010)
A. flavus	Ocimum basilicum	Leaves	Linalool and 1,8-cineol	El-Soud et al. (2015)
A. flavus	Daucus carota subsp. halophilus	Umbels	Lemicin and sabinene	Tavares et al. (2008)
A. fumigatus	*Thapsia villosa*	Aerial parts	Limonene and Methyleugenol	Pinto et al. (2017)
A. fumigatus	*Salvia officinalis*	Aerial parts	1,8-cineole and camphor	Abu-Darwish et al. (2013)
A. fumigatus	*Mentha spicata*	Aerial parts	Dextro-carvone and limonene	Kedia et al. (2014)
A. fumigatus	Thymus × viciosoi	Aerial parts	Carvacrol, thymol and p-cymene	Vale-Silva et al. (2010)

(continued)

Table 3.1 (continued)

Fungus spp	Active essential oil			References
	Plant	Part	Major compounds	
A. fumigatus	Daucus carota subsp. halophilus	Umbels	Elemicin and sabinene	Tavares et al. (2008)
A. fumigatus	Thymus villosus subsp. lusitanicus	Aerial flowering parts	Geranyl acetate and terpinen-4-ol	Pinto et al. (2013)
A. glaucus	*Mentha spicata*	Aerial parts	Dextro-carvone and limonene	Kedia et al. (2014)
A. niger	Thymus × viciosoi	Aerial parts	Carvacrol, thymol and p-cymene	Vale-Silva et al. (2010)
A. niger	Hymenocrater longiflorus	Aerial parts	α-pinene	Ahmadi et al. (2010)
A. niger	Daucus carota subsp. halophilus	Umbels	Elemicin and sabinene	Tavares et al. (2008)
A. niger	Citrus sinensis	Epicarp	Limonene	Sharma and Tripathi (2008)
A. niger	Thymus villosus subsp. lusitanicus	Aerial flowering parts	Geranyl acetate and terpinen-4-ol	Pinto et al. (2013)
A. niger	*Salvia officinalis*	Aerial parts	1,8-cineole and camphor	Abu-Darwish et al. (2013)
A. niger	*Mentha spicata*	Aerial parts	Dextro-carvone and limonene	Kedia et al. (2014)
A. niger	*Cymbopogon citratus*	Bunches	E-citral and Z-citral	Sonker et al. (2014)
A. niger	*Matricaria chamomilla*	Flowers	α-bisabolol and trans-farnesol	Tolouee et al. (2010)
A. niger	*Thapsia villosa*	Aerial parts	Limonene and Methyleugenol	Pinto et al. (2017)
A. parasiticus	Satureja hortensis	Leaves	Carvacrol and thymol	Razzaghi-Abyaneh et al. (2008)
A. ochratoxin	*Eremanthus erythropappus*	Trunks	Alpha-bisabolol	Brandão et al. (2020)
A. ochratoxin	*Cymbopogon citratus*	Bunches	E-citral and Z-citral	Sonker et al. (2014)
A. unguis	*Mentha spicata*	Aerial parts	Dextro-carvone and limonene	Kedia et al. (2014)

harmful to human health: aflatoxins (AFB1, AFB2, AFG1 and AFG2); ochratoxin A (OTA); sterygmatocystin (ST) and cyclopiazonic acid (CPA) (Priyanka et al. 2014). Due to the adverse effects of these mycotoxins, many countries established concentration limits in grains, cereals and other foods (Priyanka et al. 2014). Thus, to minimize the problems associated with *Aspergillus* fungi, new fungicides from natural products have been studied. Among these, some essential oils that are presented in Table 3.1.

Most studies on the antifungal activity of essential oils against *Aspergillus* are focused on the species *A. flavus*, because it is one of the main responsible for the deterioration of food and produces toxins that make these foods unfit for consumption Among the oils mentioned in Table 3.1 that showed activity against *A. flavus*, several had similar major compounds in their compositions, such as 1,8-cineole, thymol, linalool and eugenol. These compounds are already known for their antifungal activity (Morcia et al. 2012), and they are considered to be the main responsible for the activity observed in these essential oils.

The other *Aspergillus* species were also susceptible to the activity of essential oils with the compounds 1,8-cineole, thymol, linalool and eugenol in their majority composition, in addition to others such as caryophyllene, p-cymene, limonene and carvacrol, also with antifungal activity already reported (Karpiński 2020). These studies demonstrate that essential oils can be an alternative for food preservation, protecting against *Aspergillus* fungi, and show which essential oils may present activity for a particular species based on their chemical composition.

3.2.2   *Penicillium*

Penicillium is a genus of anamorph fungi belonging to the family Trichocomaceae and composed of 483 species accepted to date (Berbee et al. 1995; Houbraken et al. 2020). Fungi of this genus occur in a wide variety of habitats, which allows them to be distributed worldwide, in soil, air, vegetation and various food products (Visagie et al. 2014; El Hajj Assaf et al. 2020). These are also of great importance in several sectors such as the food industry, being part of the composition of cheeses (Giraud et al. 2010) and fermented sausages (Ludemann et al. 2010), in addition to applications in biotechnology (Bazioli et al. 2017; El Hajj Assaf et al. 2020). These fungi also produce a wide variety of secondary metabolites with biological activity of wide application in medicine, such as penicillin, which was the first isolated substance with antibiotic action in history and continues to be used until today (Brian 1947; Fleming 1980; Mancini et al. 2021).

Despite the importance and applicability of *Penicillium* fungi, several species are classified as pathogens and cause food spoilage (Pitt and Hocking 2009; Samson et al. 2019), in addition to producing a variety of mycotoxins (Frisvad et al. 2004). For example, the species *P. verrucosum* and *P. nordicum* produce ochratoxin A, which is a potent nephrotoxic and nephrocarcinogenic mycotoxin associated with kidney problems in humans, found in foods based on wheat, oats, rice, beer, coffee, and wine contaminated by these fungi (Ostenfeld et al. 2001; Cabañes et al. 2010). In view of the deterioration of foods caused by fungi of the *Penicillium* genus and the human health damage resulting from the toxins produced by these microorganisms, studies have been carried out to evaluate the fungicidal potential of essential oils against some species, as shown in Table 3.2.

Table 3.2 demonstrates the wide variety of species of the genus that are food contaminants and/or pathogens. The compositions of the aforementioned essential

Table 3.2 Essential oils active against *Penicillium* spp.

| Fungus spp | Active essential oil | | | References |
	Plant	Part	Major compounds	
P. aurantiogriseum	*Allium cepa*	Bulbs	Dimethyl trisulfide and methyl propyl trisulfide	Kocić-Tanackov et al. (2017)
P. brevicompactum	*Allium cepa*	Bulbs	Dimethyl trisulfide and methyl propyl trisulfide	Kocić-Tanackov et al. (2017)
P. brevicompactum	*Thymus vulgaris*	–	p-cymene, thymol and 1,8-cineole	Segvić Klarić et al. (2007)
P. chrysogenum	*Allium cepa*	Bulbs	Dimethyl trisulfide and methyl propyl trisulfide	Kocić-Tanackov et al. (2017)
P. chrysogenum	*Thymus vulgaris*	–	p-cymene, thymol and 1,8-cineole	Segvić Klarić et al. (2007)
P. citrium	*Trachyspermum ammi*	Fruit	Thymol, p-cymene and γ-terpinene	Singh et al. (2004)
P. citrinum	*Laurus nobilis*	Flowers	1,8-cineole	Mssillou et al. (2020)
P. commune	*Pistacia lentiscus*	Aerial parts	α-pinene, β-myrcene, p-cymene, and terpinen-4-ol	Barra et al. (2007)
P. digitatum	*Tagetes patula*	Capitula	Piperitone and piperitenone	Romagnoli et al. (2005)
P. digitatum	*Satureja hortensis*	Aerial parts	Carvacrol and γ-terpinene	Atrash et al. (2018)
P. expansion	*Asteriscus graveolens*	Aerial parts	6-oxocyclonerolidol and 6-hydroxycyclonerolidol	Znini et al. (2011)
P. expansum	*Aaronsohnia pubescens*	Aerial parts	(2)-beta-ocimene, myrcene and a-pinene	Makhloufi et al. (2015)
P. expansum	*Pulicaria mauritanica*	Aerial parts	Carvotanacetone	Znini et al. (2013)
P. funiculosum	*Gallesia integrifolia*	Fruit	2,8-dithianonane, dimethyl trisulfide and lenthionine	Raimundo et al. (2018)
P. funiculosum	*Salvia sclarea*	–	Linalyl acetate and linalool	Dzamic et al. (2008)
P. funiculosum	*Hyssopus officinalis*	Aerial parts	1,8-cineole, β-pinene and isopinocamphone	Dzamic et al. (2013)
P. glabrum	*Allium cepa*	Bulbs	Dimethyl trisulfide and methyl propyl trisulfide	Kocić-Tanackov et al. (2017)
P. griseofulvum	*Melaleuca alternifolia*	Aerial parts	Terpinen-4-ol	Chidi et al. (2020)
P. griseofulvum	*Thymus vulgaris*	–	p-cymene, thymol and 1,8-cineole	Segvić Klarić et al. (2007)
P. italicum	*Rosmarinus officinalis*	Flowers	1,8-cineol, a-pinene and borneol	Sofiene et al. (2019)

(continued)

Table 3.2 (continued)

Fungus spp	Active essential oil			Major compounds	References
	Plant	Part			
P. jensenii	*Aaronsohnia pubescens*	Aerial parts		(2)-beta-ocimene, myrcene and a-pinene	Makhloufi et al. (2015)
P. madriti	*Trachyspermum ammi*	Fruit		Thymol, p-cymene and γ-terpinene	Singh et al. (2004)
P. notatum	*Melaleuca alternifolia*	Leaves		α-pinene, γ-terpinene, terpinen-4-ol and limonene	Sevik et al. (2021)
P. ochrochloron	*Gallesia integrifolia*	Fruit		2,8-dithianonane, dimethyl trisulfide and lenthionine	Raimundo et al. (2018)
P. ochrochloron	*Hyssopus officinalis*	Aerial parts		1,8-cineole, β-pinene and isopinocamphone	Dzamic et al. (2013)
P. ochrochloron	*Petroselinum crispum*	Aerial parts		Apiol, Myristicin and β-Phellandrene	Linde et al. (2016)
P. ochrochloron	*Salvia sclarea*	–		Linalyl acetate and linalool	Dzamic et al. (2008)
P. purpurogenum	*Aaronsohnia pubescens*	Aerial parts		(2)-beta-ocimene, myrcene and a-pinene	Makhloufi et al. (2015)
P. verrucosum	*Melaleuca alternifolia*	Aerial parts		Terpinen-4-ol	Chidi et al. (2020)
P. viridicatum	*Trachyspermum ammi*	Fruit		Thymol, p-cymene and γ-terpinene	Singh et al. (2004)

oils that showed antifungal activity against these species are quite varied, but many contain substances already known for their antifungal activity, such as 1,8-cineol, thymol, terpinen-4-ol, α-pinene and p- cymene (Morcia et al. 2012; Nóbrega et al. 2020; Balahbib et al. 2021). Most of the mentioned species that produce essential oils with antifungal activity against *Penicillium* species belong to the families Lamiaceae (*Thymus vulgaris, Satureja hortensis, Salvia sclarea, Hyssopus officinalis* and *Rosmarinus officinalis*) and Asteraceae (*Tagetes patula, Asteriscus graveolens, Aaronsohnia pubescen* and *Mauritanian Pulicaria*). These families have been extensively studied for their antifungal potential (Zapata et al. 2010; Skendi et al. 2020; Karpiński 2020) and the data presented here reinforce the potential of essential oils from these species for application as food preservatives.

3.2.3 *Fusarium*

Fusarium is a genus of filamentous fungi, whose species are distributed in the tropical and subtropical regions of the planet and can be found in soil, water, plants, and air (Laurence et al. 2011; Gakuubi et al. 2017). There is no universally accepted concept of species in the genus *Fusarium* and therefore taxonomists disagree about the number of species. In general, species are recognized based on the concept of morphological, biological and phylogenetic characteristics or a combination of these (Watanabe et al. 2011).

Species of the genus *Fusarium* have a high adaptability and are resistant to antifungal agents (Ma et al. 2013). Most species are phytopathogenic, causing various diseases in plants, such as wilt, canker, root and stem rot, among others (Duan et al. 2016). These species are known to produce mycotoxins that can contaminate agricultural products, which can lead to reduced crop yield and make agricultural products unsuitable for consumption by humans and other animals (Zabka and Pavela 2018; Toghueo 2020), due to serious risks to health, such as anorexia, depression, gastroenteritis, immune dysfunction, and hepatotoxicity (Wan et al. 2013; Gakuubi et al. 2017). Among the species best known for producing mycotoxins are *F. graminearum, F. oxysporum, F. sporotrichioides, F. verticillioides* and *F. proliferatum* (Gakuubi et al. 2017).

The search for natural compounds with antifungal properties is growing, due to the problems that synthetic fungicides have caused to human health and the environment. Table 3.3 presents the antifungal activities of essential oils that can be used as alternatives to synthetic fungicides against species of the genus *Fusarium*.

Studies carried out by (Naveen Kumar et al. 2016) showed that the essential oil of *Curcuma longa* has antifungal activity against *F. graminearum*. The authors demonstrated that the oil was able to inhibit the production of fungal biomass and the mycotoxin zearalenone, with inhibition values of 3500 and 3000 mg/mL, respectively. (Avanço et al. 2017) showed that the essential oil of *C. longa* inhibited the growth of *F. verticillioides* in the application of 17.9 and 294.9µg/mL, with inhibition values of 56.0% and 79.3%, respectively, and decreased the production of the mycotoxin fumonisins B1 and B2 in all concentrations of oils tested. The activity of essential oils from *Curcuma longa* can be mainly attributed to its chemical composition, with the predominance of the class of terpenes, which are lipophilic molecules and can be toxic to the cellular structures of fungi.

The essential oil of *Piper divaricatum* and its main constituents methyleugenol and eugenol were tested against the fungus *F. solan*, and demonstrated to have strong antifungal activities (da Silva et al. 2014a). (Xing-dong and Hua-li 2014) studied the antifungal activity of *Zanthoxylum bungeanum* essential oil and its major constituent α-pinene, and found that both were able to inhibit the growth of *F. sulphureum*, with Minimum Inhibitory Concentration (MIC) values equal to 6.25% and 12.50%, respectively. Furthermore, this study demonstrated that the oil has greater fungicidal potential than the pure active components showed that *Zingiber officinale* oil was able to control the growth of *F. verticillioides* and subsequent production of the mycotoxin fumonisin. The essential oil exhibited an inhibitory activity, with a MIC of 2500µg/ml. In addition to these oils, others also showed antifungal potential against species of the genus *Fusarium*, especially *Cinnamomum zeylanicum, Origanum compactum, Syzygium aromaticum* (Roselló et al. 2015), *Cymbopogon martinii* (Kalagatur et al. 2018), *Eucalyptus camaldulensis* (Gakuubi et al. 2017), *Baccharis dracunculifolia* and *Pogostemon cablin* (Luchesi et al. 2020).

The essential oils described in Table 3.3 have a great antifungal potential, which is generally related to their chemical composition. These oils can be an excellent alternative to synthetic fungicides and can be used in the food industry to control *Fusarium* fungus infestation and mycotoxin contamination.

Table 3.3 Essential oils active against *Fusarium* spp.

Fusarium spp	Active essential oil			References
	Plant	Part	Major compounds	
F. culmorum	*Cinnamomum zeylanicum*	Branches	Eugenol	Roselló et al. (2015)
F. culmorum	*Origanum compactum*	Flower	Carvacrol and thymol	Roselló et al. (2015)
F. culmorum	*Syzygium aromaticum*	Leaves	Eugenol	Roselló et al. (2015)
F. graminearum	*Baccharis dracunculifolia*	Leaves	trans-nerolidol and γ-elemene	Luchesi et al. (2020)
F. graminearum	*Curcuma longa*	Rhizome	ar-turmerone	Naveen Kumar et al. (2016)
F. graminearum	*Cymbopogon martinii*	Leaves	Geraniol	Kalagatur et al. (2018)
F. graminearum	*Echinophora platyloba*	Aerial parts	Ocimene	Hashemi et al. (2016)
F. graminearum	*Pogostemon cablin*	Leaves	Patchoulol and Seichelene	Luchesi et al. (2020)
F. graminearum	*Zingiber officinale*	Rhizomes	α-Zingiberene and geranial	Ferreira et al. (2018)
F. oxysporum	*Eucalyptus camaldulensis*	Leaves	1,8-cineole and α-Pinene	Gakuubi et al. (2017)
F. oxysporum	*Mentha piperita*	Leaves	Menthone and Mentol	Moghaddam et al. (2013)
F. proliferatum	*Eucalyptus camaldulensis*	Leaves	1,8-cineole and α-Pinene	Gakuubi et al. (2017)
F. proliferatum	*Mentha arvensis*	Seed	Menthol and Menthone	Kumar et al. (2016)
F. solani	*Eucalyptus camaldulensis*	Leaves	1,8-Cineole and α-Pinene	Gakuubi et al. (2017)
F. solani	*Piper divaricatum*	Aerial parts	Methyleugenol, Eugenol	da Silva et al. (2014a)
F. solani	*Syzygium aromaticum*	Buds	Eugenol	Sameza et al. (2016)
F. subglutinans	*Eucalyptus camaldulensis*	Leaves	1,8-cineole and α-Pinene	Gakuubi et al. (2017)
F. sulphureum	*Zanthoxylum bungeanum*	Seeds	α-Pinene	Xing-dong and Hua-li (2014)
F. verticillioides	*Cinnamomum zeylanicum*	Branches	Eugenol	Roselló et al. (2015)
F. verticillioides	*Curcuma longa*	Rhizomes	α-Turmerone and β-Turmerone	Avanço et al. (2017)
F. verticillioides	*Eucalyptus camaldulensis*	Leaves	1,8-cineole, α-pinene	Gakuubi et al. (2017)
F. verticillioides	*Mentha arvensis*	Seed	Menthol and Menthone	Kumar et al. (2016)

(continued)

Table 3.3 (continued)

| *Fusarium* spp | Active essential oil | | | |
	Plant	Part	Major compounds	References
F. verticillioides	*Origanum compactum*	Flower	Carvacrol and thymol	Roselló et al. (2015)
F. verticillioides	*Rosmarinus officinalis*	Leaves	1,8-cineole	da Silva Bomfim et al. (2015)
F. verticillioides	*Syzygium aromaticum*	Leaves	Eugenol	Roselló et al. (2015)
F. verticillioides	*Zingiber officinale*	Rhizomes	α-zingiberene and citral	Yamamoto-Ribeiro et al. (2013)

3.2.4   *Alternaria*

The fungal genus *Alternaria* is known to have saprobic, endophytic and pathogenic species that are associated with a wide variety of serious plant pathogens, which are responsible for large losses, mainly in large agricultural crops (Woudenberg et al. 2013). This genus of fungi comprises around 250 species (Pinto and Patriarca 2017) and affects the growth process of vegetables, with diseases that impact seedlings, leaves, stalks, stems, flowers, and fruits. In carrot cultivation, the diseases caused by this type of fungus are called burnt leaf or alternation, which manifests a high destructive potential between temperatures of 25–32 °C, as well as a relative humidity of around 40% during the day and 95% at night (de Lima et al. 2016).

In agriculture, many species of this genus produce toxic secondary metabolites, of which some can cause allergies and even esophageal cancer. Synthetic fungicides are used to control the invasion of these phytopathogenic fungi, with azoxystrobin and difenoconazol as the most used against *Alternaria* species. However, these commercial products cause serious risks to the environment and human health, which makes it necessary to search for new friendly alternatives to control these phytopathogenic fungi, such as compounds from botanical sources that have great potential for use in crops to inhibit this fungicidal action (Muy-Rangel et al. 2017). Thus, there are studies that report the fungicidal action of natural products against fungi of the *Alternaria* genus, such as the essential oils shown in Table 3.4.

It is demonstrated the essential oils of three species of *Artemisia* (*A. lavandulae-folia*, *A. scoparia* and *A. annua*), which were respectively characterized by the major compounds eucalyptol, acenaphthene, and artemisia ketone, showed significant activity against the fungus *Alternaria solani* (Huang et al. 2019). The same was observed for the essential oils of *Lippia alba*, respectively characterized by the major compounds camphor, citral, linalool and 1,8-cineole (Tomazoni et al. 2016). These studies provide a theoretical basis for the application of these essential oils as future ecological alternatives to synthetic fungicides in disease control, especially in the tomato crop (*Lycopersicon esculentum*), which is susceptible to attack by the fungus *A. solani* (Huang et al. 2019).

Table 3.4 Essential oils active against *Alternaria* spp.

| Fungus spp | Active essential oil | | | References |
	Plant	Part	Major compounds	
A. solani	*Artemisia lavandulaefolia*	Leaves	Eucalyptol	Huang et al. (2019)
A. solani	*Artemisia scoparia*	Leaves	Acenaphthene	Huang et al. (2019)
A. solani	*Artemisia annua*	Leaves	Artemisia ketone	Huang et al. (2019)
A. tenuissima	*Allium sativum*	Bulbos	Diallyl disulphide and Diallyl sulphide	Muy-Rangel et al. (2017)
A. alternata	*Cymbopogon nardus*	Leaves	Citronellal	Chen et al. (2014)
A. solani	*Lippia alba* (chemotype 1)	Leaves	Camphor	Tomazoni et al. (2016)
A. solani	*L. alba* (chemotype 2)	Leaves	Citral	Tomazoni et al. (2016)
A. solani	*L. alba* (chemotype 3)	Leaves	Linalool	Tomazoni et al. (2016)
A. solani	*L. alba* (chemotype 4)	Leaves	Camphor/1,8-cineole	Tomazoni et al. (2016)
A. alternata	*Laurus nobilis*	Leaves	Eugenol	Xu et al. (2014)
A. alternata	*Ocimum basilicum*	Leaves	Methyl chavicol	Perveen et al. (2020)
A. radicina	*Satureja montana*	Seeds	Carvacrol	Lopez-Reyes et al. (2016)
A. alternata	*Syringa oblata*	Buds	Eugenol	Jing et al. (2018)
A. radicina	*Thymus vulgaris*	Seeds	Thymol	Lopez-Reyes et al. (2016)

In another study, *Cymbopogon nardus* essential oil presented the citronellal compound as the major compound and was investigated to verify its antifungal potential against *Alternaria alternata* (Chen et al. 2014). The results showed that the essential oil of *C. nardus* had a strong inhibiting activity against the fungus *A. alternata*, and the incidence of *Lycopersicon esculentum* (cherry tomato) disease treated with citronella oil was significantly reduced when compared to the control treatment. The authors concluded that citronella oil can significantly inhibit *A. alternata in vitro* and *in vivo* and has potential as a promising natural product for the control of black rot in cherry tomatoes (Chen et al. 2014). Other oils that showed activities against *A. alternata* were *Syringa ablata* and *Laurus nobilis*, both characterized by the major compound eugenol (Jing et al. 2018; Xu et al. 2014). The authors suggest that eugenol has fungicidal potential for use in the control of diseases caused by the phytopathogen (*A. alternata*). In addition, the essential oil of *Ocimum basilicum*, characterized by methyl chavicol as the major compound, also showed activity against the phytopathogen *Alternaria alternata* (Perveen et al. 2020).

The essential oil of garlic (*Allium sativum*) was characterized by the compounds diallyl disulphide and diallyl sulphide and showed activity against *Alternaria tenuissima* (Muy-Rangel et al. 2017). The essential oils of *Satureja montana* and *Thymus vulgaris* were tested against *Alternaria radicina* and were respectively characterized by the major compounds carvacrol and thymol, with positive results in the control of this phytopathogen that causes problems in carrot crops (Lopez-Reyes et al. 2016).

It is important to mention that *Alternaria alternata* is a critical phytopathogen that causes foodborne spoilage and produces a polyketide mycotoxin, alternariol (AOH) and its derivative alternariol monomethyl ether (AME), which are harmful to human and animal health through food chains. This is all of great concern because this phytopathogen has caused losses in agricultural production and has played a significant role in food security, due to the production of these mycotoxins (Wang et al. 2019). Thus, the oils mentioned are an excellent alternative to combat these fungi.

3.2.5 Candida

The genus *Candida* comprises a group of yeasts consisting of approximately 200 species that are found in different ecosystems in the microbiota of the human and animal body, colonizing the skin and mucous membranes of the digestive, urinary, oral, and vaginal tracts. They have a structure that includes the chitin cell wall and the phospholipid cytoplasmic membrane, composed of proteins that act as enzymes, and ergosterol. It is considered the main group of opportunistic pathogenic fungi, with just over 20 species responsible for human infections (Modrzewska and Kurnatowski 2013; da Rocha et al. 2021). Among the species of these genus, *Candida albicans* stands out as the most pathogenic (Silva et al. 2012; Yapar 2014; Kołaczkowska and Kołaczkowski 2016). *Candida albicans* is characterized by polymorphism, which occurring in many forms, namely blastospores, germ tubes, pseudohyphae, true hyphae, and chlamydospores (Williams et al. 2011).

There is growing concern about species that show resistance profiles to available antifungal agents, such as the emerging pathogen C. auris, which has been showing resistance to antifungal agents such as: fluconazole, amphotericin B and echinocandins (Meis and Chowdhary 2018). The table below presents research on the antifungal activities of essential oils against Candida species.

Table 3.5 describes more than 60 major components of essential oils with activity against Candida species. Candida albicans was the most researched species, followed by Candida tropicalis, Candida krusei and Candida glabrata. (Cavalcanti et al. 2011) in their studies against infection of the root canal system and infections of the oral microbiota, demonstrated that essential oils from Melaleuca alternifolia, Cymbopogon winterianus, Thymus vulgaris, Ocimum basilicum, Cymbopogon martinii, Rosmarinus officinalis and Cinnamomun cassia presented activity against Candida albicans, Candida krusei and Candida tropicalis strains.

In studies on the activity of essential oils from four *Cinnamomum* species (Jantan et al. 2008) and from *Cuminum cyminum, Anethum graveolens, Pimpinella anisum* and *Foeniculum vulgare* (Vieira et al. 2018) evaluating the MIC in *Candida* species, the tested yeasts showed susceptibility to the oils. (Zomorodian et al. 2018) demonstrated that the partial inhibition of biofilm formation for *Candida albicans, Candida tropicalis* and *Candida krusei* strains was achieved by applying essential oils from *Ferula assafoetida*. Essential oil nanoemulsions were also effective, such as wild

Table 3.5 Essential oils active against *Candida* spp.

| Fungus spp | Active essential oil | | | References |
	Plant	Part	Major compounds	
C. albicans	*Cinnamomum cassia*	–	Cinnamic aldehyde	Almeida et al. (2012)
C. albicans	*Cymbopogon martinii*	–	Geraniol and geranyl acetate	Almeida et al. (2012)
C. albicans	*Cymbopogon winterianus*	–	Citronellal and citronellol	Almeida et al. (2012)
C. albicans	*Thymus vulgaris*	–	Thymol and ρ-cimeno	Almeida et al. (2012)
C. albicans	*Melaleuca alternifolia*	–	Terpinen-4-ol and α-terpinen	Almeida et al. (2012)
C. albicans	*Syzygium aromaticum*	Flower buds	Eugenol	Ferrão et al. (2020)
C. glabrata	*Cinnamomum cassia*	Bark	Trans-cinnamaldehyde	Ferrão et al. (2020)
C. lusitanae	*Pelargonium graveolens*	Aerial parts	Geraniol, linalol and citronellol	Ferrão et al. (2020)
C. tropicalis	*Myristica fragrans*	Seeds	Sabinene; terpin-4-ol, miristicin, elimicin and limonene	Ferrão et al. (2020)
C. albicans	*Rosmarinus officinalis*	–	1,8-cineol, limonene, ρ-cimeno, α-pineno and canfora	Cavalcanti et al. (2011)
C. albicans	*Cymbopogon winterianus*	–	Citronellal, geraniol and citronellol	Cavalcanti et al. (2011)
C. albicans	*Melaleuca alternifolia*	–	Terpinen-4-ol, γ-terpinene, α-terpinene and 1,8-cineol	Cavalcanti et al. (2011)
C. krusei	*Rosmarinus officinalis*	–	1,8-cineol, limonene, ρ-cimene, α-pineno and canfora	Cavalcanti et al. (2011)
C. krusei	*Cymbopogon winterianus*	–	Citronellal, geraniol and citronellol	Cavalcanti et al. (2011)
C. krusei	*Melaleuca alternifolia*	–	Terpinen-4-ol, γ-terpinene, α-terpineno and 1,8-cineol	Cavalcanti et al. (2011)
C. tropicalis	*Rosmarinus officinalis*	–	1,8-cineol, limonene, ρ-cimene, α-pinene and canfora	Cavalcanti et al. (2011)
C. tropicalis	*Cymbopogon winterianus*	–	Citronellal, geraniol and citronellol	Cavalcanti et al. (2011)
C. tropicalis	*Melaleuca alternifolia*	–	Terpinen-4-ol, γ-terpinene, α-terpinene and 1,8-cineol	Cavalcanti et al. (2011)

(continued)

Table 3.5 (continued)

Fungus spp	Active essential oil			References
	Plant	Part	Major compounds	
C. glabrata	Cinnamomum zeylanicum	Leaves	Cinnamic aldehyde, cinnamyl acetate and 1,8-cineol	Almeida et al. (2017)
C. glabrata	Cinnamomum cassia	Bark	Cinnamic aldehyde, benzyl benzoate and α-pinene	Almeida et al. (2017)
C. glabrata	Cymbopogon winterianus	–	Citronellal, geraniol and citronellol	Almeida et al. (2017)
C. albicans	Cuminum cyminum	Seeds	Cuminaldehyde	Vieira et al. (2018)
C. albicans	Anethum graveolens	Seeds	Carvone	Vieira et al. (2018)
C. albicans	Pimpinella anisum	Seeds	Trans-anethole	Vieira et al. (2018)
C. albicans	Foeniculum vulgare	Leaves	Anethole	Vieira et al. (2018)
C. glabrata	Cuminum cyminum	Seeds	Cuminaldehyde	Vieira et al. (2018)
C. glabrata	Anethum graveolens	Seeds	Carvone	Vieira et al. 2018)
C. glabrata	Pimpinella anisum	Seeds	Trans-anethole	Vieira et al. (2018)
C. glabrata	Foeniculum vulgare	Leaves	Anethole	Vieira et al. (2018)
C. parapsilosis	Cuminum cyminum	Seeds	Cuminaldehyde	Vieira et al. (2018)
C. parapsilosis	Anethum graveolens	Seeds	Carvone	Vieira et al. (2018)
C. parapsilosis	Pimpinella anisum	Seeds	Trans-anethole	Vieira et al. (2018)
C. parapsilosis	Foeniculum vulgare	Leaves	Anethole	Vieira et al. (2018)
C. krusei	Cuminum cyminum	Seeds	Cuminaldehyde	Vieira et al. (2018)
C. krusei	Anethum graveolens	Seeds	Carvone	Vieira et al. (2018)
C. krusei	Pimpinella anisum	Seeds	Trans-anethole	Vieira et al. (2018)
C. krusei	Foeniculum vulgare	Leaves	Anethole	Vieira et al. (2018)
C. albicans	Cinnamomum zeylanicum	Leaves	Cinnamaldehyde	Jantan et al. (2008)
C. albicans	Cinnamomum cordatum	Leaves	Linalol and methyl cinnamate	Jantan et al. (2008)
C. albicans	Cinnamomum pubescens	Bark	Methyl cinnamate	Jantan et al. (2008)
C. albicans	Cinnamomum impressionicostatum	Branches	Methyl cinnamate	Jantan et al. (2008)
C. glabrata	Cinnamomum zeylanicum	Leaves	Cinnamaldehyde	Jantan et al. (2008)
C. glabrata	Cinnamomum cordatum	Leaves	Linalol and methyl cinnamate	Jantan et al. (2008)
C. glabrata	Cinnamomum pubescens	Bark	Methyl cinnamate	Jantan et al. (2008)

(continued)

Table 3.5 (continued)

| Fungus spp | Active essential oil | | | References |
	Plant	Part	Major compounds	
C. glabrata	Cinnamomum impressionicostatum	Branches	Methyl cinnamate	Jantan et al. (2008)
C. parapsilosis	Cinnamomum zeylanicum	Leaves	Cinnamaldehyde	Jantan et al. (2008)
C. parapsilosis	Cinnamomum cordatum	Leaves	Linalol and methyl cinnamate	Jantan et al. (2008)
C. parapsilosis	Cinnamomum pubescens	Bark	Methyl cinnamate	Jantan et al. (2008)
C. parapsilosis	Cinnamomum impressionicostatum	Branches	Methyl cinnamate	Jantan et al. (2008)
C. tropicalis	Cinnamomum zeylanicum	Leaves	Cinnamaldehyde	Jantan et al. (2008)
C. tropicalis	Cinnamomum cordatum	Leaves	Linalol and methyl cinnamate	Jantan et al. (2008)
C. tropicalis	Cinnamomum pubescens	Bark	Methyl cinnamate	Jantan et al. (2008)
C. tropicalis	Cinnamomum impressionicostatum	Branches	Methyl cinnamate	Jantan et al. (2008)
C. albicans	Ferula assafoetida	Resin gum	(E)-1-Propenyl sec-butyl disulfide, 10-epi-γ-Eudesmol and (Z)-Dissulfeto de 1 -Propenyl sec-butyl disulfide	Zomorodian et al. (2018)
C. dubliniensis	Ferula assafoetida	Resin gum	(E)-1-Propenyl sec-butyl disulfide, 10-epi-γ-Eudesmol and (Z)-Dissulfeto de 1 -Propenyl sec-butyl disulfide	Zomorodian et al. (2018)
C. glabrata	Ferula assafoetida	Resin gum	(E)-1-Propenyl sec-butyl disulfide, 10-epi-γ-Eudesmol and (Z)-Dissulfeto de 1 -Propenyl sec-butyl disulfide	Zomorodian et al. (2018)
C. krusei	Ferula assafoetida	Resin gum	(E)-1-Propenyl sec-butyl disulfide, 10-epi-γ-Eudesmol and (Z)-Dissulfeto de 1 -Propenyl sec-butyl disulfide	Zomorodian et al. (2018)

(continued)

Table 3.5 (continued)

Fungus spp	Active essential oil			References
	Plant	Part	Major compounds	
C. tropicalis	Ferula assafoetida	Resin gum	(E)-1-Propenyl sec-butyl disulfide, 10-epi-γ-Eudesmol and (Z)-Dissulfeto de 1 -Propenyl sec-butyl disulfide	Zomorodian et al. (2018)
C. albicans	Thymus pannonicus ssp. auctus	–	Germacren D, farnesol, α-terpinyl acetate and trans-nerolidol	Boz and Dunca (2018)
C. albicans	Thymus pannonicus ssp. pannonicus	–	Geranial, neral, thymol and ρ-cimene	Boz and Dunca (2018)
C. albicans	Pogostemon heyneanus	Leaves	Acetophenone; β-pinene; α-pinene, (E)-cariofilene, patchouli alcohol and α-guaiene	Adhavan et al. (2017)
C. albicans	Pogostemon plectranthoides	Leaves	Atractilone, curzerenone; guayazuleno and caryophyllene oxide	Adhavan et al. (2017)
C. albicans	Cuminum cyminum	Aerial parts	1,8-cineol	Minooeianhaghighi et al. (2017)
C. albicans	Lavandula binaludensis	Aerial parts	γ-terpinene	Minooeianhaghighi et al. (2017)
C. albicans	Thymus vulgaris	–	Thymol, carvacrol, terpine-4-ol, nerol acetate and fenchol	Jafri and Ahmad (2020)
C. tropicalis	Thymus vulgaris	–	Thymol, carvacrol, terpine-4-ol, nerol acetate and fenchol	Jafri and Ahmad (2020)
C. krusei	Lippia alba	Leaves	Geranial and neral	Mesa et al. (2009)
C. parapsilosis	Copaifera multijuga	–	β-cariofilene, α-humulene and trans-α-bergamotene	Deus et al. (2011)
C. guilliermondii	Copaifera multijuga	–	β-cariofilene, α-humulene and trans-α-bergamotene	Deus et al. (2011)
C. tropicallis	Copaifera multijuga	–	β-cariofilene, α-humulene and trans-α-bergamotene	Deus et al. (2011)
C. albicans	Ruta angustifólia	Aerial parts	2-undecanone and 2-decanone	Haddouchi et al. (2013)
C. albicans	Ruta graveolens	Aerial parts	2-undecanone, 2-nonanona, 1-noneno and α-limonene	Haddouchi et al. (2013)

(continued)

Table 3.5 (continued)

| Fungus spp | Active essential oil | | | | References |
	Plant	Part	Major compounds		
C. albicans	Ruta chalepensis	Aerial parts	2-nonanone, 2-undecanone and 1-nonene		Haddouchi et al. (2013)
C. albicans	Ruta tuberculata	Aerial parts	Piperitone, trans-ρ-menth-2-en-1-ol, cis-piperitol and cis-ρ-menth-2-en-1-ol		Haddouchi et al. (2013)
C. albicans	Pulicaria undulata	Leaves	Carvothanacetone and 2,5-dimethoxy-ρ-cymene		Ali et al. (2012)

patchouli *Pogostemon heyneanus* and *Pogostemon plectranthoides*, which exhibited inhibitory activity against *Candida albicans* (Adhavan et al. 2017). Thus, these species can act as an alternative to overcome resistance to species of the genus *Candida* present to conventional antifungal agents.

3.2.6 Cladosporium

Fungi belonging to the genus *Cladosporium* are called dematic, myelinated or black fungi, because they have a naturally brownish color due to the presence of the pigment dihydroxynaphthalenomelanin in their cell wall, which constitutes a photoprotective element and is considered a virulence factor for the fungus. Many *Cladosporium* species are plant pathogens, causing leaf spots and other lesions, or parasitizing other fungi. They are found as contaminants and spoilage agents in food or industrial products and can be isolated in various substrates, such as soil, stones, as well as paper and leather (Bensch et al. 2012; Menezes et al. 2017).

The genus *Cladosporium* is one of the largest and most heterogeneous genera of hyphomycetes, currently comprising more than 772 names (Dugan et al. 2004). Species of this genus are phytopathogenic filamentous fungi generally chosen for bioautographic tests, since they have high sensitivity and allow the detection of substances with antifungal potential, in contrast to their dark color. Currently, *Cladosporium* species of medical interest associated with human disease are *C. cladosporioides*, *C. herbarum*, *C. oxysporum* and *C. sphaerospermum* (Menezes et al. 2017). Moreover, there are other *Cladosporium* species, such as *Cladosporium fulvum*, *C. cucumerinum* and *C. carrionii* (Sharma and Tripathi 2006; Ali 2014; Menezes et al. 2016). These species can cause serious problems to human health; therefore, studies with essential oils have been developed to discover new natural fungicides for application in combating these fungi. Some of these are shown in Table 3.6.

Table 3.6 Essential oils active against *Cladosporium* spp.

Fungus spp	Active essential oil			References
	Plant	Part	Major compounds	
C. fulvum	*Citrus sinensis*	Epicarp	Limonene, linalol and myrcene	Sharma and Tripathi (2006)
C. cladosporioides	*Citrus sinensis*	Epicarp	Limonene, linalol and myrcene	Sharma and Tripathi (2006)
C. herbarum	*Ruta angustifólia*	Aerial parts	2-undecanone and 2-decanone	Haddouchi et al. (2013)
C. herbarum	*Ruta graveolens*	Aerial parts	2-undecanone, 2-nonanone, 1-nonene and α-limonene	Haddouchi et al. (2013)
C. herbarum	*Ruta chalepensis*	Aerial parts	2-nonanone, 2-undecanone and 1-nonene	Haddouchi et al. (2013)
C. herbarum	*Ruta tuberculata*	Aerial parts	Piperitone, trans-ρ-menth-2-en-1-ol, cis-piperitol and cis-ρ-menth-2-en-1-ol	Haddouchi et al. (2013)
C. cucumerinum	*Pulicaria undulata*	Leaves	Carvothanacetone and 2,5-dimethoxy-ρ-cymene	Ali et al. (2012)
C. carrionii	*Melissa officinalis*	–	Citral	Menezes et al. (2016)
C. oxysporum	*Melissa officinalis*	–	Citral	Menezes et al. (2016)
C. sphaerospermum	*Melissa officinalis*	–	Citral	Menezes et al. (2016)
C. herbarum	*Thymus capitatus*	Leaves	Thymol, Carvacrol and γ-terpinene	Goudjil et al. (2020)
C. cladosporioides	*Zhumeria majdae*	Aerial parts	Linalol and Canfora	Davari and Ezazi (2017)
C. cladosporioides	*Eucalyptus sp.*	Leaves	1,8-cineol	Davari and Ezazi (2017)
C. cladosporioides	*Laurus nobilis*	Leaves	1,8-cineol	Mssillou et al. (2020)
C. sphaerospermum	*Conchocarpus fontanesianus*	Leaves	Spathulenol and α-cadinol	Cabral et al. (2016)
C. cladosporioides	*Hedyosmum brasiliense*	Flowers	Curzerene	Murakami et al. (2017)
C. sphaerospermum	*Hedyosmum brasiliense*	Flowers	Curzerene	Murakami et al. (2017)
C. cladosporioides	*Lippia gracilis*	Leaves	Thymol, ρ-cimene and thymol methyl ether	Caroline et al. (2014)
C. sphaerospermum	*Lippia gracilis*	Leaves	Thymol, ρ-cimene and thymol methyl ether	Caroline et al. (2014)
C. cladosporioides	*Piper aduncum*	Fruit	Linalol, γ-terpinene and trans-ocimene	Morandim et al. (2010)
C. cladosporioides	*Piper tuberculatum*	Fruit	β-pinene, α-pinene and β-cariofilene	Morandim et al. (2010)

(continued)

Table 3.6 (continued)

| Fungus spp | Active essential oil | | | References |
	Plant	Part	Major compounds	
C. cladosporioides	*Piper crassinervium*	Leaves	Germacrene D, β-eudesmol and spathulenol	Morandim et al. (2010)
C. cladosporioides	*Piper solmsianum*	Leaves	E-isoelemicin	Morandim et al. (2010)
C. cladosporioides	*Piper cernuum*	Leaves	Germacrene D, β-eudesmol and spathulenol	Morandim et al. (2010)
C. sphaerospermum	*Piper aduncum*	Fruit	Linalol, γ-terpinene and trans-ocimene	Morandim et al. (2010)
C. sphaerospermum	*Piper tuberculatum*	Fruit	β-pinene, α-pinene and β-cariofilene	Morandim et al. (2010)
C. sphaerospermum	*Piper crassinervium*	Leaves	Germacrene D, β-eudesmol and spathulenol	Morandim et al. (2010)
C. sphaerospermum	*Piper solmsianum*	Leaves	E-isoelemicin	Morandim et al. (2010)
C. sphaerospermum	*Piper cernuum*	Fruit	Germacrene D, β-eudesmol and spathulenol	Morandim et al. (2010)
C. cucumerinum	*Artemisia abyssinica*	Aerial parts	Davanone, canfora and (E)-nerolidol	Ali (2014)
C. cucumerinum	*Artemisia arborescens*	Aerial parts	Artemisia ketone, canfora and α-bisabolol	Ali (2014)
C. cladosporioides	*Piper hispidum*	Aerial parts	β-cariofilene, α-humulene and δ-3-carene	da Silva et al. (2014b)
C. cladosporioides	*Piper aleyreanum*	Aerial parts	β-elemene, bicyclogermacrene and δ-elemene	da Silva et al. (2014b)
C. cladosporioides	*Piper anonifolium*	Aerial parts	Selin-11-en-4-α-ol, β-selinene and α-selinene	da Silva et al. (2014b)
C. sphareospermum	*Piper hispidum*	Aerial parts	β-cariofilene, α-humulene and δ-3-carene	da Silva et al. (2014b)
C. sphareospermum	*Piper aleyreanum*	Aerial parts	β-elemene, bicyclogermacrene and δ-elemene	da Silva et al. (2014b)
C. sphareospermum	*Piper anonifolium*	Aerial parts	Selin-11-en-4-α-ol, β-selinene and α-selinene	da Silva et al. (2014b)
C. carrionii	*Melissa officinalis*	–	Geranial, citral, trans-β-cariofilene and Germacrene D	Menezes et al. (2015)

Table 3.6 describes about 50 major components of essential oils with activity against *Cladosporium* species and *Cladosporium cladosporioides* was the most researched species, followed by *Cladosporium sphaerospermum*.

Some plant species have therapeutic properties on some types of pathogens. (Haddouchi et al. 2013) identified the activity of the essential oils of *Ruta chalepensis* and *Ruta tuberculata* against *Cladosporium herbarum*, as well as *Ruta angustifolia* and *Ruta graveolens* against *Candida albicans*, which were close to or more active than the antifungal Amphotericin B. (Morandim et al. 2010; da Franco Caroline et al. 2014) confirmed, through bioautography tests, assessed the effectiveness of essential oils from *Piper* species as a botanical antifungal against *C. sphaerospermum* and *C. cladosporioides*, similar saprophytic species, air and food contaminants, and with common occurrence. Many of these essential oils already have major compounds known for their antifungal activity, such as 1,8-cineole and thymol, and can currently be used as an alternative to commercial fungicides against species of the genus *Cladosporium*.

References

Abu-Darwish MS, Cabral C, Ferreira IV et al (2013) Essential oil of common sage (Salvia officinalis L.) from Jordan: assessment of safety in mammalian cells and its antifungal and anti-inflammatory potential. Biomed Res Int 2013:538940. https://doi.org/10.1155/2013/538940

Adhavan P, Kaur G, Princy A, Murugan R (2017) Essential oil nanoemulsions of wild patchouli attenuate multi-drug resistant gram-positive, gram-negative and Candida albicans. Ind Crops Prod 100:106–116. https://doi.org/10.1016/j.indcrop.2017.02.015

Ahmadi F, Sadeghi S, Modarresi M et al (2010) Chemical composition, in vitro anti-microbial, antifungal and antioxidant activities of the essential oil and methanolic extract of Hymenocrater longiflorus Benth., of Iran. Food Chem Toxicol 48:1137–1144. https://doi.org/10.1016/j.fct.2010.01.028

Ali N (2014) Cytotoxic and antiphytofungal activity of the essential oils from two artemisia species. World J Pharm Res 3:1350–1354

Ali NAA, Sharopov FS, Alhaj M et al (2012) Chemical composition and biological activity of essential oil from Pulicaria undulata from Yemen. Nat Prod Commun 7:257–260. https://doi.org/10.1177/1934578X1200700238

Almeida LFD, Cavalcanti YW, Castro RD, Lima EO (2012) Atividade antifúngica de óleos essenciais frente a amostras clínicas de Candida albicans isoladas de pacientes HIV positivos. Rev Bras Plantas Med 14:649–655. https://doi.org/10.1590/S1516-05722012000400012

Almeida L, Paula JF, Almeida-Marques R et al (2017) Atividade inibitória de óleos essenciais vegetais frente à candida glabrata, resistente a fluconazol. Rev Bras Ciências da Saúde 21. https://doi.org/10.4034/RBCS.2017.21.02.05

Atrash S, Ramezanian A, Rahemi M et al (2018) Antifungal effects of savory essential oil, gum arabic, and hot water in Mexican lime fruits. HortScience 53:524–530. https://doi.org/10.21273/HORTSCI12736-17

Avanço GB, Ferreira FD, Bomfim NS et al (2017) Curcuma longa L. essential oil composition, antioxidant effect, and effect on Fusarium verticillioides and fumonisin production. Food Control 73:806–813. https://doi.org/10.1016/j.foodcont.2016.09.032

Balahbib A, El Omari N, El Hachlafi N et al (2021) Health beneficial and pharmacological properties of p-cymene. Food Chem Toxicol 153. https://doi.org/10.1016/j.fct.2021.112259

Barra A, Coroneo V, Dessi S et al (2007) Characterization of the volatile constituents in the essential oil of Pistacia lentiscus L. from different origins and its antifungal and antioxidant activity. J Agric Food Chem 55:7093–7098. https://doi.org/10.1021/jf071129w

Basak S, Guha P (2017) Use of predictive model to describe sporicidal and cell viability efficacy of betel leaf (Piper betle L.) essential oil on Aspergillus flavus and Penicillium expansum and its antifungal activity in raw apple juice. LWT 80:510–516. https://doi.org/10.1016/j.lwt.2017.03.024

Bazioli JM, Amaral LDS, Fill TP, Rodrigues-Filho E (2017) Insights into Penicillium brasilianum secondary metabolism and its biotechnological potential. Molecules 22. https://doi.org/10.3390/molecules22060858

Bensch K, Braun U, Groenewald JZ, Crous PW (2012) The genus Cladosporium. Stud Mycol 72:1–401. https://doi.org/10.3114/sim0003

Berbee ML, Yoshimura A, Sugiyama J, Taylor JW (1995) Is Penicillium monophyletic? An evaluation of phylogeny in the family Trichocomaceae from 18S, 5.8S and ITS ribosomal DNA sequence data. Mycologia 87:210–222. https://doi.org/10.1080/00275514.1995.12026523

Boz I, Dunca S (2018) The study of essentials oils obtained from L. – microbiological aspects. Acta Biol Marisiensis 1:53–59. https://doi.org/10.2478/abmj-2018-0006

Brandão RM, Ferreira VRF, Batista LR et al (2020) Antifungal and antimycotoxigenic effect of the essential oil of Eremanthus erythropappus on three different Aspergillus species. Flavour Fragr J 35:524–533. https://doi.org/10.1002/ffj.3588

Brian PW (1947) Penicillin and other antibiotics. Nature 160:554–556. https://doi.org/10.1038/160554a0

Cabañes FJ, Bragulat MR, Castellá G (2010) Ochratoxin A producing species in the genus Penicillium. Toxins 2:1111–1120

Cabral RS, Suffredini IB, Young MCM (2016) Chemical composition and in vitro biological activities of essential oils from Conchocarpus fontanesianus (A. St.-Hil.) Kallunki & Pirani (Rutaceae). Chem Biodivers 13:1273–1280. https://doi.org/10.1002/cbdv.201600036

Cavalcanti Y, Almeida L, Padilha W (2011) Antifungal activity of three essential oils on Candida strains

Chen Q, Xu S, Wu T et al (2014) Effect of citronella essential oil on the inhibition of postharvest Alternaria alternata in cherry tomato. J Sci Food Agric 94:2441–2447. https://doi.org/10.1002/jsfa.6576

Chidi F, Bouhoudan A, Khaddor M (2020) Antifungal effect of the tea tree essential oil (Melaleuca alternifolia) against Penicillium griseofulvum and Penicillium verrucosum. J King Saud Univ Sci 32:2041–2045. https://doi.org/10.1016/j.jksus.2020.02.012

da Franco Caroline S, Ribeiro Alcy F, Carvalho Natale CC et al (2014) Composition and antioxidant and antifungal activities of the essential oil from Lippia gracilis Schauer. Afr J Biotechnol 13:3107–3113. https://doi.org/10.5897/AJB2012.2941

da Rocha WRV, Nunes LE, Neves MLR et al (2021) Candida genus – virulence factores, epidemiology, candidiasis and resistance mechanisms. Res Soc Dev 10:e43910414283. https://doi.org/10.33448/rsd-v10i4.14283

da Silva CJA, do Nascimento Malta DJ (2017) A importância dos fungos na biotecnologia. Cad Grad Ciências Biológicas e da Saúde Unit Pernambuco 2:49–66

da Silva J, Silva J, Nascimento S et al (2014a) Antifungal activity and computational study of constituents from Piper divaricatum essential oil against Fusarium infection in black pepper. Molecules 19:17926–17942. https://doi.org/10.3390/molecules191117926

da Silva JKR, Pinto LC, Burbano RMR et al (2014b) Essential oils of Amazon Piper species and their cytotoxic, antifungal, antioxidant and anti-cholinesterase activities. Ind Crops Prod 58:55–60. https://doi.org/10.1016/j.indcrop.2014.04.006

da Silva Bomfim N, Nakassugi LP, Faggion Pinheiro Oliveira J et al (2015) Antifungal activity and inhibition of fumonisin production by Rosmarinus officinalis L. essential oil in Fusarium verticillioides (Sacc.) Nirenberg. Food Chem 166:330–336. https://doi.org/10.1016/j.foodchem.2014.06.019

Dambolena JS, Zunino MP, López AG et al (2010) Essential oils composition of Ocimum basilicum L. and Ocimum gratissimum L. from Kenya and their inhibitory effects on growth and fumonisin production by Fusarium verticillioides. Innov Food Sci Emerg Technol 11:410–414. https://doi.org/10.1016/j.ifset.2009.08.005

Davari M, Ezazi R (2017) Chemical composition and antifungal activity of the essential oil of Zhumeria majdae, Heracleum persicum and Eucalyptus sp. against some important phytopathogenic fungi. J Mycol Med 27:463–468. https://doi.org/10.1016/j.mycmed.2017.06.001

de Lima CB, Rentschler LLA, Bueno JT, Boaventura AC (2016) Plant extracts and essential oils on the control of Alternaria alternata, Alternaria dauci and on the germination and emergence of carrot seeds (Daucus carota L.). Ciência Rural 46:764–770. https://doi.org/10.1590/0103-84 78cr20141660

Deus R, Alves C, Arruda M (2011) Avaliação do efeito antifúngico do óleo resina e do óleo essencial de copaíba (Copaifera multijuga Hayne). Rev Bras Plantas Med 13. https://doi.org/10.1590/ S1516-05722011000100001

Dong X, Liang W, Meziani MJ et al (2020) Carbon dots as potent antimicrobial agents. Theranostics 10:671–686. https://doi.org/10.7150/thno.39863

Duan C, Qin Z, Yang Z et al (2016) Identification of pathogenic Fusarium spp. causing maize ear rot and potential mycotoxin production in China. Toxins (Basel) 8:186. https://doi.org/10.3390/ toxins8060186

Dugan F, Bensch K, Braun U (2004) Check-list of Cladosporium names. Schlechtendalia 11:1–103

Dzamic AM, Soković MD, Ristić MS et al (2008) Chemical composition and antifungal activity of Salvia sclarea (Lamiaceae) essential oil. Arch Biol Sci 60:233–237

Dzamic AM, Soković MD, Novaković M et al (2013) Composition, antifungal and antioxidant properties of Hyssopus officinalis L. subsp. pilifer (Pant.) Murb. essential oil and deodorized extracts. Ind Crops Prod 51:401–407

El Hajj Assaf C, Zetina-Serrano C, Tahtah N et al (2020) Regulation of secondary metabolism in the Penicillium genus. Int J Mol Sci 21:9462

El-Soud NHA, Deabes M, El-Kassem LA, Khalil M (2015) Chemical composition and antifungal activity of Ocimum basilicum L. essential oil. Open Access Maced J Med Sci 3:374–379. https://doi.org/10.3889/oamjms.2015.082

Ferrão S, Butzge J, Mezzomo L et al (2020) Atividade antifúngica de óleos essenciais frente a Candida spp. Braz J Health Rev 3:100–113. https://doi.org/10.34119/bjhrv3n1-007

Ferreira FMD, Hirooka EY, Ferreira FD et al (2018) Effect of Zingiber officinale Roscoe essential oil in fungus control and deoxynivalenol production of Fusarium graminearum Schwabe in vitro. Food Addit Contam Part A 35:2168–2174. https://doi.org/10.1080/19440049.2018.1520397

Fleming A (1980) On the antibacterial action of cultures of a Penicillium, with special reference to their use in the isolation of B. influenzae. Rev Infect Dis 2:129–139. https://doi.org/10.1093/ clinids/2.1.129

Frisvad J, Smedsgaard J, Larsen TO, Samson R (2004) Mycotoxins and other extrolites produced by species in Penicillium subgenus Penicillium. Stud Mycol 2004:201–241

Gakuubi MM, Maina AW, Wagacha JM (2017) Antifungal activity of essential oil of Eucalyptus camaldulensis Dehnh. against selected Fusarium spp. Int J Microbiol 2017:1–7. https://doi. org/10.1155/2017/8761610

Galagan JE, Calvo SE, Cuomo C et al (2005) Sequencing of Aspergillus nidulans and comparative analysis with A. fumigatus and A. oryzae. Nature 438:1105–1115. https://doi.org/10.1038/ nature04341

Geiser DM (2009) Sexual structures in Aspergillus: morphology, importance and genomics. Med Mycol 47(Suppl 1):S21–S26. https://doi.org/10.1080/13693780802139859

Giraud F, Giraud T, Aguileta G et al (2010) Microsatellite loci to recognize species for the cheese starter and contaminating strains associated with cheese manufacturing. Int J Food Microbiol 137:204–213. https://doi.org/10.1016/j.ijfoodmicro.2009.11.014

Goudjil MB, Zighmi S, Hamada D et al (2020) Biological activities of essential oils extracted from Thymus capitatus (Lamiaceae). S Afr J Bot 128:274–282. https://doi.org/10.1016/j. sajb.2019.11.020

Haddouchi F, Chaouche TM, Zaouali Y et al (2013) Chemical composition and antimicrobial activity of the essential oils from four Ruta species growing in Algeria. Food Chem 141:253–258. https://doi.org/10.1016/j.foodchem.2013.03.007

Hashemi M, Ehsani A, Afshari A et al (2016) Chemical composition and antifungal effect of Echinophora platyloba essential oil against Aspergillus flavus, Penicillium expansum and Fusarium graminearum. J Chem Health Risks 6:91–97. https://doi.org/10.22034/ jchr.2016.544133

Hasheminejad N, Khodaiyan F, Safari M (2019) Improving the antifungal activity of clove essential oil encapsulated by chitosan nanoparticles. Food Chem 275:113–122. https://doi.org/10.1016/j.foodchem.2018.09.085

Houbraken J, Kocsubé S, Visagie CM et al (2020) Classification of Aspergillus, Penicillium, Talaromyces and related genera (Eurotiales): an overview of families, genera, subgenera, sections, series and species. Stud Mycol 95:5–169. https://doi.org/10.1016/j.simyco.2020.05.002

Hu Y, Zhang J, Kong W et al (2017) Mechanisms of antifungal and anti-aflatoxigenic properties of essential oil derived from turmeric (Curcuma longa L.) on Aspergillus flavus. Food Chem 220:1–8. https://doi.org/10.1016/j.foodchem.2016.09.179

Huang Y, Zhao J, Zhou L et al (2010) Antifungal activity of the essential oil of illicium verum fruit and its main component trans-anethole. Molecules 15:7558–7569. https://doi.org/10.3390/molecules15117558

Huang X, Chen S-y, Zhang Y et al (2019) Chemical composition and antifungal activity of essential oils from three artemisia species against Alternaria solani. J Essent Oil-Bear Plants 22:1581–1592. https://doi.org/10.1080/0972060X.2019.1708812

Jafri H, Ahmad I (2020) Thymus vulgaris essential oil and thymol inhibit biofilms and interact synergistically with antifungal drugs against drug resistant strains of Candida albicans and Candida tropicalis. J Mycol Med 30:100911. https://doi.org/10.1016/j.mycmed.2019.100911

Jantan I, Karim Moharam BA, Santhanam J, Jamal JA (2008) Correlation between chemical composition and antifungal activity of the essential oils of eight cinnamomum. Species. Pharm Biol 46:406–412. https://doi.org/10.1080/13880200802055859

Jing L, Lei Z, Li L et al (2014) Antifungal activity of citrus essential oils. J Agric Food Chem 62:3011–3033. https://doi.org/10.1021/jf5006140

Jing C, Zhao J, Han X et al (2018) Essential oil of Syringa oblata Lindl. as a potential biocontrol agent against tobacco brown spot caused by Alternaria alternata. Crop Prot 104:41–46. https://doi.org/10.1016/j.cropro.2017.10.002

Kalagatur NK, Nirmal Ghosh OS, Sundararaj N, Mudili V (2018) Antifungal activity of chitosan nanoparticles encapsulated with Cymbopogon martinii essential oil on plant pathogenic fungi Fusarium graminearum. Front Pharmacol 9. https://doi.org/10.3389/fphar.2018.00610

Karpiński T (2020) Essential oils of Lamiaceae family plants as antifungals. Biomolecules 10:103. https://doi.org/10.3390/biom10010103

Kedia A, Prakash B, Mishra PK et al (2014) Antifungal, antiaflatoxigenic, and insecticidal efficacy of spearmint (Mentha spicata L.) essential oil. Int Biodeter Biodegr 89:29–36. https://doi.org/10.1016/j.ibiod.2013.10.027

Kocić-Tanackov S, Dimić G, Mojović L et al (2017) Antifungal activity of the onion (Allium cepa L.) essential oil against Aspergillus, Fusarium and Penicillium species isolated from food. J Food Process Preserv 41:e13050. https://doi.org/10.1111/jfpp.13050

Kołaczkowska A, Kołaczkowski M (2016) Drug resistance mechanisms and their regulation in non-albicans Candida species. J Antimicrob Chemother 71:1438–1450. https://doi.org/10.1093/jac/dkv445

Krijgsheld P, Bleichrodt R, van Veluw GJ et al (2013) Development in Aspergillus. Stud Mycol 74:1–29. https://doi.org/10.3114/sim0006

Kujur A, Kumar A, Prakash B (2021) Elucidation of antifungal and aflatoxin B1 inhibitory mode of action of Eugenia caryophyllata L. essential oil loaded chitosan nanomatrix against Aspergillus flavus. Pestic Biochem Phys 172:104755. https://doi.org/10.1016/j.pestbp.2020.104755

Kumar A, Shukla R, Singh P, Dubey NK (2010) Chemical composition, antifungal and antiaflatoxigenic activities of Ocimum sanctum L. essential oil and its safety assessment as plant based antimicrobial. Food Chem Toxicol 48:539–543. https://doi.org/10.1016/j.fct.2009.11.028

Kumar P, Mishra S, Kumar A, Sharma AK (2016) Antifungal efficacy of plant essential oils against stored grain fungi of Fusarium spp. J Food Sci Technol 53:3725–3734. https://doi.org/10.1007/s13197-016-2347-0

Laurence MH, Summerell BA, Burgess LW, Liew ECY (2011) Fusarium burgessii sp. nov. representing a novel lineage in the genus Fusarium. Fungal Divers 49:101–112. https://doi. org/10.1007/s13225-011-0093-1

Linde GA, Gazim ZC, Cardoso BK et al (2016) Antifungal and antibacterial activities of Petroselinum crispum essential oil. Genet Mol Res 15. https://doi.org/10.4238/gmr.15038538

Lopez-Reyes JG, Gilardi G, Garibaldi A, Gullino ML (2016) In vivo evaluation of essential oils and biocontrol agents combined with hot water treatments on carrot seeds against Alternaria radicina. J Phytopathol 164:131–135. https://doi.org/10.1111/jph.12400

Luchesi LA, Paulus D, Busso C et al (2020) Chemical composition, antifungal and antioxidant activity of essential oils from Baccharis dracunculifolia and Pogostemon cablin against Fusarium graminearum. Nat Prod Res:1–4. https://doi.org/10.1080/14786419.2020.1802267

Ludemann V, Greco M, Rodríguez MP et al (2010) Conidial production by Penicillium nalgiovense for use as starter cultures in dry fermented sausages by solid state fermentation. LWT – Food Sci Technol 43:315–318. https://doi.org/10.1016/j.lwt.2009.07.011

Ma L-J, Geiser DM, Proctor RH et al (2013) Fusarium pathogenomics. Annu Rev Microbiol 67:399–416. https://doi.org/10.1146/annurev-micro-092412-155650

Maia LC, de Carvalho Junior AA (2010) Os fungos do Brasil. Catálogo Plantas e Fungos do Bras 1:43–48. https://doi.org/10.7476/9788560035083.0005

Makhloufi A, Ben Larbi L, Moussaoui A et al (2015) Chemical composition and antifungal activity of Aaronsohnia pubescens essential oil from Algeria. Nat Prod Commun 10:149–151

Mancini CM, Wimmer M, Schulz LT et al (2021) Association of penicillin or cephalosporin allergy documentation and antibiotic use in hospitalized patients with pneumonia. J Allergy Clin Immunol Pract 9:3060–3068.e1. https://doi.org/10.1016/j.jaip.2021.04.071

Meis J, Chowdhary A (2018) Candida auris: a global fungal public health threat. Lancet Infect Dis 18. https://doi.org/10.1016/S1473-3099(18)30609-1

Menezes C, Guerra F, Pinheiro L et al (2015) Investigation of Melissa officinalis L. essential oil for antifungal activity against Cladosporium carrionii. Int J Trop Dis Health 8:49–56. https://doi.org/10.9734/IJTDH/2015/17841

Menezes C, Pérez A, Filho A et al (2016) Antifungal activity of phytochemicals against species of Cladosporium and Cladophialophora. Int J Trop Dis Health 17:1–7. https://doi.org/10.9734/IJTDH/2016/27233

Menezes C, Perez A, Oliveira E (2017) Cladosporium spp: morfologia, infecções e espécies patogênicas. Acta Bras 1:23. https://doi.org/10.22571/Actabra1120176

Mesa A, Ramos J, Zapata B et al (2009) Citral and carvone chemotypes from the essential oils of Colombian Lippia alba (Mill.) N.E. Brown: composition, cytotoxicity and antifungal activity. Mem Inst Oswaldo Cruz 104:878–884. https://doi.org/10.1590/S0074-02762009000600010

Minooeianhaghighi MH, Sepehrian L, Shokri H (2017) Antifungal effects of Lavandula binaludensis and Cuminum cyminum essential oils against Candida albicans strains isolated from patients with recurrent vulvovaginal candidiasis. J Mycol Med 27:65–71. https://doi.org/10.1016/j.mycmed.2016.09.002

Modrzewska B, Kurnatowski P (2013) Selected pathogenic characteristics of fungi from the genus Candida. Ann Parasitol 59:57–66

Moghaddam M, Pourbaige M, Tabar HK et al (2013) Composition and antifungal activity of peppermint (Mentha piperita) essential oil from Iran. J Essent Oil Bear Plants 16:506–512. https://doi.org/10.1080/0972060X.2013.813265

Morandim A, Pin A, Pietro N et al (2010) Composition and screening of antifungal activity against Cladosporiumsphaerospermum and Cladosporiumcladosporioides of essential oils of leaves and fruits of Piper species. Afr J Biotechnol 9:6135–6139

Morcia C, Malnati M, Terzi V (2012) In vitro antifungal activity of terpinen-4-ol, eugenol, carvone, 1,8-cineole (eucalyptol) and thymol against mycotoxigenic plant pathogens. Food Addit Contam Part A Chem Anal Control Expo Risk Assess 29:415–422. https://doi.org/10.108 0/19440049.2011.643458

Mssillou I, Agour A, El Ghouizi A et al (2020) Chemical composition, antioxidant activity, and antifungal effects of essential oil from *Laurus nobilis* L. flowers growing in Morocco. J Food Qual 2020:8819311. https://doi.org/10.1155/2020/8819311

Murakami C, Cordeiro I, Scotti MT et al (2017) Chemical composition, antifungal and antioxidant activities of Hedyosmum brasiliense Mart. ex Miq. (Chloranthaceae) essential oils. Medicines (Basel) 4(3):55

Muy-Rangel MD, Osuna-Valle JR, García-Estrada RS et al (2017) Actividad antifúngica in vitro del aceite esencial de ajo (Allium sativum L.) contra Alternaria tenuissima. Rev Mex Fitopatol Mex J Phytopathol 36:162–171. https://doi.org/10.18781/r.mex.fit.1708-3

Naveen Kumar K, Venkataramana M, Allen JA et al (2016) Role of Curcuma longa L. essential oil in controlling the growth and zearalenone production of Fusarium graminearum. LWT – Food Sci Technol 69:522–528. https://doi.org/10.1016/j.lwt.2016.02.005

Negeri NG, Woldeamanuel Y, Asrat D et al (2014) Assessment of aflatoxigeinic Aspergillus species in food commodities from local market of Addis Ababa. Research 1. https://doi.org/10.13070/rs.en.1.1195

Nerilo SB, Rocha GHO, Tomoike C et al (2016) Antifungal properties and inhibitory effects upon aflatoxin production by Zingiber officinale essential oil in Aspergillus flavus. Int J Food Sci Technol 51:286–292. https://doi.org/10.1111/ijfs.12950

Nóbrega JR, de Figuerêdo Silva D, de Andrade Júnior FP et al (2020) Antifungal action of α-pinene against Candida spp. isolated from patients with otomycosis and effects of its association with boric acid. Nat Prod Res:1–4. https://doi.org/10.1080/14786419.2020.1837803

Nogueira JHC, Gonçalez E, Galleti SR et al (2010) Ageratum conyzoides essential oil as aflatoxin suppressor of Aspergillus flavus. Int J Food Microbiol 137:55–60. https://doi.org/10.1016/j.ijfoodmicro.2009.10.017

Oliveira RC, Carvajal-Moreno M, Correa B, Rojo-Callejas F (2020) Cellular, physiological and molecular approaches to investigate the antifungal and anti-aflatoxigenic effects of thyme essential oil on Aspergillus flavus. Food Chem 315:126096. https://doi.org/10.1016/j.foodchem.2019.126096

Ostenfeld LT, Anne S, Jørn S (2001) Biochemical characterization of ochratoxin A-producing strains of the genus Penicillium. Appl Environ Microbiol 67:3630–3635. https://doi.org/10.1128/AEM.67.8.3630-3635.2001

Perveen K, Bokhari NA, Al-Rashid SAI, Al-Humaid LA (2020) Chemical composition of essential oil of Ocimum basilicum L. and its potential in managing the Alternaria rot of tomato. J Essent Oil Bear Plants 23:1428–1437. https://doi.org/10.1080/0972060X.2020.1868351

Pinto VEF, Patriarca A (2017) Alternaria species and their associated mycotoxins. In: Methods in molecular biology. Humana Press, New York, pp 13–32

Pinto E, Gonçalves MJ, Hrimpeng K et al (2013) Antifungal activity of the essential oil of Thymus villosus subsp. lusitanicus against Candida, Cryptococcus, Aspergillus and dermatophyte species. Ind Crops Prod 51:93–99. https://doi.org/10.1016/j.indcrop.2013.08.033

Pinto E, Gonçalves M-J, Cavaleiro C, Salgueiro L (2017) Antifungal activity of Thapsia villosa essential oil against Candida, Cryptococcus, Malassezia, Aspergillus and Dermatophyte species. Molecules 22. https://doi.org/10.3390/molecules22101595

Pitt JI, Hocking AD (2009) Fungi and food spoilage. Springer, New York

Priyanka SR, Venkataramana M, Kumar GP et al (2014) Occurrence and molecular detection of toxigenic Aspergillus species in food grain samples from India. J Sci Food Agric 94:537–543. https://doi.org/10.1002/jsfa.6289

Raimundo KF, de Campos Bortolucci W, Glamočlija J et al (2018) Antifungal activity of Gallesia integrifolia fruit essential oil. Braz J Microbiol 49 Suppl 1:229–235. https://doi.org/10.1016/j.bjm.2018.03.006

Rana IS, Rana AS, Rajak RC (2011) Evaluation of antifungal activity in essential oil of the Syzygium aromaticum (L.) by extraction, purification and analysis of its main component eugenol. Braz J Microbiol 42:1269–1277. https://doi.org/10.1590/S1517-83822011000400004

Razzaghi-Abyaneh M, Shams-Ghahfarokhi M, Yoshinari T et al (2008) Inhibitory effects of Satureja hortensis L. essential oil on growth and aflatoxin production by Aspergillus parasiticus. Int J Food Microbiol 123:228–233. https://doi.org/10.1016/j.ijfoodmicro.2008.02.003

Rokas A (2013) Aspergillus. Curr Biol 23:R187–R188. https://doi.org/10.1016/J.CUB.2013.01.021

Romagnoli C, Bruni R, Andreotti E et al (2005) Chemical characterization and antifungal activity of essential oil of capitula from wild Indian Tagetes patula L. Protoplasma 225:57–65. https://doi.org/10.1007/s00709-005-0084-8

Roselló J, Sempere F, Sanz-Berzosa I et al (2015) antifungal activity and potential use of essential oils against Fusarium culmorum and Fusarium verticillioides. J Essent Oil Bear Plants 18:359–367. https://doi.org/10.1080/0972060X.2015.1010601

Sameza ML, Nguemnang Mabou LC, Tchameni SN et al (2016) Evaluation of clove essential oil as a mycobiocide against Rhizopus stolonifer and Fusarium solani, tuber rot causing fungi in yam (Dioscorea rotundata Poir.). J Phytopathol 164:433–440. https://doi.org/10.1111/jph.12468

Samson RA, Visagie CM, Houbraken J et al (2014) Phylogeny, identification and nomenclature of the genus Aspergillus. Stud Mycol 78:141–173. https://doi.org/10.1016/j.simyco.2014.07.004

Samson RA, Houbraken J, Thrane U et al (2019) Food and indoor fungi. Westerdijk Fungal Biodiversity Institute, Utrecht

Segvić Klarić M, Kosalec I, Mastelić J et al (2007) Antifungal activity of thyme (Thymus vulgaris L.) essential oil and thymol against moulds from damp dwellings. Lett Appl Microbiol 44:36–42. https://doi.org/10.1111/j.1472-765X.2006.02032.x

Sevik R, Akarca G, Kilinc M, Ascioglu Ç (2021) Chemical composition of tea tree (Melaleuca alternifolia) (Maiden & Betche) cheel essential oil and its antifungal effect on foodborne molds isolated from meat products. J Essent Oil Bear Plants 24:561–570. https://doi.org/10.108 0/0972060X.2021.1942232

Sharma N, Tripathi A (2006) Fungitoxicity of the essential oil of Citrus sinensis on postharvest pathogens. World J Microbiol Biotechnol 22:587–593. https://doi.org/10.1007/s11274-005-9075-3

Sharma N, Tripathi A (2008) Effects of Citrus sinensis (L.) Osbeck epicarp essential oil on growth and morphogenesis of Aspergillus niger (L.) Van Tieghem. Microbiol Res 163:337–344. https://doi.org/10.1016/j.micres.2006.06.009

Silva S, Negri M, Henriques M et al (2012) Candida glabrata, Candida parapsilosis and Candida tropicalis: biology, epidemiology, pathogenicity and antifungal resistance. FEMS Microbiol Rev 36:288–305. https://doi.org/10.1111/j.1574-6976.2011.00278.x

Singh G, Maurya S, Catalan C, de Lampasona MP (2004) Chemical constituents, antifungal and antioxidative effects of ajwain essential oil and its acetone extract. J Agric Food Chem 52:3292–3296. https://doi.org/10.1021/jf035211c

Skendi A, Katsantonis DN, Chatzopoulou P et al (2020) Antifungal activity of aromatic plants of the Lamiaceae family in bread. Foods 9:1642

Snyder AB, Churey JJ, Worobo RW (2019) Association of fungal genera from spoiled processed foods with physicochemical food properties and processing conditions. Food Microbiol 83:211–218. https://doi.org/10.1016/j.fm.2019.05.012

Sofiene BK, Hanafi M, Berhal C et al (2019) Rosmarinus officinalis essential oil as an effective antifungal and herbicidal agent. Span J Agric Res 17. https://doi.org/10.5424/sjar/2019172-14043

Soliman EA, El-Moghazy AY, El-Din MSM, Massoud MA (2013) Microencapsulation of essential oils within alginate: formulation and in vitro evaluation of antifungal activity. J Encapsulation Adsorpt Sci 03:48–55. https://doi.org/10.4236/jeas.2013.31006

Sonker N, Pandey AK, Singh P, Tripathi NN (2014) Assessment of Cymbopogon citratus (DC.) stapf essential oil as herbal preservatives based on antifungal, antiaflatoxin, and antiochratoxin activities and in vivo efficacy during storage. J Food Sci 79:M628–M634. https://doi.org/10.1111/1750-3841.12390

Stević T, Berić T, Šavikin K et al (2014) Antifungal activity of selected essential oils against fungi isolated from medicinal plant. Ind Crops Prod 55:116–122. https://doi.org/10.1016/j.indcrop.2014.02.011

Tavares AC, Gonçalves MJ, Cavaleiro C et al (2008) Essential oil of Daucus carota subsp. halophilus: composition, antifungal activity and cytotoxicity. J Ethnopharmacol 119:129–134. https://doi.org/10.1016/j.jep.2008.06.012

Tian J, Huang B, Luo X et al (2012) The control of Aspergillus flavus with Cinnamomum jensenianum Hand.-Mazz essential oil and its potential use as a food preservative. Food Chem 130:520–527. https://doi.org/10.1016/j.foodchem.2011.07.061

Toghueo RMK (2020) Bioprospecting endophytic fungi from Fusarium genus as sources of bioactive metabolites. Mycology 11:1–21. https://doi.org/10.1080/21501203.2019.1645053

Tolouee M, Alinezhad S, Saberi R et al (2010) Effect of Matricaria chamomilla L. flower essential oil on the growth and ultrastructure of Aspergillus niger van Tieghem. Int J Food Microbiol 139:127–133. https://doi.org/10.1016/j.ijfoodmicro.2010.03.032

Tomazoni EZ, Pansera MR, Pauletti GF et al (2016) In vitro antifungal activity of four chemotypes of Lippia alba (Verbenaceae) essential oils against Alternaria solani (Pleosporeaceae) isolates. An Acad Bras Cienc 88:999–1010. https://doi.org/10.1590/0001-3765201620150019

Vale-Silva LA, Gonçalves MJ, Cavaleiro C et al (2010) Antifungal activity of the essential oil of Thymus x viciosoi against Candida, Cryptococcus, Aspergillus and dermatophyte species. Planta Med 76:882–888. https://doi.org/10.1055/s-0029-1240799

Vieira J, Gonçalves C, Villarreal Villarreal J et al (2018) Chemical composition of essential oils from the apiaceae family, cytotoxicity, and their antifungal activity in vitro against candida species from oral cavity. Braz J Biol 79. https://doi.org/10.1590/1519-6984.182206

Villa TG, Veiga-Crespo P (2014) Antimicrobial compounds: current strategies and new alternatives. Springer, Berlin/Heidelberg

Visagie CM, Houbraken J, Frisvad JC et al (2014) Identification and nomenclature of the genus Penicillium. Stud Mycol 78:343–371

Wan LYM, Turner PC, El-Nezami H (2013) Individual and combined cytotoxic effects of Fusarium toxins (deoxynivalenol, nivalenol, zearalenone and fumonisins B1) on swine jejunal epithelial cells. Food Chem Toxicol 57:276–283. https://doi.org/10.1016/j.fct.2013.03.034

Wang L, Jiang N, Wang D, Wang M (2019) Effects of essential oil citral on the growth, mycotoxin biosynthesis and transcriptomic profile of alternaria alternata. Toxins (Basel) 11. https://doi.org/10.3390/toxins11100553

Watanabe M, Yonezawa T, Lee K et al (2011) Molecular phylogeny of the higher and lower taxonomy of the Fusarium genus and differences in the evolutionary histories of multiple genes. BMC Evol Biol 11:322. https://doi.org/10.1186/1471-2148-11-322

Williams DW, Kuriyama T, Silva S et al (2011) Candida biofilms and oral candidosis: treatment and prevention. Periodontol 2000 55:250–265. https://doi.org/10.1111/j.1600-0757.2009.00338.x

Woudenberg JHC, Groenewald JZ, Binder M, Crous PW (2013) Alternaria redefined. Stud Mycol 75:171–212. https://doi.org/10.3114/sim0015

Xing-dong L, Hua-li X (2014) Antifungal activity of the essential oil of Zanthoxylum bungeanum and its major constituent on Fusarium sulphureum and dry rot of potato tubers. Phytoparasitica 42:509–517. https://doi.org/10.1007/s12600-014-0388-3

Xu S, Yan F, Ni Z et al (2014) In vitro and in vivo control of Alternaria alternata in cherry tomato by essential oil from Laurus nobilis of Chinese origin. J Sci Food Agric 94:1403–1408. https://doi.org/10.1002/jsfa.6428

Yamamoto-Ribeiro MMG, Grespan R, Kohiyama CY et al (2013) Effect of Zingiber officinale essential oil on Fusarium verticillioides and fumonisin production. Food Chem 141:3147–3152. https://doi.org/10.1016/j.foodchem.2013.05.144

Yapar N (2014) Epidemiology and risk factors for invasive candidiasis. Ther Clin Risk Manag 10:95–105. https://doi.org/10.2147/TCRM.S40160

Ye CL, Dai DH, Hu WL (2013) Antimicrobial and antioxidant activities of the essential oil from onion (Allium cepa L.). Food Control 30:48–53. https://doi.org/10.1016/j.foodcont.2012.07.033

Zabka M, Pavela R (2018) Review chapter: Fusarium genus and essential oils. In: Mérillon JM, Riviere C. (eds) Natural Antimicrobial Agents. Sustainable Development and Biodiversity. Springer cham. New York City. p. 95–120

Zapata B, Durán C, Stashenko E et al (2010) Actividad antimicótica y citotóxica de aceites esenciales de plantas de la familia Asteraceae. Rev Iberoam Micol 27:101–103. https://doi.org/10.1016/j.riam.2010.01.005

Znini M, Cristofari G, Majidi L et al (2011) Antifungal activity of essential oil from Asteriscus graveolens against postharvest phytopathogenic fungi in apples. Nat Prod Commun 6:1763–1768

Znini M, Cristofari G, Majidi L et al (2013) Essential oil composition and antifungal activity of Pulicaria mauritanica Coss., against postharvest phytopathogenic fungi in apples. LWT – Food Sci Technol 54:564–569. https://doi.org/10.1016/j.lwt.2013.05.030

Zomorodian K, Saharkhiz J, Pakshir K et al (2018) The composition, antibiofilm and antimicrobial activities of essential oil of Ferula assa-foetida oleo-gum-resin. Biocatal Agric Biotechnol 14:300–304. https://doi.org/10.1016/j.bcab.2018.03.014

Chapter 4
Combination of Essential Oil and Food Packaging

Jian Ju, Chang Jian Li, Yang Deng, and Mi Li

4.1 Introduction

With the continuous development of economic and trade globalization, we have the opportunity to come into contact with food from all over the world (Ju et al. 2017). As a result, the importance of maintaining the long shelf life and nutritional quality of food is increasing. In order to achieve satisfactory fresh-keeping effect, preservatives are usually added directly to food, but this may lead to overuse of preservatives (Batiha et al. 2021; Kaderides et al. 2021; Tong et al. 2021). And as we known, some chemical synthetic preservatives even pose a potential threat to human health (Ju et al. 2019a; Zhu et al. 2020). The food industry is forced to develop new methods and technologies to meet the needs of consumers. At the same time, antibacterial packaging arises at the historic moment. The main purpose of antibacterial packaging is to prolong the shelf life of food and maintain the original quality of food (Jin 2017; Liu et al. 2021a). In the antibacterial packaging system, antimicrobial agents achieve long-term antibacterial effect through slow release. This not only avoids the direct contact between antimicrobial agents and food, but also prolongs the effect of antimicrobial agents (Ju et al. 2019b; Zhang et al. 2021).

At present, there are many kinds of antimicrobial agents that can be used in food preservation, including organic synthetic antimicrobial agents, inorganic antimicrobial agents and natural antimicrobial agents (Ju et al. 2020a; Ls et al. 2021). The

J. Ju (✉)
College of Food Science and Engineering, Qingdao Agricultural University, Qingdao, China

C. J. Li · M. Li
State Key Laboratory of Food Science and Technology, Jiangnan University,
Wuxi, Jiangsu, China

Y. Deng
College of Food Science and Engineering, Qingdao Agricultural University, and Qingdao
Special Food Research Institute, Qingdao, China   .

Table 4.1 Types and characteristics of some common antibacterial agents

Type	Component	Characteristic
Natural antibacterial agent	Botanical antibacterial agents: plant extracts, herbs and essential oils.	Natural non-toxic, high safety, good antibacterial effect, poor heat resistance and difficult processing.
Organic antibacterial agent	Alcohols, phenols and ethers; Aldehydes and ketones; Esters; Organic metals, such as methylmercury, polyquaternary ammonium salt, polyquaternary phosphine salt, polyorganotin, polyhaloamines, polyguanidine salt, chitosan and its derivatives.	Fast sterilization, stable chemical properties, convenient processing, poor heat resistance and easy to produce microbial drug resistance.
Inorganic antibacterial agent	Antibacterial agents for metal elements: silver, copper and zinc.	
Photocatalytic antibacterial agents: mainly nano TiQ2, ZnO and Cao.	It has wide antibacterial range, good heat resistance and low toxicity, and will not lead to microbial drug resistance, but it has high cost and is easy to oxidize and discolor.	
Compound antibacterial agent	Different types of antibacterial agents are used in combination.	It has the advantages of various antibacterial agents and produces synergistic effects, but the preparation method is complex.

types and characteristics of some common antibacterial agents are described in Table 4.1. Natural antimicrobial agents can be divided into animal-derived antimicrobial agents (such as chitosan, fish essence, protein propolis), microbial antimicrobial agents (such as lysozyme, nisin, natamycin) and botanical antimicrobial agents (such as essential oils, tea polyphenols, Chinese herbal medicine) (Gyawali and Ibrahim 2014; Onaolapo and Onaolapo 2018; Ong et al. 2021; Yu and Shi 2021). Because natural antimicrobials are extracted from natural raw materials, they are usually considered to be safe antimicrobials. Especially in recent years, as people pay more and more attention to food safety, the demand for "green", "natural" and "safe" antimicrobials is increasing (Gutiérrez et al. 2018; Kai et al. 2020; Seyedeh et al. 2021).

Because essential oils are not only in line with the current development direction of food additives, but also have many different biological properties, such as antibacterial, anti-oxidation, anti-tumor, analgesia, insecticidal, anti-diabetes, anti-inflammation, et al. (Ju et al. 2018b; Zhu et al. 2020). Therefore, essential oil has been widely concerned by people in recent years. Generally speaking, essential oils usually contain several main active components, but some other small molecular compounds can also contribute to the biological activity of essential oils and show synergism to some extent (Ju et al. 2020b; Tak and Isman 2017).

One of the main trends in the field of food packaging is the use of active antibacterial packaging. In this regard, the EOAP has shown great application potential in the field of food preservation. Therefore, this chapter analyzes the application status and technical strategy of essential oil in antibacterial packaging in detail.

4.2 Antibacterial Packaging System

At present, most food packaging on the market is mainly divided into two forms: packaging/food and packaging/headspace/food (Fig. 4.1) (Yao 2016). The first form of packaging is usually suitable for solid or liquid food, which can contact with the packaging material directly without gaps. The antimicrobial agents in this kind of antibacterial packaging can be added directly to food surface or combine the antimicrobial agents with the packaging materials, and then play an antibacterial role by diffusion (Ayana and Turhan 2010; Devlieghere et al. 2004). The second form of packaging is usually that there is a certain gap between the packaging material and the food, so that the volatile antimicrobial agents in the packaging material can be evenly distributed between the packaging material and the food by evaporation. Antimicrobials can also produce antibacterial effects through packaging and contact with food (Quintavalla and Vicini 2002). During this period, the main factors affecting the antibacterial effect of packaging are the type and concentration of antimicrobial agents and the application of antimicrobial agents in packaging. In addition, the types of packaging materials and microorganisms, as well as environmental factors such as temperature, pH and gas environment, are also factors that cannot be ignored (Ju et al. 2019a). Table 4.2 lists the food additives currently approved in the United States, Europe, Australia, New Zealand and other countries and regions. At the same time, these substances may also be used as antimicrobial agents in food packaging (Cooksey 2001; Han 2003).

4.3 Migration of Active Compounds from Packaging to Food

The migration of antimicrobials from packaging materials to food is a challenge for the development of active food packaging, as compounds transferred from packaging to food may be toxic (Guilbert and Gontard 2005). The main factors affecting the transfer rate of antimicrobial agents from packaging to food are the affinity

Fig. 4.1 Schematic diagram of active food packaging system (Han 2003)

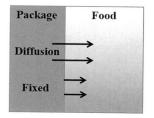
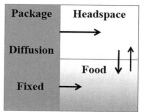

Table 4.2 Types of antibacterial agents that can be used in food packaging

Name	Authorized food additive number			Nam	Authorized food additive number		
	USA[a]	Europe[b]	Australia/New Zealand[c]		USA[a]	Europe[b]	Australia/New Zealand[c]
Citric acid	GRAS	E330	330	Citral	GRAS	–	–
Benzoic acid	GRAS	E210	210	Carvacrol	FA	–	–
Malic acid	GRAS	E296	296	Butylated hydroxytoluene	GRAS	E321	321
Sorbic acid	GRAS	E200	200	Butyl hydroxyanisole	GRAS	E320	320
Succinic acid	GRAS	E363	–	Linalool	GRAS	–	–
Tartaric acid	GRAS	E334	334	Lysozyme	GRAS	E1105	1105
Phosphoric acid	GRAS	E338	338	Methyl p-hydroxybenzoate	–	E218	218
Propionic acid	GRAS	E280	280	Natamycin	FA	E235	235
Acetic acid	GRAS	E260	260	Phosphate	GRAS	E452	–
Lactic acid	GRAS	E270	270	Nisin	GRAS	E234	234
Lauric acid	FA	–	–	Potassium sorbate	GRAS	E202	202
Konjac glucomannan	GRAS	E425	–	Propyl p-hydroxybenzoate	GRAS	E216	216
Cyclohexamethylene tetramine	–	E239	–	Sodium benzoate	GRAS	E211	211
Glucose oxidase	GRAS	–	1102	Sulfur dioxide	GRAS	E220	220
Geraniol	GRAS	–	–	Tartaric acid	GRAS	E334	334
Eugenol	GRAS	–	–	Tert butyl hydroquinone	FA	–	319
Ethyl p-hydroxybenzoate	GRAS	E214	–	Terpineol	FA	–	–
Ethanol	GRAS	E1510	–	Thymol	FA	–	–
Artemisia brain	GRAS	–	–	Ethylenediamine tetraethylamine	FA	–	–
P-methylphenol	FA	–	–	–			

Note: [a]The US Food and Drug Administration (FDA) classifies substances used in food production into three categories: food additives (FA) are food ingredients and substances generally considered safe; [b]Figures starting with "E" indicate that the additive is approved by the European Union (EC) Food Scientific Committee (SCF). [c]Figures show that the additive is considered safe in food by the Australian Food Administration (ANZFA) and the New Zealand Food Safety Authority (ANZFSC) between antimicrobial agents and packaging materials, solubility of antimicrobial agents, storage temperature, et al. (Ju et al. 2019a). In general, the higher the temperature is, the greater the interaction between molecules is. In addition, food composition is also a factor that cannot be ignored. For example, food ingredients such as water and fat increase the release of phenolic compounds into food in active packaging. If the antibacterial agent is highly soluble in food, the migration profile will follow an unconstrained free diffusion, and an antibacterial agent with low solubility will produce a unilateral system (as shown in Fig. 4.2). Therefore, the migration experiment is necessary to ensure the quality of packaging and food safety (Pablo et al. 2015). However, chromatography may be an effective measurement method. For example, some researchers used this method to measure the migration of essential oil active components in whey protein film to food. The results showed that the migration rate of eucalyptol was the highest compared with other active components (Ribeiro-Santos et al. 2017b). In addition, some authors have shown that the higher the temperature is, the greater the migration rate of the compounds in the film is. For example, temperature increases the release rate of carvacrol, thymol and linalool from the starch film to the food simulator (Sharma et al. 2020b). Thus it can be seen that the migration experiment is a very important link in the process of designing active packaging.

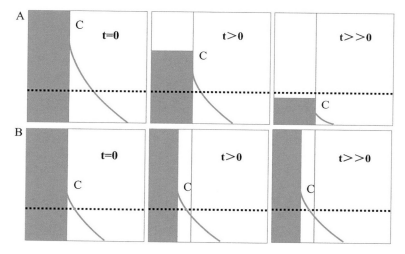

Fig. 4.2 The migration curve of antimicrobial agents. (**a**) Release of soluble antimicrobial agents through free diffusion. (**b**) Release of antimicrobial agents from monolithic system (Ju et al. 2019b)

4.4 Effect of Essential Oil on Antibacterial Property of Packaging

Food microorganism is the main factor leading to food corruption (Ju et al. 2017). The growth of bacteria and fungi during storage will lead to the degradation or decomposition of nutrients in the food, thus changing the appearance, smell and taste of the food. More importantly, some corrupt microorganisms can produce a large number of toxins, and the existence of these toxins pose a serious threat to human health (Sharma et al. 2020a, b). However, essential oils containing active packaging can prolong the shelf life of food, thereby reducing waste and economic losses. For example, Ju et al. (2020c) prepared an active antibacterial sachet containing a combination of essential oils, the result of which showed that it can effectively inhibit the growth of fungi in baked goods during storage (Ju et al. 2020c). The study also from the research group of Ju et al. found that applying cinnamaldehyde and eugenol to the inside of the bag could effectively prolong the shelf life of bread (Ju et al. 2018a, b). In addition, the methylcellulose membrane containing clove essential oil and oregano essential oil reduced the number of yeast and mold in sliced bread within 15 days (Otoni et al. 2014).

4.5 Effect of Essential Oil on Antioxidant Performance of Packaging

The oxidation of food is another important factor leading to food deterioration and corruption. The oxidation of fat can not only lead to discoloration and decay of food, but also cause unpleasant smell and even produce toxins (Wang et al. 2019). Therefore, it is very important to use natural antioxidants in active packaging to prevent the oxidation of food. Essential oils contain a lot of polyphenols and flavonoids. Among them, polyphenols can directly quench singlet oxygen, prevent free radical chain reaction, and transfer active hydrogen atoms to free radicals, turn free radicals into less active substances and then scavenge them (Fig. 4.3). Zheng et al. (2019) proved this point. Eugenol was used in acorn starch and chitosan-based edible membranes. It was found that the addition of eugenol significantly increased the antioxidant activity of edible films (about 86.77%) (Zheng et al. 2019). Similarly, Miao et al. (2019) prepared a biodegradable film with antioxidant and antibacterial properties by adding fennel essential oil to polylactic acid (PLA) and polyhydroxybutyrate (PHB). The determination of pH value, TVB- N value and free amino acid value showed that the film could significantly prolong the shelf life of oyster (Miao et al. 2019).

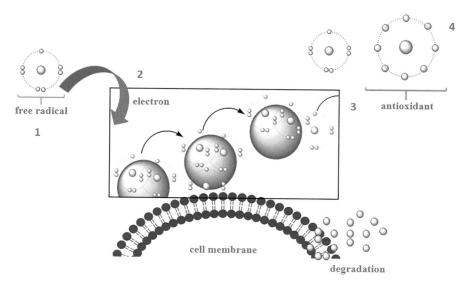

free radical

1

2

electron

3

antioxidant

4

cell membrane

degradation

Fig. 4.3 Schematic diagram of the possible effect of essential oil on scavenging free radicals

4.6 Effect of Essential Oil on Microstructure of Packaging Materials

Because the essential oil is volatile, it can be used in the antibacterial packaging system. At present, the commonly used film-forming materials are polymer materials (EVHO,PVA,LDPE) and edible packaging materials (protein, polysaccharide, cellulose) (Ribeiro-Santos et al. 2017a, b). In the antibacterial packaging system of essential oil, small molecules of essential oil can be gradually released through the micropores in the film, and then come into contact with food, so as to achieve the purpose of prolonging the shelf life of food (Ju et al. 2019a). The addition of essential oil to polymer materials will have an important impact on the physical and chemical properties of packaging materials.

The extensibility of food packaging materials is an important index to measure the physical properties of packaging materials. The addition of essential oil into packaging materials can have an important effect on the tensile properties of packaging materials, which has been confirmed in many studies. The tensile properties of packaging materials are mainly determined by the interaction between the properties of polymers and essential oil molecules (Ojagh et al. 2010). For example, phenolic compounds in essential oils can interact with different protein sites, resulting in enhanced tensile strength of edible films containing proteins (Atarés and Chiralt 2016). In contrast, the tensile strength of cassava starch film containing cinnamon essential oil decreased with the addition of essential oil (Zhou et al. 2021).

Secondly, the barrier performance of food packaging materials is another important index to measure the physical properties of packaging materials, because the barrier performance of packaging materials will have an important impact on the quality of products during storage (Lka et al. 2021). Usually, the hydrophobicity or hydrophilicity of the packaging material is measured by the moisture resistance or water vapor permeability of the packaging material (Pires et al. 2013; Tcv et al. 2021). The hydrophilicity or hydrophobicity of the material can be adjusted by adding different substances to the packaging material. Because the essential oil is a hydrophobic small molecular compound, the addition of essential oil to the packaging material will improve the hydrophobicity and barrier of the material. For example, the addition of *Mosla chinensis Maxim* essential oils, *Artemisia dracunculus L* essential oils and thyme essential oils to cod protein can significantly reduce the water vapor permeability of the protein film (Pires et al. 2013). However, some authors have shown that the addition of *Zataria multiflora* essential oil to potato starch and pectin film can increase the water vapor permeability of the film, and the reason for this different phenomenon needs to be further confirmed (Sani et al. 2021).

Thirdly, the color and transparency of food packaging are important external factors affecting consumers' desire to buy. Of course, in some cases, the color of food packaging will also affect the shelf life and storage stability of food. In the active antibacterial packaging containing essential oils, the color of the packaging mainly depends on the type and concentration of essential oils and whether the essential oils will react with polymer materials (Abdur et al. 2020). The addition of rosemary and thyme essential oils to the polylactic acid film can change the color of the polylactic acid film. At the same time, the color change intensity of the film depends on the concentration of essential oil (Yahyaoui et al. 2016). In addition, it was found that the addition of cinnamon essential oil to starch film make the color of the film deepened and the yellowness increased (Arezoo et al. 2019). In contrast, the addition of clove essential oil and thyme essential oil to the Polylactide polybutylene terephthalate (PLA-PBAT) film had no significant effect on the optical properties of the film (Sharma et al. 2020a). In some cases, the addition of essential oils to polymer materials may lead to an increase in the surface roughness of packaging materials. The reason for this phenomenon may be that the dispersion of essential oil droplets on the surface of the solution reduces the specular reflectance.

4.7 Construction form of Essential Oil in Antibacterial Packaging System

4.7.1 The Essential Oil Is Directly Mixed into the Matrix Material of the Package

At present, the kind of antibacterial packaging most widely studied is to mix essential oil directly into polymer materials (Fig. 4.4a). Recently published literature showed that the research and application of this method in food-related fields is growing rapidly, such as the EOAP and antibacterial paper containing essential oils

Fig. 4.4 The realization way of EO as an additive in antibacterial packaging

(Chan et al. 2010; Rodríguez et al. 2007; Tankhiwale and Bajpai 2009). Adding thyme essential oil with different concentration to konjac glucomannan matrix can inhibit the growth of *Listeria monocytogenes*, *Staphylococcus aureus* and Escherichia coli (Liu et al. 2021a, b). Similarly, the addition of cinnamon, oregano and clove essential oils to polypropylene (PP) and polyethylene/vinyl alcohol copolymer (PE/EVOH) packaging films can make the packaging materials have better antifungal activity (Suppakul et al. 2011). Other antimicrobial agents, such as various natural phenols, catechins, enzymes and peptides with antibacterial activity, also have good antibacterial activity when added to the polymer (Hotchkiss 1997). Because a single antibacterial agent sometimes cannot achieve the ideal antibacterial effect, two or more kinds of antibacterial agents with synergistic antibacterial effect can be added to the polymer. The principle of adding antimicrobial agents to polymers to protect food is mainly based on the fact that most of the harmful microorganisms in food are concentrated on the surface of food. When the antibacterial packaging materials come into contact with the food surface, it can inhibit the growth and reproduction of microorganisms on the food surface. Normally, the mass fraction of antimicrobial agents in the polymer is 0.1–5% (Paola and Joseph 2002).

4.7.2 To Coat or Adsorb Essential Oils on Packaging Materials

Because the essential oil is unstable and volatile when exposed to heat, the packaging material can be coated after the packaging material is formed (Fig. 4.4b and c). There are several successful applications for coating essential oils on packaging bags. Ju and his colleagues successfully extended the shelf life of the bread by coating cinnamaldehyde and eugenol on the inside of the bread bag (Ju et al. 2018a, b). And it is worth mentioning that this method has no adverse effect on the sensory properties of bread. Qi et al. (2020) also confirmed that adding essential oil inside the bag could effectively inhibit the fungal growth of bread during storage and effectively prolong the shelf life of bread (Qi et al. 2020). Similarly, some authors have confirmed that paper packaging coated with cinnamon essential oil can prevent the infiltration of water vapor and effectively prevent fungal contamination of food in the package (RodriGuez et al. 2008).

4.7.3 Add a Carrier Containing Essential Oil to the Package

Adding antibacterial small packages or fresh-keeping cards containing essential oils to the packaging is one of the most successful strategies for the application of antibacterial packaging at present (Fig. 4.4d). Among them, ethanol is the first to be used in this kind of antibacterial packaging, but the disadvantage of using ethanol as an antibacterial agent in packaging is that it will produce the smell of ethanol, thus affecting the sensory quality of food (Brody et al. 2001). Recently, Ju and his colleagues prepared an antibacterial sachet containing the active ingredient of essential oil and applied it to extend the shelf life of bread. The results showed that the antibacterial sachet could not only effectively prolong the shelf life of bread, but also had no significant effect on the sensory properties of bread (Ju et al. 2020b). In addition, the liner containing organic acids and surfactants has also been successfully applied to the preservation of meat products (Hansen et al. 1989).

4.7.4 Add Essential Oil to the Package in the Form of Gas

The application of essential oil in modified atmosphere packaging in the form of gas may be a very potential method for food preservation (Fig. 4.4e). In this way, the essential oil can achieve a balanced distribution in the food packaging (Nielaen and Rios 2000; Serrano et al. 2005). However, the current research on this type of active food packaging is relatively scarce. Earlier, some scholars investigated the inhibitory activity of the vapor produced by the combination of cinnamon essential oil and clove essential oil on microorganisms, and the results showed that the combination had antagonistic effect on the growth of *Escherichia coli* when the minimum inhibitory concentration was used. However, when the maximum inhibitory concentration was used, the combination had a synergistic inhibitory effect on *Listeria monocytogenes*, *Bacillus cereus* and *Y. enterocolitica* (Miroslava et al. 2019).

4.7.5 The Essential Oil Is Encapsulated and then Added to the Packaging Matrix

Because the essential oil is unstable, and easy to oxidize and decompose, especially under the condition of heating, it is easy to deactivate. Besides, its compatibility with packaging materials is poor, so it is often limited in the preparation process of active food packaging. In order to overcome these problems, the physical and chemical stability of essential oil can be well maintained by embedding essential oil with microcapsule and nano-encapsulation technology (Fig. 4.4f). For example, clove and oregano essential oils are first encapsulated in coarse emulsions and nanoemulsions and then mixed with carboxymethyl cellulose matrix to prepare active antibacterial packaging. This method can reduce the amount of yeast and mold in sliced bread within 15 days (Otoni et al. 2014).

4.8 Conclusion

This paper gives a comprehensive review of the active packaging system containing essential oil. Firstly, two common antibacterial packaging systems were introduced and the migration law of essential oil from packaging to food was analyzed. On this basis, the effects of essential oil mixed with polymer matrix on the antibacterial properties, antioxidant properties and microstructure of packaging were analyzed. Finally, the construction strategy of essential oil in antibacterial packaging system was summarized. The research content of this paper will provide important guiding significance for the application of essential oil in active food packaging.

At present, although some achievements have been made in the research on antibacterial packaging of essential oil, there are still many problems to be solved. Although many essential oils and their active ingredients have been approved as food flavor additives or have been listed as harmless components, whether all essential oils can be used in the food industry needs to be carefully evaluated. In the next work, it is necessary to explore the cytotoxicity of essential oils and provide a list of essential oils that can be used in the food industry. In addition, it is not clear whether the interaction between essential oils and polymer materials will produce other harmful compounds.

Acknowledgments The work described in this article is supported by the China Scholarship Council.

Conflict of Interest The authors declare no competing interests.

References

Abdur R, Seid M, Rana M, Shahid M (2020) Development of active food packaging via incorporation of biopolymeric nanocarriers containing essential oils. Trends Food Sci Technol 101:206–121
Atarés L, Chiralt A (2016) Essential oils as additives in biodegradable films and coatings for active food packaging. Trends Food Sci Technol 48:51–62

Arezoo E, Mohammadreza E, Maryam M, Abdorreza MN (2019) The synergistic effects of cinnamon essential oil and nano TiO2 on antimicrobial and functional properties of sago starch films. Int J Biol Macromol 157:743–751

Ayana B, Turhan KN (2010) Use of antimicrobial methylcellulose films to control staphylococcus aureus during storage of kasar cheese. Packag Technol Sci 22(8):461–469

Batiha ES, Hussein DE, Algammal AM, George TT, Cruz-Martins N (2021) Application of natural antimicrobials in food preservation: recent views. Food Control 126:108066. https://doi.org/10.1016/j.foodcont.2021.108066

Brody AL, Strupinsky EP, Kline LR (2001) Active packaging for food applications. CRC Press LLC, New York, p 11

Chan HL, An DS, Park HJ, Dong SL (2010) Wide-spectrum antimicrobial packaging materials incorporating nisin and chitosan in the coating. Packag Technol Sci 16(3):99–106

Cooksey K (2001) Antimicrobial food packaging. Addit Polym 8:6–10

Devlieghere F, Vermeiren L, Debevere J (2004) New preservation technologies: possibilities and limitations. Int Dairy J 14(4):273–285

Gutiérrez DR, Fernández J, Lombó F (2018) Plant nutraceuticals as antimicrobial agents in food preservation: terpenoids, polyphenols and thiols. Int J Antimicrob Agents 52(3):309–315

Guilbert S, Gontard N (2005) Agro-polymers for edible and biodegradable films. Review of agricultural polymeric materials, physical and mechanical characteristics. Innovations Food Packag 2005:263–276

Gyawali R, Ibrahim SA (2014) Natural products as antimicrobial agents. Food Control 46:412–429

Han JH (2003) Antimicrobial food packaging. Novel Food Packag Tech 3:50–70

Hansen RE, Rippl CG, Midkiff DG, Neuwirth JG (1989) Antimicrobial absorbent food pad. U.S. Patent 4,865,855

Hotchkiss JH (1997) Food-packaging interactions influencing quality and safety. Food Addit Contam 14(7):601–607

Jin TZ (2017) Current state of the art and recent innovations for antimicrobial food packaging. Microbial Control Food Preserv 16:349–372

Ju J, Wang C, Qiao Y, Li D, Li W (2017) Effects of tea polyphenol combined with nisin on the quality of weever (Lateolabrax japonicus) in the initial stage of fresh-frozen or chilled storage state. J Aquat Food Prod Technol 26(5):543–552

Ju J, Xie Y, Guo Y, Cheng Y, Qian H, Yao W (2018a) The inhibitory effect of plant essential oils on foodborne pathogenic bacteria in food. Crit Rev Food Sci Nutr 59(20):3281–3292

Ju J, Xu X, Xie Y, Guo Y, Cheng Y, Qian H (2018b) Inhibitory effects of cinnamon and clove essential oils on mold growth on baked foods. Food Chem 240:850–855

Ju J, Chen X, Xie Y, Yu H, Yao W (2019a) Application of essential oil as a sustained release preparation in food packaging. Trends Food Sci Technol 92:22–32

Ju J, Xie Y, Guo Y, Cheng Y, Qian H, Yao W (2019b) Application of edible coating with essential oil in food preservation. Crit Rev Food Sci Nutr 59(15):2467–2480

Ju J, Xie Y, Guo Y, Cheng Y, Qian H, Yao W (2020a) Major components in lilac and litsea cubeba essential oils kill Penicillium roqueforti through mitochondrial apoptosis pathway. Ind Crop Prod 149:112349

Ju J, Xie Y, Yu H, Guo Y, Yao W (2020b) Synergistic interactions of plant essential oils with antimicrobial agents: a new antimicrobial therapy. Crit Rev Food Sci Nutr 2:1–12

Ju J, Xie Y, Yu H, Guo Y, Yao W (2020c) A novel method to prolong bread shelf life: sachets containing essential oils components. LWT Food Sci Technol 131:109744

Kai C, Min Z, Bhesh B, Arun S, Mujumdar D (2020) Edible flower essential oils: a review of chemical compositions, bioactivities, safety and applications in food preservation. Food Res Int 139:109809

Kaderides K, Kyriakoudi A, Mourtzinos I, Goula AM (2021) Potential of pomegranate peel extract as a natural additive in foods. Trends Food Sci Technol 115(3):380–390

Lka B, Fei L, Bsc C, Ab C, Mh C, Fang Z (2021) Controlled release of antioxidants from active food packaging: a review. Food Hydrocoll 120:106992

Liu Y, Ahmed S, Sameen DE, Wang Y, Qin W (2021a) A review of cellulose and its derivatives in biopolymer-based for food packaging application. Trends Food Sci Technol 112(2):532–546

Liu Z, Lin D, Shen R, Zhang R, Yang X (2021b) Konjac glucomannan-based edible films loaded with thyme essential oil: physical properties and antioxidant-antibacterial activities. Food Packag Shelf Life 29(1):100700

Ls A, Fx A, Hsab C (2021) Bio-synthesis of food additives and colorants-a growing trend in future food. Biotechnol Adv 47:107694

Miroslava C, Hleba L, Medo J, Taninová D, Klouek P (2019) The in vitro and in situ effect of selected essential oils in vapour phase against bread spoilage toxicogenic aspergilli. Food Control 110:107007

Miao L, Walton WC, Wang L, Li L, Wang Y (2019) Characterization of polylactic acids-polyhydroxybutyrate based packaging film with fennel oil, and its application on oysters. Food Packag Shelf Life 22:100388

Nielaen PV, Rios R (2000) Inhibition of fungal growth on bread by volatile components from-spices and herbs, and the possible application in active packaging, with specialem phases on mustard essential oil. Int J Food Microbiol 60(2–3):219–229

Ojagh SM, Rezaei M, Razavi SH, Hosseini SMH (2010) Development and evaluation of a novel biodegradable film made from chitosan and cinnamon essential oil with low affinity toward water. Food Chem 122(1):161–166

Ong G, Kasi R, Subramaniam R (2021) A review on plant extracts as natural additives in coating applications. Prog Org Coat 151:106091

Onaolapo AY, Onaolapo OJ (2018) Food additives, food and the concept of 'food addiction': is stimulation of the brain reward circuit by food sufficient to trigger addiction. Pathophysiology 25(4):263–276

Pablo RS, Cristian MO, Yanina SM, Luciana DG, Adriana NM (2015) Edible films and coatings containing bioactives. Curr Opin Food Sci 5:86–92

Paola A, Joseph HH (2002) Review of antimicrobial food packaging. Innovative Food Sci Emerg Technol 3:113–126

Pires C, Ramos C, Teixeira B, Batista I, Nunes ML, Marques A (2013) Hake proteins edible films incorporated with essential oils: physical, mechanical, antioxidant and antibacterial properties. Food Hydrocoll 30(1):224–231

Qi S, Li J, Sun Y, Chen Q, Le T (2020) The antifungal effects of cinnamaldehyde against aspergil-lus Niger and its application in bread preservation. Food Chem 317:126405

Quintavalla S, Vicini L (2002) Antimicrobial food packaging in meat industry. Meat Sci 62(3):373–380

Otoni CG, Pontes S, Medeiros E, Soares N (2014) Edible films from methylcellulose and nano-emulsions of clove bud (syzygium aromaticum) and oregano (origanum vulgare) essential oils as shelf life extenders for sliced bread. J Agric Food Chem 62(22):5214–5219

Ribeiro-Santos R, Andrade M, Sanches-Silva A (2017b) Application of encapsulated essential oils as antimicrobial agents in food packaging. Curr Opin Food Sci 14:78–84

Ribeiro-Santos R, Andrade M, de Melo NR, Sanches-Silva A (2017a) Use of essential oils in active food packaging: recent advances and future trends. Trends Food Sci Technol 61:132–140

RodriGuez A, NeriN C, Batlle R (2008) New cinnamon-based active paper packaging against rhizopusstolonifer food spoilage. J Agric Food Chem 56(15):6364–6369

Rodríguez A, Batlle R, Nerín C (2007) The use of natural essential oils as antimicrobial solutions in paper packaging. Part ii. Prog Org Coat 60(1):33–38

Sani IK, Geshlaghi SP, Pirsa S, Asdagh A (2021) Composite film based on potato starch/apple peel pectin/zro 2 nanoparticles/microencapsulated zataria multiflora essential oil; investigation of physicochemical properties and use in quail meat packaging. Food Hydrocoll 117:106719

Serrano M, Martínez-Romero D, Castillo S, Guillén F, Valero D (2005) The use of natural antifun-gal compounds improves the beneficial effect of map in sweet cherry storage. Innovative Food Sci Emerg Technol 6(1):115–123

Seyedeh NJ, Elham A, Jianguo F, Seid MJ (2021) Natural antimicrobial-loaded nanoemulsions for the control of food spoilage/pathogenic microorganisms. Adv Colloid Interf Sci 295:102504

Sharma S, Barkauskaite S, Jaiswal AK, Jaiswal S (2020a) Essential oils as additives in active food packaging. Food Chem 343(8):128403

Sharma S, Barkauskaite S, Jaiswal S, Duffy B, Jaiswal AK (2020b) Development of essential oil incorporated active film based on biodegradable blends of poly (lactide)/poly (butylene adipate-co-terephthalate) for food packaging application. J Packag Technol Res 4(3):235–245

Suppakul P, Sonneveld K, Bigger SW (2011) Loss of AM additives from antimicrobial films during storage. J Food Eng 105(2):270–276

Tankhiwale R, Bajpai SK (2009) Graft copolymerization onto cellulose-based filter paper and its further development as silver nanoparticles loaded antibacterial food-packaging material. Colloids Surf B 69(2):164–168

Tak JH, Isman MB (2017) Enhanced cuticular penetration as the mechanism of synergy for the major constituents of thyme essential oil in the cabbage looper, trichoplusia ni. Ind Crops Prod 101:29–35

Tcv A, Com A, Lmj B, Sai C, Rpv A (2021) Essential oils as additives in active starch-based food packaging films: a review. Int J Biol Macromol 182:1803–1891

Tong J, Zhang Z, Wu Q, Huang Z, Zhao Y (2021) Antibacterial peptides from seafood: a promising weapon to combat bacterial hazards in food. Food Control 125(2):108004

Wang ZC, Lu Y, Yan Y, Nisar T, Chen DW (2019) Effective inhibition and simplified detection of lipid oxidation in tilapia (oreochromis niloticus) fillets during ice storage. Aquaculture 511:634183

Yahyaoui M, Gordobil O, Herrera DR, Abderrabba M, Labidi J (2016) Development of novel antimicrobial films based on poly(lactic acid) and essential oils. React Funct Polym 109:1–8

Yao Z (2016) Antimicrobial food packaging. Food Technol 54:56–65

Yu L, Shi H (2021) Recent advances in anti-adhesion mechanism of natural antimicrobial agents on fresh produce. Curr Opin Food Sci 42(1):8–14

Zhang W, Jiang H, Rhim JW, Cao J, Jiang W (2021) Effective strategies of sustained release and retention enhancement of essential oils in active food packaging films/coatings. Food Chem 1:130671

Zheng K, Xiao S, Li W, Wang W, Chen H, Yang F, Qin C (2019) Chitosanacorn starch-eugenol edible film: Physico-chemical, barrier, antimicrobial, antioxidant and structural properties. Int J Biol Macromol 135:344–352

Zhu Y, Li C, Cui H, Lin L (2020) Encapsulation strategies to enhance the antibacterial properties of essential oils in food system. Food Control 123(2):107856

Zhou Y, Wu X, Chen J, He J (2021) Effects of cinnamon essential oil on the physical, mechanical, structural and thermal properties of cassava starch-based edible films. Int J Biol Macromol 184(6):574–583

Chapter 5
Combination of Essential Oil, and Food Additives

Jian Ju, Chang Jian Li, Yang Deng, and Mi Li

5.1 Introduction

Diseases caused by foodborne pathogens remain a major health problem in the world. Despite new improvements in food hygiene and food production technology, food safety is still a common public health problem (Mittal et al. 2018; Weng et al. 2021). At present, the common foodborne pathogens include *Salmonella*, *Staphylococcus aureus*, *Bacillus cereus*, *Clostridium perfringens*, *Botox*, *Campylobacter jejuni*, *Escherichia coli O157:H7* and *Listeria monocytogenes*. Due to the indiscriminate overuse of antibiotics, most bacteria develop drug resistance (Ju et al. 2018a; Guo et al. 2014).

At present, some researchers have tested the antibacterial activity of some plant extracts or compounds against drug-resistant microorganisms. The authors found that herbs and some phytochemicals are effective against almost all targets (Ayaz et al. 2019). For example, Tellimagrandin and Corilagin can suppress PBP2a. Gallic acid, thymol and carvol can destroy the permeability of cell membrane. Epigallocatechin gallate (EGCG) can inhibit β-lactamases. Reserpine, isopimarane, EGCG and sage acid can inhibit bacterial efflux pump (Abreu et al. 2012; Hemaiswarya et al. 2008).

Figure 5.1 shows the mechanism of antibiotic resistance and the targets of natural products.

J. Ju (✉)
College of Food Science and Engineering, Qingdao Agricultural University, Qingdao, China

C. J. Li · M. Li
State Key Laboratory of Food Science and Technology, Jiangnan University, Wuxi, China

Y. Deng
College of Food Science and Engineering, Qingdao Agricultural University, and Qingdao Special Food Research Institute, Qingdao, China

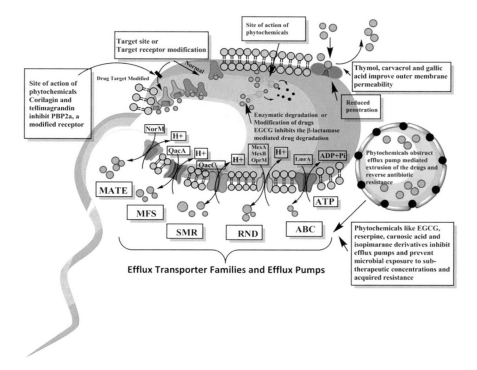

Fig. 5.1 Mechanisms of antibiotic resistance and target sites of natural products (Ayaz et al. 2019)

Phytochemicals in combination with existing antibiotics can synergistically act to neutralize different resistance mechanisms, including over expression of efflux pumps, expression of drug inactivating and target site modifying enzymes and modification of permeability barriers. MATE: Multidrug and Aoxic Compound Extrusion super-family, MFS: Major Facilitator Super-family, SMR: Small Multidrug Resistance super-family, RND: Resistance-nodulation-cell Division super-family, ABC: ATP-binding Cassette super-family.

At present, consumers' concern about the negative perception of chemical preservatives has prompted the food industry to pursue and develop natural preservatives (Ju et al. 2017). In addition, natural preservatives are favored by consumers because of their wide range of sources and biodegradability. In recent years, plant essential oils and some plant extracts have attracted more and more attention by food scientists as natural antimicrobial agents in the food system (Ju et al. 2018b; Chouhan et al. 2017). In particular, plant essential oil, as a new antibacterial agent, has been developed. What is more recognized about the antibacterial mechanism of essential oil is that the hydrophobicity of essential oil can make the small molecules of essential oil act on the cell membrane, which leads to the increase of cell membrane permeability, the leakage of cell inclusions, and finally lead to cell apoptosis (Ju et al. 2018c, 2020a). For example, eugenol and citral small molecules can destroy the cell membrane permeability of *Penicillium roqueforti* and *Aspergillus niger*, resulting in the leakage of intracellular genetic material (Ju et al. 2020b, c). Similarly, red pepper essential oil can damage the permeability and integrity of *Staphylococcus aureus* and increase the concentration of protein, sugar and nucleic acid in the cytoplasm (Dannenberg et al. 2018).

The chemical components of essential oils are varied, and different active components often show different interactions (Nidhi et al. 2020; Baj et al. 2018). There may be four kinds of effects among the active components of essential oil: irrelevant effect, additive effect, antagonistic effect and synergistic effect (Ju et al. 2020a). Therefore, this chapter mainly summarized the main factors affecting the antibacterial activity of essential oils in food system, focusing on the synergistic bacteriostatic effect of different active components of essential oils, the synergistic bacteriostatic effect of essential oils and food additives, and the synergistic bacteriostatic effect of essential oils and antibiotics. Our main purpose is to reduce the dosage of essential oil in food preservation and the effect of essential oil on food sensory properties through synergistic effect.

5.2 Factors Affecting the Antibacterial Activity of Essential Oils in Food System Microbial Species

In general, gram-positive bacteria are more sensitive to essential oils than gram-negative bacteria. The main reason why Gram-negative bacteria are not sensitive to the active components of essential oils is that they have an outer membrane around the cell wall, which is almost impermeable to hydrophobic compounds, thus preventing the infiltration of various active components in the essential oil (Burt 2004; Klancnik et al. 2011; Ju et al. 2018b). In addition, there are hydrolases in the periplasmic space of Gram-negative bacteria, and these hydrolases also promote the hydrolysis of antibacterial components, thus further reducing the sensitivity of Gram-negative bacteria to drugs (Al-Reza et al. 2009). On the contrary, Gram-positive bacteria do not have this natural barrier, so the hydrophobic small molecules in essential oil can directly reach the cell membrane and come into contact with the phospholipid bilayer in the cell membrane, resulting in increased membrane permeability and even ion leakage (Gao et al. 2010). However, not all studies have shown that Gram-positive bacteria are more sensitive than Gram-negative bacteria. In fact, a large number of studies have confirmed that hydrophilic bacteria (Gram-negative) seem to be one of the most sensitive specie (Stecchini et al. 1993; Hao et al. 1998; Wan et al. 1998). There are some porins in the outer membrane of Gram-negative bacteria, and these porins produce channels large enough to allow small molecules of essential oils to pass through, so that essential oils can still act on Gram-negative bacteria (Holley and Patel 2005; Kotzekidou et al. 2008).

5.2.1 Food Composition

The antibacterial activity of essential oils in food matrix is usually higher than the theoretical value obtained under laboratory conditions. The main reason for this phenomenon is the effect of food components on the antibacterial activity of essential oils (Seow et al. 2014; Klancnik et al. 2011). Compared with the laboratory culture medium, the nutrients in food are more comprehensive and more available, which enables the damaged bacteria to repair quickly in the food matrix (Gill et al. 2002). It is generally believed that high levels of fat or protein in foods can reduce the sensitivity of bacteria to essential oils. For example, studies have shown that fat in whole milk can protect bacterial cells from antibiotics (Klancnik et al. 2011). The mixture of thymol and carvacrol with lactic acid and acetic acid against Staphylococcus aureus in meat products was not as effective as in broth (de Oliveira et al. 2010). In addition, the reaction between carvanol and protein has been considered to be a limiting factor in anti-Bacillus cereus activity in milk (Pol et al. 2001). Phenolic compounds interact with peptide chains of proteins to form complexes through hydrogen bonding and hydrophobic interaction. Nitrogenous compounds and fats in meat products will react with phenolic compounds, seriously affecting their antibacterial properties (de Oliveira et al. 2010). However, carbohydrates in food do not seem to protect bacteria from essential oils as well as fats and proteins (Shelef et al. 1984).

5.2.2 Acidity and Alkalinity

In general, the sensitivity of microorganisms to essential oils seems to increase with the decrease of pH in food system (Gutierrez et al. 2009). Therefore, the combination of low pH and suitable essential oils may play a synergistic role in reducing the number of microorganisms. For example, studies have confirmed that when pH value 5, it can increase the sensitivity of Listeria monocytogenes to essential oils, while higher than this value, it will increase the growth rate of bacteria (Gutierrez et al. 2008). This may be due to the fact that at low pH, essential oils are not dissociated and their hydrophobicity increases, which make it easier to dissolve in cell membrane lipids (Rivas et al. 2010). Therefore, the antibacterial activity of essential oils in high-protein and high-fat foods is often lower than that in vegetable products. Based on this, some authors also suggested that acetic acid can be added to oregano essential oil to improve the antibacterial activity of carvanol. However, this regular is not invariable. In many cases, the antibacterial activity of essential oils also depends on the types of microorganisms, especially for acid-tolerant microorganisms, such as Escherichia coli O157:H7 in apple juice and vegetable juice (Friedman et al. 2004).

5.2.3   *Sodium Chloride*

As the main component of food seasoning, sodium chloride can play different roles in combination with essential oils under different conditions. Some researchers have shown that the combination of sodium chloride and clove essential oil has an effective inhibitory effect on *Escherichia coli* at low concentration (Angienda and Hill 2011). In addition, sodium chloride and peppermint essential oil had synergistic inhibitory effect on *Salmonella* and *Listeria monocytogenes* (Tassou et al. 1995). In contrast, some studies have shown that the combination of sodium chloride and carvanol has antagonistic effect on *Bacillus cereus* (Ultee et al. 1995). Similar studies also showed that the addition of 4% w/v sodium chloride to agar medium did not improve the antibacterial activity of cinnamaldehyde (Moleyar and Narasimham 1992).

5.2.4   *Bacterial Concentration, Temperature and Solubility*

In general, the higher the concentration of bacteria is, the higher the minimum inhibitory concentration (MIC) of essential oil is (Lambert et al. 2001; Seow et al. 2014). However, at present, there is not much analysis on the correlation between the antibacterial activity of essential oil and bacterial concentration. Earlier, researchers have found that Prigon essential oil can better inhibit the growth of *Erwinia* (Scortchini and Rossi 1991) when the concentration of microbial cells is low (3 log cfu/mL). Of course, it also depends on different types of essential oils, because different essential oils contain different proportions or contents of active ingredients. In addition, temperature also has a certain effect on antibacterial activity of essential oil under certain conditions. However, the effect of temperature on antibacterial activity of essential oil is still controversial. This is mainly because different microbial species usually have different optimum growth temperature. In addition, different kinds of essential oils have different solubility or volatility at different temperatures.

The solubility of essential oil in food medium is also the main factor affecting the antibacterial activity of essential oil. In general, the higher the solubility is, the better the antibacterial activity is. Therefore, various emulsifiers are used to improve the solubility of essential oils, such as ethanol, Tween-80, acetone, polyethylene glycol, propylene glycol, n-hexane, dimethyl sulfoxide and agar (Sokoic et al. 2009). Some related studies have shown that the combination of essential oil and emulsifier can improve the antibacterial activity of essential oil (Hammer et al. 1999; Kim et al. 1995). However, some related studies have confirmed that the combination of essential oil and emulsifier cannot improve the antibacterial activity of essential oil (Sokoic et al. 2009; Chalova et al. 2010). Therefore, at present, the above research conclusions are still controversial.

5.3 Synergistic Bacteriostatic Effect of Different Essential Oils

The use of different combinations or active compounds of essential oils can produce four possible effects: synergistic, additive, irrelevant or antagonistic. The evaluation of the possible effect is mainly calculated according to fractional inhibitory concentration (FIC) formula. FIC = (MIC of A in combination/MIC of A) + (MIC of B in combination/MIC of B). The interaction was defined as synergistic if the FIC index was 0.5 or less, as additive if the FIC index was between 0.5 and 1, as no interaction if the FIC index was between 1 and 2, and as antagonistic if the FIC index was greater than 2. The chessboard method can be used to calculate the FIC value, as shown in Fig. 5.2 (Ju et al. 2020d). Of course, in addition to this, there are some other commonly used evaluation methods in the field of biomedicine. In Table 5.1, we summarize and analyze these commonly used methods and their advantages and disadvantages. Researchers can choose one or more methods according to the needs of the experiment.

At present, some researchers have evaluated the antibacterial efficacy of some essential oil combinations. The synergism or antagonism of essential oil combinations depends on the type of essential oils or microorganisms. Different essential oils or microorganisms may have different effects (Ju et al. 2020e). It is reported that thymol and carvol have synergistic inhibitory effects on *Escherichia coli*

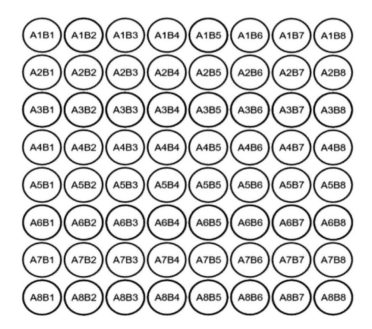

Fig. 5.2 Schematic diagram of chessboard method (Ju et al. 2020d)

Table 5.1 The evaluation method of drug combination and its advantages and disadvantages (Ju et al. 2020a)

Evaluation method	Advantages	Disadvantages	References
Algebraic sum	The method is very simple, and it is easy to judge the nature of the combination of drugs.	It can only be used to judge the combination of drugs with simple linear relationship.	Levine (1978)
Q value method	This method can be directly used to compare the original dose effect level and the operation is simple.	The amount of information is small, only qualitative.	Abt and Paul (1972)
Burgi formula method	This method is commonly used to measure the final effect of combined drugs.	Because it does not need to consider the dose-response relationship and the mode of action of the combination of drugs, its application is limited.	Guo et al. (2005)
Chou-Talalay	It can be used in combination of multiple drugs or in combination of unsteady dose ratio, and can be described qualitatively and quantitatively.	This method cannot give the drug map with non-constant ratio, only the equivalent line map after standardization.	Chou and Rideout (1987)
Finney Harmonic average method	When the combined action of the two drugs is similar, the equivalent line can be derived by this method.	It may be too simple to determine the experimental results by probability operation, unless a large number of experiments are carried out, the results are not very reliable.	Finney (1952)
Webb Fractional product method	It is the simplest method and is widely used at present.	It is only suitable for the additive calculation of non repellent drugs.	Ribo and Rogers (2010)
Reaction surface method	It can display two-dimensional and three-dimensional atlas with reliable results. According to the spectrum, the best joint mode can be obtained.	The mathematical model is complex and the workload is heavy. It needs relevant statistical knowledge and S-curve relationship of drug effect.	Minto et al. (2000)
Relative effect method	The combined effect is determined by the value of the two drugs acting alone and the actual value of the single drug.	The confidence interval can not be calculated. In most cases, it lacks reliability and can only be analyzed qualitatively.	Pradhan and Kim (2014)
Mapping analysis	This method does not need to consider the type of action between the two drugs. It is a relatively general analysis method.	A large amount of data is needed to draw dose-response curve, and a fixed ratio of drug combination is needed, so it will be limited in practical application.	Zheng and Sun (2000)

(continued)

Table 5.1 (continued)

Evaluation method	Advantages	Disadvantages	References
Parametric method	Data can be analyzed systematically. Abundant data and reliable results.	It needs high statistical knowledge and a lot of work. It is suitable for data that can be fitted by Hill equation.	Wang et al. (2014)
Logistic Regression model analysis	This method needs to design appropriate experimental factors, and the method is not mature, and the relevant reports of using this method are rare.	With the help of computer, it is easy to judge the nature of combined drug use.	Gennings et al. (2002)
Weight matching method	This method can be used to judge the interaction between six drugs and six concentrations, and can show the intensity of the interaction between drugs.	It is necessary to carry out pre experiment. When the effective dose range of the drug is small, the experiment design should be changed.	Qing and Rui (1999)
Orthogonal t-value method	The principle is clear, easy to understand, simple to calculate, and easy to analyze the synergistic or antagonistic effect between the two drugs.	The combination of drugs can be made according to or without drugs, and then the nature of the combination can be judged. It is limited in practical application.	Jin and Gong (2011)

O157:H7, Staphylococcus aureus, Listeria monocytogenes, Saccharomyces cerevisiae and *Aspergillus niger* (Guarda et al. 2011). However, another study showed that the combination of cinnamon essential oil and clove essential oil had an antagonistic effect on the growth of *Escherichia coli*, but the combination showed a synergistic effect on *Listeria monocytogenes, Bacillus cereus* and *Yersinia enterocolitica* under the same conditions (Moleyar and Narasimham 1992). Therefore, it is difficult to directly predict the antimicrobial efficacy of essential mixtures. In general, some studies have concluded that when mixed in proportion, the whole essential oil shows stronger antibacterial activity than its main components. For example, basil essential oil has a stronger inhibitory effect on *Lactobacillus campylobacter* and *Saccharomyces cerevisiae* than its main component linalool or methyl piperol (Lachowicz et al. 2010). The antibacterial activity of conifer essential oil against *Listeria monocytogenes* was significantly higher than that of its main active components (Mourey and Canillac 2002). This shows that due to the synergistic effect, the trace components in essential oil may be very important against microbial activity.

Cinnamon essential oils can be used as food flavor or preservative in food processing, and its inhibitory effect on mold is better than many essential oils, such as clove essential oil, thyme essential oil, rosemary essential oil, oregano essential oil and litsea cubeba essential oil. However, cinnamon essential oils have obvious and strong aroma, which limits its application in food processing (Wang et al. 2017). The addition of linalool to cinnamaldehyde can significantly improve the antimicrobial activity of cinnamaldehyde and reduce its dosage in food preservation (Dorman and Deans 2000; Marino et al. 2001). Similarly, the combination of cinnamaldehyde and citral had significant synergistic inhibitory effect on microorganisms (Liu et al. 2014). Hyegeun et al. (2018) studied the synergistic bacteriostatic effect of 97 kinds of essential oils. The results showed that cinnamon essential oils and citronella essential oil had good inhibitory effect on *Penicillium pariformis* in gas phase (Hyegeun et al. 2018). At the same time, some researchers compared the fungal inhibitory effect of compound essential oils with propionic acid. For example, Yi (2017) prepared the compound essential oils by compounding cineole and terpineol, and found that the inhibitory effect of the compound essential oils on *Aspergillus flavus* and *Penicillium citrus* was significantly better than that of propionic acid (Yi 2017). In addition, Wang et al. (2018) obtained the compound essential oils by compounding cinnamaldehyde, citral, eugenol and menthol. The results showed that the inhibitory effect of the compound essential oils on mold growth in corn was better than that of propionic acid (Wang et al. 2018). Due to the synergism of essential oils, the application of compound essential oils in food processing has greater potential, which can reduce the use of essential oils and reduce the negative effects of essential oil odors on the sensory quality of food. Based on this, Ju et al. (2020e) successfully applied the combination of eugenol and citral to the preservation of bread. The combination could significantly prolong the shelf life of bread at 25 °C and 35 °C without affecting the original sensory quality of bread (Fig. 5.3) (Ju et al. 2020e).

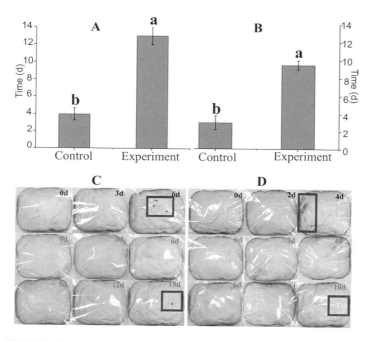

Fig. 5.3 Shelf life of bread. (**a, c**) represent the shelf life and physical changes of bread stored at 25 °C, respectively. (**b, d**) represent shelf life and physical changes of bread stored at 35 °C, respectively (Ju et al. 2020e)

5.4   Synergistic Bacteriostatic Effect of Essential Oils and Food Additives

Compared with using a larger dose of single preservative, the combined use of multiple preservatives may be a more effective way to keep fresh. This will increase the targets of preservatives on microorganisms, which makes microorganisms more sensitive (Gill et al. 2002; Ju et al. 2020a). For example, sodium chloride and peppermint essential oils have synergistic inhibitory effects on *Salmonella enteritis* and *Listeria monocytogenes* (Tassou et al. 1995). In addition, the combination of sodium chloride and eugenol could completely inhibit the growth and histamine production of *Pseudomonas aerogenes*. The reason for this joint action mechanism may be that eugenol increases the permeability of cell membrane and promotes the absorption of sodium chloride, which can act on intracellular enzymes (Wendakoon and Sakaguchi 1993). In another study, soy sauce was shown to have a synergistic bacteriostatic effect with carvanol. However, this synergistic bacteriostatic effect is also offset by the presence of salt (Ultee et al. 1995). Of course, the combined use of essential oils and food preservatives can not only increase the target of food preservatives, but also help to reduce the effect of bad flavor of essential oils on the original quality of food, and reduce the dosage and cost of essential oils in food preservation. For example, Ju et al. (2020f) investigated the effect of antibacterial packaging containing eugenol and citral on the shelf life of bread. The results showed that the antibacterial package could significantly prolong the shelf life of bread and had no significant effect on bread flavor (Ju et al. 2020f).

Similarly, eucalyptus and peppermint essential oils have a synergistic effect on *Pseudomonas aeruginosa* when combined with methyl p-hydroxybenzoate (Patrone et al. 2010). The combination of compound plant essential oils (thymol and cinnamaldehyde) with citric acid can significantly inhibit the growth of Escherichia coli, *Staphylococcus aureus* and *Salmonella* (Fig. 5.4). It can be seen from Fig. 5.4 that the different addition ratio of citric acid has a certain effect on the bacteriostatic effect of microorganisms (Zhang et al. 2018). When nisin combined with carvol or thymol, the inhibitory activity of the combination against *Bacillus cereus* was significantly higher than that of nisin alone (Periago et al. 2001). Similarly, the combination of oregano essential oil and sodium nitrite had a synergistic inhibitory effect on the growth and toxin production of *Botox* in broth. This synergistic inhibition mechanism is based on the fact that oregano essential oil can reduce the number of spore germination, while sodium nitrite can inhibit spore growth (Ismaiel and Pierson 1990). Similarly, thyme essential oil showed a synergistic inhibitory effect on *Listeria monocytogenes* when used in combination with nisin (Solomakos et al. 2008). All of these studies show that the combination of essential oils and food additives can be a good substitute for chemical synthetic preservatives.

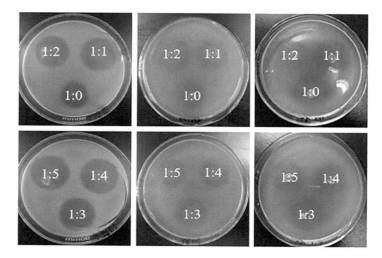

Fig. 5.4 Inhibitory effect of plant essential oil combined with citric acid on microorganisms in different proportions (Zhang et al. 2018)

5.5 Synergistic Bacteriostatic Effect of Essential Oils and Antibiotics

At present, due to the indiscriminate use of antibiotics, the drug resistance of microorganisms is increasing. Many studies have shown that essential oils are effective against drug-resistant bacteria at minimal inhibitory concentration (MIC), so essential oils can be combined with antibiotics to minimize the potential toxicity of antibiotics. Based on the above research strategies, a large number of studies have confirmed the effectiveness of this method. For example, the combination of cinnamon essential oil and ampicillin or chloramphenicol has a synergistic inhibitory effect on *Staphylococcus aureus* (Yassine et al. 2019). The *Myrtus communis* essential oil and amphotericin B have synergistic inhibitory effect on *Candida albicans* and *Aspergillus* (Mahboubi and Ghazian 2010b). The Rosa mandshurica essential oil can enhance the activity of vancomycin, and this combination has a synergistic effect on the inhibition of *Staphylococcus aureus* (Mahboubi et al. 2010). In addition, Daniel et al. (2018) studied the inhibitory effect of the combination of Eucalyptus globulus essential oil and ampicillin on *Staphylococcus aureus* by two-dimensional chessboard method. The results showed that the combination had significant synergistic inhibitory effect on *Staphylococcus aureus*, and the synergistic effect was 32 times higher than that of *Eucalyptus globulus* essential oil alone (Daniel et al. 2018).

Similar to pure essential oil mixtures, additive effects rather than synergistic effects of essential oils and antibiotics are sometimes observed. For example, the combination of thyme essential oil and fennel seed essential oil with methanol extract showed an additive effect on nine Gram-positive and Gram-negative pathogens, especially on *Pseudomonas aeruginosa* (Al-Bayati 2008). Similarly, when Schelz et al. (2006) studied the interaction of peppermint essential oil with four antibiotics (ampicillin, oxytetracycline, erythromycin and gentamicin), they found that only the mixture of peppermint oil and oxytetracycline and the mixture of menthol and oxytetracycline had additive effect, while the other antibiotics showed different interactions with essential oils (Schelz et al. 2006).

5.6 Conclusion

As discussed earlier, the synergistic bacteriostatic effect between different essential oil components, essential oils and antibiotics or food additives may be a new strategy to prevent microorganisms from developing drug resistance. Therefore, in order to reduce the drug resistance of pathogenic bacteria to antibiotics, it is necessary to look for high-efficiency and low-toxic combined drugs. This requires that we must establish relevant experimental parameters to provide a scientific basis for the subsequent use of plant extraction.

In addition, the antibacterial activities of essential oils are varies from different food substrates. Further study on the role of these natural products in different food substrates may be a fruitful research field. In order to avoid the influence of essential oil on food flavor or other sensory properties, the appropriate essential oil must be selected according to different food types. In general, the investigation and collection of information about the synergistic antibacterial activity of essential oils with antibiotics and food additives is helpful to identify natural antistaling agents suitable for food preservative systems.

Acknowledgments The work described in this article is supported by the China Scholarship Council.

Conflict of Interest The authors declare no competing interests.

References

Abt K, Paul F (1972) Acute toxicity of drug combinations: a method for evaluating the interaction of the active components. Excerpta Medica 10(2):192–195
Abreu AC, McBain AJ, Simoes (2012) Plants as sources of new antimicrobials and resistance-modifying agents. Nat Prod Rep 29(9):1007–1021
Al-Reza SM, Rahman A, Kang SC (2009) Chemical composition and inhibitory effect of essential oil and organic extracts of Cestrum nocturnum L. on food-borne pathogens. Int J Food Sci Technol 44(6):1176–1182
Al-Bayati FA (2008) Synergistic antibacterial activity between Thymus vulgaris and Pimpinella anisum essential oils and methanol extracts. J Ethnopharmacol 116(3):403–406

Angienda PO, Hill DJ (2011) The effect of sodium chloride and pH on the antimicrobial effectiveness of essential oils against pathogenic and foodspoilage bacteria: implications in food safety. Int J Biol Sci 8(4):222–227

Ayaz M, Ullah F, Sadiq A, Ullah F, Devkota HP (2019) Synergistic interactions of phytochemicals with antimicrobial agents: potential strategy to counteract drug resistance. Chem Biol Interact 308:294–303

Baj T, Baryluk A, Sieniawska E (2018) Application of mixture design for optimum antioxidant activity of mixtures of essential oils from ocimum basilicum l. origanum majorana l. and rosmarinus officinalisl. Indus Crops Products 115:52–61

Burt SA (2004) Essential oils: their antibacterial properties and potential applications in foods—a review. Int J Food Microbiol 94(3):223–253

Chalova VI, Crandall PG, Ricke SC (2010) Microbial inhibitory and radical scavenging activities of cold-pressed terpeneless Valencia orange (Citrus sinensis) oil in different dispersing agents. J Sci Food Agr 90(5):870–876

Chouhan S, Sharma K, Guleria S (2017) Antimicrobial activity of some essential oils—present status and future perspectives. Medicines 4(3):58–69

Dannenberg G, Funck GD, Silva W, Fiorentini NM (2018) Essential oil from pink pepper (schinus terebinthifolius raddi): chemical composition, antibacterial activity and mechanism of action. Food Control 95:115–120

Daniel B, Buchamer AV, Marchetti ML, Florencia A, Arnaldo B, Nora M (2018) Combination of cloxacillin and essential oil of melaleuca armillaris as an alternative against staphylococcus aureus. Front Veter Sci 5:177–185

de Oliveira CEV, Stamford TLM, Neto NJG, de Souza EL (2010) Inhibition of staphylococcus aureus in broth and meat broth using synergies of phenolics and organic acids. Int J Food Microbiol 137(2–3):312–316

Dorman HJD, Deans SG (2000) Antimicrobial agents from plants:antibacterial acfivity of plant volatile oils. J Appl Microbiol 88:308–316

Finney DJ (1952) Probit analysis, 2nd edn. Cambridge University Press, Cambridge, pp 146–153

Friedman M, Henika PR, Levin CE, Mandrell RE (2004) Antibacterial activities of plant essential oils and their components against Escherichia coli O157:H7 and Salmonella enterica in apple juice. J Agric Food Chem 52(19):6042–6048

Gao HY, Zhao C, Liu C (2014) Pathogenic microorganisms in response to environmental stress of VBNC state and its potential impact on risk assessment. Microbiology 41(1):169–177

Gao C, Tian C, Lu Y, Xu J, Luo J, Guo X (2010) Essential oil composition and antimicrobial activity of Sphallerocarpus gracilis seeds against selected food-related bacteria. Food Control 22(4):517–522

Gennings C, Carter WH, Campain JA, Bae DS, Yang RSH (2002) Statistical analysis of interactive cytotoxicity in human epidermal keratinocytes following exposure to a mixture of four metals. J Agric Biol Environ Stat 7(1):58–73

Chou TC, Rideout D (1987) Synergism and antagonism in chemotherapy, 37–64. Academic, New York

Gill AO, Delaquis P, Russo P, Holley RA (2002) Evaluation of antilisterial action of cilantro oil on vacuum packed ham. Int J Food Microbiol 73:83–92

Guo JY, Huo HR, Jiang TL (2005) Evaluation of research methods to measure the effect of combination drugs. Pharmacol Clin Trad Chinese Med 21(3):60–64

Gutierrez J, Barry-Ryan C, Bourke P (2009) Antimicrobial activity of plant essential oils using food model media: efficacy, synergistic potential and interactions with food components. Food Microbiol 26(2):142–150

Guarda A, Rubilar JF, Miltz J, Galotto MJ (2011) The antimicrobial activity of microencapsulated thymol and carvacrol. Int J Food Microbiol 146(2):144–150

Gutierrez J, Barry-Ryan C, Bourke P (2008) The antimicrobial efficacy of plantessential oil combinationsand interactions with foodingredients. Int J Food Microbiol 124(1):91–97

Hao YY, Brackett RE, Doyle MP (1998) Efficacy of plant extracts in inhibiting Aeromonas hydrophila and Listeria monocytogenes in refrigerated cooked poultry. Food Microbiol 15:367–378

Hammer KA, Carson CF, Riley TV (1999) Influence of organic matter, cations and surfactants on the antimicrobial activity of Melaleuca alternifolia (tea tree) oil in vitro. J Appl Microbiol 86(3):446–452

Hemaiswarya S, Kruthiventi AK, Doble M (2008) Synergism between natural products and antibiotics against infectious diseases. Phytomedicine 15(8):639–652

Holley RA, Patel D (2005) Improvement in shelf-life and safety of perishable foods by plant essential oils and smoke antimicrobials. Food Microbiol 22(4):273–292

Hyegeun H, Kim L, Beuchat R, Jee H (2018) Synergistic antimicrobial activities of essential oil vapours against penicillium corylophilum on a laboratory medium and beef jerky. Int J Food Microbiol 291:104–110

Ismaiel AA, Pierson MD (1990) Effect of sodium nitrite and origanum oil on growth and toxin production of Clostridium botulinum in TYG broth and ground pork. J Food Prot 53(11):958–960

Jin W, Gong W (2011) Disintegration of orthogonal t value method on prescription of anti-fatigue health wine. Ra Harmaal J:120–131

Ju J, Xu X, Xie Y, Guo Y, Cheng Y, Qian H, Yao W (2018a) The inhibitory effect of plant essential oils on foodborne pathogenic bacteria in food. Crit Rev Food Sci Nutr 59(20):3281–3292

Ju J, Xu X, Xie Y, Guo Y, Cheng Y, Qian H (2017) Inhibitory effects of cinnamon and clove essential oils on mold growth on baked foods. Food Chem 240(1):850–855

Ju J, Xu X, Xie Y, Guo Y, Cheng Y, Qian H (2018b) Application of starch microcapsules containing essential oil in food preservation. Crit Rev Food Sci Nutr 60(17):2825–2836

Ju J, Xie Y, Guo Y, Cheng Y, Qian H, Yao W (2018c) Application of edible coating with essential oil in food preservation. Crit Rev Food Sci Nutr:01–62

Ju J, Xie Y, Yu H, Guo Y, Yao W (2020a) Synergistic interactions of plant essential oils with antimicrobial agents: a new antimicrobial therapy. Crit Rev Food Sci Nutr (2):1–12

Ju J, Xie Y, Yu H, Guo Y, Yao W (2020b) Analysis of the synergistic antifungal mechanism of eugenol and citral. LWT- Food Sci Technol 123:109128

Ju J, Xie Y, Yu H, Guo Y, Yao W (2020c) Major components in lilac and litsea cubeba essential oils kill penicillium roqueforti through mitochondrial apoptosis pathway. Ind Crop Prod 149:112349

Ju J, Xie Y, Yu H, Guo Y, Yao W (2020d) Synergistic properties of citral and eugenol for the inactivation of foodborne molds in vitro and on bread. LWT-Food Sci Technol 122:109063

Ju J, Xie Y, Yu H, Guo Y, Cheng Y, Zhang R (2020e) Synergistic inhibition effect of citral and eugenol against aspergillus niger and their application in bread preservation. Food Chem 310(Apr.25):125974.1–125974.7

Ju J, Xie Y, Yu H, Guo Y, Cheng Y, Zhang R (2020f) A novel method to prolong bread shelf life: Sachets containing essential oils components. LWT Food Sci Technol 131:109744

Kim JM, Marshall MR, Wei CI (1995) Antibacterial activity of some essential oil components against five foodborne pathogens. J Agric Food Chem 43(11):2839–2845

Klancnik A, Piskernik S, Mozina SS, Jersek GL (2011) Investigation of some factors affecting the antibacterial activity of rosemary extracts in food models by a food microdilution method. Int J Food Sci Technol 46(2):413–420

Kotzekidou P, Giannakidis P, Boulamatsis A (2008) Antimicrobial activity of some plant extracts and essential oils against foodborne pathogens in vitro and on the fate of inoculated pathogens in chocolate. LWT- Food Sci Technol 41(1):119–127

Lambert RJW, Skandamis PN, Coote PJ, Nychas GJE (2001) A study of the minimum inhibitory concentration and mode of action of oregano essential oil, thymol and carvacrol. J Appl Microbiol 91(3):453–462

Lachowicz KJ, Jones GP, Briggs DR, Bienvenu FE, Coventry MJ (2010) The synergistic preservative effects of the essential oils of sweet basil (ocimum basilicum l.) against acid-tolerant food microflora. Lett Appl Microbiol 26(3):209–214

Levine RR (1978) Pharmacology: drug actions & reactions, 2nd ed., 284–5. Boston: Little Brown Co

Liu H, Xia GH, Wen CY (2014) Inhibitory effect of compound essential oil on Botrytis cinerea after harvest. J Food Indus Technol 35(021):115–118

Marino M, Bersani C, Couli G (2001) Impedance measurements to study the aatimierobial activity of essential oils from Lam/acea and Compos/tae. Int J Food Microbiol 67:187–195

Mahboubi M, Ghazian BF (2010a) Antistaphylococcal activity of zataria multiflora essential oil and its synergy with vancomycin. Phytomedicine 17(7):548–550

Mahboubi M, Ghazian BF (2010b) In vitro synergistic efficacy of combination of amphotericin B with Myrtus communis essential oil against clinical isolates of Candida albicans. Phytomedicine 17(10):771–774

Minto CF, Schnider TW, Short TG, Gregg KM, Gentilini A, Shafer SL (2000) Response surface model for anesthetic drug interactions. Anesthesiology 92(6):1603–1616

Mittal RP, Rana A, Jaitak V (2018) Essential oils: an impending substitute of synthetic antimicrobial agents to overcome antimicrobial resistance. Curr Drug Targets 19(6):605–624

Moleyar V, Narasimham P (1992) Antibacterial activity of essential oil components. Int J Food Microbiol 16(4):337–342

Mourey A, Canillac N (2002) Anti-listeria monocytogenes activity of essential oils components of conifers. Food Control 13(4–5):289–292

Nidhi P, Rolta R, Kumar V, Dev K, Sourirajan A (2020) Synergistic potential of citrus aurantiuml. essential oil with antibiotics against candida albicans. J Ethnopharmacol 262:113135

Patrone V, Campana R, Vittoria E, Baffone W (2010) In vitro synergistic activities of essential oils and surfactants in combination with cosmetic preservatives against Pseudomonas aeruginosa and Staphylococcus aureus. Curr Microbiol 60(4):237–241

Periago PM, Palop A, Fernandez PS (2001) Combined effect of nisin, carvacrol and thymol on the viability of Bacillus cereus heat-treated vegetative cells. Food Sci Technol Int 7(6):487–492

Pol IE, Mastwijk HC, Slump RA, Popa ME, Smid EJ (2001) Influence of food matrix on inactivation of Bacillus cereus by combinations of nisin, pulsed electric field treatment and carvacrol. J Food Prot 64(7):1012–1018

Pradhan AMS, Kim YT (2014) Relative effect method of landslide susceptibility zonation in weathered granite soil: a case study in deokjeok-ricreek, South Korea. Nat Hazards 72(2):1189–1217

Qing SZ, Rui YS (1999) Quantitative design of drug compatibility by weighted modification method. Acta Pharmacol Sin 20(11):1043–1051

Ribo JM, Rogers F (2010) Toxicity of mixtures of aquatic contaminants using the luminescent bacteria bioassay. Environ Toxicol 5(2):135–152

Rivas L, McDonnell MJ, Burgess CM, Brien M, Navarro-Villa A, Fanning S, Duffy G (2010) Inhibition of verocytotoxigenic Escherichia coli in model broth and rumen systems by carvacrol and thymol. Int J Food Microbiol 139(2):70–78

Scortichini M, Rossi MP (1991) In vitro susceptibility of Erwinia amylovora (Burrill) to geraniol and citronellol. J Appl Microbiol 71(2):113–118

Schelz Z, Molnar J, Hohmann J (2006) Antimicrobial and antiplasmid activities of essential oils. Fitoterapia 77(4):279–285

Seow YX, Yeo CR, Chung HL, Yuk HG (2014) Plant essential oils as active antimicrobial agents. Crit Rev Food Sci Nutr 54(5):625–644

Shelef LA, Jyothi EK, Bulgarelli MA (1984) Growth of enteropathogenic and spoilage bacteria in sage-containing broth and foods. J Food Sci 49(737–740):809–817

Sokovic MD, Vukojevic J, Marin PD, Brkic DD, Vajs V, Van LJ (2009) Chemical composition of essential oils of Thymus and Mentha species and their antifungal activities. Molecules 14(1):238–249

Solomakos N, Govaris A, Koidis P, Botsoglou N (2008) The antimicrobial effect of thyme essential oil, nisin, and their combination against Listeria monocytogenes in minced beef during refrigerated storage. Food Microbiol 25(1):120–127

Stecchini ML, Sarais I, Giavedoni P (1993) Effect of essential oils on Aeromonas hydrophila in a culture medium and in cooked pork. J Food Prot 56(5):406–409

Tassou CC, Drosinos EH, Nychas GJE (1995) Effects of essential oil from mint (Mentha piperita) on Salmonella enteritidis and Listeria monocytogenes in model food systems at 4 °C and 10 °C. J Appl Microbiol 78(6):593–600

Wan J, Wilcock A, Coventry MJ (1998) The effect of essential oils of basil on the growth of Aeromonas hydrophila and Pseudomonas fluorescens. J Appl Microbiol 84:152–158

Wang Y, Ma EWM, Chow TWS, Tsui KL (2014) A two-step parametric method for failure prediction in hard disk drives. IEEE Transac Indus Inform 10(1):419–430

Wang J, Cao XS, Song L, Ding ZQ, Tang F, Yue YD (2017) Analysis of essential oil components and antimicrobial activity of Cinnamomum camphora leaves from different habitats by Purge and trap thermal desorption gas chromatography-mass spectrometry. J Food Sci (12):131–136

Wang LM, Xing FG, Lv C, Liu Y (2018) Control effect of compound plant essential oil fungicide on corn mold and mycotoxin. J Nucl Agric Sci 32(4):732–739

Weng X, Zhang C, Jiang H (2021) Advances in microfluidic nanobiosensors for the detection of foodborne pathogens. LWT- Food Sci Technol 151(3):112172

Wendakoon CN, Sakaguchi M (1993) Combined effect of sodium chloride and clove on growth and biogenic amine formation of Enterobacter aerogenes in mackerel muscle extract. J Food Prot 56(5):410–413

Yassine E, Atki IA, Fatima I (2019) Antibacterial activity of cinnamon essential oils and their synergistic potential with antibiotics. J Adv Pharm Technol Res 10(2):63–76

Yi M (2017) Inhibitory effect of Eucalyptus terpineol compound essential oil on 6 kinds of molds. Flavour Fragrance Cosmetics 05v.31;(224):17–22

Zhang HR, Liu JS, Zhang LL, Yang CM (2018) Study on antibacterial effect and synergistic effect of plant essential oil and organic acid. J Feed Indus 39(18):52–56

Zheng QS, Sun RY (2000) Analysis of drug interaction in combined drug therapy by flection method. Acta Pharmacol Sin 21(2):183–187

Chapter 6
Encapsulation of Essential Oils by Spray-Drying: Antimicrobial Activity, and Applications in Food Preservation

Lidiane Diniz do Nascimento, Kauê Santana da Costa, Márcia Moraes Cascaes, and Eloisa Helena de Aguiar Andrade

6.1 Introduction

Food safety is one of the main issues in the food industry and is directly linked to the capacity of preservation. The concept of food preservation was introduced centuries ago, as it was necessary to find ways to keep food fresh and edible. To inhibit the growth of microorganisms and ensure the shelf life of foods, different strategies can be used, such as the addition of chemical preservatives, the decrease of water activity, and thermal processes (Khorshidian et al. 2018; Hertrich and Niemira 2021; Sridhar et al. 2021).

Due to the need to preserve food, control the growth of microorganisms, enhance the antioxidant and antimicrobial activities, and extend the shelf-life of foodstuffs, substances as sodium benzoate, sodium nitrite, and sulfur dioxide were synthesized (Hassoun et al. 2020). However, some synthetic antimicrobials approved by regulatory agencies and used as food preservatives have shown health risks to the consumer (Gutiérrez-del-Río et al. 2018; Falleh et al. 2020). Pisoschi et al. (2018) reported that some synthetic food preservatives as nitrates, sorbates, sulfites, formaldehyde, and benzoates were described for their life-threatening side effects. Some synthetic preservatives are reported by their potential health damage, being also related to neurodegenerative and cardiovascular diseases, and cancers (Beya et al. 2021). The nitrite, for example, added in meat products, may have carcinogenic potential due to nitrosamine production (Radünz et al. 2020).

L. D. do Nascimento (✉) · E. H. de Aguiar Andrade
Coordenação de Botânica, Museu Paraense Emílio Goeldi, Belém, Pará, Brazil
e-mail: lidianenascimento@museu-goeldi.br

K. S. da Costa
Instituto de Biodiversidade, Universidade Federal do Oeste do Pará, Santarém, Pará, Brazil

M. M. Cascaes
Universidade Federal do Pará, Programa de Pós-Graduação em Química, Belém, Pará, Brazil

© The Author(s), under exclusive license to Springer Nature Switzerland AG 2022
M. Santana de Oliveira (ed.), *Essential Oils*, https://doi.org/10.1007/978-3-030-99476-1_6

These factors have driven the search for natural substances capable of replacing or reducing the use of synthetic preservatives. In addition, more consumers have been searching for products that are less processed and able to provide health benefits. Several natural preservatives derived from plants (essential oils and extracts), animals (chitosan), and microorganisms have been used in food products (Donsì and Ferrari 2016; Hassoun et al. 2020). In this context, the replacement of synthetic by natural preservatives in foods has gained prominence, with special attention to the applications of essential oils as food preservatives (Pisoschi et al. 2018; Falleh et al. 2020).

According to Falleh et al. (2020), a variety of natural and safe essential oils have been used in food (e.g.: meat, meat products, vegetables, and fruits), representing a new alternative to control or eliminate pathogens from a specific food matrix. Chang et al. (2017) reported that "essential oils are classified as GRAS (generally recognized as safe) by the Food and Drug Administration (FDA) and as food additives by the European agencies". Essential oils are natural plant-based products that contain low-molecular-weight and volatile compounds which are described for different biological activities, including antioxidant, antifungal, repellent, and antibacterial (Siqueira et al. 2015; Da Costa et al. 2019; Silva et al. 2019; Ma et al. 2020; Nascimento et al. 2021). The antibacterial activity associated with several essential oils and their chemical constituents is related to their ability to destabilize the bacterial lipid bilayer promoting cell membrane degeneration which causes bacterial death (Radünz et al. 2020).

However, essential oils are volatile, lipophilic, characterized by intense aroma, and also susceptible to oxidation when exposed to light, oxygen, and moisture. These drawbacks can compromise the biological activity associated with essential oil and its components. Therefore, adding these pure natural products to food matrices can be a big challenge.

The microencapsulation of essential oils has been proposed as an alternative to these drawbacks due to its action in the protection of volatile constituents of essential oil from external factors and the reduction of volatilization loss. Melo et al. (2020) highlighted that the microencapsulation permits a gradual release of volatile compounds into the product, corroborating with increased shelf life.

This chapter presents a review of recent studies using essential oils microencapsulated by spray drying as an alternative to control microbial growth and discusses the influences of these microcapsules on food preservation.

6.2 Food Conservation: Applications of Essential Oils and Their Compounds in the Food Industry

Preservative compounds applied in the food industry must show some desired biological properties, such as antioxidant and antimicrobial activities, as well as the absence or low toxicity (Bhavaniramya et al. 2019; Nieto 2020; Maurya et al. 2021).

Essential oils have been investigated as an alternative to synthetic preservatives in foods (Ribeiro-Santos et al. 2017; Khorshidian et al. 2018; Maurya et al. 2021), due to their reported antimicrobial and antioxidant activities (Nascimento et al. 2020b).

Some synthetic additives of foods, such as tartrazine, propyl gallate, butylated hydroxytoluene, and butylated hydroxyanisole have been reported by their toxicity to living organisms (Hirose et al. 1981; El-Wahab and Moram 2013; Kamal and Fawzia 2018; Mizobuchi et al. 2021). In addition, some of these compounds are known due to their reduced biodegradability, low environmental sustainability, and reduced bio-incompatibility (Maurya et al. 2021).

Recent growing concerns about health promotion through quality foods have changed the eating habits of consumers as well as have driven the food industries to replace synthetic conservatives and artificial aromas in their products with natural compounds or plant-based derivatives, such as extracts and essential oils with similar activities (Ridgway et al. 2019; Knorr and Watzke 2019). Based on these assumptions, the research for natural compounds with antioxidant and antimicrobial activities, as well as less toxicity when compared to the currently applied synthetic compounds have been the main aim of some scientific efforts (Higueras et al. 2014; Beya et al. 2021; Ng et al. 2021).

Some bioactive natural compounds, as well as essential oils, have been reported with antioxidant and antimicrobial activities against some pathogens found in foods (Escobar et al. 2020). *Ocimum basilicum* (basil), *Origanum vulgare* (oregano), *Pipper nigrum* (black pepper), *Thymus vulgaris* (thyme), and *Rosmarinus officinalis* (rosemary) are examples of some spice plants with essential oils that are applied as preservatives in the food industry (Aljabeili et al. 2018; Veenstra and Johnson 2019; Vinciguerra et al. 2019; Nascimento et al. 2020b; Escobar et al. 2020).

Rosemary essential oil has been widely investigated as a food preservative due to its potent antioxidant activity, antibacterial, and antifungal activities (Wang et al. 2008; Rašković et al. 2014; Bajalan et al. 2017). The strong antioxidant activity of rosemary essential oil has been reported due to its major constituents that include 1,8-cineole, camphor, and α-pinene (Bajalan et al. 2017).

Some of the chemical classes most related to the antimicrobial activity of essential oils are terpenes, aliphatic alcohols, aldehydes, and phenolics. Some phenolic molecules, as carvacrol, thymol, and eugenol, can inhibit or prevent the growth of spoilage and pathogenic microorganisms (Ribeiro-Santos et al. 2017; Pisoschi et al. 2018; Beya et al. 2021). Moreover, carvacrol and thymol are examples of natural bioactive compounds obtained from essential oils extracted from spice plants and are investigated as food conservatives due to their antioxidant and antimicrobial activities against microbial pathogens (Veenstra and Johnson 2019; Bhavaniramya et al. 2019; de Souza et al. 2021). Khorshidian et al. (2018) reported thymol, carvacrol, carvone, eugenol, and cinnamaldehyde as some of the mains chemical constituents responsible for exerting antimicrobial activity in cheeses.

Nevertheless, it is important to highlight that some well-known biological activities of these plant-based products are reduced when applied directly to food due to the presence of compounds with high volatility and low solubility, as well as interactions with other components of the food matrix that could inhibit their mode of action. These characteristics have limited their direct uses in food (Aguiar et al. 2016). Thus, different delivery strategies, such as micro and nanoencapsulation have been developed to effectively address these challenges and improve the controlled release of these plant-derivatives products and conserve the food quality (Zanetti et al. 2018; Nguyen et al. 2021).

6.3 Some Aspects About Essential Oils

Essential oils are derived from plants and are composed of a complex blend of terpenes and terpenoids, which confer the aromatic characteristics of each plant. They have a low-molecular-weight and can be precursors, intermediates, and final products of a biosynthetic route. Its contents may vary according to the time of the year, stage of plant development, geographical location, plant organ, and the extraction technique applied (Ribeiro et al. 2014; Cook and Lanaras 2016; Nascimento et al. 2019, 2020b; Ferreira et al. 2020; Cascaes et al. 2021).

The essential oils play an important role in the protection of the plants, acting as antibacterial, antifungals, antivirals, and also against herbivores (Bakkali et al. 2008; Puškárová et al. 2017; Ilić et al. 2019; Karpiński 2020; Abd Rashed et al. 2021). Several essential oil constituents are important insect repellents, being effective against mosquitoes, ticks, fleas, and flies (Cook and Lanaras 2016; Santana et al. 2021).

Essential oils can be extracted through a variety of techniques, with emphasis on hydrodistillation and steam-distillation, which have been applied in research laboratories and industrial processes (Cook and Lanaras 2016; Moradi et al. 2018; Nascimento et al. 2020a, 2021; Silva Júnior et al. 2021). In recent years, environmentally friendly alternatives have been sought, capable of reducing the energy consumption and CO_2 emissions, which has strengthened the use of green techniques, such as microwave-assisted extraction and supercritical CO_2 extraction (Cardoso-Ugarte et al. 2013; Khalili et al. 2018; Nascimento et al. 2020a; Martínez-Abad et al. 2020; Silva et al. 2021). The technique to being used varies according to the plant organ and the application of the essential oil, which is directly linked to the demand of the essential oils market, safety, and costs of the process (Nascimento et al. 2019; Maes et al. 2019; Beya et al. 2021).

Essential oils contain many bioactive compounds, which provide flavor and a wide range of biological and specific properties, which makes them of great interest to use in the food industry (Cook and Lanaras 2016; Khorshidian et al. 2018; Hernández-Nava et al. 2020). Shortly, it is estimated that the main applications of essential oils will be in the areas of food and beverage, personal care and cosmetics, and aromatherapy. The global essential oils market demand was estimated at 247.08 kilotons in 2020 and it is expected to grow at a compound annual growth rate of 7.5% from 2020 to 2027 (Grand View Research 2020).

6.4   Spray-Drying as an Alternative to Microencapsulation of Essential Oils

Essential oils are extremely volatile and susceptible to degradation due to exposure to light, heat, oxygen, or due to their interactions with other components present in complex formulations. These factors need to be taken into account during storage and that may compromise or limit their biological activities (Gonçalves et al. 2017; Benjemaa et al. 2018). Zhu et al. (2021) highlighted that volatility, aromatic odor and insolubility still are the main factors that restrict the application of essential oils in the food industry.

Based on these limitations, microencapsulation is an alternative to avoid the degradation and evaporation of the bioactive constituents present in essential oils. In addition, microencapsulation can favor the solubility of essential oils in aqueous solutions, minimize undesirable flavors and aromas, as well as optimize storage conditions, since the final product will be in powder form. These characteristics are of great interest to the food industry.

The essential oils microencapsulation can be conducted by several methodologies, as coacervation, lyophilization, extrusion, molecular inclusion, ionic gelation, casting, microfluidization, and spray-drying (Salvia-trujillo et al. 2015; Kujur et al. 2017; Lucía et al. 2017; Riquelme et al. 2017; Nascimento et al. 2019). The process of spray-drying microencapsulation consists of transforming a solution, suspension, or emulsion from a liquid state to a solid state and then creating a protective coating around the substance of interest. The microencapsulated product has some advantages compared to its original form concerning transport, handling, and use in food matrices (Botrel et al. 2015; Mohammed et al. 2020).

The spray-drying technique allows continuous operation and can be easily adapted to industrial levels. The use of the spray-drying technique allows obtaining products with lower volume and weight when compared to particles in liquid or gel form, making them easier to store and transport. The main steps involved in the spray-drying process are atomization, liquid contact - hot air, water evaporation, and the separation of dry product from moist air (Gharsallaoui et al. 2007; Asbahani et al. 2015; Botrel et al. 2015).

The main advantages of the spray-drying microencapsulation process are low operating cost, good yield, fast solubilization, and high stability capsules. However, some disadvantages are lack of uniformity between the microcapsules produced, restrictions regarding the choice of wall material, production of very fine powders, and possible loss of heat-sensitive components, such as aroma and other volatile compounds (Madene et al. 2006). Moreover, almost all spray drying processes in the food industry are carried out with aqueous formulations, so the wall material must be soluble in water, at an acceptable level (Leimann 2008). Some of the main wall materials used in spray-drying processes are the Arabic gum, whey protein, maltodextrin, chitosan, and starch, which can be used in isolation or combinations. Despite the higher cost, Arabic gum is a versatile encapsulating material due to its water solubility, low viscosity, emulsification capacity, and good retention of volatile compounds (Madene et al. 2006; Gharsallaoui et al. 2007; Al-Ismail et al. 2015; Maes et al. 2019; da Silva et al. 2020).

6.5   Essential Oils Microencapsulated by Spray-Drying as Antimicrobial Agents in Food Preservation

The microencapsulation of active components in powders is adequate for food ingredients as well as for chemicals, drugs, and cosmetics purposes (Fuchs et al. 2006). The use of bioactive compounds from essential oils in food has attracted the interest of industries, due to their potential health benefits, so the microencapsulated essential oils have emerged as useful alternatives for the food industry (Holkem et al. 2015; Fernandes et al. 2017; Nascimento et al. 2019; Dávila-Rodríguez et al. 2020). In addition, according to Mehran et al. (2020), increasing consumer demand for functional ingredients is a key factor in the development of microencapsulation research for applications in the food industry.

It is noticed that essential oils in their free form are already added in food matrices, for different purposes, whether adding aroma, flavor or potentiating the characteristics of the final product. However, as previously mentioned, working with essential oils in their free form can bring disadvantages, mainly due to volatilization loss and low solubility (Radünz et al. 2020; Dávila-Rodríguez et al. 2020).

Different authors have applied microencapsulated essential oils in food matrices, to evaluate the preservative potential of these natural products. In this work, we emphasize the studies that employed essential oils microencapsulated by the spray-drying technique and that evaluated the antimicrobial viability and potential of microcapsules containing essential oil as antimicrobial agents (Table 6.1).

Chang et al. (2017) microencapsulated the essential oil of *O. vulgare* (oregano) in polyvinyl alcohol and evaluated the antimicrobial activity of the microcapsules against *Dickeya Chrysanthemi*, a bacillus that causes maladies to fresh vegetables. The authors evaluated 18 different essential oils, among them white thyme, clove, cinnamon, and tea tree, but the oregano essential oil produced the largest growth inhibition zone against *D. chrysanthemi*. The authors applied a sachet containing the encapsulated oregano essential oil at 20 °C and 85% relative humidity for five days and the controls consisted of five inoculated pieces of lettuce in containers with no oregano sachets. They noted that the sachets showed an antimicrobial effect, exhibiting controlled release at high humidity and that the sachets could be useful for antimicrobial packaging systems development, which could increase the microbiological safety and extend the shelf life of fresh vegetables without, affect the texture and color characteristics of the lettuce. They also described the release profiles of oregano essential oil from the sachets and concluded that high temperatures and relative-humidity induced faster release of oregano essential oil from the microcapsules because when the water interacted with the polar groups of the hydrophilic

Table 6.1 Some Essential oil microencapsulated by spray-drying using different wall material and their application as antimicrobials in foods

Essential oil used as core material	Major compounds in the essential oil	Encapsulation matrix	Microorganisms and their respective zone of inhibition or minimum inhibition concentration	The methodology used for the evaluation of antimicrobial activity	Application	Reference
Coriandrum sativum (coriander)	Linalool (83%)	Chitosan	Bacteria Gram (−) *Aeromonas hydrophila* (9.0 mm) *Escherichia coli* (9.0 mm) *Escherichia coli* O157 (9.0 mm) *Klebsiella pneumoniae* (ineffective) *Pseudomonas aeruginosa* (ineffective) *Salmonella typhimurium* (9.0 mm) *Yersinia enterocolitica* (8.0 mm) Gram (+) *Bacillus cereus* (8.0 mm) *Listeria monocytogenes* (ineffective) Yeast *Candida albicans* (ineffective)	Agar Diffusion method.	Natural antioxidant and antimicrobial agent.	Duman and Kaya (2016)

(continued)

Table 6.1 (continued)

Essential oil used as core material	Major compounds in the essential oil	Encapsulation matrix	Microorganisms and their respective zone of inhibition or minimum inhibition concentration	The methodology used for the evaluation of antimicrobial activity	Application	Reference
Cymbopogon flexuosus (lemongrass)	α-citral (36.2) and β-citral (22.42%)	Arabic gum and maltodextrin	Total coliforms Thermotolerant coliforms and Coagulase-positive *Staphylococcus*	*In situ* antimicrobial activity. Microencapsulated oil was added to cheese.	Natural preservative in Coalho cheese.	Melo et al. (2020)
Origanum vulgare (oregano)	Not informed	polyvinyl alcohol (PVA)	Bacterial Gram (−) *Dickeya chrysanthemi* (44.7 ± 1.6 mm)	Vapor Diffusion assay and sachets containing the microencapsulated essential oil.	Antimicrobial packaging system.	Chang et al. (2017)
O. vulgare (oregano)	Carvacrol (>60%)	whey protein isolate	Filamentous fungi and yeast	Microencapsulated oil was added to cheese.	Parmesan cheese conservation.	Fernandes et al. (2018)
O. vulgare (oregano)	Carvacrol (85.89%)	Arabic gum, maltodextrin, and modified starch	Bacteria Gram (+) *Staphylococcus aureus* (6.75–9.25 mm) Gram (−) *Escherichia coli* (7.75–10.75 mm) *Proteus mirabilis* (6.20–10.25 mm) *Klebsiella* sp (7.40–9.50 mm)	Disc diffusion.	Tablets with microencapsulated essential oil as antimicrobial agents.	Partheniadis et al. (2019)
Rosmarinus officinalis (rosemary)	1,8-cineole (40.8%) and camphor (28.8%)	Whey protein isolate and inulin	Mesophilic Bacteria	Microencapsulated oil was added to cheese.	Preservative in Minas frescal cheese	Fernandes et al. (2017)

Essential oil used as core material	Major compounds in the essential oil	Encapsulation matrix	Microorganisms and their respective zone of inhibition or minimum inhibition concentration	The methodology used for the evaluation of antimicrobial activity	Application	Reference
Schinus terebinthifolia (pink pepper)	α-pinene (35.9%), β-pinene (15.6%) and δ-3-carene (13.1%)	Soy protein isolate (SPI), high methoxyl pectin (HMP) and maltodextrin	Bacteria Gram (+) *Staphylococcus aureus* (10.6 to 19.8 mm (SPI) and 13.1 to 22.2 mm (SPI/HMP)) *Bacillus subtilis* (8.8 to 15.0 mm (SPI) and 11.0 to 16.1 (SPI/HMP)) *Listeria monocytogenes* (0.0 to 12.0 mm (SPI) and 0.0 to 14.6 (SPI/HMP)) *Listeria innocua* (0.0 to 14.0 and 0.0 to 15.4 mm (SPI/HMP)) Gram (−) *Escherichia coli* (no inhibitory activity) *Salmonella typhimurium* (no inhibitory activity)	Diffusion and added to milk	Natural preservative	Locali-Pereira et al. (2020)

(continued)

Table 6.1 (continued)

Essential oil used as core material	Major compounds in the essential oil	Encapsulation matrix	Microorganisms and their respective zone of inhibition or minimum inhibition concentration	The methodology used for the evaluation of antimicrobial activity	Application	Reference
Thymus vulgaris (thyme)	Thymol (36%) and *p*-cymene (26.2%)	Casein and maltodextrin	Bacteria Gram (+) *Staphylococcus aureus* (0.1 mg·mL⁻¹) *Listeria monocytogenes* (0.1 mg·mL⁻¹) Gram (−) *Salmonella Typhimurium* (0.1 mg·mL⁻¹) *Escherichia coli O157* (0.1 mg·mL⁻¹)	Disc diffusion and *in situ* antimicrobial activity (microencapsulated oil was added to hamburger-like meat products).	Conservation of hamburger-like meat products.	Radünz et al. (2020)
T. vulgaris (thyme)	Thymol (472 mg/g)	Starch and agave fructans	Fungi *Fusarium pseudocircinatum* *Alternaria alternata* *Neofusicocum kwambonambiense* *seudocladosporioides* *Colletotrichum gloeosporioides*	Sachets with the microcapsules.	Antifungal packaging.	Esquivel-Chávez et al. (2021)

polyvinyl alcohol (wall material) it resulted in a displacement of the volatile compound from the interior of the capsule to the headspace.

Another study also evaluated the use of sachets containing essential oils microencapsulated by spray-drying as an alternative to control the growth of microorganisms. The authors investigated the use of microcapsules sachets prepared with *T. vulgaris* (thyme) essential oil and starch/agave fructans as an alternative to the control of phytopathogens associated with mango decay. To evaluate the inhibition of the sachets in the growth of the microorganism strains, the authors filled the sachets with 0.10, 0.15, and 0.20 g of active microcapsules, which were fixed with double-sided tape to the inner side of the plate lids. Each plate contained the potato dextrose agar medium and the inoculated mycelia disc (6 mm of the diameter) of the actively growing test phytopathogen and so it was incubated at 18 °C for 12 days. Furthermore, they evaluated the effect of active packaging on the quality attributes of mango. In this step, the fruits were artificially wounded, immersed in a conidial suspension of 1×10^6 conidia/mL for 1 min, dried, and stored in a humidity chamber (9 days at 20 ± 2 °C), where the mangos were positioned in PVC rings. For each fruit, there was a sachet positioned at the center of the lids of the plastic box to avoid contact between the fruit and the sachet. According to the results, at 20 °C, the sachets with the microcapsules of thyme essential oil were an efficient packing alternative to control the microorganisms associated with mango decay and it controlled the growth of *C. gloeosporioide* in mango. The authors highlighted that the antifungal activity of the packaging system was performed only at 20 °C and so other temperatures should be also evaluated (Esquivel-Chávez et al. 2021).

The essential oil of pink pepper (*Schinus terebinthifolia*), rich in α-pinene and β-myrcene, was microencapsulated by spray-drying of single-layer emulsions, stabilized by soy protein isolate (SPI), and of double-layer emulsions, stabilized by soy protein isolate/high methoxyl pectin (SPI/HMP). The pure essential oil and the microcapsules were evaluated against six bacteria, four Gram-positive (*Staphylococcus aureus, Bacillus subtilis, Listeria monocytogenes*, and *Listeria innocua*) and two Gram-negative (*Escherichia coli* and *Salmonella typhimurium*). Furthermore, they evaluated the stability and the antioxidant activity of the microcapsules during storage and the *in vitro* and *in situ* antimicrobial activity of free and microencapsulated pink pepper oil, using milk as a food model. According to the results of the antibacterial activity, it was observed inhibition zones only for the Gram-positive bacteria evaluated, which could be related to the composition of the cell wall, once that the Gram-negative, present an outer membrane composed of two layers of lipopolysaccharide, but the Gram-positive bacteria cells are coated with only one layer of peptidoglycan. After the spray-drying, the single-layer particles (SPI) showed high losses of the compounds when compared to the double-layer microcapsules, which best preserved the volatile composition of the pure oil. The authors concluded that the barrier created by wall materials allowed a more gradual release of volatiles. Both microcapsules reduced the bacterial growth in milk, whereas non-encapsulated oil showed no satisfactory inhibition. The faster reduction of microbial growth in milk was observed for SPI/HMP microcapsules (Locali-Pereira et al. 2020).

Some authors have applied essential oils microencapsulated by spray-drying as alternatives to cheese conservation. A study evaluated the use of microencapsulated *R. officinalis* (rosemary) essential oil in the conservation of Minas cheese. For the preparation of microcapsules, inulin and whey protein isolate were used as wall material, in a proportion of 20%, while the amount of rosemary essential oil used was 25% of the mass of the wall materials. The conditions used during the spray-drying were inlet temperature of 170 °C and feed rate of 0.9 L/h. The authors produced the Minas frescal cheese following the traditional manufacturing techniques and used pasteurized milk. Then, the pure and the microencapsulated essential oil of rosemary were added to the cheese at concentrations of 0.5% and after that, it was conducted the total counting of mesophilic bacteria and the enumeration of coliforms in Minas frescal cheese. The free and the microencapsulated essential oil were characterized by 1,8-cineole, with levels equal to 40.8% and 44.8%, respectively. From the analysis, the coliform development in cheese was not affected by any treatment, so it was recommended to perform more experiments, with higher concentrations of essential oil. Despite that, the microencapsulated rosemary essential oil was able to control the proliferation of mesophilic bacteria in Minas frescal cheese and could be used as a potential preservative in Minas frescal cheese (Fernandes et al. 2017).

In another study, the essential oil of *O. vulgare* (oregano) was microencapsulated by spray-drying and the microcapsules were tested as an alternative to inhibit the growth of fungi in grated Parmesan cheese. To obtain the microcapsules, whey protein isolate was used as coating material and the conditions used during the microencapsulation were inlet temperature of 170 °C, atomizing airflow 40 L/min, and feed rate of 0.9 L/h. Carvacrol was the major compound found in the free oregano essential oil and in the oil extracted from the microcapsules. To evaluate the antifungal activity of the microencapsulated essential oil, the microcapsules with 0,1% and 0,5% of essential oil were combined with potassium sorbate and homogeneously mixed to the grated cheese. The different treatments were placed in polypropylene bags stored in a chamber at 25 °C. After 45 days of storage, only the treatment containing microcapsules with 0.5% of oregano essential oil remained with undetectable counting, being considered the most effective treatment in the control of filamentous fungi and yeast growth in grated Parmesan cheese. Hence, the results confirmed the antimicrobial activity of oregano oil and indicated that the use of the microcapsules obtained by spray-drying (product in powder form) can be an alternative to the conservation of grated cheese package (Fernandes et al. 2018).

Recently, Melo et al. (2020) evaluated the viability of *Cymbopogon flexuosus* (lemongrass) essential oil microencapsulated as a natural alternative to the conservation of Coalho cheese. The microcapsules were prepared using Arabic gum and maltodextrin as wall material and during the microencapsulation by spray-drying, the authors also reported an inlet temperature of 170 °C and a feed flow rate of 0.9 L/h. The lemongrass essential oil was compound mostly by α-citral and its percentages varied from 32.62% to 35.88% in the microcapsules prepared with Arabic Gum/maltodextrin (1:1) and only Arabic gum, respectively. The chemical constituent β-citral also presented high percentages and varying from 22.42% (in the pure oil) to 20,7% (microcapsules prepared with Arabic Gum/maltodextrin, in the 1:1 proportion). To evaluate the antimicrobial activity of the lemongrass essential oil in cheese, only the microcapsules prepared with Arabic gum/maltodextrin (3:1) were chosen due to their better oil retention and solubility. The Coalho cheese was prepared and three treatments were applied to evaluate the growth of the microorganisms: treatment 1 (control with no addition of essential oil); treatment 2 (pure essential oil, 0.25%), and treatment 3 (microencapsulated lemongrass essential oil, 0.25%). The total coliforms, thermotolerant coliforms, and coagulase-positive *Staphylococcus* analysis for Coalho cheese were performed at 6 °C during 21 days of storage. From the results, regarding total coliforms, the microencapsulated lemongrass essential oil was efficient during the Coalho cheese storage, furthermore, the microencapsulation process did not compromise the bioactivity of the lemongrass essential oil when added to the cheese. Thus, the microcapsules extended the shelf life of the product and could also be used in the production of other products as well as for biodegradable packaging (Melo et al. 2020).

Khorshidian et al. (2018) also noticed that the concentration of essential oils and their major compounds applied in cheeses should be considered carefully because of their possible negative impacts on organoleptic properties. Those points can be extended to other foodstuffs. Thus, regardless of the method employed, microencapsulation can be an effective alternative to minimize aromas and flavors and still maintain or even improve the bioactivity of essential oils and their components with the combination of different wall materials, some already described for their antimicrobial properties, such as chitosan.

A study applied the spray-drying technique for the encapsulation of *Coriandrum sativum*, using chitosan (obtained from the waste shells of crayfish) as wall material. The authors evaluated the antimicrobial activity of coriander essential oil (83% of linalool), crayfish chitosan, and the obtained microcapsule against some bacteria food pathogens. The crayfish chitosan showed the best results for the antimicrobial activities with inhibition zones ranging from 20 mm (*Yersinia enterocolitica*) to 41 mm (*P. aeruginosa*), however, the pure coriander essential oil was not effective during the antimicrobial tests (no inhibition zones). The microcapsules did not present inhibitions zones for *C. albicans*, *L. monocytogenes*, and *Pseudomonas aeruginosa*, but for the other analyzed microorganisms, the results were similar and varied from 8 to 9 mm (inhibition zones). Hence, the wall material was more effective against the microorganisms when compared to the essential oil and the microcapsules (Duman and Kaya 2016).

An emulsion composed of the oregano essential oil, Arabic gum, maltodextrin, and modified starch was spray-dried at a feed rate of 5 mL/min, inlet air temperature 180 °C, and outlet 117 °C, and airflow 600 L/h. The powder (microcapsules) obtained was converted into tablets, which were evaluated by its antimicrobial activity against *S. aureus, E. coli, Proteus mirabilis*, and *Klebsiella* sp. For comparative purposes, antimicrobial activity was also carried out using only the free essential oil. According to the authors, spray-dried powders are compressible and appropriate to be processed into tablets. Also, the encapsulating wall can form a dense and coherent structure able to resist stresses during compression, and thus preventing the essential oil volatilization. The tablets were prepared using powders with 10% and 20% w/w of essential oil, and containing 5% w/w of addition of cros-carmellose sodium, which was added as disintegrants. The results showed that it was possible to prepare tablets with the oregano microcapsules in a powder form, without loss of the essential oil (compression range 60–100 MPa). The release profile of the tablets was similar to the profile obtained from spray-dried powder with 20% of essential oil. Furthermore, the reconstituted emulsion from oregano essential oil tablets showed excellent antimicrobial activity, similar to pure essential oil. The authors also highlighted that the selected wall materials did not alter the antimicrobial activity (Partheniadis et al. 2019).

Radünz et al. (2020) evaluated the microcapsules prepared with the essential oil of thyme, casein, and maltodextrin as alternatives to the conservation of hamburger-like meat products. The antimicrobial potential of the microcapsules was evaluated *in vitro* (against *S. aureus, L. monocytogenes, S. typhimurium, and E. coli*), and *in situ* (against thermotolerant coliforms and *E. coli*), in which hamburger-like meat products were used as a test food. In this analysis, the authors established different conditions: treatments 1 (a standard meat product), treatment 2 (a sodium nitrite-added meat product), treatment 3 (a meat product with the addition of 0.1 g/100 g of unencapsulated thyme essential oil), treatment 4 (meat product with the addition of 1 g/100 g of encapsulated thyme essential oil), and treatment 5 (control capsule without oil addition). Then, the hamburger-like meat products were refrigerated at 4 °C for 14 days and the count of thermotolerant coliforms and *E. coli* was performed at 0, 7, and 14 days of storage. The GC-MS analysis showed that thymol was the major compound identified in the essential oil before and after the encapsulation, with 36.0 and 58.5%, respectively. Concerning the number of thermotolerant coliforms, after 14 days of refrigerated storage, the treatment with the addition of unencapsulated essential oil decreased slightly, and treatment with encapsulated essential oil considerably reduced the concentration of thermotolerant coliforms. Furthermore, the authors also reported that after 14 days, the concentrations in the treatments 1 and 5 increased exponentially, but the sodium nitrite was zero. Therefore, Radünz et al. (2020) concluded that the thyme essential oil microcapsules controlled the growth of the microorganisms during periods of up to 14 days and that those microcapsules could be used as a natural preservative in food.

Beya et al. (2021) published a review on the application of natural antimicrobials in meat. The authors highlighted that several studies reported only the *in vitro* tests regarding the antimicrobial activities of natural products. However, it is important to evaluate the *in situ* bioactivity and understand the possible different outcomes. They pointed two aspects that are indispensable when using natural preservatives in food systems: (1) the change of sensory attributes of food when the natural preservative is added; (2) and interactions of the natural preservative with other food ingredients in the system.

The molecular mechanism of action of essential oils regarding their antimicrobial activity remains not completely understood and there are several suggested mechanisms, once that these natural products are composed of different chemical constituents that act in synergy and/or antagonism (Blasa et al. 2011; Sundararajan et al. 2018; Hassoun et al. 2020; Beya et al. 2021). Nevertheless, there is a consensus that antimicrobial activity is usually related to the hydrophobic character of essential oil and that its lipophilic compounds can stimulate damage to the cell membrane and alterations of microbial cell permeability (Sundararajan et al. 2018; Pisoschi et al. 2018; Hassoun et al. 2020).

Falleh et al. (2020) reported that essential oils rich in phenolic compounds such as thymol, carvacrol, and eugenol were associated with high activities against foodborne pathogens. Phenolics interfere with cell membrane function as they interact with membrane proteins inducing their structure and function alteration. Eugenol and carvacrol disrupt the cell membrane and inhibit ATPase activity, while carvacrol and thymol increase membrane permeability by dissolution into the phospholipid bilayer (Pisoschi et al. 2018). When comparing the influence of the antibacterial activity of essential oils, the Gram-negative bacteria are more resistant to the action of the essential oils due to the lipopolysaccharide cell wall, once that this layer prevents components from entering the membrane (Zanetti et al. 2018; Melo et al. 2020).

6.6 Antimicrobial Packaging Systems in Food

It is unquestionable that the consumer market is constantly changing, thus growing the demands for less processed products. In parallel, the industry has invested more in technologies and innovations compatible with minimally processed products, such as natural foods that require certain properties and special characteristics of packaging (Donsì and Ferrari 2016; Ribeiro-Santos et al. 2017; Zanetti et al. 2018). Antimicrobial packaging is one of the innovative technology concepts of active food packaging and is an alternative to extending the shelf life of the product, inhibiting and/or retarding the proliferation of undesirable microorganisms in foodstuff. Studies have been developed to find approaches to include essential oils, reported by their natural antimicrobials. Microencapsulation can be an alternative for developing antimicrobial packaging systems and can be applied to natural trap compounds, such as essential oils, to be used in food packaging (Ribeiro-Santos et al. 2017; Chang et al. 2017).

Essential oils and their isolated compounds, known for their antibacterial, antifungal, and antioxidant activities, can be microencapsulated and added into the packaging. Microencapsulation is an alternative capable of avoiding the characteristic drawbacks of the use of free essential oils (volatility, low solubility in water, and susceptibility for oxidation) and therefore a path to production of active antimicrobial packaging. Esquivel-Chávez et al. (2021) highlighted that the active compound can be incorporated by two mechanisms: inside the packaging material during its production or in the headspace of the packaging as a sachet during food packaging. Hence, the use of essential oils in active packing is a strategy used for food preservation (Hassoun et al. 2020).

6.7   Limitations of the Use of Essential Oils

Despite the advantages presented when using essential oils as antimicrobial agents in food, the chemical variability and availability of these natural products may be limitations in the application of these natural products in large-scale processes. Furthermore, Falleh et al. (2020) highlighted that the flavor of the essential oils may alter the organoleptic and sensory characteristics of the food products and so it remains one of the main limitations of their use as food preservatives. Another factor is the concern about possible contamination by chemical products such as pesticides. Beya et al. (2021) also listed the importance of unified legislation about the use of natural food preservatives.

6.8   Final Considerations

Spray-drying can provide particles in the form of powder on micro and nanoscales, is characterized by a good cost-benefit relationship, and is a very well-established industrial process, involving not only the food industry but also the pharmaceutical, chemical, and cosmetic industries. Essential oils are a source of bioactive molecules, reported for their antimicrobial and antioxidant activities. These bioactivities may be the key to the use of essential oils in food conservation, especially nowadays, where there is a demand for food products with less addition of synthetic compounds.

The addition of different essential oils microencapsulated by spray-drying has been effective in controlling microorganisms that compromise the quality and shelf-life of food products. Studies developed *in situ* are fundamental for a better understanding of the efficacy and limitations associated with the use of microencapsulated essential oils as antimicrobial agents in food. According to the present review, the use of microcapsules with essential oils was effective to control the growth of fungi and bacteria in vegetables, milk, cheeses, fruits, and hamburger-like meat products. The contact of microcapsules with food occurred through the addition to the product (as in the case of studies involving cheeses), preparation of tablets, or through the elaboration of controlled release sachets. Moreover, the essential oils most described for their antimicrobial activities in food matrices are usually extracted from spicy plants.

Certainly, studies describing the influence of the addition of microencapsulated essential oils on organoleptic characteristics, as well as tests under more experimental conditions, evaluating the influence of temperature, humidity, essential oil concentration, and storage time, are essential to ensure the quality of the final product.

References

Abd Rashed A, Rathi D-NG, Ahmad Nasir NAH, Abd Rahman AZ (2021) Antifungal properties of essential oils and their compounds for application in skin fungal infections: conventional and nonconventional approaches. Molecules 26:1093. https://doi.org/10.3390/molecules26041093

Aguiar J, Estevinho BN, Santos L (2016) Microencapsulation of natural antioxidants for food application – the specific case of coffee antioxidants – a review. Trends Food Sci Technol 58:21–39. https://doi.org/10.1016/j.tifs.2016.10.012

Al-Ismail KM, Mehyar G, Al-Khatib HS, Al-Dabbas M (2015) Effect of microencapsulation of cardamom's essential oil in gum Arabic and whey protein isolate using spray drying on its stability during storage. Qual Assur Saf Crop Foods 7:613–620. https://doi.org/10.3920/QAS2014.0422

Aljabeili HS, Barakat H, Abdel-Rahman HA (2018) Chemical composition, antibacterial and antioxidant activities of thyme essential oil (Thymus vulgaris). Food Nutr Sci 09:433–446. https://doi.org/10.4236/fns.2018.95034

Asbahani AE, Miladi K, Badri W et al (2015) Essential oils : from extraction to encapsulation. Int J Pharm 483:220–243. https://doi.org/10.1016/j.ijpharm.2014.12.069

Bajalan I, Rouzbahani R, Pirbalouti AG, Maggi F (2017) Antioxidant and antibacterial activities of the essential oils obtained from seven Iranian populations of Rosmarinus officinalis. Ind Crop Prod 107:305–311. https://doi.org/10.1016/j.indcrop.2017.05.063

Bakkali F, Averbeck S, Averbeck D, Idaomar M (2008) Biological effects of essential oils – a review. Food Chem Toxicol 46:446–475. https://doi.org/10.1016/j.fct.2007.09.106

Benjemaa M, Neves MA, Falleh H et al (2018) Nanoencapsulation of Thymus capitatus essential oil : formulation process , physical stability characterization and antibacterial e efficiency monitoring. Ind Crop Prod 113:414–421. https://doi.org/10.1016/j.indcrop.2018.01.062

Beya MM, Netzel ME, Sultanbawa Y et al (2021) Plant-based phenolic molecules as natural preservatives in comminuted meats: a review. Antioxidants 10:263. https://doi.org/10.3390/antiox10020263

Bhavaniramya S, Vishnupriya S, Al-Aboody MS et al (2019) Role of essential oils in food safety: antimicrobial and antioxidant applications. Grain Oil Sci Technol 2:49–55. https://doi.org/10.1016/j.gaost.2019.03.001

Blasa M, Angelino D, Gennari L, Ninfali P (2011) The cellular antioxidant activity in red blood cells (CAA-RBC): a new approach to bioavailability and synergy of phytochemicals and botanical extracts. Food Chem 125:685–691. https://doi.org/10.1016/j.foodchem.2010.09.065

Botrel DA, Fernandes RVB, Borges SV (2015) Microencapsulation of essential oils using spray drying technology. In: SAGIS LMC (ed) Microencapsulation and microspheres for food applications. Academic, San Diego, pp 235–251

Cardoso-Ugarte GA, Juárez-Becerra GP, SosaMorales ME, López-Malo A (2013) Microwave-assisted extraction of essential oils from herbs. J Microw Power Electromagn Energy 47:63–72. https://doi.org/10.1080/08327823.2013.11689846

Cascaes MM, Silva SG, Cruz JN et al (2021) First report on the Annona exsucca DC. Essential oil and in silico identification of potential biological targets of its major compounds. Nat Prod Res:1–4. https://doi.org/10.1080/14786419.2021.1893724

Chang Y, Choi I, Cho AR, Han J (2017) Reduction of Dickeya chrysanthemi on fresh-cut iceberg lettuce using antimicrobial sachet containing microencapsulated oregano essential oil. LWT - Food Sci Technol 82:361–368. https://doi.org/10.1016/j.lwt.2017.04.043

Cook CM, Lanaras T (2016) Essential oils : isolation , production and uses. In: Caballero B, Finglas PM, Toldrá F (eds) Encyclopedia of food and health. Elsevier Ltd., pp 552–557

Da Costa KS, Galúcio JM, Da Costa CHS et al (2019) Exploring the potentiality of natural products from essential oils as inhibitors of Odorant-binding proteins: a structure- and ligand-based virtual screening approach to find novel mosquito repellents. ACS Omega. https://doi.org/10.1021/acsomega.9b03157

da Silva DA, Aires GCM, Pena R da S (2020) Gums—characteristics and applications in the food industry. In: Innovation in the food sector through the valorization of food and agro-food by-products. IntechOpen, pp 1–24

Dávila-Rodríguez M, López-Malo A, Palou E et al (2020) Essential oils microemulsions prepared with high-frequency ultrasound: physical properties and antimicrobial activity. J Food Sci Technol 57:4133–4142. https://doi.org/10.1007/s13197-020-04449-8

de Melo AM, Turola Barbi RC, de Souza WFC et al (2020) Microencapsulated lemongrass (Cymbopogon flexuosus) essential oil: a new source of natural additive applied to Coalho cheese. J Food Process Preserv 44. https://doi.org/10.1111/jfpp.14783

de Souza GH, de A, dos Santos Radai JA, Mattos Vaz MS, et al (2021) In vitro and in vivo antibacterial activity assays of carvacrol: a candidate for development of innovative treatments against KPC-producing Klebsiella pneumoniae. PLoS One 16:e0246003. https://doi.org/10.1371/journal.pone.0246003

do Nascimento LD, Almeida LQ, de EMP Sousa, et al (2020a) Microwave-assisted extraction: an alternative to extract Piper aduncum essential oil. Brazilian J Dev 6:40619–40638. https://doi.org/10.34117/bjdv6n6-558

do Nascimento LD, de Moraes AAB, Costa KS et al (2020b) Bioactive natural compounds and antioxidant activity of essential oils from spice plants : new findings and potential applications. Biomol Ther 10:988. https://doi.org/10.3390/biom10070988

Donsì F, Ferrari G (2016) Essential oil nanoemulsions as antimicrobial agents in food. J Biotechnol 233:106–120. https://doi.org/10.1016/j.jbiotec.2016.07.005

dos Ferreira GKS, Margalho JF, Almeida LQ, et al (2020) Seasonal and circadian evaluation of the essential oil of Piper divaricatum G. Mey. (Piperaceae) Leaves. Brazilian J Dev 6:41356–41369. https://doi.org/10.34117/bjdv6n6-612

Duman F, Kaya M (2016) Crayfish chitosan for microencapsulation of coriander (Coriandrum sativum L .) essential oil. Int J Biol Macromol 92:125–133. https://doi.org/10.1016/j.ijbiomac.2016.06.068

El-Wahab HMFA, Moram GSE-D (2013) Toxic effects of some synthetic food colorants and/or flavor additives on male rats. Toxicol Ind Health 29:224–232. https://doi.org/10.1177/0748233711433935

Escobar A, Pérez M, Romanelli G, Blustein G (2020) Thymol bioactivity: a review focusing on practical applications. Arab J Chem 13:9243–9269. https://doi.org/10.1016/j.arabjc.2020.11.009

Esquivel-Chávez F, Colín-Chávez C, Virgen-Ortiz JJ et al (2021) Control of mango decay using antifungal sachets containing of thyme oil/modified starch/agave fructans microcapsules. Futur Foods 3:100008. https://doi.org/10.1016/j.fufo.2020.100008

Falleh H, Ben Jemaa M, Saada M, Ksouri R (2020) Essential oils: a promising eco-friendly food preservative. Food Chem 330:127268. https://doi.org/10.1016/j.foodchem.2020.127268

Fernandes DEB, Costa I, Aes G et al (2017) Microencapsulated rosemary (Rosmarinus officinalis) essential oil as a biopreservative in Minas cheese. J Food Proc Preserv Proc 41:1–9. https://doi.org/10.1111/jfpp.12759

Fernandes RVB, Botrel DA, Monteiro PS et al (2018) Microencapsulated oregano essential oil in grated Parmesan cheese conservation. Int Food Res J 25:661–669

Fuchs M, Turchiuli C, Bohin M et al (2006) Encapsulation of oil in powder using spray drying and fluidised bed agglomeration. J Food Eng 75:27–35. https://doi.org/10.1016/j.jfoodeng.2005.03.047

Gharsallaoui A, Roudaut G, Chambin O et al (2007) Applications of spray-drying in microencapsulation of food ingredients : an overview. Food Res Int 40:1107–1121. https://doi.org/10.1016/j.foodres.2007.07.004

Gonçalves ND, Pena FL, Sartoratto A et al (2017) Encapsulated thyme (Thymus vulgaris) essential oil used as a natural preservative in bakery product. Food Res Int 96:154–160. https://doi.org/10.1016/j.foodres.2017.03.006

Grand View Research (2020) Essential oils market size, share & trends analysis report by application (food & beverages, Spa & relaxation), by product (orange, peppermint), by sales channel, and segment forecasts, 2020–2027. https://www.grandviewresearch.com/industry-analysis/essential-oils-market

Gutiérrez-del-Río I, Fernández J, Lombó F (2018) Plant nutraceuticals as antimicrobial agents in food preservation: terpenoids, polyphenols and thiols. Int J Antimicrob Agents 52:309–315. https://doi.org/10.1016/j.ijantimicag.2018.04.024

Hassoun A, Carpena M, Prieto MA et al (2020) Use of spectroscopic techniques to monitor changes in food quality during application of natural preservatives: a review. Antioxidants 9:882. https://doi.org/10.3390/antiox9090882

Hernández-Nava R, López-Malo A, Palou E et al (2020) Encapsulation of oregano essential oil (Origanum vulgare) by complex coacervation between gelatin and chia mucilage and its properties after spray drying. Food Hydrocoll 109:106077. https://doi.org/10.1016/j.foodhyd.2020.106077

Hertrich SM, Niemira BA (2021) Advanced processing techniques for extending the shelf life of foods. In: Food safety and quality-based shelf life of perishable foods. Springer International Publishing, Cham, pp 91–103

Higueras L, López-Carballo G, Hernández-Muñoz P et al (2014) Antimicrobial packaging of chicken fillets based on the release of carvacrol from chitosan/cyclodextrin films. Int J Food Microbiol 188:53–59. https://doi.org/10.1016/j.ijfoodmicro.2014.07.018

Hirose M, Shibata M, Hagiwara A et al (1981) Chronic toxicity of butylated hydroxytoluene in Wistar rats. Food Cosmet Toxicol 19:147–151. https://doi.org/10.1016/0015-6264(81)90350-3

Holkem AT, Codevilla C, Menezes C (2015) Emulsification / internal ionic gelation : alternative for microencapsulation bioactive compounds. Ciência e Nat 37:116–124. https://doi.org/1 0.5902/2179-460X19739

Ilić DP, Stanojević LP, Troter DZ et al (2019) Improvement of the yield and antimicrobial activity of fennel (Foeniculum vulgare Mill.) essential oil by fruit milling. Ind Crop Prod 142:111854. https://doi.org/10.1016/j.indcrop.2019.111854

Kamal AA, Fawzia SE-S (2018) Toxicological and safety assessment of tartrazine as a synthetic food additive on health biomarkers: a review. African J Biotechnol 17:139–149. https://doi.org/10.5897/AJB2017.16300

Karpiński TM (2020) Essential oils of lamiaceae family plants as antifungals. Biomol Ther 10. https://doi.org/10.3390/biom10010103

Khalili G, Mazloomifar A, Larijani K et al (2018) Solvent-free microwave extraction of essential oils from Thymus vulgaris L. and Melissa officinalis L. Ind Crop Prod 119:214–217. https://doi.org/10.1016/j.indcrop.2018.04.021

Khorshidian N, Yousefi M, Khanniri E, Mortazavian AM (2018) Potential application of essential oils as antimicrobial preservatives in cheese. Innov Food Sci Emerg Technol 45:62–72. https://doi.org/10.1016/j.ifset.2017.09.020

Knorr D, Watzke H (2019) Food processing at a crossroad. Front Nutr 6. https://doi.org/10.3389/fnut.2019.00085

Kujur A, Kiran S, Dubey NK, Prakash B (2017) LWT - Food Science and Technology Microencapsulation of Gaultheria procumbens essential oil using chitosan-cinnamic acid microgel : improvement of antimicrobial activity , stability and mode of action. LWT - Food Sci Technol 86:132–138. https://doi.org/10.1016/j.lwt.2017.07.054

Leimann FV (2008) MICROENCAPSULAÇÃO DE ÓLEO ESSENCIAL DE CAPIM LIMÃO UTILIZANDO O PROCESSO DE COACERVAÇÃO SIMPLES. Universidade Federal de Santa Catarina

Locali-Pereira AR, Lopes NA, Menis-Henrique MEC et al (2020) Modulation of volatile release and antimicrobial properties of pink pepper essential oil by microencapsulation in single- and double-layer structured matrices. Int J Food Microbiol 335:108890. https://doi.org/10.1016/j.ijfoodmicro.2020.108890

Lucía C, Marcela F, Ainhoa L (2017) Encapsulation of almond essential oil by co-extrusion / gelling using Chitosan as wall material:67–74. https://doi.org/10.4236/jeas.2017.71004

Ma S, Jia R, Guo M et al (2020) Insecticidal activity of essential oil from Cephalotaxus sinensis and its main components against various agricultural pests. Ind Crop Prod 150:112403. https://doi.org/10.1016/j.indcrop.2020.112403

Madene A, Jacquot M, Scher J, Desobry S (2006) Flavour encapsulation and controlled release – a review. Int J Food Sci Technol 41:1–21. https://doi.org/10.1111/j.1365-2621.2005.00980.x

Maes, Bouquillon, Fauconnier (2019) Encapsulation of essential oils for the development of biosourced pesticides with controlled release: a review. Molecules 24:2539. https://doi.org/10.3390/molecules24142539

Martínez-Abad A, Ramos M, Hamzaoui M et al (2020) Optimisation of sequential microwave-assisted extraction of essential oil and pigment from lemon peels waste. Foods 9:1493. https://doi.org/10.3390/foods9101493

Maurya A, Prasad J, Das S, Dwivedy AK (2021) Essential oils and their application in food safety. Front Sustain Food Syst 5. https://doi.org/10.3389/fsufs.2021.653420

Mehran M, Masoum S, Memarzadeh M (2020) Microencapsulation of Mentha spicata essential oil by spray drying: optimization, characterization, release kinetics of essential oil from microcapsules in food models. Ind Crop Prod 154:112694. https://doi.org/10.1016/j.indcrop.2020.112694

Mizobuchi M, Ishidoh K, Kamemura N (2021) A comparison of cell death mechanisms of antioxidants, butylated hydroxyanisole and butylated hydroxytoluene. Drug Chem Toxicol 1–8. https://doi.org/10.1080/01480545.2021.1894701

Mohammed NK, Tan CP, Manap YA et al (2020) Spray drying for the encapsulation of oils—a review. Molecules 25:3873. https://doi.org/10.3390/molecules25173873

Moradi S, Fazlali A, Hamedi H (2018) Microwave-assisted hydro-distillation of essential oil from rosemary: comparison with traditional distillation. Avicenna J Med Biotechnol 10:22–28

Nascimento LD, Cascaes MM, Kauê SC et al (2019) Essential oils microencapsulation: concepts and applications. In: Voigt CL (ed) A produção do conhecimento na engenharia química. Atena Editora, Ponta Grossa, pp 22–35

Nascimento LD, Silva SG, Cascaes MM et al (2021) Drying effects on chemical composition and antioxidant activity of Lippia thymoides essential oil, a natural source of thymol. Molecules 26:2621. https://doi.org/10.3390/molecules26092621

Ng KR, Lin Lee JJ, Lyu X, Chen WN (2021) Yeast-derived plant phenolic emulsions as novel, natural, and sustainable food preservatives. ACS Food Sci Technol 1:326–337. https://doi.org/10.1021/acsfoodscitech.0c00062

Nguyen TTT, Le TVA, Dang NN et al (2021) Microencapsulation of essential oils by spray-drying and influencing factors. J Food Qual 2021:1–15. https://doi.org/10.1155/2021/5525879

Nieto G (2020) A review on applications and uses of thymus in the food industry. Plan Theory 9:961. https://doi.org/10.3390/plants9080961

Partheniadis I, Vergkizi S, Lazari D et al (2019) Formulation, characterization and antimicrobial activity of tablets of essential oil prepared by compression of spray-dried powder. J Drug Deliv Sci Technol 50:226–236. https://doi.org/10.1016/j.jddst.2019.01.031

Pisoschi AM, Pop A, Georgescu C et al (2018) An overview of natural antimicrobials role in food. Eur J Med Chem 143:922–935. https://doi.org/10.1016/j.ejmech.2017.11.095

Puškárová A, Bučková M, Kraková L et al (2017) The antibacterial and antifungal activity of six essential oils and their cyto/genotoxicity to human HEL 12469 cells. Sci Rep 7:8211. https://doi.org/10.1038/s41598-017-08673-9

Radünz M, dos Santos Hackbart HC, Camargo TM et al (2020) Antimicrobial potential of spray drying encapsulated thyme (Thymus vulgaris) essential oil on the conservation of hamburger-like meat products. Int J Food Microbiol 330:108696. https://doi.org/10.1016/j.ijfoodmicro.2020.108696

Rašković A, Milanović I, Pavlović N et al (2014) Antioxidant activity of rosemary (Rosmarinus officinalis L.) essential oil and its hepatoprotective potential. BMC Complement Altern Med 14:225. https://doi.org/10.1186/1472-6882-14-225

Ribeiro AF, Andrade EHA, Salimena FRG, Maia JGS (2014) Circadian and seasonal study of the cinnamate chemotype from Lippia origanoides Kunth. Biochem Syst Ecol 55:249–259. https://doi.org/10.1016/j.bse.2014.03.014

Ribeiro-Santos R, Andrade M, Sanches-Silva A (2017) Application of encapsulated essential oils as antimicrobial agents in food packaging. Curr Opin Food Sci 14:78–84. https://doi.org/10.1016/j.cofs.2017.01.012

Ridgway E, Baker P, Woods J, Lawrence M (2019) Historical developments and paradigm shifts in public health nutrition science, guidance and policy actions: a narrative review. Nutrients 11:531. https://doi.org/10.3390/nu11030531

Riquelme N, Lidia M, Matiacevich S (2017) Food and Bioproducts Processing Active films based on alginate containing lemongrass essential oil encapsulated : effect of process and storage conditions. Food Bioprod Process 104:94–103. https://doi.org/10.1016/j.fbp.2017.05.005

Salvia-trujillo L, Rojas-graü A, Soliva-fortuny R, Martín-belloso O (2015) Physicochemical characterization and antimicrobial activity of food-grade emulsions and nanoemulsions incorporating essential oils. Food Hydrocoll 43:547–556. https://doi.org/10.1016/j.foodhyd.2014.07.012

Santana K, do Nascimento LD, Lima e Lima A et al (2021) Applications of virtual screening in bioprospecting: facts, shifts, and perspectives to explore the chemo-structural diversity of natural products. Front Chem 9. https://doi.org/10.3389/fchem.2021.662688

Silva Júnior OS, de Jesus Pereira Franco C, Barbosa de Moraes AA et al (2021) In silico analyses of toxicity of the major constituents of essential oils from two ipomoea L. species. Toxicon. https://doi.org/10.1016/j.toxicon.2021.02.015

Silva S, Da Costa RA, De Oliveira MS et al (2019) Chemical profile of lippia thymoides, evaluation of the acetylcholinesterase inhibitory activity of its essential oil, and molecular docking and molecular dynamics simulations. PLoS One 14:1–17. https://doi.org/10.1371/journal.pone.0213393

Silva SG, de Oliveira MS, Cruz JN et al (2021) Supercritical CO_2 extraction to obtain Lippia thymoides Mart. Schauer (Verbenaceae) essential oil rich in thymol and evaluation of its antimicrobial activity. J Supercrit Fluids 168:105064. https://doi.org/10.1016/j.supflu.2020.105064

Siqueira CAT, Serain AF, Pascoal ACRF et al (2015) Bioactivity and chemical composition of the essential oil from the leaves of Guatteria australis A.St.-Hil. Nat Prod Res 29:1966–1969. https://doi.org/10.1080/14786419.2015.1015017

Sridhar A, Ponnuchamy M, Kumar PS, Kapoor A (2021) Food preservation techniques and nanotechnology for increased shelf life of fruits, vegetables, beverages and spices: a review. Environ Chem Lett 19:1715–1735. https://doi.org/10.1007/s10311-020-01126-2

Sundararajan B, Moola AK, Vivek K, Kumari BDR (2018) Formulation of nanoemulsion from leaves essential oil of Ocimum basilicum L. and its antibacterial, antioxidant and larvicidal activities (Culex quinquefasciatus). Microb Pathog 125:475–485. https://doi.org/10.1016/j.micpath.2018.10.017

Veenstra JP, Johnson JJ (2019) Oregano (Origanium Vulgare) extract for food preservation and improving gastrointestinal health. Int J Nutr 3:43–52. https://doi.org/10.14302/issn.2379-7835.ijn-19-2703

Vinciguerra V, Rojas F, Tedesco V et al (2019) Chemical characterization and antifungal activity of Origanum vulgare , Thymus vulgaris essential oils and carvacrol against Malassezia furfur. Nat Prod Res 33:3273–3277. https://doi.org/10.1080/14786419.2018.1468325

Wang W, Wu N, Zu YG, Fu YJ (2008) Antioxidative activity of Rosmarinus officinalis L. essential oil compared to its main components. Food Chem 108:1019–1022. https://doi.org/10.1016/j.foodchem.2007.11.046

Zanetti M, Carniel TK, Dalcanton F et al (2018) Use of encapsulated natural compounds as antimicrobial additives in food packaging: a brief review. Trends Food Sci Technol 81:51–60. https://doi.org/10.1016/j.tifs.2018.09.003

Zhu Y, Li C, Cui H, Lin L (2021) Encapsulation strategies to enhance the antibacterial properties of essential oils in food system. Food Control 123:107856. https://doi.org/10.1016/j.foodcont.2020.107856

Chapter 7
Safety Assessment of Essential Oil as a Food Ingredient

Fernando Almeida-Souza, Isadora F. B. Magalhães, Allana C. Guedes, Vanessa M. Santana, Amanda M. Teles, Adenilde N. Mouchrek, Kátia S. Calabrese, and Ana Lúcia Abreu-Silva

7.1 Introduction

It has long been recognized that some essential oils have been shown to exhibit various biological properties such as analgesic, antioxidant, antispasmodic, carminative, analgesic, anti-inflammatory, antiviral, antimycotic, antitoxic, antiparasitic, antifungal, and insecticide (Bhagat et al. 2018; Blowman et al. 2018). Over the past two hundred years, the development of Chemistry was helpful to produce standardized plant extracts and isolate their active compounds (Brnawi et al. 2018; Govindarajan et al. 2018). Despite the obscure beginnings of the use of aromatic plants in prehistoric times to prevent, palliate or cure diseases, analyzes of pollen from Stone Age settlements indicate the use of aromatic plants that can be dated back to 10,000 BC (Kubeczka 2015).

In ancient Egypt, there was a medical document that was called the Ebers Papyrus, which contained about 700 formulas and remedies, including aromatic

F. Almeida-Souza (✉)
Pós-graduação em Ciência Animal, Universidade Estadual do Maranhão, São Luís, MA, Brazil

Laboratório de Imunomodulação e Protozoologia, Instituto Oswaldo Cruz, Fiocruz, Rio de Janeiro, Brazil
e-mail: fernandosouza@professor.uema.br

I. F. B. Magalhães · A. C. Guedes · V. M. Santana · A. L. Abreu-Silva
Pós-graduação em Ciência Animal, Universidade Estadual do Maranhão, São Luís, MA, Brazil

A. M. Teles · A. N. Mouchrek
Universidade Federal do Maranhão, São Luís, Maranhão, Brazil

K. S. Calabrese
Laboratório de Imunomodulação e Protozoologia, Instituto Oswaldo Cruz, Fiocruz, Rio de Janeiro, Brazil

plants and plant products. Theophrastus von Hohenheim, known by the name of Paracelsus, was a 15th-century physician and alchemist that, defined the role of alchemy by developing medicines and medicinal plant extracts. His works were about distillation, in which it is released the most desirable part of the plant, the Fifth essence or quintessence, by separating the "essential" from the "non essential" part. The essential term oil comes up referring to its quintessence theory (Kubeczka 2015).

Essential oils are aromatic oily liquids that can be extracted by a process of distillation, compression or extraction using solvents from various plant organs, namely flowers, leaves, seeds, buds, shoots, roots, among others, and are supplied in secretory cells, cavities, channels, epidermal or glandular cells (Bakkali et al. 2008; Burt 2004). They are a mixture of several compounds, including terpenes, alcohols, acids, aldehydes, and sulfides, which are likely produced by plants in response to physiological stress, pathogen attack, and ecological factors (Calo et al. 2015).

Characterized by having a small molecular weight, they are very volatile and evaporate easily. Essential oils are natural, complex liquids that have a strong odor, and are sometimes colored., Their production is made by plants as secondary metabolites, liposoluble and soluble in organic solvents with a density generally lower than that of water (Bakkali et al. 2008).

Essential oils have been used in foods as flavorings and preservatives due to their antimicrobial agents and antioxidant properties. The main active components that generate this are: thymol, carvacrol, eugenol, cinnamaldehyde and linalool (Kuorwel et al. 2011). Although its mechanisms of action are still poorly understood, the use of essential oil is justified since the food microbial is a major concern for consumers, regulatory agencies and food industries (Burt 2004; Calo et al. 2015).

Food products can be subject to microbial contamination mainly caused by bacteria, yeasts and fungi. Many of these microorganisms can end up causing undesirable reactions that lead to food deterioration and thus altering the taste, odor, color, sensory and textural properties of the food (Gutierrez et al. 2008). Therefore, the food industry aims to produce food that has a longer shelf-life in the markets and is free from damage, concerning the presence of pathogenic microorganisms and their toxins. Also, the new consumer bias and food legislation have triggered an increasingly urgent and necessary change in food production. Consumers look for good fresh quality food, with little amount of salt, sugar, fat, and acids, among others, and free of artificial preservatives and minimally processed, but with a long shelf life (Moubarac et al. 2014).

Essential oils as foods additives represent a natural compound and produce antioxidant and antimicrobial effects, reducing the use of synthetic preservatives, and thus, are suitable for organic foods. They allow the use of clean food labels, and they are in line with the propensity of the so-called green consumerism, in which the consumer, in addition to seeking better quality and price, includes the environmental variable, giving preference to products and services that do not harm the environment in production, distribution, consumption and final disposal. Due to their extensive history of use in culinary and consumption by humans, they are generally recognized as safe (Kuorwel et al. 2011). In addition to their use in human food, essential oils can be applied as food supplements for animal production, such as ruminants, modifying their metabolism in order to reduce methane and ammonia emissions (Cobellis et al. 2016a).

However, the use of essential oils in food has limitations and the main ones are causing sensory changes in food, due to its strong odor and flavor, and in some cases its color. Essential oils can show a high variability in quality and quantity of bioactive constituents. Also, bioactive compounds are potentially lost or reduced by many food processing techniques or even essential oil extraction techniques (Kuorwel et al. 2011; Negi 2012). Furthermore, people often assume that, as essential oils are natural products, "natural" means safe, but there are many natural compounds and chemicals that are not safe. All these factors need to be analyzed before including essential oil in foods.

In this chapter, we searched in the literature for the main essential oil used in food industry, and carry out a detailed description of its chemical composition and of its biological activities relevant for its use as a food constituent, such as antimicrobial and antioxidant properties. Then, we summarized the studies related to the safety use of essential oils in food. Together, these data provide tools for an adequate assessment of the safety of essential oil as a food ingredient.

7.2   Chemical Composition of Essential Oils Used as Food

Food and Drug Association (FDA) defined a list of essential oils from medicinal or aromatic plants that can be Generally Recognized as Safe (GRAS) for addition in food (Laranjo et al. 2019). A word cloud of Latin name of the genus or plant species was created based on the number of citations in PubMed database using "essential oil" and "food" as keywords. The four essential oils with the highest research numbers were *Origanum* spp. (461), *Thymus vulgaris* L. (138), *Citrus aurantium* L. (113), and *Rosmarinus officinalis* L. (109) (Fig. 7.1).

The chemical composition of an essential oil can receive different influences. It can vary depending on the climate, extraction method, part of the plant to be used and location. As we will see in Table 7.1, these factors can generate different chemotypes for the same species.

Artemisia dracunculus SL, also known as estragon or tarragon, is a typical seasoning of French cuisine used to enhance the flavor of certain ingredients and foods. Besides posses different morphtypes, different studies have found it to be rich in compounds such as terpenes and terpenoids, aromatic and aliphatic compounds (Azizkhani et al. 2021), phenylpropanoids (Bedini et al. 2017), and methyleugenol and estragole as one of its major compounds (Meepagala et al. 2002; Szczepanik et al. 2018). The presence of some constituents ends up making the oil unfeasible for food use, a decision by the scientific committee for Food of the Directorate-General for Health and Consumer Protection. Estragole and methyleugenol are two compounds, often described in essential oils used as food ingredients, that possess carcinogenic and genotoxic properties (SCF 2001a, b).

Fig. 7.1 Word cloud of essential oil species generally classified as Generally Recognized as Safe (GRAS) by Food and Drug Association (FDA), associated with the number of results on PubMed

Another widely researched genus in the food area is *Brassica* spp. (Reyes-Jurado et al. 2016), which identified allyl-isotocyanate (98.4%) in *Brassica nigra* oil. Saka et al. (2017) researched the production of essential oil from *Brassica rapa* collected in different locations and extracted by microwave-assisted hydrodistillation and hydrodistillation techniques and, regardless of the technique used, what influenced the yield was its geographic location. Its chemical composition only varied in the percentage, remaining the same major compounds (Methyl-5-hexenenitrile, 2-Phenylethanol and Allyl isothiocyanate) in the three sampling regions. Usami et al. (2014) found (E)-1,5-heptadiene (40.3%), 3-methyl-3-butenenitrile (26.0%), and 3-phenylpropanenitrile (12.4%) as majorities of *B. rapa*.

The same can be verified when researching parts of the same plant that may still be influenced by the extraction technique used. Değirmenci and Hatice (2020) verified that the major chemical compounds of flowers in *Citrus aurantium* L oil was linalool L (14.12%), squalene (6.77%) and d-limonene (5.8%). Navaei Shoorvarzi et al. (2020) observed similar major compounds by GC-MS analysis, these being linalyl acetate (22.9%), limonene (7.3%) and α-terpineol (6.9%).

Essential oils have a variety of volatile compounds, mostly hydrocarbons, and are typically less than 5% of the plant product. Terpenes has a strong presence in *C. aurantium* and this occurs in most citrus fruits as well, with the presence of monoterpenes, sesquiterpenes and other hydrocarbons. Teneva et al. (2019) studying *C. aurantium* bark essential oil verified forty-eight compounds, in which the majority were d-limonene (85.22%), β-myrcene (4.30%) and pinene (1.29%). The

Table 7.1 Chemical composition of essential used as food ingredient

Species	Parts	Extraction Method; Extraction Time; Yield	Place of Collection	Major Components	References
Artemisia dracunculus	Aerial parts	Hydrophilic-lipophilic	Kashan, Iran	Estragole (81.89%), beta- cis-ocimene (4.62%), beta-trans-ocimene (3.44%)	Azizkhani et al. (2021)
	Aerial parts	Hydrodistillation; 0.40%	Urbino, Italy	Methyl chavicol (73.3%), camphor (16.9%), artemisia alcohol (5.9%)	Bedini et al. (2017)
	Aerial parts	Steam-distilled; 4%	Albany, Oregon	5-Phenyl-1,3-pentadiyne, methyleugenol, and capillarin	Meepagala et al. (2002)
	Aerial parts	Hydrodistillation; 0.5% - 1%	Kotayk, Armenia	Estragole (84.9%), linalool (5.09%), beta-ocimen (4%)	Sahakyan et al. (2021)
	Aerial parts	Hydrodistillation	Gostyń, Polony	Methyleugenol (31.4%), elemicin (26.7%), (E)-isoelemicin (15.0%).	Szczepanik et al. (2018)
Brassica spp.	Aeed	Pressing	Jalisco, Mexico	Allyl isothiocyanate- AITC (98.4%), allyl trisulfide (0.2%), allyl disulfide (0.1%)	Reyes-Jurado et al. (2016)
	Leaves and roots	Hydrodistillation and microwave-assisted hydrodistillation	Algeria: Bouira, Mostagane, Sétif	Allyl isothiocyanate, allyl disulfide, methyl-5-hexenenitrile	Saka et al. (2017)
	Aerial parts	Hydrodistillation	Yamagata, Japan	1,5-Heptadiene (40.27%), 3-methyl-3-butenenitrile (25.97%), 3-phenylpropanenitrile (12.41%)	Usami et al. (2014)

(continued)

Table 7.1 (continued)

Species	Parts	Extraction Method; Extraction Time; Yield	Place of Collection	Major Components	References
Citrus aurantium	Flowers	Hydrosol and ethanol; 16.38%	Güzelyurt, Cyprus	Linalool (15.72%), α—terpineol (4.87%), hotrienol (1.60%)	Değirmenci and Hatice (2020)
	Dried bloom	Hydrodistillation	Mashhad, Iran	Linalyl acetate (22.9%), limonene (7.3%), and α-terpineol (6.9%)	Navaei Shoorvarzi et al. (2020)
	Fresh zest	Steam distillation; 4 h	Bulgaria	Limonene (85.22%), β-myrcene (4.3%), and α-pinene (1.29%)	Teneva et al. (2019)
	Flowers	Steam distillation	Nabeul, Tunisia	Limonene (27.5%), e-nerolidol (17.5%), α-terpineol (14%)	Hsouna et al. (2013)
	Dried blossoms	Steam distillation	Guangzhou, China	Linalool (64.6 ± 0.04%), α-terpineol (7.61 ± 0.03%), (R)-limonene (6.15 ± 0.04%)	Shen et al. (2017)

Species	Parts	Extraction Method; Extraction Time; Yield	Place of Collection	Major Components	References
Ocimum basilicum	Dried plants	Hydrodistillation; 3 h	Turquia	p-allyl-anisole (5.65–17.90%), nerol (6.69–16.11%), linalool (5.10–10.81%)	Yaldiz et al. (2019)
	Aerial parts	Hydrodistillation; 1.05%	Algerian Saharan Atlas (Laghouat region)	h linalool (52.1%), linalyl acetate (19.1%), α-terpineol (5.7%)	Rezzoug et al. (2019)
	Dried leaves and aerial parts	Hydrodistillation; 2 h	Urmia, Iran	Methylchavicol, linalool, 1,8-cineol	Mandoulakani et al. (2017)
	Aerial parts	Hydrodistillation	Maragheh, Iran	Methyl chavicol (43.09–69.91%), linalool (4.8–17.9%), cadinol (1.5–3.2%)	Gohari et al. (2020a)
	Aerial parts	Hydrodistillation	East Azerbaijan provience, Iran	Chavicol (26.2%), linalool (12.4%), germacrene D (4.26%)	Gohari et al. (2020b)
	Leaves	Hydrodistillation	South Africa	Estragole (41.40%), 1,6-octadien-3-ol, 3,7-dimethyl (29.49%), trans-, α-bergamotene (5.32%)	Falowo et al. (2019)
	Leaves	Hydrodistillation	Cairo, Egypt	linalool (48.4%), 1,8-cineol (12.2%), eugenol (6.6%)	Abou El-Soud et al. (2015)
	Fresh leaves	Hydrodistillation	Pisa, Italy	Linalool (48.8%), 1,8-cineole (13%), trans-α-bergamotene (7.3%)	Kiferle et al. (2019)

(continued)

Table 7.1 (continued)

Species	Parts	Extraction Method; Extraction Time; Yield	Place of Collection	Major Components	References
Origanum spp.	–	–	Guangzhou, China	Carvacrol (58.13%), p-cymene (17.85%), thymol (8.15%)	Xie et al. (2019)
	–	–	Milan, Italy	Carvacrol (35.95%), thymol (25.2%), p-cymene (21.54%)	Avola et al. (2020)
	Aerial parts	Hydrodistillation; 60 min; 5.3%	Putre, Chile	Thymol (15.9%), Z-sabinene hydrate (13.4%), γ-terpinene (10.6%)	Simirgiotis et al. (2020)
	Aerial parts	Hydrodistillation	Kashmir Himalayas	Carvacrol (52.99–91.18%), β- cariofileno (0.04–1.87%), terpinen-4-ol (0.02–0.32%)	Jan et al. (2020)
	leaves and flowers	hydrodistillation	Crete (Greece)	Carvacrol (52.2%), γ-terpinene (8.4%), p-cymene (6.1%)	Mitropoulou et al. (2015)
	flowering aerial parts	–	–	Phenol carvacrol (50.32%), thymol (14.8%), γ-terpinene (13.6%) and p-cymene (8.40%)	López et al. (2018)
	leaves	Hydrodistillation	Chapecó, Brazil	γ-Terpinene (46.3%), terpinolene (21.2%), p-cymene (15.7%)	Badia et al. (2020)
	Aerial parts	Hydrodistillation	Montecorice, Italy	Carvacrol (77.8%), p- cimeno (5.3%) and γ – terpineno (4.9%)	Della Pepa et al. (2019)
	Aerial parts	Hydrodistillation	Montecorice, Italy	Terpinen-4-ol (29.6%), δ-2-careno (20.1%), canfeno (13.4%)	Della Pepa et al. (2019)
	–	–	Milan, Italy	Carvacrol (36%), thymol (25%) and p-cymene (22%)	Kapustová et al. (2021)
	Aerial parts	Hydrodistillation	SE Aegean, Greece	Carvacrol (66.0%), p-cymene (7.9%) and γ-terpinene (4.9%)	Vanti et al. (2021)

Species	Parts	Extraction Method; Extraction Time; Yield	Place of Collection	Major Components	References
Mentha piperita	Leaves	Steam distillation	São Paulo, Brazil	Menthol (43.75%), isomenthone 291 (27.71%), menthone (9.37%)	de Melo, et al. (2020)
	–	–	Kayseri, Turkey	Menthol (20,31%), p-menthone (14,89%), limonene (9,50%)	Yilmaztekin et al. (2019)
	Fruit	Hydrodistillation	Paraíba, Brazil	Menthol (41.34%), isomenthone (23.47%), cismenthone (10.84%)	de Oliveira et al. (2017)
	Fruit	Steam distillation	São Paulo, Brazil	Menthol (56.85%), isomenthone (21.13%), menthyl acetate (4.62%)	de Sousa Guedes et al. (2016)
	Fresh leaves	Hydrodistillation; 2 h	Prešov, Slovak Republic	Menthol (74.95%), menthyl acetate (15.18%), menthone (6.89%)	Grulova et al. (2015)
	Leaves	Hydrodistillation	Paraíba, Brazil	Menthol (30.31%), isomenthone (26.70%), menthol acetate (8.52%)	Guerra et al. (2015)
	Leaves	Hydrodistillation	Mysore, India	Menthol (24.96%), l-menthone (22.18%), pulegone (19.33%)	Rachitha et al. (2017)
Rosa spp.	–	–	Bulgaria and Turkey	Citronellol (29.6–36.2%), geraniol (16.7–13.3%), nerol (8.0–7.3%),	Krupčík et al. (2015)
	Fresh flowers	Distillation	Kashan, Iran	Nonadecane (24.72%), heneicosane (19.325%), oleic acid (17.63%)	Ghavam, et al. (2021)
	Fresh flowers	Hydrodistillation	Vallauris, France	2-Phenyl etanol (25%), Citronellol (20.9%), Geraniol (21.2%)	Labadie et al. (2015)
	Fresh flowers	Hydrodistillation	Siagne, France	2-Phenylethanol (42%), citronellol (22%), geraniol (14%).	Labadie et al. (2016)
	Flowers	Steam distillation	Shiraz, Iran	Phenyl ethyl alcohol (74.6%), hexadecane (8.5%), methyl eugenol (4.1%)	Mahboubifar et al. (2021)
	Fresh fruit	Hydrodistillation	Guangzhou, China	n-Hexadecanoic acid (16.06%), octadecane (8.16%), octadecatrien-1-ol (6.66%)	Liu et al. (2016)

(continued)

Table 7.1 (continued)

Species	Parts	Extraction Method; Extraction Time; Yield	Place of Collection	Major Components	References
Rosmarinus officinalis	–	Steam distillation	Minas Gerais, Brazil	Eucalyptol (1,8-cineole) (35.75%), camphor (28.7%) and limonene (24.88%)	de Medeiros Barbosa et al. (2019)
	Dried leaves	Hydrodistillation	Paraná, Brazil	1,8-Cineole (eucalyptol, 52.2%), camphor (15.2%) and α-pinene (12.4%)	da Silva Bomfim et al. (2020)
	Aerial parts	Hydrodistillation	North-West of Morocco	1,8-Cineole (23.673%), camphor (18.743%), borneol (15.46%)	Bouyahya et al. (2017)
	Aerial parts	Hydrodistillation	Pančić, Belgrade	1,8-Cineole (43.77%), camphor (12.53%) and α-pinene (11.51%)	Rašković et al. (2014)
	Fresh leaves	Hydrodistillation	Ghamsar, Iran	Verbenone (20.29%), 1, 8-cineole (15.56%), a-pinene (7.58%)	Akhbari et al. (2018)
	Leaves	Steam distillation	Rio Grande do Sul, Brazil	α-Pinene (37.26%), 1,8-Cineole (26.24%), Verbenone (5.53%)	Silvestre et al. (2019)
	–	–	Firenze, Italy	1,8-Cineole (41.2%), camphor (11.7%), α-pinene (9.8%)	Bedini et al. (2020)
	–	–	Italy	1,8-Cineole (45.27%), borneol (12.94%), α-pinene (11.39%)	Iseppi et al. (2018)
	Flowering aerial parts	Hydrodistillation	Camerino (central Italy)	1,8-Cineole (36.2%), camphor (16.4%), α-pinene (11.7%),	Sirocchi et al. (2017)
	–	–	Saanichton, BC, Canada	1,8-Cineole (37.6%), (±)-camphor, (+)-α-pinene	Tak et al. (2016)
	Aerial parts	Hydrodistillation	Region of Fez, Morocco	α-Pinene (15,4%), camphene (9.16%), para-cymene (4.15%)	Elyemni et al. (2019)

Species	Parts	Extraction Method; Extraction Time; Yield	Place of Collection	Major Components	References
Salvia officinalis	Flowers, leaves and stems	Hydrodistillation; 2.5 h; 0,11%	Niš, Serbia	α-Thujone (28.2%), camphor (27.5%), 1,8-cineole (8.3%)	Radulović et al. (2017)
	Aerial part	Hydrodistillation	Marrakech, Morocco	Trans-thujone (14.10% and 29.84%), 1,8-cineole (5.10% and 16.82%), camphor (4.99% and 9.14%)	Bouajaj et al. (2013)
	Aerial part	Hydrodistillation; 3 h	Spain	α-Thujone (22.8–41.7%), camphor (10.7–19.8%), 1,8-cineole (4.7–15.6%)	Cutillas et al. (2017)
	Leaves	Hydrodistillation	East of Tunisia	Camphor (25.14%), α-thujone (18.83%), 1,8-cineole (14.14%)	Khedher et al. (2017)
	Leaves	Hydrodistillation	Morocco	Camphor (24.14%), α-Thujone (21.45%), 1,8-Cineole (16.46%)	Ed-Dra et al. (2020)
	–		Firenze, Italy	α-Thujone (22.2%), camphor (16.2%), 1,8-cineole (11.9%)	Bedini et al. (2020)
	Aerial parts	Hydrodistillation	Romania	Caryophyllene (25.364%), camphene (14.139%), eucalyptol (13.902%)	Alexa et al. (2018)
	Fresh leaves	Hydrodistillation	Croatia	β-Thujone (23%), camphor (17%), borneol (8%)	Miljanović et al. (2020)

(continued)

Table 7.1 (continued)

Species	Parts	Extraction Method; Extraction Time; Yield	Place of Collection	Major Components	References
Thymus vulgaris	–	–	Emilia Romagna Apennines, Italy	Thymol (35.84–41.15%), p-cymene (17.50–21.73%), c-terpinene (15.06–18.42%)	Tardugno et al. (2020)
	Leaves	Steam distillation	Fars province (Meymand region), Iran	Thymol (57%), p-cymene (15%), γ- terpineol (12%)	Almasi et al. (2021)
	–	Steam distillation	São Paulo, Brazil	Thymol (43.19%), p-cymene 268 (28.55%), γ-terpinene (6.36%)	De Carvalho et al. (2015)
	Fresh leaves	Steam distillation	Parana, Brazil	CVL (45.5%), α-terpineol (22.9%), and endo-Borneol (14.3%)	Fachini-Queiroz et al. (2012)
	–	–	St. Louis, MO, EUA	Thymol (43.52%), p-cymene (31.65%), linalool (5.38%)	Lazarević et al. (2020)

– not informed

research carried out by Hsouna et al. (2013) found nine compounds, of which limonene (39.74%), β-pinene (25.44%) and α-terpineol (7.30%) were the major compounds. Shen et al. (2017) found as major compounds linalool (64.6%), α-terpineol (7.61%), and (R)-limonene (6.15%) in *C. aurantium* flowers. The chemical composition variation in essential oils above depended on the geographic location, and environmental conditions such as temperature, precipitation, altitude and hours of sunshine.

A plant species that is widely used in cooking is *Ocimum basilicum* L. Mandoulakani et al. (2017) investigated the cultivation of *O. basilicum* under water stress and whether the compounds methyl chavicol, methyleugenol, eugenol, bergamothene, b-myrcene and linalool would be affected. They observed that water stress significantly affected the content of all components except linalool, making it evident that seasonal variation, climate change, plant growth regulators and environmental stresses during cultivation contribute to the extraction of chemical compounds. Not all stress can negatively influence the chemical composition as observed by Gohari et al. (2020a) studying basil essential oil under salt stress, identified 46 components, revealing greater amounts of methyl chavicol, linalool and epi-α-cadinol than other metabolites.

Ocimum basilicum has an extremely variable composition due to the existence of several chemotypes (linalool, methyl chavicol, eugenol, methyl eugenol and neral), in which the presence of its constituents can determine the specific aroma, color of the plant and also a variety of flavors when consumed. Abou El-Soud et al. (2015), described that chemotype directly influences the biological activity of the plant product. In their research with *O. basilicum*, linalool, 1,8-cineol, eugenol, methyl cinnamate and α-cubeben were among the main components.

Several species of the genus *Origanum* represent a common spice for culinary uses and widely use in traditional and modern medicine, as well as in food and cosmetics. Many of them are referred to with the common name of "oregano" (Lombrea et al. 2020) and presents the same influence and variability such as described to *O. basilicum*.

Xie et al. (2019), when searching for *Origanum vulgare*, found as major compounds carvacrol (58.13%), β-myrcene (17.85%), and thymol (8.15%), while Avola et al. (2020) founded carvacrol (35.95%) followed by thymol (25.2%), p-cemene (21.54%) as major compounds, probably due to environmental factors, extraction methods or genetic differences.

The genus *Mentha* belongs to the Lamiaceae family and consists of eighteen species, including peppermint (*Mentha piperita* L.). *Mentha* spp. are mainly used as flavoring agents in foods and beverages and are commonly exploited for their rich composition in biological activities (De Sousa Guedes et al. 2016). It is one of the most homogenous plant genus in terms of chemical composition. The main compound it's the menthol, but even its quantity may vary according to different factors.

As exposed by all the studies mentioned above, variations in the amount of the main compounds and in the chemical composition of essential oils depends of a variety of variables such as the plant's genetics, harvest time, climatic and geographic conditions, light, seasons of the year and the extraction methods. And this variation should also interfere in both desirable and indesirable compounds quantities and presence/absence.

7.3 Essential Oils in Food

The ideal essential oil for food industry needs to combine a preservative, antibacterial, antifungal, and antioxidant effect, that together will result in less food deterioration. Another important parameter is that the oil should not induce severe organoleptic changes in the product, such as changes in taste and odor. The safe use of these oils must always be in tune with the beneficial effects on food, one cannot exist without the other (Fig. 7.2).

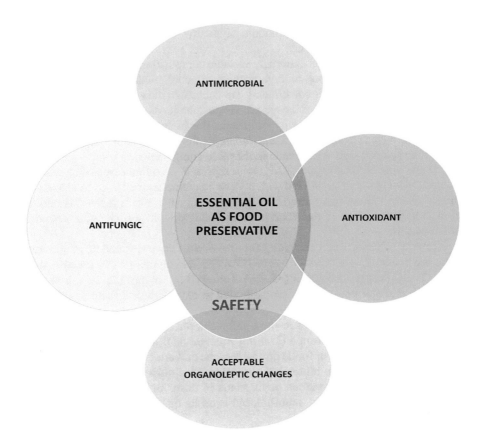

Fig. 7.2 Illustrative image demonstrating that the ideal essential oil for food preservation must be safe and combine an antimicrobial, antifungal and antioxidant effect, and only produce acceptable organoleptic changes

7.3.1   Antibacterial and Preservative Effect of Essential Oils in Food

Food contamination can occur at various stages of food production, including post-harvest processing, transport and storage, and may involve bacteria, fungi, fungal toxins, among others (Mutlu-Ingok et al. 2020).

Food-borne pathogens are major microorganisms that can interfere in food safety, being an important cause of human illness worldwide as a result of contaminated food. Bacteria such as *Campylobacter* spp., *Staphylococcus aureus*, *Escherichia coli*, *Salmonella* spp., and *Listeria monocytogenes* play a central role in these circumstances, causing mainly gastrointestinal symptoms including abdominal cramps, vomiting, nausea and diarrhea. A wide variety of foods are related to foodborne illness, especially those with animal origin, including eggs, meat, and dairy products, due to the role of animals as reservoirs for zoonotic pathogenic microorganisms (Abebe et al. 2020).

Natural products are a source of bioactive molecules and among them, essential oils have attracted attention due to their antibacterial activities (De Jesus et al. 2020). Oils such as thyme, *Thymus vulgaris*, tea tree, *Melaleuca alternifolia*, cinnamon, *Cinnamomum* spp., clove, *Syzygium aromaticum*, and lemon, *Citrus limon*, have showed antimicrobial activities with consequences in shelf lifes and food safety. Compounds such as aromatic volatile compounds and terpenes seems to be related to their antimicrobial effect (Bhavaniramya et al. 2019).

The use of essential oils as food preservatives has gained greater recognition as a way to replace numerous chemicals and additives, adopting a "healthier" form of conservation, as they have antioxidant and antimicrobial bioactive compounds that can increase the shelf life of the product. Despite the demonstrated potential of these oils, their use as preservatives in food has been limited by the requirement of high concentrations to fulfill sufficient antimicrobial activity (Hyldgaard et al. 2012).

Listeria monocytogenes is a threat to the food chain. In humans it causes listeriosis, a disease with 20–30% mortality range. Read-to-eat foods seems to be implicated with this agent because they don't go through heat treatment (Jordan and McAuliffe 2018). The anti-*Listeria* effect of several oils has been evaluated in many studies with food (Table 7.2).

Oregano oil is one of the most promising plant-derived products that can be used to develop antibacterial agents. Thyme and oregano oils have antilisterial activity of wich increased CO_2 levels and colder storage temperatures can potentialize this effect (Scollard et al. 2009). The correct use of heat serves a potent antimicrobial agent and essential oils can help to increase their efficiency, such as oregano associated with citric acid in *sous-vide* processed salmon (Dogruyol et al. 2020).

Like any other product obtained from plants, seasonal variations may interfere in chemical composition of essential oils. Factors such as composition, concentration, pH, storage temperature and type of food can affect essential oils antimicrobial activity. Treatment with mint, *Mentha piperita*, in tzatziki, taramosalata and *pâté* as food models proved this variation (Tassou et al. 1995).

Table 7.2　Essential oils with antibacterial effect in food

Plant	Part of the plant	Bacteria	Food	References
Bunium persicum	Leaf	*S. enteritidis,* and *L. monocytogenes*	Turkey meat	Keykhosravy et al. (2020)
Carum carvi	Fruit	*S. enteritidis*, *S. aureus*, and *Bacillus subtilis*	Baby carrots	Gniewosz et al. (2013)
Citrus medica	Leaf	*S. enteritidis*, *E. coli,* and *L. monocytogenes*	Ready-to-eat fruit salads	Belletti et al. (2008)
Citrus limon	Flower	*L. monocytogenes*	Minced beef	Ben Hsouna et al. (2017)
Citrus sinensis	Fruit	*Salmonella* and *Listeria*	Tomatoes	Das et al.,2020)
Coriandrum sativum	Seed	*Salmonella* spp., *E. coli* and *L. monocytogenes*	Cooked pork sausages	Šojić et al. (2019)
Cuminum cyminum	Seed	*B. cereus*	Barley soup	Pajohi et al. (2011)
Illicium verum	Seed	*Glutamicibacter* and *Aequorivita*	Grass carp fillets	Huang et al. (2018)
Litsea cubeba	Fruit	Enterohemorrhagic *E. coli* O157:H7	Vegetable juices	Dai et al. (2021)
Melaleuca alternifolia	Leaf	*L.monocytogenes*	Ground beef	de Sá Silva et al. (2019)
Mentha piperita	Leaf	*S. enteritidis*	Tzatziki, taramosalata and pâté	Tassou et al. (1995)
Metasequoia glyptostroboides	Leaf	*L.monocytogenes*	Milk	Bajpai et al. (2014)
Origanum vulgare	Leaf	*L.monocytogenes*	Minced beef	Hulankova et al. (2013)
		L.monocytogenes	Sous-vide processed salmon	Dogruyol et al. (2020)
		Salmonella enterica	Cherry	Kwon et al. (2017)
		Escherichia coli and *S. aureus*	Pate of Chicken	Moraes-Lovison et al. (2017)
		Aeromonas spp	Grass carp fillets	Huang et al. (2018)
		Lactic acid bacteria	Tuscan sausage	Badia et al. (2020)
		Lactic acid bacteria	Ham	Menezes et al. (2018)
		Salmonella Enteritidis, L. monocytogenes and *S. aureus*	Fermented meat sausage	Carvalho et al. (2019)
		Pseudomonas fluorescens	Mozzarella cheese	Rossi et al. (2018)

(continued)

Table 7.2 (continued)

Plant	Part of the plant	Bacteria	Food	References
Rosmarinus officinalis	Leaf	Lactic acid bacteria	Tuscan sausage	Badia et al. (2020)
		Yersinia enterocolitica, L. monocytogenes, E. coli Pseudomonas spp. and *S. enteritidis*	Ready-to-eat vegetables	Iseppi et al. (2018)
		Pseudomonas spp. count, Enterobacteriaceae count, Lactic acid bacteria, *S. aureus* count, *L. monocytogenes*, and *E. coli* O157:H7	Lamb meat	Sani et al. (2017)
Satureja montana	Leaf	*S. Typhimurium*	Mini-carrots	Ndoti-Nembe et al. (2015)
Sinapis alba	Seed	*Salmonella*	Ground chicken	Porter et al. (2020)
Thymus mongolicus Ronn	Leaf	*Glutamicibacter* and *Aequorivita*	Grass carp fillets	Huang et al. (2018)
Thymus vulgaris	Leaf	*Y. enterocolitica, L. monocytogenes, E. coli Pseudomonas spp.* and *S. enteritidis*	Ready-to-eat vegetables	Iseppi et al. (2018)
		L. monocytogenes	Tofu	Liu and Yang (2012)
		Vibrio spp.	Oysters	Liu and Yang (2012)
		L. plantarum	Orange-milk beverage	Liu and Yang (2012)
Thymus daenensis	Leaf	*E. coli* O157:H7	Cherry tomatoes	He et al. (2021)
Zataria Multiflora	Aerial parts	*S. enteritidis*, and *L. monocytogenes*	Turkey meat	Keykhosravy et al. (2020)
Zataria Multiflora	Aerial parts	*Pseudomonas spp.*, lactic acid bacteria, and psychrotrophic bacteria	Chicken breast meat	Bazargani-Gilani et al. (2015)

Some compounds from oils seem to play a bigger role in antimicrobial effect, such as phenols and aldehydes, while other like monoterpenes and ketones can have a downregulation effect (Bagheri et al. 2020). One of the richest sources of phenolic compounds, such as eugenol, eugenol acetate and gallic acid, is *S. aromaticum*, known as clove, and used as food preservative and spice for centuries (Cortés-Rojas et al. 2014). Their oil has a strong odor that can interfere in the use as food ingredient, and the encapsulation can disguise this, promoting an even stronger antimicrobial inhibition against *S. aureus*, *E. coli*, *L. monocytogenes*, and *Salmonella typhimurium* (Radünz et al. 2019).

There must be an adaptation between the oils dose used to inhibit bacteria in vitro, and the one necessary to achieve the same goal in food. While a concentration between 0.2 and 10µL/mL is sufficient in laboratory in vitro studies against *S. aureus*, *S. typhimurium*, *E. coli* O157:H7, *Bacillus cereus*, *L. monocytogenes*, and *Shigella dysenteria*, a higher concentration is needed to have a similar result in foods. Around 0.5–20µL/g seems to be the effective dosage for an antimicrobial effect of oils in fresh meat, milk, meat products, cooked rice, dairy products, and fish (Burt 2004). This higher dose increases the unpleasant smell from oils causing the main limitation for their use in fresh food (Iseppi et al. 2018).

The cell membrane seems to be a major target for their action in bacteria cells by interfere in membrane potential, transport of nutrients and ions, and permeability of the cell. They can also interfere through intracellular mechanisms, targeting molecules related to biosynthesis or energy generation. Both extracellular e intracellular mechanisms can exist depending on the type of oil (Hyldgaard et al. 2012) and some of them have higher effect against Gram-positive bacteria than Gram-negative bacteria (Diao et al. 2013) due to the structure of cell membrane that allows hydrophobic molecules to easily penetrate (Nazzaro et al. 2013).

Their interference in bacteria membrane can be explained by the natural hydrophobicity of oils, which facilitates their partition with lipids from the membrane (Devi et al. 2010) and attachment, becoming difficult to separate them from the bacterial membrane. This process increases permeability which affects the energy status of the cell, metabolic regulation and other vital processes to the bacteria, which leads to death (Nazzaro et al. 2013).

The genus *Citrus* includes different fruits such as lemons, grapefruits and orange and their oil are a growing interest in the food industry for preservative, antioxidant and flavorist effect (Mustafa 2015). Their action in bacteria can happen through increasing cell permeability, such as *Citrus reticulata* against *S. aureus* (Song et al. 2020), and inhibition of biofilm formation, such as *Citrus paradisi* in *Pseudomonas aeruginosa* (Luciardi et al. 2019). They also can reduce the tensile strength and elongation of food films (Do Evangelho et al. 2019).

Microorganisms are usually not free and can produce a matrix called biofilm, where a group of microorganisms of the same or a different species are attached to surfaces. Biofilms can be a barrier against antimicrobial agents and are widely distributed in the environment, including industrial surfaces, where can be a source of food contamination. Many essential oils have shown effect in biofilms, therefore can be used for the formulation of sanitizers for contaminated surfaces. When isolated from mozzarella cheese and treated with *O. vulgare* oil, *Pseudomonas fluorescens* biofilm formation was reduced by promoting the detachment of bacteria cells, and so could be used as an alternative for dairy food industry (Rossi et al. 2018).

Nanoemulsions can help essential oils to improve their volatility, water solubility, organoleptic characteristics (Prakash et al. 2018), stability, and potential against pathogens (Chouhan et al. 2017). In foods, nanoemulsion increase the dispersibility of essential oils in areas that pathogens can proliferate, minimizes the effects in product quality (Donsì and Ferrari 2016), and increase bioactivity due to increased bioavailability in the food matrix (Basak and Guha 2018). This technology works in several oils, such as clove encapsulated in chitosan nanoparticles (Hadidi et al. 2020), and *Pepper fragrant* functionalized nanoparticles (Jin et al. 2019).

Oregano oil encapsulated with nanoemulsion, using a temperature phase inversion method, had greater antibacterial effect against *E. coli*, and the incorporation of this nanoemulsion in the pate of Chicken did not change the physicochemical characteristics of the product, proving it is suitable for incorporation into food formulations to prevent and control microbial growth and extend its shelf life (Moraes-Lovison et al. 2017).

An edible coating for tomatoes made with nanoemulsion of sweet orange essential oil and sodium alginate effectively eradicate sessile and biofilms forms of *Salmonella* and *Listeria* (Das et al. 2020). The synergistic effects of the combination between ultrasound and thyme, *Thymus daenensis* oil nanoemulsion, decontaminated *E. coli* O157:H7 from the surface of cherry tomatoes without affecting its firmness and color. It also reduced *E. coli* in wastewater, providing an anti cross-contamination effect (He et al. 2021). The association of thyme oil nanoemulsion with ultrasound altered *E. coli* O157:H7 cell membrane permeability lead to a possible new form of food pasteurization (Guo et al. 2020). Its association with chitosan nanoparticles and nanocapsules is also effective against *S. aureus* and *B. cereus* (Sotelo-Boyás et al. 2017).

The direct addition of essentials oils to the food matrix has limitations associated with low water solubility, high volatility, low stability and strong odor (Fernández-López and Viuda-Martos 2018). An alternative form to combat that is to add them in active packaging instead as ingredients of the product itself. Oils can be encapsulated in edible and biodegradable polymer coatings or sachets that provide slow release to the food surface, and increase their stability (Prakash et al. 2018). A pullulan-based film containing rockrose, *Cistus ladanifer* oil, has antibacterial activity, which indicate their potential to develop films to pack foods, improving their shelf life (Luís et al. 2020).

7.3.2   Antifungic Effect of Essential Oils in Food

Food safety is at the frontal stage in food production, processing and distribution. The presence of aflatoxigenic fungi and mycotoxins in foods can have health implications and directly affect their safety (Ayofemi Olalekan Adeyeye 2019). As a way to prevent fungi and intoxications related to mycotoxins, natural products are being sought specially with a growing negative view between consumers about synthetic food additives (Redondo-Blancos et al. 2019).

Table 7.3 Essential oils with antifungal effect in food

Species	Plant part	Fungi	Food	References
Apium graveolens	Seed	*Aspergillus flavus AFLHPR14*	Rice seeds	Das et al. (2019)
Artemisia nilagirica	Shade-dried parts (leaf, rhizome, shoot)	*Aspergillus flavus, A. niger* and *A. ochraceus*	Grapes	Sonker et al. (2014)
Carum carvi	Fruit	Saccharomyces cerevisiae, or Aspergillus niger	Baby carrot	Gniewosz et al. (2013)
Cinnamomum zeylanicum	Leaf	*Aspergillus carbonarius*	Pears and apples	Kapetanakou et al. (2019)
Citrus sinensis	Peel	*Aspergillus niger, Mucor wutungkiao, Penicillium funiculosum*, and *Rhizopus oryzae*	Potato slices	Shi et al. (2018)
Lippia alba	Leaf	*Aspergillus flavus*	Gram seeds	Pandey et al. (2016)
Lippia sidoides	Leaf	*Rhizopus stolonifera*	Strawberry	Parisi et al. (2019)
Mentha cardiaca	Aerial parts	*A. favus*	Dry fruits	Dwivedy et al. (2017)
Origanum virens	Aerial parts	*Aspergillus flavus*	Maize	García-Díaz et al. (2019)
Origanum vulgare	Leaf	*Cladosporium* sp., *Fusarium* sp., and *Penicillium* sp	Minas Padrão cheese	Bedoya-Serna et al. (2018)
			Tomatoes	Rodriguez-Garcia et al. (2016)
Pimenta dioica	Fruits	*A. flavus*	Maize cob	Chaudhari et al. (2020)
Satureja montana	Aerial parts	*Aspergillus flavus*	Maize	García-Díaz et al. (2019)

Several oils are effective against fungi in vitro and in food model systems (Valdivieso-Ugarte et al. 2019) (Table 7.3). This antifungal action can happen through changes in fungal membrane with ergosterol content reduction and inhibition of aflatoxin inducers. This is the mechanism of action of *Apium graveolens* essential oil against *Aspergillus flavus* AFLHPR14, a very toxigenic strain isolated from contaminated rice seed, demonstrating the potential of *A. graveolens* essential oil as anti-contaminant of stored commodities (Das et al. 2019).

Mycotoxins contaminate approximately 25% of the world food supply per annum (Patil et al. 2010). Aflatoxins and fumonisins are foodborne mycotoxins and their presence in foods is associated with human aflatoxicosis, neural tube defects and many types of primary cancers (Sun et al. 2011). This toxin can be produced by *Fusarium verticillioides* and *F. proliferatum* in foods such as corn and corn-based products. Cinnamon oil is a promising target to mitigate their occurrence and is able to reduce fumonisin B1 (FB1), a highly toxic fumonisin to human and animal (Xing et al. 2014). Rats sub-chronically exposed to FB1 and or aflatoxin B1 (AFB1) orally treated with this oil demonstrated its protective effect against single or combined exposure (Abdel-Wahhab et al. 2018). When used to coat pears and apples, it also reduced the growth of *Aspergillus carbonarius* (Kapetanakou et al. 2019).

In maize, aflatoxin, a mycotoxin produced by *Aspergillus* spp., is a common problem in the food chain and the use of natural products are being sought to prevent this. *Satureja montana* and *Origanum virens* oils are good targets with antifungal properties lasting until 75 days after the first application (García-Díaz et al. 2019).

The action of *Mentha cardiaca* oil in dry fruits seems to be associated with an effect in fungal plasma membrane against *A. flavus* and antiaflatoxigenic activity against AFB1 (Dwivedy et al. 2017). In potato slices, spoilage fungi *Aspergillus niger*, *Mucor wutungkiao*, *Penicillium funiculosum*, and *Rhizopus oryzae,* were inhibit by *Citrus sinensis* (L.) Osbeck oil with d-limonene as the major component (Shi et al. 2018).

Oregano essential oil can have multiple antifungal effects, therefore is an important target for a natural compound that can be incorporated in food chain. *Cladosporium* sp., *Fusarium* sp., and *Penicillium* sp. isolated from Minas Padrão cheese were inhibit by nanoemulsions encapsulating *O. vulgare* oil (Bedoya-Serna et al. 2018). The association between this oil and edible coating protects against fungi contamination without altering aroma acceptability of tomatoes (Rodriguez-Garcia et al. 2016). Another species of oregano, *Lippia berlandieri* Schauer, known as Mexican oregano, can also be incorporated to edible films to control mold formed by *A. niger, Penicillium* spp. (Avila-Sosa et al. 2010), and *A. flavus* (Gómez-Sánchez et al. 2011).

Pimenta dioica is a tree that belongs to the Myrtaceae family and is used as culinary spice since immemorial time. It is commonly named pimento or allspice in reference to an aroma that seems like a mixture from many other spices (Rao et al. 2012). In maize cob slices, essential oil from fruits of *P. dioica* prevented *A. flavus* growth without interfering in seed germination (Chaudhari et al. 2020).

Aspergillus spp. and *Penicillium* spp. can spread in raw or processed food materials and produce mycotoxins that are implicated in several diseases (Basak and Guha 2018). To control fruit-rotting fungi in grapes, *Artemisia nilagirica* essential oil can be used as an alternative fungistatic and fungicidal for *Aspergillus* species (Sonker et al. 2014).

When considering the use of essential oils, not only its antifungal action should be considered, but also any sensory changes it may cause in foods. Although clove oil has a stronger antifungal effect than cinnamon in minced meat, the odor of cinnamon is more palatable (Saad et al. 2015).

7.3.3 Antioxidant Effect of Essential Oils in Food

Oxidation in food is related to production of free radicals and reactive oxygen precursors. It can induce several changes in foods such as alteration in nutrients, flavors, loss of color and toxic compounds production. The lipids from foods are very susceptible to oxidation, especially triglycerides and phospholipids, and this alters the quality of products and limits shelf life (Ahmed et al. 2016). Natural antioxidants can delay this process and are less toxic and safer in general than synthetics (Kaur et al. 2019).

Some beneficial effects of essential oils can be attributed to prooxidant effects on cells (Bakkali et al. 2008) with an effective dose range of 0.01–10 mg/mL (Valdivieso-Ugarte et al. 2019). They can inhibit lipid peroxidation in food, and have radical scavenging capacity, and this effect is related with chemical composition of terpenoids with phenolic groups such as eugenol, carvacrol, methyl chavicol, and thymol (Mimica-Dukic et al. 2016). Their action doesn't always depend on their main component because it can be a consequence of a synergism effect between more components (Dawidowicz and Olszowy 2014).

In research, the effect of these oils is generally compared to reference antioxidant compounds such as butylated hydroxyl toluene (BHT), 6-hydroxy-2,5,7,8-tetramethylchroman-2-carboxylic acid (Trolox), butylated hydroxyanisole (BHA), L-ascorbic acid (vitamin C) and 3,4,5-trihydroxybenzoic acid (gallic acid).The antioxidant activity of the oil from aerial parts of *Salvia lanigera* Poir., a plant belonging to Lamiaceae family, was higher than L-ascorbic acid, BHT and gallic acid (Tenore et al. 2011). Trans-geraniol, α-citral and β-citral are the major compounds of *Thymus bovei*, an oil with a potential antioxidant activity almost equal to Trolox standard (Jaradat et al. 2016).

Clove oil from *S. aromaticum* has strong antioxidant profile by powerfully inhibition of reactive oxygen species (ROS) production in human neutrophils stimulated by an inducer of endogenous superoxide production (Pérez-Rosés et al. 2016). It also demonstrated high antioxidant effect in 2,2-diphenyl-1-picrylhydrazyl (DPPH) assay, a standardized test to evaluate compounds antioxidant effects, and low hydroxyl radical inhibition (Radünz et al. 2019). This oil demonstrated some degree of antioxidant activity in an egg yolk-based thiobarbituric acid reactive substances (TBARS) assay (Dorman et al. 2000).

A strong antioxidant activity assures the efficacy of *Zanthoxylum alatum* Roxb. oil (Prakash et al. 2012), a plant with fruits traditionally used as a spice for several food preparations (Alam and Us Saqib 2017). The oil from *Clausena anisata* (Willd.) Hook. F. ex Benth, with main components E-ocimenone, Z-ocimenone, gamma-terpinene and germacrene D had a smaller antiradical effect than the reference compound BHT (Yaouba et al. 2011).

Allium cepa, onion, belongs to Liliaceae family and its a spice used worldwide. The waste of onion can be enormous and affect the environment, so establishing new products derived from this plant as natural oxidants can reduce this waste and create new ingredients (Roldán et al. 2008). As it was confirmed, *A. cepa* essential oil can serve as an antioxidant and antimicrobial agent in food systems (Ye et al. 2013).

The addition of essential oils directly to food or indirectly through edible packaging or coatings can replace the use of synthetic products harmful to the human body such as BHA and BHT and prolong shelf life (Amorati et al. 2013). *Zataria multiflora* oil for example can be combined with resveratrol to create a new material to pack food (Hashemi et al. 2019).

Despite studies confirming antioxidants activities, there are still areas in this field in need of further researches (Mutlu-Ingok et al. 2020) to fully understand the mechanism of action and future perspectives in essential oils use. The high antioxidant activity of bitter cumin, *Cuminum cyminum*, for example, seems to be correlated to the high phenolic content (Allahghadri et al. 2010), a common correlation found between essential oils (Valdivieso-Ugarte et al. 2019). Essential oils can also inhibit oxidation stronger than plant extracts, such as *Ageratum conyzoides* L that showed greater lipid peroxidation inhibition than methanol extract (Patil et al. 2010).

The method of extraction can also interfere with the antioxidant activity. The oil from *Piper nigrum L.*, with main compounds β-caryophyllene, limonene, sabinene, 3-carene, β-pinene and α-pinene, was a more effective antioxidant extract by a supercritical carbon dioxide technique than by hydro-distillation technique (Bagheri et al. 2014). The antioxidant activity can also be affected by the location where the plant was collected. This variation in chemical composition can have environmental influence too, such as propolis oil collected from different locations in China (Chi et al. 2019).

Already used as food preservative, rosemary oil is a promise in the search for new antioxidant agents (Rašković et al. 2014). Other oils have also shown promising results such as *Curcuma aromatica* Salisb (Al-Reza et al. 2010), *Apium graveolens* (Das et al. 2019), *Bunium persicum* (Nickavar et al. 2014), and *Mentha cardiaca* (Dwivedy et al. 2017).

7.4 Safety Assessment of Essential Oils in Food

The Expert Panel of the Flavor and Extract Manufacturers Association (FEMA) evaluates natural flavor compounds, which include essential oils since 2015. In that year, a guide considered chemical characterization and also composition variability as tools to evaluate the safety of essential oils used as flavor ingredient. These variables are affected by factors such as species, subspecies, part of the plant used, method of isolation, geographical location, and harvest time. The chemical compounds found can be organized into congeneric groups and then evaluated in accordance with data relating to toxicology, metabolism and absorption. If there is not already data available about a compound, researches about a similar chemotaxonomy is considered. With this guide is possible to

Table 7.4 Steps from Flavor and Extract Manufacturers Association (FEMA) to evaluate the safety of essential oils used as flavor ingredient

Steps	Description
1	Data review, analysis of consumption as a flavoring in relation to intake as a food additive
2–6	Application of the limit-to-toxicological concern (TTC) approach, analysis of data on toxicity and metabolism of congeneric groups
7–12	Toxicity assessment with inclusion of the genotoxic effects of unidentified compounds
13–14	Consider interactions between constituents and assess overall safety

Adapted from Cohen et al. (2020)

assure a no reasonably possible significant risk (Smith et al. 2005). A re-evaluation of the guide was performed in 2018 and it includes 14 steps (Table 7.4).

One of the weaknesses in the use of essential oils in food is the lack of information about their metabolism in the body and its toxicity in the short, medium or long term. Because FDA considers some oils to be GRAS, further research about their toxic effects on the body are discouraged, which makes their interaction with organs and tissues not fully known (Horky et al. 2019).

The assessment of oils toxicity is a basic premise for their safety use by humans and animals. In vivo experiments must be made to assure that these highly concentrated oils are no harmful to health (Hashemi et al. 2017). Laboratory animals such as mice, rat and rabbit are widely used in toxicology studies to determine the dose that is both safe and effective. Pregnant animals treated with compounds can also help to evaluate negative reproductive effects (Brent 2004), and in vitro study with living cells can determinate possible genotoxic effects by interference in mitochondrial dysfunction and intracellular redox (Bakkali et al. 2008).

Brazilian cherry tree, *Eugenia uniflora L.*, has been very explored by the food industry. Acute administration of this oil by oral route did not cause lethality or toxicological effects in mice (Victoria et al. 2012). A similar result was found with *Mentha cardiaca* (Dwivedy et al. 2017).

Lavandula angustifolia is a natural preservative and GRAS with no toxicity demonstrated in oral treatment in mice (Mekonnen et al. 2019). *Lavandula stoechas* subsp. *luisieri* essential oil, with oxygenated monoterpenes as main components, showed low toxicity after oral treatment in rats, therefore is also a good source for food supplement (Arantes et al. 2016).

Cuminum cyminum, cumin, is a very popular and traditional spice from Middle Ages (Singh et al. 2017). Their oil increased high-density lipoprotein (HDL) levels on mice, but it also changed blood parameters by increment of hemoglobin concentration, hematocrit, and platelet count (Allahghadri et al. 2010). Toxicity of cumin oil was explored in Wistar rats and at 1000 mg/kg/day dose no adverse effects or mortality was observed, as well as no alterations on clinical signs, histology in lungs, spleen, kidneys and liver, body weight, hematology and biochemistry after 23 days and 45 days treatment (Taghizadeh et al. 2017).

The use of coriander, *Cinnamomum glaucescens*, essential oil as a flavor ingredient added to food is considered safe, not irritative and without any adverse effects to humans (Burdock and Carabin 2009). The oil was also not toxic after oral treatment in mice and increased good cholesterol (HDL) levels, which can stimulate its use in cooking (Prakash et al. 2013).

Cinnamon bark, *Cinnamomum* spp., and spearmint, *Mentha spicata L.*, oils are common agents for flavoring chewing gums, toothpastes and mouthwashes, but there are more reported oral adverse reactions for cinnamon than for spearmint. Most cinnamon reactions are allergic, and mainly due to repeated exposure (Tisserand and Young 2014). In mice, after oral administration with *Cinnamomum zeylanicum* essential oil for 2 weeks, there was no significant toxicity (Mahmoudvand et al. 2017).

To be recommended for human consumption, essential oils should go through more severe toxicity testing. Some essential oils come from plants that are already widely used in food, therefore studies tend to neglect their possible adverse effects. An example of this is *B. persicum* oil, which doesn't stimulate mayor concern about the toxicity due to its large use as flavoring compound in people's diet (Hassanzadazar et al. 2018), and although their low-toxic in rats, histological changes in kidney, lung and liver were demonstrated in a sub-acute assay (Tabarraei et al. 2019).

Some essential oils can interfere with pregnancy or fetal development (Dosoky and Setzer 2021). Therefore, animal testing can help to find oils that are not suitable for pregnant women or that can be safely used by them. *Mentha x villosa* Hudson is used as a food spice, and although the treatment didn't cause any malformations in fetus of pregnant rats, there were mild hemorrhagic points at brain, kidney, liver and blood vessels which needs to be investigated before their safe use by pregnant women (Da Silva et al. 2012).

Citrus aurantium, bitter orange, belongs to the Rutaceae family and the oil has chemical composition mainly of limonene, linalool, and β-myrcene. Long-term studies are needed in order to affirm its safety, especially regarding the proper dosage (Suntar et al. 2018). An acute toxicity analysis in mice treated with this oil extracted from *C. aurantium* flowers indicated safety with no mortality or signs of toxicity up to 2000 mg/kg dose (Almalki 2021). In a reproductive assay in pregnant rats, this oil didn't induce alteration in maternal reproductive performance and was not teratogenic, showing no toxicological effect, no changes in ossification sites, and no malformation (Volpato et al. 2015).

Median lethal dose (LD50) is another parameter for the safety evaluation of a chemical compound. It represents the lethal dose of a compound per unit weight which kills 50% of a population submitted to the test. A high LD50 value indicates that the species has a high tolerance to that compound. The LD50 of rosemary oil, *R. officinalis* L., after oral administration is greater than 2000 mg/kg in murine model (Faria et al. 2011; Mengiste et al. 2018). This species belongs to the Lamiaceae family and is widely used as a spice (Singletary 2016). After a 72 h treatment at 50, 100, 200, 500, 1000 and 2000 mg/kg doses, this oil didn't induce any deaths or symptoms associated with toxicity in rat (Faria et al. 2011), a similar result to a 28 days treatment in mice that didn't alter biochemical parameters, body weights or induced macroscopic changes in liver and kidneys (Mengiste et al. 2018).

Genotoxicity corresponds to the destructive effect that a given agent can have on the cells genetic material. Oral administration of oil from *Curcuma longa* L with Ar-turmerone as major compound didn't show any mutagenicity or genotoxicity in rats (Liju et al. 2013). Although it was slightly toxic orally and moderately toxic intraperitoneally in mice (Oyemitan et al. 2017).

A potential substitute for synthetic food preservatives, lemongrass, *Cymbopogon citratus*, showed safe results in acute and subacute toxicity in mice and rabbits (Lulekal et al. 2019). A genotoxicity assay in mice revealed no significant changes in body weight and biochemistry or urinalysis exams (Costa et al. 2011), a good indicative for its use in food.

No genetic toxicity was attributed for *Litsea cubeba* oil although a slightly toxicity was observed in murine model (Luo et al. 2005). The subchronic oral administration of ginger, *Zingiber officinale* Roscoe, essential oil in rats during 13 weeks was not toxic (Jeena et al. 2011).

The essential oil extracted from the rhizomes of *Ligusticum chuanxiong* Hort., a plant used as a food ingredient in China, was declared safe by an acute toxicity research in mice for short term application (Zhang et al. 2012). *Lippia origanoides* essential oil in an acute and chronic toxicity assay in rats, revealed no sign of toxicity, with no change in biochemical and hematological parameters (Andrade et al. 2014).

A moderate toxic effect was observed in mice after treatment with oil from lemon balm, *Melissa officinalis* L., with main compounds citronellal, neral, and geranial. The animals exhibit changes in behavior, interference in liver and kidney functions with depletion in antioxidant capacities (Stojanović et al. 2019).

Wistar rats orally treated with oregano essential oil, *O. vulgare*, during 90 days didn't show any side effects in food or water consumption, body weight, hematology, biochemistry, necropsy, organ weight, histopathology or mortality (Llana-Ruiz-Cabello et al. 2017). It also didn't show any genotoxicity effect, fulfilling the requirements of the European Food Safety Authority (EFSA) for food packaging (Llana-Ruiz-Cabello et al. 2018). Several safe concentrations have already been established for other animals such as ruminants, horses, birds and fish according to the European Commission, Panel on Additives and Products or Substances Used in Animal Feed (FEEDAP) (Bampidis et al. 2019). Even though this oil has many positive effects, an embryotoxic effect was detected in pregnant mice (Domaracký et al. 2007).

Although lime, *Citrus aurantifolia* essential oil, is acknowledged as safe (GRAS), a mild toxicity in rats with elevated levels of lymphocytes and liver enzymes and low levels of hemoglobin was observed (Adokoh et al. 2019).

The investigation of side effects after ingestion of essential oil is crucial for the natural compounds be incorporated in food industry. In Australia, oils investigations are increasing and there are several cases of intoxication associated with ingestion of eucalyptus, tea tree, lavender, clove and peppermint oils. Children seem to be a population associated with greater risk in these cases (Lee et al. 2019). A case report shows a coma produced by clove oil, extracted from flowers of *Eugenia caryophyllata*, after consumption of 5–10 mL by a 2-year-old child (Hartnoll et al. 1993).

In albino rats, oil from eucalyptus, *Eucalyptus globules* L., and clove, *Eugenia caryophyllus*, exhibited mild effect on kidney function, decrease in hemoglobin concentration and platelets count and moderate pathological changes in the liver (Shalaby et al. 2011).

Tymol, geraniol, linalool, borneol, sabinene hydrate and carvacrol are among the main compounds of thyme, *T. vulgaris*, a very common natural agent used for food flavoring (Satyal et al. 2016). This oil is considered as GRAS but an acute assay reveled a moderate oral toxicity in rats after 28 days of treatment (Rojas-Armas et al. 2019), and some evidence also indicate that it may not be safe for food preservation (Eisenhut 2007). In murine model, LD50 has a high value of 4000 mg/kg with no side effects during the test (Grespan et al. 2014) and no detectable effects on embryo development (Domaracký et al. 2007).

While LD50 is the treatment with a single dose of a compound that kills 50% of the animals, LC50 is the lethal concentration in which 50% of the test animals are killed with a single exposure. The LC50 of thyme oil is 7142.85μL/kg body weight in mice, reaffirming its non-toxicity to mammals and its potential to preserve food and control spoilage (Kumar et al. 2008). Different places of plant extraction may influence *T. vulgaris* toxicity, as it happened with this oil from two regions of Northwestern Algeria. In an acute toxicity assay in mice, the oil from Tlemcen had toxic effect at a 4500 mg/kg dosis and the oil from Mostaganem had no toxicity even at 5000 mg/kg (Abdelli et al. 2017).

Although animal tests are considered the gold standard in toxicity and safety assessment, other experiments may replace them, such as in vitro tests with cell cultures (Lanzerstorfer et al. 2020). Other animal models can help to substitute murine, especially inferior species such as nematode *Caenorhabditis elegans*, a *Drosophila melanogaster* fly and fishes such as zebrafish, *Danio rerio* (Gosslau 2016). The brine shrimp lethality (BSLT) test is an in vivo, simple, practical, and inexpensive test for the preliminary assessment of compounds extracted from plants toxicity. It is based on the ability of the tested agent to induce death in *Artemia salina*, a microcrustacean (Subhan et al. 2008). In this test, *T. vulgaris* L. essential oil was considered as toxic (Niksic et al. 2021).

Some concerns about adverse effects on the liver led to a study in rats that revealed that at high doses, higher than 1 g/kg, oral and intraperitoneal treatment with essential oil from *C. aurantium L.* flowers lead to mild liver toxicity, causing a significant increase in alanine aminotransferase (ALT), aspartate aminotransferase (AST), and lactate dehydrogenase (LDH) (Hamedi 2020).

Widely used to enhance the flavor of foods, mustard can be used in various ways as seasonings and sauces. The black mustard, *B. nigra*, gives a spicier flavor

(Palle-Reisch et al. 2013). Only at higher doses of 5000 mg/kg body weight the oil from seeds of *B. nigra* was lethal in a on acute toxicity study in rats (Kumar et al. 2013).

The species *O. basilicum*, commonly known as basil, is also widely used in food as season. Wistar rat orally treated with this oil for 14 consecutive days suffered from damages on stomach and liver only at high doses, higher than 1500 mg/kg body weight (Fandohan et al. 2008). The LD50 of this oil is 532 mg/kg body weight (Venâncio et al. 2010).

7.5   Use of Essential Oils in Animal Dietary Supplementation

Essential oils have been used as herbal medicines for a long time, with antimicrobials, antifungal, antibacterial, and relaxing properties. Their use in animal diet aims a nutritional improvement, ensuring benefits to the production, animal performance, and often being an alternative to some conventional treatments. Thus, the search for oils in animal diet has been widely studied in veterinary medicine.

The oregano essential oil has been widely reported as a dietary supplement in animal diets (Abdel-Latif et al. 2020; Ding et al. 2020; Feng et al. 2021; Forte et al. 2017; Gordillo et al. 2021; Migliorini et al. 2019a, b; Mizuno et al. 2018; Mohiti-Asli and Ghanaatparast-Rashti 2015; Ruan et al. 2021; Zhang et al. 2021). A practical prevention strategy for the ectoparasitic flagellate *Ichthyobodo salmonis* and ciliate *Trichodina truttae* was reported with good results in *Oncorhynchus keta*, a salmon fish, using dietary supplementation with oregano oil, and suggested that its anti-parasitic and antimicrobial effect is possibly attributable to the carvacrol, a major component from this oil (Mizuno et al. 2018). The use of herbal oils in aquafeeds is an important approach to maintaining fish health status, and oregano can increase the antioxidant rest and immune responses of common carp, *Cyprinus carpio* (Abdel-Latif et al. 2020).

Zebrafish diet was supplemented with oil of mastic, *Pistacia lentiscus* Var. *chia*, which had an immunomodulatory action, leading to an increase in proinflammatory cytokine, with possible prebiotic effects (Serifi et al. 2019). Inclusion of essential oil constituted by eucalyptol, carvacrol and thymol on dietary to rainbow trout *Oncorhynchus mykiss* had good results by the increase of actin stability and preservation of muscle protein solubility and water holding capacity, increasing fish meat quality and shelf life during frozen storage, through a selective-antioxidant effect on muscle proteins (Santos et al. 2019). Similar results were found, but with an oil blend from aromatic plants eucalyptus, oregano, thyme and sweet orange diluted in fish oil and citric acid (Ceppa et al. 2017).

Savory essential oil, *Sartureja hortensis*, used in angelfish *Pterophyllum scalare* is a beneficial dietary supplement to improve growth performance, stress resistance, and innate immune (Ghafari Farsani et al. 2018). Menthol oil activates the immunity, antioxidative, and anti-inflammatory responses of Nile tilapia under toxicity by chlorpyrifos, an insecticide of the organophosphate class, known to be a water pollutant (Dawood et al. 2020). Clove essential oil has an antioxidant role by increasing antioxidant enzyme activities and antagonizing lipid peroxidation, and by an immune-stimulant effect in response to *Streptococcus iniae* infection of Nile Tilapia (Abdelkhalek et al. 2020).

The oil from *C. aurantium* can be used as a dietary supplement in the diet of common carp *Cyprinus carpio*, helping the growth and the immune response of these animals (Acar et al. 2021). In addition, the use of this oil in the diet of the silver catfish, *Rhamdia quelen,* showed a good growth of the fish, although altered liver biochemistry (Lopes et al. 2018).

In livestock, a diet supplemented with essential oils adds value to the final product and oregano oil can be used to assess performance, oxidative status, quality characteristics and sensory properties of pork. This diet can be effective in outdoor rearing, improving performance, increasing the oxidative stability of the meat without changing the meat quality (Forte et al. 2017).

Concern about antibiotic drug residues and resistant bacteria prompted researchers to look for more natural options in the search for essential oils as food additives to improve production performance (Dhama et al. 2015). Oregano oil improves antioxidant capacity and intestinal defense of ducks, causing growth and performance improvements and production of natural antibodies of broilers and chickens. Therefore, can help to keep enteric health without growth-promoting antibiotics (Ruan et al. 2021; Zhang et al. 2021).

A blend from essential oils of *R. officinalis, T. vulgaris, M. piperita* and *Anethum graveolent* associated with fish oil can be used as an additive of a dietary for the improvement of intestinal health, in addition to improving the immune response of laying hens (Mousavi et al. 2018). Thyme oil encapsulated in chitosan nanoparticles is also an alternative for antibiotic therapy for broiler chickens in a dietary supplementation, and resulted in improved animal performance (Hosseini and Meimandipour 2018).

The correct doses of supplementation with thyme oil in the broiler diet is important for a good performance, as the lowest concentrations were found in the liver and muscles and the largest in plasma and kidney of these animals, it is important to establish sufficient concentration of this oil for the diet (Ocel'ová et al. 2016). This oil can be used in broiler food supplementation to increase an antioxidant effect, but due to the low amount in muscles, not affected the composition of fatty acids and lipid oxidation (Placha et al. 2019). The blend of thyme, oregano, rosemary and star anise essential oils and saponin Quillaja to supplement broilers diet also helped growth performance, improved amino acids values and crude protein (Reyer et al. 2017).

In the dietary supplementation of goats with rosemary essential oil, it is possible to consider good results in milk production, without alteration in the plasmatic concentration of metabolites. In addition, the use of rosemary essential oil in a supplemented diet of dairy ewes made feeding more palatable, improving food intake and colostrum production (Smeti et al. 2014; Smeti et al. 2015). Rosemary and clove essential oils were studied in the diet of growing rabbits with satisfactory results in increasing hemoglobin values and potential antioxidant effect, but the immunological parameters did not change significantly (El-Gindy et al. 2021). The essential oil of rosemary can be used to supplement small ruminants, such as sheep, with modulation of the ruminal microbiota, which increased amount of *Fibrobacter succinogenes*, a rumen probiotic bacterium (Cobellis et al. 2016b).

Supplementation with oregano oil in diet, at the dose of 500 ppm, had beneficial effect on prevention of coccidiosis in broilers (Mohiti-Asli and Ghanaatparast-Rashti 2015), reduced oocysts in litter material and altered branched-chain fatty acids in cecal digesta of broiler chicken (Gordillo et al. 2021).

The use of coconut oil in broiler feed improved growth performance and villi histology, even with the presence of coccidial oocysts (Hafeez et al. 2020). The microencapsulated feed additive composed by garlic, carvacrol and thymol essential oils, have antiparasitic properties against ectoparasite *Sparicotyle chrysophrii* (Firmino et al. 2020). The same oil promoted skin immunity, decreasing its susceptibility to pathogenic bacteria, to gilthead seabream, *Sparus aurata*, by feed supplementation (Firmino et al. 2021).

To reduce lipid peroxidation in yolks and increase shelf life in eggs, researchers used oregano oil in the diet of laying hens. The reduction of lipid peroxidation in egg yolk is beneficial to consumer health by reducing levels of free radicals, also in laying hens in winter might be useful for maintaining egg quality and for prolonging shelf life (Migliorini et al. 2019a, b). Dietary using oregano oil enhanced digestive enzyme activity, improved intestinal morphology, favoring feed efficiency and eggshell quality of late-phase laying hens (Feng et al. 2021). Also, with a dietary supplementation of star anise oil, there is enhanced laying performance and overall antioxidant status of laying hens in a dose-dependent manner (Yu et al. 2018).

In murine model, the use of orange essential oil by intragastric administration increased good bacterial flora of the intestine, *Lactobacillus*, influencing the microflora of these animals (Wang et al. 2019). Thyme oil included on the diet of Japanese quail chicks caused growth inhibition response of gram-positive bacteria and *E. coli*, therefore, being a suitable alternative for antibiotic growth promoters (Dehghani et al. 2019).

The ingestion of *Eupatorium buniifolium* oil on nurse honeybees shows that this ingestion can impact the composition of cuticular hydrocarbons by a dose-dependent effect, and this could affect the signaling process mediated by pheromone compounds (Rossini et al. 2020). The combination of essential oils of cinnamon, oregano, clove and thyme has strong activities against furunculosis in salmonid fish (*Salmonidae*), which is caused by *Aeromonas salmonicida* subsp. *Salmonicida* strains (Hayatgheib et al. 2020).

A commercial essential oil blend containing cinnamaldehyde, thymol and feed grade carrier for supplementation in post-weaned pigs diet has natural antimicrobial properties that improved the growth process, with good fecal digestibility performance, and improved antioxidant activity (Tian 2019). A blend of essential oil composed by anise, cinnamon, garlic, rosemary and thyme, was used for supplementation in milk replacer for dairy heifers, and contributed to ruminal manipulation in pre-weaning and carry-over effects, immunity improvement, and decreased morbidity of neonatal diarrhea (Palhares Campolina et al. 2021).

A combination of oregano and thyme essential oil and prebiotic formulates a colostrum-based liquid for newborn calves, that when administrated after birth improved IgA titers, contributing to the immune status of the new born calf to fight off potential diseases and pathogens (Swedzinski et al. 2019).

7.6 Final Considerations

Natural products can be a source of compounds to reduce synthetic additives and encourage a healthier lifestyle that attracts consumers from all over the world. Essential oils can be extracted, mainly through hydrodistillation process, from various plants and represent an alternative for the food industry. These oils can act by several ways such as reducing or eliminating bacteria and fungi from food, or by reducing oxidative processes, which will increase food safety and shelf life. The safety is fundamental for the use of essential oils as food additives. Many oils considered GRAS by FDA have confirmed their safety in more accurate research, although there is a lack of studies for many of these oils and some tests with animal models showed a certain level of toxicity. Due to the extraction of many essential oils from plants that are already used in cooking, there is a false belief that they will always be safe. One of the problems with this logic is that it ignores the natural high concentration of these oils and also the higher concentrations needed to achieve an effective preservative effect in foods. The lack of studies in humans can mask side effects, since they are not being studied in more specific research. Few studies report poisoning by oral ingestion, although they are caused by accidental ingestion not related to food. The essential oils have huge potential as a natural alternative for increasing food shelf life, and improvement in the processes of clinical and toxicological studies regulation is necessary to assure essential oils efficacy and safety. All these factors need to be analyzed before including essential oils in foods, nevertheless, essential oils have a long history of use in food and still have a relevant place in the present and future of food science and technology.

Acknowledgements This study was financed in part by the Coordenação de Aperfeiçoamento de Pessoal de Nível Superior – Brasil (CAPES) – Finance Code 001. Dr. Fernando Almeida-Souza is research fellow of CAPES. Dr. Katia da Silva Calabrese and Dr. Ana Lúcia Abreu-Silva are research fellow of National Council for Scientific and Technological Development – CNPq.

References

Abdel-Latif HMR, Abdel-Tawwab M, Khafaga AF, Dawood MAO (2020) Dietary origanum essential oil improved antioxidative status, immune-related genes, and resistance of common carp (*Cyprinus carpio* L.) to *Aeromonas hydrophila* infection. Fish Shellfish Immunol 104:1–7. https://doi.org/10.1016/j.fsi.2020.05.056

Abdel-Wahhab MA, El-Nekeety AA, Hassan NS, Gibriel AAY, Abdel-Wahhab KG (2018) Encapsulation of cinnamon essential oil in whey protein enhances the protective effect against single or combined sub-chronic toxicity of fumonisin B1 and/or aflatoxin B1 in rats. Environ Sci Pollut Res 25:29144–29161. https://doi.org/10.1007/s11356-018-2921-2

Abdelli W, Bahri F, Romane A, Höferl M, Wanner J, Schmidt E, Jirovetz L (2017) Chemical composition and anti-inflammatory activity of Algerian *Thymus vulgaris* essential oil. Nat Prod Commun 12:611–614

Abdelkhalek NK, Risha E, El-Adl MA, Salama MF, Dawood MAO (2020) Antibacterial and antioxidant activity of clove oil against *Streptococcus iniae* infection in Nile tilapia (*Oreochromis niloticus*) and its effect on hepatic hepcidin expression. Fish Shellfish Immunol 104:478–488. https://doi.org/10.1016/j.fsi.2020.05.064

Abebe E, Gugsa G, Ahmed M (2020) Review on major food-borne zoonotic bacterial pathogens. J Trop Med 2020:1–19. https://doi.org/10.1155/2020/4674235

Abou El-Soud NH, Deabes M, Abou El-Kassem L, Khalil M (2015) Chemical composition and antifungal activity of *Ocimum basilicum* L. essential oil. Open Access Maced J Med Sci 3:374. https://doi.org/10.3889/oamjms.2015.082

Acar Ü, Kesbiç OS, Yılmaz S, İnanan BE, Zemheri-Navruz F, Terzi F, Fazio F, Parrino V (2021) Effects of essential oil derived from the bitter Orange (*Citrus aurantium*) on growth performance, histology and gene expression levels in common carp juveniles (*Cyprinus carpio*). Animals 11:1431. https://doi.org/10.3390/ani11051431

Adokoh CK, Asante DB, Acheampong DO, Kotsuchibashi Y, Armah FA, Sirikyi IH, Kimura K, Gmakame E, Abdul-Rauf S (2019) Chemical profile and in vivo toxicity evaluation of unripe *Citrus aurantifolia* essential oil. Toxicol Rep 6:692–702. https://doi.org/10.1016/j.toxrep.2019.06.020

Ahmed M, Pickova J, Ahmad T, Liaquat M, Farid A, Jahangir M (2016) Oxidation of lipids in foods. Sarhad J Agric 32:230–238. https://doi.org/10.17582/journal.sja/2016.32.3.230.238

Akhbari M, Masoum S, Aghababaei F, Hamedi S (2018) Optimization of microwave assisted extraction of essential oils from Iranian *Rosmarinus officinalis* L. using RSM. J Food Sci Technol 55:2197–2207. https://doi.org/10.1007/s13197-018-3137-7

Al-Reza SM, Rahman A, Sattar MA, Rahman MO, Fida HM (2010) Essential oil composition and antioxidant activities of *Curcuma aromatica* Salisb. Food Chem Toxicol 48:1757–1760. https://doi.org/10.1016/j.fct.2010.04.008

Alam F, Us Saqib QN (2017) Evaluation of Zanthoxylum armatum Roxb for in vitro biological activities. J Traditional Complementary Med 7:515–518. https://doi.org/10.1016/j.jtcme.2017.01.006

Alexa E, Sumalan RM, Danciu C, Obistioiu D, Negrea M, Poiana MA, Rus C, Radulov I, Pop G, Dehelean C (2018) Synergistic antifungal, allelopatic and anti-proliferative potential of *Salvia officinalis* L., and *Thymus vulgaris* L. essential oils. Molecules 23:185. https://doi.org/10.3390/molecules23010185

Allahghadri T, Rasooli I, Owlia P, Nadooshan MJ, Ghazanfari T, Taghizadeh M, Astaneh SDA (2010) Antimicrobial property, antioxidant capacity, and cytotoxicity of essential oil from cumin produced in Iran. J Food Sci 75:H54–H61. https://doi.org/10.1111/j.1750-3841.2009.01467.x

Almalki WH (2021) Citrus aurantium flowers essential oil protects liver against ischemia/reperfusion injury. S Afr J Bot 142:325–334. https://doi.org/10.1016/j.sajb.2021.06.041

Almasi L, Radi M, Amiri S, Torri L (2021) Fully dilutable *Thymus vulgaris* essential oil: acetic or propionic acid microemulsions are potent fruit disinfecting solutions. Food Chem 343:128411. https://doi.org/10.1016/j.foodchem.2020.128411

Amorati R, Foti MC, Valgimigli L (2013) Antioxidant activity of essential oils. J Agric Food Chem 61:10835–10847. https://doi.org/10.1021/jf403496k

Andrade VA, Almeida AC, Souza DS, Colen KGF, Macêdo AA, Martins ER, Fonseca FSA, Santos RL (2014) Antimicrobial activity and acute and chronic toxicity of the essential oil of *Lippia origanoides*. Pesqui Veterinária Bras 34:1153–1161. https://doi.org/10.1590/s0100-736x2014001200002

Arantes S, Candeias F, Lopes O, Lima M, Pereira M, Tinoco T, Cruz-Morais J, Martins M (2016) Pharmacological and toxicological studies of essential oil of *Lavandula stoechas* subsp. luisieri. Planta Med 82:1266–1273. https://doi.org/10.1055/s-0042-104418

Avila-Sosa R, Hernández-Zamoran E, López-Mendoza I, Palou E, Munguía MTJ, Nevárez-Moorillón GV, López-Malo A (2010) Fungal inactivation by Mexican oregano (*Lippia berlandieri* Schauer) essential oil added to Amaranth, chitosan, or starch edible films. J Food Sci 75:M127–M133. https://doi.org/10.1111/j.1750-3841.2010.01524.x

Avola R, Granata G, Geraci C, Napoli E, Graziano ACE, Cardile V (2020) Oregano (*Origanum vulgare* L.) essential oil provides anti-inflammatory activity and facilitates wound healing in a human keratinocytes cell model. Food Chem Toxicol 144:111586. https://doi.org/10.1016/j.fct.2020.111586

Ayofemi Olalekan Adeyeye S (2019) Aflatoxigenic fungi and mycotoxins in food: a review. Crit Rev Food Sci Nutr 60:709–721. https://doi.org/10.1080/10408398.2018.1548429

Azizkhani M, Kiasari FJ, Tooryan F, Shahavi MH, Partovi R (2021) Preparation and evaluation of food-grade nanoemulsion of tarragon (*Artemisia dracunculus* L.) essential oil: antioxidant and antibacterial properties. J Food Sci Technol 58:1341–1348. https://doi.org/10.1007/s13197-020-04645-6

Badia V, de Oliveira MSR, Polmann G, Milkievicz T, Galvão AC, da Silva RW (2020) Effect of the addition of antimicrobial oregano (*Origanum vulgare*) and rosemary (*Rosmarinus officinalis*) essential oils on lactic acid bacteria growth in refrigerated vacuum-packed Tuscan sausage. Braz J Microbiol 51:289–301. https://doi.org/10.1007/s42770-019-00146-7

Bagheri H, Abdul Manap MYB, Solati Z (2014) Antioxidant activity of *Piper nigrum* L. essential oil extracted by supercritical CO2 extraction and hydro-distillation. Talanta 121:220–228. https://doi.org/10.1016/j.talanta.2014.01.007

Bagheri L, Khodaei N, Salmieri S, Karboune S, Lacroix M (2020) Correlation between chemical composition and antimicrobial properties of essential oils against most common food pathogens and spoilers: in-vitro efficacy and predictive modelling. Microb Pathog 147:104212. https://doi.org/10.1016/j.micpath.2020.104212

Bajpai VK, Yoon JI, Bhardwaj M, Kang SC (2014) Anti-listerial synergism of leaf essential oil of *Metasequoia glyptostroboides* with nisin in whole, low and skim milks. Asian Pac J Trop Med 7:602–608. https://doi.org/10.1016/s1995-7645(14)60102-4

Bakkali F, Averbeck S, Averbeck D, Idaomar M (2008) Biological effects of essential oils – a review. Food Chem Toxicol 46:446–475. https://doi.org/10.1016/j.fct.2007.09.106

Bampidis V, Azimonti G, Bastos MD, Christensen H, Kouba M, Kos Durjava M, López-Alonso M, López Puente S, Marcon F, Mayo B (2019) Safety and efficacy of an essential oil from *Origanum vulgare* ssp. hirtum (Link) Ietsw. for all animal species. EFSA J 17:e5909. https://doi.org/10.2903/j.efsa.2019.5909

Basak S, Guha P (2018) A review on antifungal activity and mode of action of essential oils and their delivery as nano-sized oil droplets in food system. J Food Sci Technol 55:4701–4710. https://doi.org/10.1007/s13197-018-3394-5

Bazargani-Gilani B, Aliakbarlu J, Tajik H (2015) Effect of pomegranate juice dipping and chitosan coating enriched with *Zataria multiflora* Boiss essential oil on the shelf-life of chicken meat during refrigerated storage. Innovative Food Sci Emerg Technol 29:280–287. https://doi.org/10.1016/j.ifset.2015.04.007

Bedini S, Flamini G, Cosci F, Ascrizzi R, Echeverria MC, Guidi L, Landi M, Lucchi A, Conti B (2017) Artemisia spp. essential oils against the disease-carrying blowfly *Calliphora vomitoria*. Parasites Vectors 10:80. https://doi.org/10.1186/s13071-017-2006-y

Bedini S, Guarino S, Echeverria MC, Flamini G, Ascrizzi R, Loni A, Conti B (2020) *Allium sativum, Rosmarinus officinalis*, and *Salvia officinalis* essential oils: a spiced shield against blowflies. Insects 11:143. https://doi.org/10.3390/insects11030143

Bedoya-Serna CM, Dacanal GC, Fernandes AM, Pinho SC (2018) Antifungal activity of nanoemulsions encapsulating oregano (*Origanum vulgare*) essential oil: in vitro study and application in Minas Padrão cheese. Braz J Microbiol 49:929–935. https://doi.org/10.1016/j.bjm.2018.05.004

Belletti N, Lanciotti R, Patrignani F, Gardini F (2008) Antimicrobial efficacy of citron essential oil on spoilage and pathogenic microorganisms in fruit-based salads. J Food Sci 73:M331–M338. https://doi.org/10.1111/j.1750-3841.2008.00866.x

Ben Hsouna A, Ben Halima N, Smaoui S, Hamdi N (2017) Citrus lemon essential oil: chemical composition, antioxidant and antimicrobial activities with its preservative effect against *Listeria monocytogenes* inoculated in minced beef meat. Lipids Health Dis 16:146. https://doi.org/10.1186/s12944-017-0487-5

Bhagat M, Sangral M, Arya K, Rather RA (2018) Chemical characterization, biological assessment and molecular docking studies of essential oil of *Ocimum viride* for potential antimicrobial and anticancer activities. BioRxiv 2018:390906. https://doi.org/10.1101/390906

Bhavaniramya S, Vishnupriya S, Al-Aboody MS, Vijayakumar R, Baskaran D (2019) Role of essential oils in food safety: antimicrobial and antioxidant applications. Grain Oil Sci Technol 2:49–55. https://doi.org/10.1016/j.gaost.2019.03.001

Blowman K, Magalhães M, Lemos MFL, Cabral C, Pires IM (2018) Anticancer properties of essential oils and other natural products. Evid Based Complement Alternat Med 2018:1–12. https://doi.org/10.1155/2018/3149362

Bouajaj S, Benyamna A, Bouamama H, Romane A, Falconieri D, Piras A, Marongiu B (2013) Antibacterial, allelopathic and antioxidant activities of essential oil of *Salvia officinalis* L. growing wild in the Atlas Mountains of Morocco. Nat Prod Res 27:1673–1676. https://doi.org/10.1080/14786419.2012.751600

Bouyahya A, Et-Touys A, Bakri Y, Talbaui A, Fellah H, Abrini J, Dakka N (2017) Chemical composition of *Mentha pulegium* and *Rosmarinus officinalis* essential oils and their antileishmanial, antibacterial and antioxidant activities. Microb Pathog 111:41–49. https://doi.org/10.1016/j.micpath.2017.08.015

Brent RL (2004) Utilization of animal studies to determine the effects and human risks of environmental toxicants (drugs, chemicals, and physical agents). Pediatrics 113:984–995

Brnawi WI, Hettiarachchy NS, Horax R, Kumar-Phillips G, Seo H-S, Marcy J (2018) Comparison of cinnamon essential oils from leaf and bark with respect to antimicrobial activity and sensory acceptability in strawberry shake. J Food Sci 83:475–480. https://doi.org/10.1111/1750-3841.14041

Burdock GA, Carabin IG (2009) Safety assessment of coriander (*Coriandrum sativum* L.) essential oil as a food ingredient. Food Chem Toxicol 47:22–34. https://doi.org/10.1016/j.fct.2008.11.006

Burt S (2004) Essential oils: their antibacterial properties and potential applications in foods—a review. Int J Food Microbiol 94:223–253. https://doi.org/10.1016/j.ijfoodmicro.2004.03.022

Calo JR, Crandall PG, O'Bryan CA, Ricke SC (2015) Essential oils as antimicrobials in food systems – a review. Food Control 54:111–119. https://doi.org/10.1016/j.foodcont.2014.12.040

Carvalho MIP, Albano HCP, Teixeira PCM (2019) Influence of oregano essential oil on the inhibition of selected pathogens in "Alheira" during storage. Acta Scientiarum Polonorum Technologia Alimentaria 18:13–23. https://doi.org/10.17306/j.afs.0624

Ceppa F, Faccenda F, De Filippo C, Albanese D, Pindo M, Martelli R, Marconi P, Lunelli F, Fava F, Parisi G (2017) Influence of essential oils in diet and life-stage on gut microbiota and fillet quality of rainbow trout (*Oncorhynchus mykiss*). Int J Food Sci Nutr 69:318–333. https://doi. org/10.1080/09637486.2017.1370699

Chaudhari AK, Singh VK, Das S, Deepika PJ, Dwivedy AK, Dubey NK (2020) Improvement of in vitro and in situ antifungal, AFB1 inhibitory and antioxidant activity of *Origanum majorana* L. essential oil through nanoemulsion and recommending as novel food preservative. Food Chem Toxicol 143:111536. https://doi.org/10.1016/j.fct.2020.111536

Chi Y, Luo L, Cui M, Hao Y, Liu T, Huang X, Guo X (2019) Chemical composition and antioxidant activity of essential oil of Chinese Propolis. Chem Biodivers 17:e1900489. https://doi. org/10.1002/cbdv.201900489

Chouhan S, Sharma K, Guleria S (2017) Antimicrobial activity of some essential oils—present status and future perspectives. Medicines 4:58. https://doi.org/10.3390/medicines4030058

Cobellis G, Trabalza-Marinucci M, Yu Z (2016a) Critical evaluation of essential oils as rumen modifiers in ruminant nutrition: a review. Sci Total Environ 545–546:556–568. https://doi. org/10.1016/j.scitotenv.2015.12.103

Cobellis G, Yu Z, Forte C, Acuti G, Trabalza-Marinucci M (2016b) Dietary supplementation of *Rosmarinus officinalis* L. leaves in sheep affects the abundance of rumen methanogens and other microbial populations. J Anim Sci Biotechnol 7:27. https://doi.org/10.1186/s40104-016-0086-8

Cohen SM, Eisenbrand G, Fukushima S, Gooderham NJ, Guengerich FP, Hecht SS, Rietjens IMCM, Bastaki M, Davidsen JM, Harman CL, McGowen MM, Taylor SV (2020) FEMA GRAS assessment of natural flavor complexes: mint, buchu, dill and caraway derived flavoring ingredients. Food Chem Toxicol 135:110870. https://doi.org/10.1016/j.fct.2019.110870

Cortés-Rojas DF, de Souza CRF, Oliveira WP (2014) Clove (*Syzygium aromaticum*): a precious spice. Asian Pac J Trop Biomed 4:90–96. https://doi.org/10.1016/s2221-1691(14)60215-x

Costa CARA, Bidinotto LT, Takahira RK, Salvadori DMF, Barbisan LF, Costa M (2011) Cholesterol reduction and lack of genotoxic or toxic effects in mice after repeated 21-day oral intake of lemongrass (*Cymbopogon citratus*) essential oil. Food Chem Toxicol 49:2268–2272. https://doi.org/10.1016/j.fct.2011.06.025

Cutillas AB, Carrasco A, Martinez-Gutierrez R, Tomas V, Tudela J (2017) *Salvia officinalis* L. essential oils from Spain: determination of composition, antioxidant capacity, antienzymatic, and antimicrobial bioactivities. Chem Biodivers 14:e1700102. https://doi.org/10.1002/cbdv.201700102

da Silva Bomfim N, Kohiyama CY, Nakasugi LP, Nerilo SB, Mossini SAG, Romoli JCZ, Mikcha JMG, Abreu Filho BA, Machinski M Jr (2020) Antifungal and antiaflatoxigenic activity of rosemary essential oil (L.) against *Aspergillus flavus*. Food Addit Contam Part A Chem Anal Control Expo Risk Assess 37:153–161. https://doi.org/10.1080/19440049.2019.1678771

Da Silva BGKS, Silva RLC, Souza Maia MB, Schwarz A (2012) Embryo and fetal toxicity of *Mentha* x *villosa* essential oil in Wistar rats. Pharm Biol 50:871–877. https://doi.org/10.3109/13880209.2011.641024

Dai J, Li C, Cui H, Lin L (2021) Unraveling the anti-bacterial mechanism of *Litsea cubeba* essential oil against *E. coli* O157:H7 and its application in vegetable juices. Int J Food Microbiol 338:108989. https://doi.org/10.1016/j.ijfoodmicro.2020.108989

Das S, Singh VK, Dwivedy AK, Chaudhari AK, Upadhyay N, Singh A, Deepika DNK (2019) Antimicrobial activity, antiaflatoxigenic potential and in situ efficacy of novel formulation comprising *Apium graveolens* essential oil and its major component. Pestic Biochem Physiol 160:102–111. https://doi.org/10.1016/j.pestbp.2019.07.013

Das S, Vishakha K, Banerjee S, Mondal S, Ganguli A (2020) Sodium alginate-based edible coating containing nanoemulsion of *Citrus sinensis* essential oil eradicates planktonic and sessile cells of food-borne pathogens and increased quality attributes of tomatoes. Int J Biol Macromol 162:1770–1779. https://doi.org/10.1016/j.ijbiomac.2020.08.086

Dawidowicz A, Olszowy M (2014) Does antioxidant properties of the main component of essential oil reflect its antioxidant properties? The comparison of antioxidant properties of essen-

tial oils and their main components. Nat Prod Res 28:1–12. https://doi.org/10.1080/1478641
9.2014.918121

Dawood MAO, El-Salam Metwally A, Elkomy AH, Gewaily MS, Abdo SE, Abdel-Razek MAS,
Soliman AA, Amer AA, Abdel-Razik NI, Abdel-Latif HMR, Paray BA (2020) The impact of
menthol essential oil against inflammation, immunosuppression, and histopathological alterations induced by chlorpyrifos in Nile tilapia. Fish Shellfish Immunol 102:316–325. https://doi.
org/10.1016/j.fsi.2020.04.059

Carvalho RJ, Souza GT, Honório VG, Sousa JP, Conceição ML, Maganani M, Souza EL (2015)
Comparative inhibitory effects of *Thymus vulgaris* L. essential oil against *Staphylococcus
aureus*, *Listeria monocytogenes* and mesophilic starter co-culture in cheese-mimicking models. Food Microbiol 52:59–65. https://doi.org/10.1016/j.fm.2015.07.003

De Jesus GS, Micheletti AC, Padilha RG, De Souza de Paula J, Alves FM, CRB L, Garcez FR,
Garcez WS, Yoshida NC (2020) Antimicrobial potential of essential oils from cerrado plants
against multidrug−resistant foodborne microorganisms. Molecules 25:3296. https://doi.
org/10.3390/molecules25143296

Barbosa IM, Almeida ETC, Castellano LRC, Souza EL (2019) Influence of stressing conditions
caused by organic acids and salts on tolerance of *Listeria monocytogenes* to *Origanum vulgare*
L. and *Rosmarinus officinalis* L. essential oils and damage in bacterial physiological functions.
Food Microbiol 84:103240. https://doi.org/10.1016/j.fm.2019.103240

de Oliveira KAR, Berger LRR, Araújo SA, Câmara MPS, Souza EV (2017) Synergistic mixtures
of chitosan and *Mentha piperita* L. essential oil to inhibit *Colletotrichum* species and anthracnose development in mango cultivar Tommy Atkins. Food Microbiol 66:96–103. https://doi.
org/10.1016/j.fm.2017.04.012

de Sousa Guedes JP, Medeiros JAC, Souza E, Silva RS, Sousa JMB, Conceição ML, Souza EL
(2016) The efficacy of *Mentha arvensis* L. and *M. piperita* L. essential oils in reducing pathogenic bacteria and maintaining quality characteristics in cashew, guava, mango, and pineapple
juices. Int J Food Microbiol 238:183–192. https://doi.org/10.1016/j.ijfoodmicro.2016.09.005

Değirmenci H, Hatice E (2020) Relationship between volatile components, antimicrobial and antioxidant properties of the essential oil, hydrosol and extracts of *Citrus aurantium* L. flowers. J
Infect Public Health 13:58–67. https://doi.org/10.1016/j.jiph.2019.06.017

Dehghani N, Afsharmanesh M, Salarmoini M, Ebrahimnejad H (2019) In vitro and in vivo evaluation of thyme (*Thymus vulgaris*) essential oil as an alternative for antibiotic in quail diet1. J
Anim Sci 97:2901–2913. https://doi.org/10.1093/jas/skz179

Della Pepa T, Elshafie HS, Capasso R, De Feo V, Camele I, Nazzaro F, Scognamiglio MR, Caputo
L (2019) Antimicrobial and phytotoxic activity of *Origanum heracleoticum* and *O. majorana*
essential oils growing in Cilento (Southern Italy). Molecules 24:2576. https://doi.org/10.3390/
molecules24142576

Devi KP, Nisha SA, Sakthivel R, Pandian SK (2010) Eugenol (an essential oil of clove) acts
as an antibacterial agent against *Salmonella typhi* by disrupting the cellular membrane. J
Ethnopharmacol 130:107–115. https://doi.org/10.1016/j.jep.2010.04.025

Dhama K, Latheef SK, Mani S, Samad HA, Karthik K, Tiwari R, Khan RU, Alagawany M, Farag
MR, Alam GM, Laudadio V, Tufarelli V (2015) Multiple beneficial applications and modes of
action of herbs in poultry health and production-a review. Int J Pharmacol 11:152–176. https://
doi.org/10.3923/ijp.2015.152.176

Diao WR, Hu QP, Feng SS, Li WQ, Xu JG (2013) Chemical composition and antibacterial activity
of the essential oil from green Huajiao (*Zanthoxylum schinifolium*) against selected foodborne
pathogens. J Agric Food Chem 61:6044–6049. https://doi.org/10.1021/jf4007856

Ding X, Wu X, Zhang K, Bai S, Wang J, Peng H, Xuan Y, Su Z, Zeng Q (2020) Dietary supplement
of essential oil from oregano affects growth performance, nutrient utilization, intestinal morphology and antioxidant ability in Pekin ducks. J Anim Physiol Anim Nutr 104:1067–1074.
https://doi.org/10.1111/jpn.13311

Do Evangelho JA, Da Silva Dannenberg G, Biduski B, el Halal SLM, Kringel DH, Gularte MA,
Fiorentini AM, da Rosa ZE (2019) Antibacterial activity, optical, mechanical, and barrier

properties of corn starch films containing orange essential oil. Carbohydr Polym 222:114981. https://doi.org/10.1016/j.carbpol.2019.114981

Dogruyol H, Mol S, Cosansu S (2020) Increased thermal sensitivity of *Listeria monocytogenes* in sous-vide salmon by oregano essential oil and citric acid. Food Microbiol 90:103496. https://doi.org/10.1016/j.fm.2020.103496

Domaracký M, Rehák P, Juhás Š, Koppel J (2007) Effects of selected plant essential oils on the growth and development of mouse preimplantation embryos in vivo. Physiol Res 56(1):97–104. https://doi.org/10.33549/physiolres.930929

Donsì F, Ferrari G (2016) Essential oil nanoemulsions as antimicrobial agents in food. J Biotechnol 233:106–120. https://doi.org/10.1016/j.jbiotec.2016.07.005

Dorman HJD, Figueiredo AC, Barroso JG, Deans SG (2000) In vitro evaluation of antioxidant activity of essential oils and their components. Flavour Fragr J 15:12–16. https://doi.org/10.1002/(SICI)1099-1026(200001/02)15:1<12::AID-FFJ858>3.0.CO;2-V

Dosoky NS, Setzer WN (2021) Maternal reproductive toxicity of some essential oils and their constituents. Int J Mol Sci 22:2380. https://doi.org/10.3390/ijms22052380

Dwivedy AK, Prakash B, Chanotiya CS, Bisht D, Dubey NK (2017) Chemically characterized *Mentha cardiaca* L. essential oil as plant based preservative in view of efficacy against biodeteriorating fungi of dry fruits, aflatoxin secretion, lipid peroxidation and safety profile assessment. Food Chem Toxicol 106:175–184. https://doi.org/10.1016/j.fct.2017.05.043

Ed-Dra A, Filali RF, Presti VL, Zekkori B, Nalbone L, Bouymajane A, Trabelsi N, Lamberta F, Bentayeb A, Giuffrida A, Giarratana F (2020) Chemical composition, antioxidant capacity and antibacterial action of five Moroccan essential oils against *Listeria monocytogenes* and different serotypes of *Salmonella enterica*. Microb Pathog 149:104510. https://doi.org/10.1016/j.micpath.2020.104510

Eisenhut M (2007) The toxicity of essential oils. Int J Infect Dis 11:365. https://doi.org/10.1016/j.ijid.2006.07.004

El-Gindy YM, Zahran SM, Ahmed MA-R, Salem AZM, Misbah TR (2021) Influence of dietary supplementation of clove and rosemary essential oils or their combination on growth performance, immunity status, and blood antioxidant of growing rabbits. Trop Anim Health Prod 53:1–11. https://doi.org/10.1007/s11250-021-02906-w

Elyemni M, Louaste B, Nechad I, Elkamli T, Bouia A, Taleb M, Chaouch M, Eloutassi N (2019) Extraction of essential oils of *Rosmarinus officinalis* L. by two different methods: Hydrodistillation and microwave assisted hydrodistillation. Sci World J 2019:3659432. https://doi.org/10.1155/2019/3659432

Fachini-Queiroz FC, Kummer R, Estevão-Silva CF, Carvalho MDB, Cunha JM, Grespan R, Bersani-Amado CA, Cuman RKN (2012) Effects of thymol and carvacrol, constituents of *Thymus vulgaris* L. essential oil, on the inflammatory response. Evid Based Complement Alternat Med 2012:657026. https://doi.org/10.1155/2012/657026

Falowo AB, Mukumbo FE, Idamokoro EM, Afolayan AJ, Muchenje V (2019) Phytochemical constituents and antioxidant activity of sweet basil (*Ocimum basilicum* L.) essential oil on ground beef from boran and nguni cattle. Int J Food Sci 2019:2628747. https://doi.org/10.1155/2019/2628747

Fandohan P, Gnonlonfin B, Laleye A, Gbenou JD, Darboux R, Moudachirou M (2008) Toxicity and gastric tolerance of essential oils from *Cymbopogon citratus*, *Ocimum gratissimum* and *Ocimum basilicum* in Wistar rats. Food Chem Toxicol 46:2493–2497. https://doi.org/10.1016/j.fct.2008.04.006

Faria LRD, Lima CS, Perazzo FF, Carvalho JCT (2011) Anti-inflammatory and antinociceptive activities of the essential oil from Rosmarinus officinalis L.(Lamiaceae). Int J Pharm Sci Rev Res 7:1–8

Feng J, Lu M, Wang J, Zhang H, Qiu K, Qi G, Wu S (2021) Dietary oregano essential oil supplementation improves intestinal functions and alters gut microbiota in late-phase laying hens. J Anim Sci Biotechnol 12:72. https://doi.org/10.1186/s40104-021-00600-3

Fernández-López J, Viuda-Martos M (2018) Introduction to the special issue: application of essential oils in food systems. Foods 7:56. https://doi.org/10.3390/foods7040056

Firmino JP, Fernández-Alacid L, Vallejos-Vidal E, Salomón R, Sanahuja I, Tort L, Ibarz A, Reyes-López FE, Gisbert E (2021) Carvacrol, thymol, and garlic essential oil promote skin innate immunity in gilthead seabream (*Sparus aurata*) through the multifactorial modulation of the secretory pathway and enhancement of mucus protective capacity. Front Immunol 12:559. https://doi.org/10.3389/fimmu.2021.633621

Firmino JP, Vallejos-Vidal E, Sarasquete C, Ortiz-Delgado JB, Balasch JC, Tort L, Estevez A, Reyes-López FE, Gisbert E (2020) Unveiling the effect of dietary essential oils supplementation in *Sparus aurata* gills and its efficiency against the infestation by *Sparicotyle chrysophrii*. Sci Rep 10:17764. https://doi.org/10.1038/s41598-020-74625-5

Forte C, Ranucci D, Beghelli D, Branciari R, Acuti G, Todini L, Cavallucci C, Trabalza-Marinucci M (2017) Dietary integration with oregano (*Origanum vulgare* L.) essential oil improves growth rate and oxidative status in outdoor-reared, but not indoor-reared, pigs. J Anim Physiol Anim Nutr 101:e352–e361. https://doi.org/10.1111/jpn.12612

García-Díaz M, Patiño B, Vázquez C, Gil-Serna J (2019) A novel niosome-encapsulated essential oil formulation to prevent *Aspergillus flavus* growth and aflatoxin contamination of maize grains during storage. Toxins 11:646. https://doi.org/10.3390/toxins11110646

Ghafari Farsani H, Gerami MH, Farsani MN, Rashidiyan G, Mehdipour N, Ghanad M, Faggio C (2018) Effect of different levels of essential oils (*Satureja hortensis*) in diet on improvement growth, blood biochemical and immunity of Angelfish (*Pterophyllum scalare* Schultze, 1823). Nat Prod Res 2018:1–6. https://doi.org/10.1080/14786419.2018.1434635

Gniewosz M, Kraśniewska K, Woreta M, Kosakowska O (2013) Antimicrobial activity of a pullulan-caraway essential oil coating on reduction of food microorganisms and quality in fresh baby carrot. J Food Sci 78:M1242–M1248. https://doi.org/10.1111/1750-3841.12217

Gohari G, Alavi Z, Esfandiari E, Panahirad S, Hajihoseinlou S, Fotopoulos V (2020a) Interaction between hydrogen peroxide and sodium nitroprusside following chemical priming of *Ocimum basilicum* L. against salt stress. Physiol Plant 168:361–373. https://doi.org/10.1111/ppl.13020

Gohari G, Safai F, Panahirad S, Akbari A, Rasouli F, Dadpour MR, Fotopoulos V (2020b) Modified multiwall carbon nanotubes display either phytotoxic or growth promoting and stress protecting activity in *Ocimum basilicum* L. in a concentration-dependent manner. Chemosphere 249:126171. https://doi.org/10.1016/j.chemosphere.2020.126171

Gómez-Sánchez A, Palou E, López-Malo A (2011) Antifungal activity evaluation of Mexican oregano (*Lippia berlandieri* Schauer) essential oil on the growth of aspergillus flavus by gaseous contact. J Food Prot 74:2192–2198. https://doi.org/10.4315/0362-028x.jfp-11-308

Gordillo JFX, Kim D-H, Lee SH, Kwon S-K, Jha R, Lee K-W (2021) Role of oregano and citrus species-based essential oil preparation for the control of coccidiosis in broiler chickens. J Anim Sci Biotechnol 12:47. https://doi.org/10.1186/s40104-021-00569-z

Gosslau A (2016) Assessment of food toxicology. Food Sci Human Wellness 5:103–115. https://doi.org/10.1016/j.fshw.2016.05.003

Govindarajan M, Vaseeharan B, Alharbi NS, Kadaikunnan S, Khaled JM, Al-anbr MN, Alyahya SA, Maggi F, Benelli G (2018) High efficacy of (Z)-γ-bisabolene from the essential oil of *Galinsoga parviflora* (Asteraceae) as larvicide and oviposition deterrent against six mosquito vectors. Environ Sci Pollut Res 25:10555–10566. https://doi.org/10.1007/s11356-018-1203-3

Grespan R, Aguiar RP, Giubilei FN, Fuso RR, Damião MJ, Silva EL, Mikcha JG, Hernandes L, Bersani Amado C, Cuman RKN (2014) Hepatoprotective effect of pretreatment with *Thymus vulgaris* essential oil in experimental model of acetaminophen-induced injury. Evid Based Complement Alternat Med 2014:1–8. https://doi.org/10.1155/2014/954136

Grulova D, Martino L, Mancini E, Salamon I, De Feo V (2015) Seasonal variability of the main components in essential oil of *Mentha× piperita* L. J Sci Food Agric 95:621–627. https://doi.org/10.1002/jsfa.6802

Guerra ICD, de Oliveira PDL, Pontes ALS, Carneiro Lúcio ASS, Tavares JF, Barbosa-Filho JM, Madruga MS, de Souza EL (2015) Coatings comprising chitosan and *Mentha piperita* L. or *Mentha× villosa* Huds essential oils to prevent common postharvest mold infections and maintain the quality of cherry tomato fruit. Int J Food Microbiol 214:168–178. https://doi. org/10.1016/j.ijfoodmicro.2015.08.009

Guo M, Zhang L, He Q, Arabi SA, Zhao H, Chen W, Ye X, Liu D (2020) Synergistic antibacterial effects of ultrasound and thyme essential oils nanoemulsion against *Escherichia coli* O157:H7. Ultrason Sonochem 66:104988. https://doi.org/10.1016/j.ultsonch.2020.104988

Gutierrez J, Barry-Ryan C, Bourke P (2008) The antimicrobial efficacy of plant essential oil combinations and interactions with food ingredients. Int J Food Microbiol 124:91–97. https://doi. org/10.1016/j.ijfoodmicro.2008.02.028

Hadidi M, Pouramin S, Adinepour F, Haghani S, Jafari SM (2020) Chitosan nanoparticles loaded with clove essential oil: characterization, antioxidant and antibacterial activities. Carbohydr Polym 236:116075. https://doi.org/10.1016/j.carbpol.2020.116075

Hafeez A, Ullah Z, Khan RU, Ullah Q, Naz S (2020) Effect of diet supplemented with coconut essential oil on performance and villus histomorphology in broiler exposed to avian coccidiosis. Trop Anim Health Prod 52:2499–2504. https://doi.org/10.1007/s11250-020-02279-6

Hamedi A (2020) Investigation of the effect of essential oil from *Citrus aurantium* L flowers on liver health parameters in a laboratory animal model. KAUMS J (FEYZ) 24:38–47

Hartnoll G, Moore D, Douek D (1993) Near fatal ingestion of oil of cloves. Arch Dis Child 69:392–393. https://doi.org/10.1136/adc.69.3.392

Hashemi M, Hashemi M, Amiri E, Hassanzadazar H, Daneshamooz S, Aminzare M (2019) Evaluation of the synergistic antioxidant effect of resveratrol and *Zataria multiflora* Boiss essential oil in sodium alginate bioactive films. Curr Pharm Biotechnol 20:1064–1071. https:// doi.org/10.2174/1389201020666190719143910

Hashemi SM, Mousavi Khaneghah A, de Souza Sant'Ana A (eds) (2017) Essential oils in food processing: chemistry, safety and applications. Wiley. https://doi.org/10.1002/9781119149392

Hassanzadazar H, Taami B, Aminzare M, Daneshamooz S (2018) *Bunium persicum* (Boiss.) B. Fedtsch: an overview on phytochemistry, therapeutic uses and its application in the food industry. J Appl Pharm Sci 8:150–158. https://doi.org/10.7324/JAPS.2018.81019

Hayatgheib N, Fournel C, Calvez S, Pouliquen H, Moreau E (2020) In vitro antimicrobial effect of various commercial essential oils and their chemical constituents on *Aeromonas salmonicida* subsp. salmonicida. J Appl Microbiol 129:137–145. https://doi.org/10.1111/jam.14622

He Q, Guo M, Jin TZ, Arabi SA, Liu D (2021) Ultrasound improves the decontamination effect of thyme essential oil nanoemulsions against *Escherichia coli* O157: H7 on cherry tomatoes. Int J Food Microbiol 337:108936. https://doi.org/10.1016/j.ijfoodmicro.2020.108936

Horky P, Skalickova S, Smerkova K, Skladanka J (2019) Essential oils as a feed additives: pharmacokinetics and potential toxicity in Monogastric animals. Animals 9:352. https://doi. org/10.3390/ani9060352

Hosseini SA, Meimandipour A (2018) Feeding broilers with thyme essential oil loaded in chitosan nanoparticles: an efficient strategy for successful delivery. Br Poult Sci 59:669–678. https://doi. org/10.1080/00071668.2018.1521511

Hsouna AB, Hamdi N, Halima NB, Abdelkafi S (2013) Characterization of essential oil from *Citrus aurantium* L. flowers: antimicrobial and antioxidant activities. J Oleo Sci 62:763–772. https://doi.org/10.5650/jos.62.763

Huang Z, Liu X, Jia S, Zhang L, Luo Y (2018) The effect of essential oils on microbial composition and quality of grass carp (*Ctenopharyngodon idellus*) fillets during chilled storage. Int J Food Microbiol 266:52–59. https://doi.org/10.1016/j.ijfoodmicro.2017.11.003

Hulankova R, Borilova G, Steinhauserova I (2013) Combined antimicrobial effect of oregano essential oil and caprylic acid in minced beef. Meat Sci 95:190–194. https://doi.org/10.1016/j. meatsci.2013.05.003

Hyldgaard M, Mygind T, Meyer RL (2012) Essential oils in food preservation: mode of action, synergies, and interactions with food matrix components. Front Microbiol 3:12. https://doi.org/10.3389/fmicb.2012.00012

Iseppi R, Sabia C, de Niederhäusern S, Pellati F, Benvenuti S, Tardugno R, Bondi M, Messi P (2018) Antibacterial activity of *Rosmarinus officinalis* L. and *Thymus vulgaris* L. essential oils and their combination against food-borne pathogens and spoilage bacteria in ready-to-eat vegetables. Nat Prod Res 33:3568–3572. https://doi.org/10.1080/14786419.2018.1482894

Jan S, Rashid M, Abd Allah EF, Ahmad P (2020) Biological efficacy of essential oils and plant extracts of cultivated and wild ecotypes of *Origanum vulgare* L. Biomed Res Int 2020:8751718. https://doi.org/10.1155/2020/8751718

Jaradat N, Adwan L, K'aibni S, Shraim N, Zaid AN (2016) Chemical composition, anthelmintic, antibacterial and antioxidant effects of *Thymus bovei* essential oil. BMC Complement Altern Med 16:1–7. https://doi.org/10.1186/s12906-016-1408-2

Jeena K, Liju VB, Kuttan R (2011) A preliminary 13-week oral toxicity study of ginger oil in male and female wistar rats. Int J Toxicol 30:662–670. https://doi.org/10.1177/1091581811419023

Jin L, Teng J, Hu L, Lan X, Xu Y, Sheng J, Song Y, Wang M (2019) Pepper fragrant essential oil (PFEO) and functionalized MCM-41 nanoparticles: formation, characterization, and bactericidal activity. J Sci Food Agric 99:5168–5175. https://doi.org/10.1002/jsfa.9776

Jordan K, McAuliffe O (2018) *Listeria monocytogenes* in foods. Adv Food Nutr Res 86:181–213. https://doi.org/10.1016/bs.afnr.2018.02.006

Kapetanakou AE, Nestora S, Evageliou V, Skandamis PN (2019) Sodium alginate–cinnamon essential oil coated apples and pears: variability of *Aspergillus carbonarius* growth and ochratoxin A production. Food Res Int 119:876–885. https://doi.org/10.1016/j.foodres.2018.10.072

Kapustová M, Granata G, Napoli E, Puškárová A, Bučková M, Pangallo D, Geraci C (2021) Nanoencapsulated essential oils with enhanced antifungal activity for potential application on Agri-food, material and environmental fields. Antibiotics 10:31. https://doi.org/10.3390/antibiotics10010031

Kaur N, Chahal KK, Kumar A, Singh R, Bhardwaj U (2019) Antioxidant activity of *Anethum graveolens* L. essential oil constituents and their chemical analogues. J Food Biochem 43:e12782. https://doi.org/10.1111/jfbc.12782

Keykhosravy K, Khanzadi S, Hashemi M, Azizzadeh M (2020) Chitosan-loaded nanoemulsion containing *Zataria Multiflora* Boiss and *Bunium persicum Boiss* essential oils as edible coatings: its impact on microbial quality of Turkey meat and fate of inoculated pathogens. Int J Biol Macromol 150:904–913. https://doi.org/10.1016/j.ijbiomac.2020.02.092

Khedher MRB, Khedher SB, Chaieb I, Tounsi S, Hammami M (2017) Chemical composition and biological activities of *Salvia officinalis* essential oil from Tunisia. EXCLI J 16:160. https://doi.org/10.17179/excli2016-832

Kiferle C, Ascrizzi R, Martinelli M, Gonzali S, Mariotti L, Pistelli L, Flamini G, Perata P (2019) Effect of iodine treatments on Ocimum basilicum L.: biofortification, phenolics production and essential oil composition. PLoS One 15:e0229016. https://doi.org/10.1371/journal.pone.0226559

Krupčík J, Gorovenko R, Špánik I, Sandra P, Armstrong DW (2015) Enantioselective comprehensive two-dimensional gas chromatography. A route to elucidate the authenticity and origin of Rosa damascena miller essential oils. J Sep Sci 38:3397–3403. https://doi.org/10.1002/jssc.201500744

Kubeczka KH (2015) History and sources of essential oil research. Essential 2:5–42. https://doi.org/10.1201/b19393-3

Kumar A, Shukla R, Singh P, Prasad CS, Dubey NK (2008) Assessment of *Thymus vulgaris* L. essential oil as a safe botanical preservative against post harvest fungal infestation of food commodities. Innovative Food Sci Emerg Technol 9:575–580. https://doi.org/10.1016/j.ifset.2007.12.005

Kumar M, Sharma S, Vasudeva N (2013) In vivo assessment of antihyperglycemic and antioxidant activity from oil of seeds of brassica nigra in streptozotocin induced diabetic rats. Adv Pharm Bull 3:359–365. https://doi.org/10.5681/apb.2013.058

Kuorwel KK, Cran MJ, Sonneveld K, Miltz J, Bigger SW (2011) Essential oils and their principal constituents as antimicrobial agents for synthetic packaging films. J Food Sci 76:R164–R177. https://doi.org/10.1111/j.1750-3841.2011.02384.x

Kwon SJ, Chang Y, Han J (2017) Oregano essential oil-based natural antimicrobial packaging film to inactivate *Salmonella enterica* and yeasts/molds in the atmosphere surrounding cherry tomatoes. Food Microbiol 65:114–121. https://doi.org/10.1016/j.fm.2017.02.004

Labadie C, Ginies C, Guinebretiere MH, Renard CMGC, Cerutti C, Carlin F (2015) Hydrosols of orange blossom (*Citrus aurantium*), and rose flower (*Rosa damascena* and *Rosa centifolia*) support the growth of a heterogeneous spoilage microbiota. Food Res Int 76:576–586. https://doi.org/10.1016/j.foodres.2015.07.014

Labadie C, Cerutti C, Carlin F (2016) Fate and control of pathogenic and spoilage micro-organisms in orange blossom (*Citrus aurantium*) and rose flower (*Rosa centifolia*) hydrosols. J Appl Microbiol 121:1568–1579. https://doi.org/10.1111/jam.13293

Lanzerstorfer P, Sandner G, Pitsch J, Mascher B, Aumiller T, Weghuber J (2020) Acute, reproductive, and developmental toxicity of essential oils assessed with alternative in vitro and in vivo systems. Arch Toxicol 95:673–691. https://doi.org/10.1007/s00204-020-02945-6

Laranjo M, Fernández-León AM, Agulheiro-Santos AC, Potes ME, Elias M (2019) Essential oils of aromatic and medicinal plants play a role in food safety. J Food Process Preserv 2019:e14278. https://doi.org/10.1111/jfpp.14278

Lazarević J, Jevremović S, Kostić I, Kostić M, Vuleta A, Jovanović SM, Jovanović DS (2020) Toxic, oviposition deterrent and oxidative stress effects of *Thymus vulgaris* essential oil against *Acanthoscelides obtectus*. Insects 11:563. https://doi.org/10.3390/insects11090563

Lee KA, Harnett JE, Cairns R (2019) Essential oil exposures in Australia: analysis of cases reported to the NSW Poisons Information Centre. Med J Aust 212:132–133. https://doi.org/10.5694/mja2.50403

Liju VB, Jeena K, Kuttan R (2013) Acute and subchronic toxicity as well as mutagenic evaluation of essential oil from turmeric (*Curcuma longa* L). Food Chem Toxicol 53:52–61. https://doi.org/10.1016/j.fct.2012.11.027

Liu MH, Zhang Q, Zhang YH, Lu XY, Fu WM, He JY (2016) Chemical analysis of dietary constituents in *Rosa roxburghii* and *Rosa sterilis* fruits. Molecules 21:1204. https://doi.org/10.3390/molecules21091204

Liu TT, Yang T-S (2012) Antimicrobial impact of the components of essential oil of *Litsea cubeba* from Taiwan and antimicrobial activity of the oil in food systems. Int J Food Microbiol 156:68–75. https://doi.org/10.1016/j.ijfoodmicro.2012.03.005

Llana-Ruiz-Cabello M, Maisanaba S, Puerto M, Pichardo S, Jos A, Moyano R, Cameán AM (2017) A subchronic 90-day oral toxicity study of *Origanum vulgare* essential oil in rats. Food Chem Toxicol 101:36–47. https://doi.org/10.1016/j.fct.2017.01.001

Llana-Ruiz-Cabello M, Puerto M, Maisanaba S, Guzmán-Guillén R, Pichardo S, Cameán AM (2018) Use of micronucleus and comet assay to evaluate evaluate the genotoxicity of oregano essential oil (*Origanum vulgare* l. Virens) in rats orally exposed for 90 days. J Toxic Environ Health A 81:525–533. https://doi.org/10.1080/15287394.2018.1447522

Lombrea A, Antal D, Ardelean F, Avram S, Pavel IZ, Vlaia L, Mut AM, Diaconeasa Z, Dehelean CA, Soica C, Danciu C (2020) A recent insight regarding the Phytochemistry and bioactivity of *Origanum vulgare* L. essential oil. Int J Mol Sci 21:9653. https://doi.org/10.3390/ijms21249653

Lopes JM, de Freitas SC, Saccol EMH, Pavanato MA, Antoniazzi A, Rovani MT, Heinzmann BM, Baldisserotto B (2018) *Citrus x aurantium* essential oil as feed additive improved growth performance, survival, metabolic, and oxidative parameters of silver catfish (*Rhamdia quelen*). Aquac Nutr 25:310–318. https://doi.org/10.1111/anu.12854

López V, Cascella M, Benelli G, Maggi F, Gómez-Rincón C (2018) Green drugs in the fight against Anisakis simplex-larvicidal activity and acetylcholinesterase inhibition of Origanum compactum essential oil. Parasitol Res 117:861–867. https://doi.org/10.1007/s00436-018-5764-3

Luciardi MC, Blázquez MA, Alberto MR, Cartagena E, Arena ME (2019) Grapefruit essential oils inhibit quorum sensing of *Pseudomonas aeruginosa*. Food Sci Technol Int 26:231–241. https://doi.org/10.1177/1082013219883465

Luís Â, Ramos A, Domingues F (2020) Pullulan films containing rockrose essential oil for potential food packaging applications. Antibiotics (Basel) 9:681. https://doi.org/10.3390/antibiotics9100681

Lulekal E, Tesfaye S, Gebrechristos S, Dires K, Zenebe T, Zegeye N, Feleke G, Kassahun A, Shiferaw Y, Mekonnen A (2019) Phytochemical analysis and evaluation of skin irritation, acute and sub-acute toxicity of *Cymbopogon citratus* essential oil in mice and rabbits. Toxicol Rep 6:1289–1294. https://doi.org/10.1016/j.toxrep.2019.11.002

Luo M, Jiang L-K, Zou G-L (2005) Acute and genetic toxicity of essential oil extracted from *Litsea cubeba* (Lour.) Pers. J Food Prot 68:581–588. https://doi.org/10.4315/0362-028x-68.3.581

Mahboubifar M, Hemmateenejad B, Jassbi AR (2021) Evaluation of adulteration in distillate samples of *Rosa damascena* Mill using colorimetric sensor arrays, chemometric tools and dispersive liquid–liquid microextraction-GC-MS. Phytochem Anal 32:1027–1038. https://doi.org/10.1002/pca.3044

Mahmoudvand H, Mahmoudvand H, Oliaee R, Kareshk A, Mirbadie S, Aflatoonian M (2017) In vitro protoscolicidal effects of *Cinnamomum zeylanicum* essential oil and its toxicity in mice. Pharmacogn Mag 13:652. https://doi.org/10.4103/pm.pm_280_16

Mandoulakani BA, Eyvazpour E, Ghadimzadeh M (2017) The effect of drought stress on the expression of key genes involved in the biosynthesis of phenylpropanoids and essential oil components in basil (*Ocimum basilicum* L.). Phytochemistry 139:1–7. https://doi.org/10.1016/j.phytochem.2017.03.006

Meepagala KM, Sturtz G, Wedge DE (2002) Antifungal constituents of the essential oil fraction of *Artemisia dracunculus* L. var. dracunculus. J Agric Food Chem 50:6989–6992. https://doi.org/10.1021/jf020466w

Mekonnen A, Tesfaye S, Christos SG, Dires K, Zenebe T, Zegeye N, Shiferaw Y, Lulekal E (2019) Evaluation of skin irritation and acute and subacute oral toxicity of *Lavandula angustifolia* essential oils in rabbit and mice. J Toxicol 2019:1–8. https://doi.org/10.1155/2019/5979546

Menezes NMC, Martins WF, Longhi DA, de Aragão GMF (2018) Modeling the effect of oregano essential oil on shelf-life extension of vacuum-packed cooked sliced ham. Meat Sci 139:113–119. https://doi.org/10.1016/j.meatsci.2018.01.017

Mengiste B, Dires K, Lulekal E, Arayaselassie M, Zenebe T (2018) Acute skin irritation, acute and sub-acute oral toxicity studies of *Rosmarinus officinalis* essential oils in mice and rabbit. Afr J Pharm Pharmacol 12:389–396. https://doi.org/10.5897/AJPP2018.4957

Migliorini MJ, Boiago MM, Roza LF, Barreta M, Arno A, Robazza WS, Galvão AC, Galli GM, Machado G, Baldissera MD, Wagner R, Stefani LCM, Silva ASD (2019a) Oregano essential oil (*Origanum vulgare*) to feed laying hens and its effects on animal health. An Acad Bras Cienc 91(1):e20170901. https://doi.org/10.1590/0001-3765201920170901

Migliorini MJ, Boiago MM, Stefani LM, Zampar A, Roza LF, Barreta M, Arno A, Robazza WS, Giuriatti J, Galvão AC, Boscatto C, Paiano D, Da Silva AS (2019b) Oregano essential oil in the diet of laying hens in winter reduces lipid peroxidation in yolks and increases shelf life in eggs. J Therm Biol 85:102409. https://doi.org/10.1016/j.jtherbio.2019.102409

Miljanović A, Bielen A, Grbin D, Marijanović Z, Andlar M, Rezić T, Roca S, Jerković I, Vikić-Topić D, Dent M (2020) Effect of enzymatic, ultrasound, and reflux extraction pretreatments on the yield and chemical composition of essential oils. Molecules 25:4818. https://doi.org/10.3390/molecules25204818

Mimica-Dukic N, Orčić D, Lesjak M, Šibul F (2016) Essential oils as powerful antioxidants: misconception or scientific fact? In: Medicinal and aromatic crops: production, phytochemistry, and utilization. American Chemical Society, pp 187–208. https://doi.org/10.1021/bk-2016-1218.ch012

Mitropoulou G, Fitsiou E, Stavropoulou E, Papavassilopoulou E, Vamvakias M, Pappa A, Oreopoulou A, Kourkoutas Y (2015) Composition, antimicrobial, antioxidant, and antiproliferative activity of *Origanum dictamnus* (dittany) essential oil. Microb Ecol Health Dis 26:26543. https://doi.org/10.3402/mehd.v26.26543

Mizuno S, Urawa S, Miyamoto M, Hatakeyama M, Sasaki Y, Koide N, Tada S, Ueda H (2018) Effects of dietary supplementation with oregano essential oil on prevention of the ectoparasitic protozoans *Ichthyobodo salmonis* and *Trichodina truttae* in juvenile chum salmon *Oncorhynchus keta*. J Fish Biol 93:528–539. https://doi.org/10.1111/jfb.13681

Mohiti-Asli M, Ghanaatparast-Rashti M (2015) Dietary oregano essential oil alleviates experimentally induced coccidiosis in broilers. Prev Vet Med 120:195–202. https://doi.org/10.1016/j.prevetmed.2015.03.014

Moraes-Lovison M, Marostegan LFP, Peres MS, Menezes IF, Ghiraldi M, Rodrigues RAF, Fernandes AM, Pinho SC (2017) Nanoemulsions encapsulating oregano essential oil: production, stability, antibacterial activity and incorporation in chicken pâté. LWT 77:233–240. https://doi.org/10.1016/j.lwt.2016.11.061

Moubarac J-C, Parra DC, Cannon G, Monteiro CA (2014) Food classification systems based on food processing: significance and implications for policies and actions: a systematic literature review and assessment. Curr Obes Rep 3:256–272. https://doi.org/10.1007/s13679-014-0092-0

Mousavi A, Mahdavi AH, Riasi A, Soltani-Ghombavani M (2018) Efficacy of essential oils combination on performance, ileal bacterial counts, intestinal histology and immunocompetence of laying hens fed alternative lipid sources. J Anim Physiol Anim Nutr 102:1245–1256. https://doi.org/10.1111/jpn.12942

Mustafa NEM (2015) Citrus essential oils: current and prospective uses in the food industry. Recent Pat Food Nutr Agric 7:115–127. https://doi.org/10.2174/2212798407666150831144239

Mutlu-Ingok A, Devecioglu D, Dikmetas DN, Karbancioglu-Guler F, Capanoglu E (2020) Antibacterial, antifungal, antimycotoxigenic, and antioxidant activities of essential oils: an updated review. Molecules 25:4711. https://doi.org/10.3390/molecules25204711

Navaei Shoorvarzi S, Shahraki F, Shafaei N, Karimi E, Oskoueian E (2020) *Citrus aurantium* L. bloom essential oil nanoemulsion: synthesis, characterization, cytotoxicity, and its potential health impacts on mice. J Food Biochem 44:e13181. https://doi.org/10.1111/jfbc.13181

Nazzaro F, Fratianni F, De Martino L, Coppola R, De Feo V (2013) Effect of essential oils on pathogenic bacteria. Pharmaceuticals 6:1451–1474. https://doi.org/10.3390/ph6121451

Ndoti-Nembe A, Vu KD, Han J, Doucet N, Lacroix M (2015) Antimicrobial effects of Nisin, essential oil, and γ-irradiation treatments against high load of *Salmonella typhimurium* on mini-carrots. J Food Sci 80:M1544–M1548. https://doi.org/10.1111/1750-3841.12918

Negi PS (2012) Plant extracts for the control of bacterial growth: efficacy, stability and safety issues for food application. Int J Food Microbiol 156:7–17. https://doi.org/10.1016/j.ijfoodmicro.2012.03.006

Nickavar B, Adeli A, Nickavar A (2014) Analyses of the essential oil from Bunium persicum fruit and its antioxidant constituents. J Oleo Sci 63:741–6. https://doi.org/10.5650/jos.ess13168

Niksic H, Becic F, Koric E, Gusic I, Omeragic E, Muratovic S, Miladinovic B, Duric K (2021) Cytotoxicity screening of *Thymus vulgaris* L. essential oil in brine shrimp nauplii and cancer cell lines. Sci Rep 11:13178. https://doi.org/10.1038/s41598-021-92679-x

Ocel'ová V, Chizzola R, Pisarčíková J, Novak J, Ivanišinoviá O, Faix Š (2016) Effect of thyme essential oil supplementation on thymol content in blood plasma, liver, kidney and muscle in broiler chickens. Nat Prod Commun 11:1545–1550. https://doi.org/10.1177/1934578X1601101031

Oliveira J, Parisi MCM, Baggio JS, Silva PPM, Paviani B, Spoto MHF, Gloria EM (2019) Control of *Rhizopus stolonifer* in strawberries by the combination of essential oil with carboxymethylcellulose. Int J Food Microbiol 292:150–158. https://doi.org/10.1016/j.ijfoodmicro.2018.12.014

Oyemitan IA, Elusiyan CA, Onifade AO, Akanmu MA, Oyedeji AO, McDonald AG (2017) Neuropharmacological profile and chemical analysis of fresh rhizome essential oil of *Curcuma longa* (turmeric) cultivated in Southwest Nigeria. Toxicol Rep 4:391–398. https://doi. org/10.1016/j.toxrep.2017.07.001

Pajohi MR, Tajik H, Farshid AA, Hadian M (2011) Synergistic antibacterial activity of the essential oil of *Cuminum cyminum* L. seed and nisin in a food model. J Appl Microbiol 110:943–951. https://doi.org/10.1111/j.1365-2672.2011.04946.x

Palhares Campolina J, Gesteira Coelho S, Belli AL, Samarini Machado FR, Pereira LG et al (2021) Effects of a blend of essential oils in milk replacer on performance, rumen fermentation, blood parameters, and health scores of dairy heifers. PLoS One 16:e0231068. https://doi. org/10.1371/journal.pone.0231068

Palle-Reisch M, Wolny M, Cichna-Markl M, Hochegger R (2013) Development and validation of a real-time PCR method for the simultaneous detection of black mustard (*Brassica nigra*) and brown mustard (*Brassica juncea*) in food. Food Chem 138:348–355. https://doi.org/10.1016/j. foodchem.2012.10.055

Pandey AK, Sonker N, Singh P (2016) Efficacy of some essential oils against *Aspergillus flavus* with special reference to *Lippia alba* oil an inhibitor of fungal proliferation and aflatoxin B1 production in green gram seeds during storage. J Food Sci 81:M928–M934. https://doi. org/10.1111/1750-3841.13254

Patil RP, Nimbalkar MS, Jadhav UU, Dawkar VV, Govindwar SP (2010) Antiaflatoxigenic and antioxidant activity of an essential oil from *Ageratum conyzoides* L. J Sci Food Agric 90:608–614. https://doi.org/10.1002/jsfa.3857

Pérez-Rosés R, Risco E, Vila R, Peñalver P, Cañigueral S (2016) Biological and nonbiological antioxidant activity of some essential oils. J Agric Food Chem 64:4716–4724. https://doi. org/10.1021/acs.jafc.6b00986

Placha I, Ocelova V, Chizzola R, Battelli G, Gai F, Bacova K, Faix S (2019) Effect of thymol on the broiler chicken antioxidative defense system after sustained dietary thyme oil application. Br Poult Sci 60:589–596. https://doi.org/10.1080/00071668.2019.1631445

Porter JA, Morey A, Monu EA (2020) Antimicrobial efficacy of white mustard essential oil and carvacrol against Salmonella in refrigerated ground chicken. Poult Sci 99:5091–5095. https:// doi.org/10.1016/j.psj.2020.06.027

Prakash A, Baskaran R, Paramasivam N, Vadivel V (2018) Essential oil based nanoemulsions to improve the microbial quality of minimally processed fruits and vegetables: a review. Food Res Int 111:509–523. https://doi.org/10.1016/j.foodres.2018.05.066

Prakash B, Singh P, Mishra PK, Dubey NK (2012) Safety assessment of *Zanthoxylum alatum* Roxb. essential oil, its antifungal, antiaflatoxin, antioxidant activity and efficacy as antimicrobial in preservation of *Piper nigrum* L. fruits. Int J Food Microbiol 153:183–191. https://doi. org/10.1016/j.ijfoodmicro.2011.11.007

Prakash B, Singh P, Yadav S, Singh SC, Dubey NK (2013) Safety profile assessment and efficacy of chemically characterized *Cinnamomum glaucescens* essential oil against storage fungi, insect, aflatoxin secretion and as antioxidant. Food Chem Toxicol 53:160–167. https://doi. org/10.1016/j.fct.2012.11.044

Rachitha P, Krupashree K, Jayashree GV, Gopalan N, Khanum F (2017) Growth inhibition and morphological alteration of *Fusarium sporotrichioides* by *Mentha piperita* essential oil. Pharm Res 9:74. https://doi.org/10.4103/0974-8490.199771

Radulović NS, Genčić MS, Stojanović NM, Randjelović PJ, Stojanović-Radić ZZ, Stojiljković NI (2017) Toxic essential oils. Part V: behaviour modulating and toxic properties of thujones and thujone-containing essential oils of *Salvia officinalis* L., *Artemisia absinthium* L., *Thuja occidentalis* L. and *Tanacetum vulgare* L. Food Chem Toxicol 105:355–369. https://doi. org/10.1016/j.fct.2017.04.044

Radünz M, da Trindade MLM, Camargo TM, Radünz AL, Borges CD, Gandra EA, Helbig E (2019) Antimicrobial and antioxidant activity of unencapsulated and encapsulated clove

(*Syzygium aromaticum*, L.) essential oil. Food Chem 276:180–186. https://doi.org/10.1016/j. foodchem.2018.09.173

Rao PS, Navinchandra S, Jayaveera K (2012) An important spice, *Pimenta dioica* (Linn.) Merill: a review. Int Curr Pharm J 1:221–225. https://doi.org/10.3329/icpj.v1i8.11255

Rašković A, Milanović I, Pavlović N, Ćebović T, Vukmirović S, Mikov M (2014) Antioxidant activity of rosemary (*Rosmarinus officinalis* L.) essential oil and its hepatoprotective potential. BMC Complement Altern Med 14:225. https://doi.org/10.1186/1472-6882-14-225

Redondo-Blancos S, Fernández J, López-Ibáñez S, Miguéleze ME, Villar CJ, Lombó F (2019) Plant phytochemicals in food preservation: antifungal bioactivity: a review. J Food Prot 83:163–171. https://doi.org/10.4315/0362-028x.jfp-19-163

Reyer H, Zentek J, Männer K, Youssef IMI, Aumiller T, Weghuber J, Wimmers K, Mueller AS (2017) Possible molecular mechanisms by which an essential oil blend from star anise, rosemary, thyme, and oregano and Saponins increase the performance and ileal protein digestibility of growing broilers. J Agric Food Chem 65:6821–6830. https://doi.org/10.1021/acs.jafc.7b01925

Reyes-Jurado F, Lopez-Malo A, Palou E (2016) Antimicrobial activity of individual and combined essential oils against foodborne pathogenic bacteria. J Food Prot 79:309–315. https://doi.org/10.4315/0362-028X.JFP-15-392

Rezzoug M, Bakchiche B, Gherib A, Roberta A, Guido F, Kilinçarslan Ö, Mammadov R, Bardaweel SK (2019) Chemical composition and bioactivity of essential oils and Ethanolic extracts of *Ocimum basilicum* L. and *Thymus algeriensis* Boiss. & Reut. from the Algerian Saharan Atlas. BMC Complement Altern Med 19:1–10. https://doi.org/10.1186/s12906-019-2556-y

Rodriguez-Garcia I, Cruz-Valenzuela MR, Silva-Espinoza BA, Gonzalez-Aguilar GA, Moctezuma E, Gutierrez-Pacheco MM, Tapia-Rodriguez MR, Ortega-Ramirez LA, Ayala-Zavala JF (2016) Oregano (*Lippia graveolens*) essential oil added within pectin edible coatings prevents fungal decay and increases the antioxidant capacity of treated tomatoes. J Sci Food Agric 96:3772–3778. https://doi.org/10.1002/jsfa.7568

Rojas-Armas J, Arroyo-Acevedo J, Ortiz-Sánchez M, Palomino-Pacheco M, Castro-Luna A, Ramos-Cevallos N, Justil-Guerrero H, Hilario-Vargas J, Herrera-Calderón O (2019) Acute and repeated 28-day oral dose toxicity studies of *Thymus vulgaris* L. essential oil in rats. Toxicol Res 35:225–232. https://doi.org/10.5487/TR.2019.35.3.225

Roldán E, Sánchez-Moreno C, de Ancos B, Cano MP (2008) Characterization of onion (*Allium cepa* L.) by-products as food ingredients with antioxidant and antibrowning properties. Food Chem 108:907–916. https://doi.org/10.1016/j.foodchem.2007.11.058

Rossi C, Chaves-López C, Serio A, Anniballi F, Valbonetti L, Paparella A (2018) Effect of *Origanum vulgare* essential oil on biofilm formation and motility capacity of *Pseudomonas fluorescens* strains isolated from discoloured Mozzarella cheese. J Appl Microbiol 124:1220–1231. https://doi.org/10.1111/jam.13707

Rossini C, Rodrigo F, Davyt B, Umpiérrez ML, González A, Garrido PM, Cuniolo A, Porrini LP, Eguaras MJ, Porrini MP (2020) Sub-lethal effects of the consumption of *Eupatorium buniifolium* essential oil in honeybees. PLoS One 15:e0241666. https://doi.org/10.1371/journal.pone.0241666

Ruan D, Fan Q, Fouad AM, Sun Y, Huang S, Wu A, Lin C, Kuang Z, Zhang C, Jiang S (2021) Effects of dietary oregano essential oil supplementation on growth performance, intestinal antioxidative capacity, immunity, and intestinal microbiota in yellow-feathered chickens. J Anim Sci 99(2):skab033. https://doi.org/10.1093/jas/skab033

Saad S, Salem R, Amin R, Abu Zaid K (2015) The using of essential oils in improving mycological status of some meat products. Benha Vet Med J 29:85–96. https://doi.org/10.21608/bvmj.2015.31797

Sahakyan N, Andreoletti P, Cherkaoui-Malki M, Petrosyan M, Trchounian A (2021) *Artemisia dracunculus* L. essential oil phytochemical components trigger the activity of cellular antioxidant enzymes. J Food Biochem 45:e13691. https://doi.org/10.1111/jfbc.13691

Saka B, Djouahri A, Djerrad Z, Terfi S, Aberrane S, Sabaou N, Baaliouamer A, Boudarene L (2017) Chemical variability and biological activities of *Brassica rapa* var. rapifera parts essential oils depending on geographic variation and extraction technique. Chem Biodivers 14:e1600452. https://doi.org/10.1002/cbdv.201600452

Sani MA, Ehsani A, Hashemi M (2017) Whey protein isolate/cellulose nanofibre/TiO2 nanoparticle/rosemary essential oil nanocomposite film: its effect on microbial and sensory quality of lamb meat and growth of common foodborne pathogenic bacteria during refrigeration. Int J Food Microbiol 251:8–14. https://doi.org/10.1016/j.ijfoodmicro.2017.03.018

Santos HMC, Méndez L, Secci G, Parisi G, Martelli R, Medina I (2019) Pathway-oriented action of dietary essential oils to prevent muscle protein oxidation and texture deterioration of farmed rainbow trout. Animal 13:2080–2091. https://doi.org/10.1017/s1751731119000016

Satyal P, Murray B, McFeeters R, Setzer W (2016) Essential oil characterization of Thymus vulgaris from various geographical locations. Foods 5:70. https://doi.org/10.3390/foods5040070

Scollard J, Francis G, O'beirne D (2009) Effects of essential oil treatment, gas atmosphere, and storage temperature on *Listeria monocytogenes* in a model vegetable system. J Food Prot 72:1209–1215. https://doi.org/10.4315/0362-028x-72.6.1209

Serifi I, Tzima E, Bardouki H, Lampri E, Papamarcaki T (2019) Effects of the essential oil from Pistacia lentiscus Var. chia on the lateral line system and the gene expression profile of zebrafish (*Danio rerio*). Molecules 24:3919. https://doi.org/10.3390/molecules24213919

Shalaby SEM, El-Din MM, Abo-Donia SA, Mettwally M, Attia ZA (2011) Toxicological affects of essential oils from eucalyptus *Eucalyptus globules* and Clove *Eugenia caryophyllus* on albino rats. Pol J Environ Stud 20:429–434

Shen CY, Jiang JG, Zhu W, Ou-Yang Q (2017) Anti-inflammatory effect of essential oil from *Citrus aurantium* L. var. amara Engl. J Agric Food Chem 65:8586–8594. https://doi.org/10.1021/acs.jafc.7b02586

Shi Y, Huang S, He Y, Wu J, Yang Y (2018) Navel orange peel essential oil to control food spoilage molds in potato slices. J Food Prot 81:1496–1502. https://doi.org/10.4315/0362-028x.jfp-18-006

de Sá Silva C, de Figueiredo HM, Stamford TL, da Silva LH (2019) Inhibition of *Listeria monocytogenes* by *Melaleuca alternifolia* (tea tree) essential oil in ground beef. Int J Food Microbiol 293:79–86. https://doi.org/10.1016/j.ijfoodmicro.2019.01.004

Scientific Committee on Food (SCF) (2001a) Opinion of the Scientific Committee on Food on Estragole (1-allyl-4-methoxybenzene) European Commission Health and Consumer Protection Directorate-General, Brussel, pp. 1–10

Scientific Committee on Food (SCF) (2001b) Opinion of the Scientific Committee on Food on Methyleugenol (1-allyl-1,2-dimethoxybenzene) European Commission Health & Consumer Protection Directorate-General, Brussel, pp. 1–10

Silvestre WP, Medeiros FR, Agostini F, Toss D, Pauletti GP (2019) Fractionation of rosemary (*Rosmarinus officinalis* L.) essential oil using vacuum fractional distillation. J Food Sci Technol 56:5422–5434. https://doi.org/10.1007/s13197-019-04013-z

Simirgiotis MJ, Burton D, Parra F, López J, Muñoz P, Escobar H, Parra C (2020) Antioxidant and antibacterial capacities of *Origanum vulgare* L. Essential oil from the arid Andean region of Chile and its chemical characterization by GC-MS. Meta 10:414. https://doi.org/10.3390/metabo10100414

Singh RP, Gangadharappa H, Kenganora M (2017) *Cuminum cyminum* – a popular spice: an updated review. Pharm J 9:292–301. https://doi.org/10.5530/pj.2017.3.51

Singletary K (2016) Rosemary. Nutr Today 51:102–112. https://doi.org/10.1097/nt.0000000000000146

Sirocchi V, Devlieghere F, Peelman N, Sagratini G, Maggi F, Vittori S, Ragaert P (2017) Effect of *Rosmarinus officinalis* L. essential oil combined with different packaging conditions to extend the shelf life of refrigerated beef meat. Food Chem 221:1069–1076. https://doi.org/10.1016/j.foodchem.2016.11.054

Smeti S, Hajji H, Bouzid K, Abdelmoula J, Muñoz F, Mahouachi M, Atti N (2014) Effects of *Rosmarinus officinalis* L. as essential oils or in form of leaves supplementation on goat's production and metabolic statute. Trop Anim Health Prod 47:451–457. https://doi.org/10.1007/s11250-014-0721-3

Smeti S, Joy M, Hajji H, Alabart JL, Muñoz F, Mahouachi M, Atti N (2015) Effects of *Rosmarinus officinalis* L. essential oils supplementation on digestion, colostrum production of dairy ewes and lamb mortality and growth. Anim Sci J 86:679–688. https://doi.org/10.1111/asj.12352

Smith RL, Cohen SM, Doull J, Feron VJ, Goodman JI, Marnett LJ, Portoghese PS, Waddell WJ, Wagner BM, Hall RL, Higley NA, Lucas-Gavin C, Adams TB (2005) A procedure for the safety evaluation of natural flavor complexes used as ingredients in food: essential oils. Food Chem Toxicol 43:345–363. https://doi.org/10.1016/j.fct.2004.11.007

Šojić B, Pavlic B, Ikonić P, Tomovic V, Ikonić B, Zekovic Z, Kocić-Tanackov S, Jokanović M, Škaljac S, Ivić M (2019) Coriander essential oil as natural food additive improves quality and safety of cooked pork sausages with different nitrite levels. Meat Sci 157:107879. https://doi.org/10.1016/j.meatsci.2019.107879

Song X, Liu T, Wang L, Liu L, Li X, Wu X (2020) Antibacterial effects and mechanism of mandarin (*Citrus reticulata* L.) essential oil against *Staphylococcus aureus*. Molecules 25:4956. https://doi.org/10.3390/molecules25214956

Sonker N, Pandey AK, Singh P (2014) Efficiency of *Artemisia nilagirica*(Clarke) Pamp. essential oil as a mycotoxicant against postharvest mycobiota of table grapes. J Sci Food Agric 95:1932–1939. https://doi.org/10.1002/jsfa.6901

Sotelo-Boyás M, Correa-Pacheco Z, Bautista-Baños S, Gómez y Gómez Y (2017) Release study and inhibitory activity of thyme essential oil-loaded chitosan nanoparticles and nanocapsules against foodborne bacteria. Int J Biol Macromol 103:409–414. https://doi.org/10.1016/j.ijbiomac.2017.05.063

Stojanović NM, Randjelović PJ, Mladenović MZ, Ilić IR, Petrović V, Stojiljković N, Ilić S, Radulović NS (2019) Toxic essential oils, part VI: acute oral toxicity of lemon balm (*Melissa officinalis* L.) essential oil in BALB/c mice. Food Chem Toxicol 133:110794. https://doi.org/10.1016/j.fct.2019.110794

Subhan N, Alam MA, Ahmed F, Shahid IJ, Nahar L, Sarker SD (2008) Bioactivity of *Excoecaria agallocha*. Rev Bras 18:521–526. https://doi.org/10.1590/s0102-695x2008000400004

Sun G, Wang S, Hu X, Su J, Zhang Y, Xie Y, Zhang H, Tang L, Wang J-S (2011) Co-contamination of aflatoxin B1and fumonisin B1in food and human dietary exposure in three areas of China. Food Addit Contam Part A 28:461–470. https://doi.org/10.1080/19440049.2010.544678

Suntar I, Khan H, Patel S, Celano R, Rastrelli L (2018) An overview on *Citrus aurantium* L.: its functions as food ingredient and therapeutic agent. Oxidative Med Cell Longev 2018:1–12. https://doi.org/10.1155/2018/7864269

Swedzinski C, Froehlich KA, Abdelsalam KW, Chase C, Greenfield TJ, Koppien-Fox J, Casper DP (2019) Evaluation of essential oils and a prebiotic for newborn dairy calves1. Transl Anim Sci 4:75–83. https://doi.org/10.1093/tas/txz150

Szczepanik M, Walczak M, Zawitowska B, Michalska-Sionkowska M, Szumny A, Wawrzeńczyk C, Brzezinska MS (2018) Chemical composition, antimicromicrobial activity and insecticidal activity against the lesser mealworm *Alphitobius diaperinus* (Panzer) (Coleoptera: Tenebrionidae) of *Origanum vulgare* L. ssp. hirtum (Link) and *Artemisia dracunculus* L. essential oils. J Sci Food Agric 98:767–774. https://doi.org/10.1002/jsfa.8524

Tabarraei H, Hassan J, Parvizi MR, Golshahi H, Keshavarz-Tarikhi H (2019) Evaluation of the acute and sub-acute toxicity of the black caraway seed essential oil in Wistar rats. Toxicol Rep 6:869–874. https://doi.org/10.1016/j.toxrep.2019.08.010

Taghizadeh M, Ostad SN, Asemi Z, Mahboubi M, Hejazi S, Sharafati-Chaleshtori R, Rashidi A, Akbari H, Sharifi N (2017) Sub-chronic oral toxicity of *Cuminum cyminum* L.'s essential oil in female Wistar rats. Regul Toxicol Pharmacol 88:138–143. https://doi.org/10.1016/j.yrtph.2017.06.007

Tak JH, Jovel E, Isman MB (2016) Comparative and synergistic activity of *Rosmarinus offici-nalis* L. essential oil constituents against the larvae and an ovarian cell line of the cabbage looper, *Trichoplusia ni* (Lepidoptera: Noctuidae). Pest Manag Sci 72:474–480. https://doi.org/10.1002/ps.4010

Tardugno R, Serio A, Purgatorio C, Savini V, Paparella A, Benvenuti S (2020) *Thymus vulgaris* L. essential oils from Emilia Romagna Apennines (Italy): phytochemical composition and antimicrobial activity on food-borne pathogens. Nat Prod Res 27:1–6. https://doi.org/10.1080/14786419.2020.1798666

Tassou CC, Drosinos EH, Nychas GJE (1995) Effects of essential oil from mint (*Mentha piperita*) on *Salmonella enteritidis* and *Listeria monocytogenes* in model food systems at 4° and 10°C. J Appl Bacteriol 78:593–600. https://doi.org/10.1111/j.1365-2672.1995.tb03104.x

Teneva D, Denkova-Kostova R, Goranov B, Hristova-Ivanova Y, Slavchev A, Denkova Z, Kostov G (2019) Chemical composition, antioxidant activity and antimicrobial activity of essential oil from *Citrus aurantium* L zest against some pathogenic microorganisms. Z Naturforsch C Biosci 74:105–111. https://doi.org/10.1515/znc-2018-0062

Tenore GC, Ciampaglia R, Arnold NA, Piozzi F, Napolitano F, Rigano D, Senatore F (2011) Antimicrobial and antioxidant properties of the essential oil of *Salvia lanigera* from Cyprus. Food Chem Toxicol 49:238–243. https://doi.org/10.1016/j.fct.2010.10.022

Tian P (2019) Essential oil blend could decrease diarrhea prevalence by improving antioxidative capability for weaned pigs. Animals 9:847. https://doi.org/10.3390/ani9100847

Tisserand R, Young R (2014) Essential oil safety, 2nd edn. Churchill Livingstone, Edinburgh

Usami A, Motooka R, Takagi A, Nakahashi H, Okuno Y, Miyazawa M (2014) Chemical composition, aroma evaluation, and oxygen radical absorbance capacity of volatile oil extracted from *Brassica rapa* cv. "Yukina" used in Japanese traditional food. J Oleo Sci 63:723–730. https://doi.org/10.5650/jos.ess14033

Valdivieso-Ugarte M, Gomez-Llorente C, Plaza-Díaz J, Gil Á (2019) Antimicrobial, antioxidant, and immunomodulatory properties of essential oils: a systematic review. Nutrients 11:2786. https://doi.org/10.3390/nu11112786

Vanti G, Tomou EM, Stojković D, Ćirić A, Bilia AR, Skaltsa H (2021) Nanovesicles loaded with *Origanum onites* and *Satureja thymbra* essential oils and their activity against food-borne pathogens and spoilage microorganisms. Molecules 26(8):2124. https://doi.org/10.3390/molecules26082124

Venâncio A, Onofre A, Lira A, Alves P, Blank A, Antoniolli Â, Marchioro M, dos Santos EC, Santos de Araujo B (2010) Chemical composition, acute toxicity, and antinociceptive activity of the essential oil of a plant breeding cultivar of basil (*Ocimum basilicum* L.). Planta Med 77:825–829. https://doi.org/10.1055/s-0030-1250607

Victoria FN, Lenardão EJ, Savegnago L, Perin G, Jacob RG, Alves D et al (2012) Essential oil of the leaves of *Eugenia uniflora* L.: antioxidant and antimicrobial properties. Food Chem Toxicol 50:2668–2674. https://doi.org/10.1016/j.fct.2012.05.002

Volpato GT, Luis AD, Francia-Farje LAD, Damasceno DC, Oliveira RV, Hiruma-Lima CA, Kempinas WG (2015) Effect of essential oil *from Citrus aurantium* in maternal reproductive outcome and fetal anomaly frequency in rats biomedical sciences. An Acad Bras Cienc 87:407–415. https://doi.org/10.1590/0001-3765201520140354

Wang L, Zhang Y, Fan G, Ren J, Zhang L, Pan S (2019) Effects of orange essential oil on intestinal microflora in mice. J Sci Food Agric 99:4019–4028. https://doi.org/10.1002/jsfa.9629

Xie Y, Huang Q, Rao Y, Hong L, Zhang D (2019) Efficacy of *Origanum vulgare* essential oil and carvacrol against the housefly, *Musca domestica* L. (Diptera: Muscidae). Environ Sci Pollut Res 26:23824–23831. https://doi.org/10.1007/s11356-019-05671-4

Xing F, Hua H, Selvaraj JN, Yuan Y, Zhao Y, Zhou L, Liu Y (2014) Degradation of fumonisin B1 by cinnamon essential oil. Food Control 38:37–40. https://doi.org/10.1016/j.foodcont.2013.09.045

Yaldiz G, Camlica M, Ozen F (2019) Biological value and chemical components of essential oils of sweet basil (*Ocimum basilicum* L.) grown with organic fertilization sources. J Sci Food Agric 99:2005–2013. https://doi.org/10.1002/jsfa.9468

Yaouba A, Tatsadjieu LN, Dongmo PM, Etoa FX, Mbofung CM, Zollo PH, Menut C (2011) Evaluation of *Clausena anisata* essential oil from Cameroon for controlling food spoilage fungi and its potential use as an antiradical agent. Nat Prod Commun 6:1367–1371

Ye CL, Dai DH, Hu WL (2013) Antimicrobial and antioxidant activities of the essential oil from onion (*Allium cepa* L.). Food Control 30:48–53. https://doi.org/10.1016/j.foodcont.2012.07.033

Yilmaztekin M, Lević S, Kalušević A, Cam M, Bugarski B, Rakić V, Pavlović V, Nedović V (2019) Characterisation of peppermint (*Mentha piperita* L.) essential oil encapsulates. J Microencapsul 36:109–119. https://doi.org/10.1080/02652048.2019.1607596

Yu C, Wei J, Yang C, Yang Z, Yang W, Jiang S (2018) Effects of star anise (*Illicium verum* Hook.f.) essential oil on laying performance and antioxidant status of laying hens. Poult Sci 97:3957–3966. https://doi.org/10.3382/ps/pey263

Zhang H, Han T, Yu C-H, Jiang Y-P, Peng C, Ran X, Qin L-P (2012) Analysis of the chemical composition, acute toxicity and skin sensitivity of essential oil from rhizomes of *Ligusticum chuanxiong*. J Ethnopharmacol 144:791–796. https://doi.org/10.1016/j.jep.2012.10.010

Zhang LY, Peng QY, Liu YR, Ma QG, Zhang JY, Guo YP, Xue Z, Zhao LH (2021) Effects of oregano essential oil as an antibiotic growth promoter alternative on growth performance, antioxidant status, and intestinal health of broilers. Poult Sci 100:101163. https://doi.org/10.1016/j.psj.2021.101163

Chapter 8
Essential Oil: Source of Antioxidants and Role in Food Preservation

Himani, Sonu Kumar Mahawer, Sushila Arya, Ravendra Kumar, and Om Prakash

8.1 Introduction

Essential oils are highly concentrated, aromatic plant oils derived without heating using steam distillation, dry distillation, hydro-diffusion, or other appropriate mechanical processes. In aromatherapy literature, they are referred to as "plant essences," and the process of extraction is crucial in classifying an aromatic ingredient as an essential oil (Manion and Widder 2017). Chemically, they are a blend of various terpenes or terpenoids, which are isoprene polymers. Essential oils are produced in the cytoplasm and are often found as minute droplets between cells. These are water insoluble, lipophilic, and soluble in organic solvents, as well as volatile and fragrant (Tongnuanchan and Benjakul 2014). Specific fragrant and chemical composition of essential oils serve a variety of important functions for plants, including attraction of beneficial insects and other pollinators, protection against biotic (insect pest and diseases) and abiotic (cold, heat etc.) stresses (Burt 2004; Dhifi et al. 2016).

After the food industries, the uses of synthetic food preservatives are considerably increased to preserve the food long lastingly from microbial as well as oxidative spoilage. As synthetic and chemical food preservatives are known to have several health issues in human being such as intoxication, cancer, and many other degenerative illnesses, researchers are searching for less toxic alternatives for these (Prakash and Kiran 2016). In this respect essential oils from plant origin are considered as most suitable alternative because of their strong antimicrobial, antioxidant properties along with several advantages over synthetic preservatives like very less or no toxicity, botanical in origin, eco-friendly etc. other than the antimicrobial actions of essential oils, their antioxidant potential is also considered as important

Himani · S. K. Mahawer · S. Arya · R. Kumar (✉) · O. Prakash
Department of Chemistry, College of Basic Sciences and Humanities, G.B. Pant University of Agriculture and Technology, Nagar, Uttarakhand, India

in food preservation to preserve processed as well as non-processed food products by the reduction of their spoilage due to production of reactive oxygen species. The antioxidant capacity of essential oils is largely determined by their chemical compositions. The significant antioxidant action of essential oils is due to the binding of phenolic and other secondary metabolites with double bonds.

Over 3000 varieties of EOs have been recognized, but out of them approximately 300 are of industrial significance for uses in the food industry, primarily in the flavours and fragrances sector. Because of its high volatility, ephemerality, and biodegradability, essential oils are well accepted by consumers (Bakkali et al. 2008; Falleh et al. 2020). For example, Mediterranean food products like meat and products, preserved with essential oils are well liked and of added value products by consumers (Laranjo et al. 2017). However, there are some lacunas in the use of essential oils as food preservatives at industrial level such as less water solubility, strong distinctive aroma, less stability etc. By the consideration of these lacuna, essential oils can be effectively used in food preservation at the place of synthetical chemicals. In this chapter we are discussing about the current knowledge available regarding the role of essential oils in food preservations.

8.2   Antioxidant Action of Essential Oils in Food Preservation

In the food industry, an antioxidant can be defined as any compound or molecule that is able to react with radicals or capable to slowing down or inhibiting the oxidation of any easily oxidisable material like the polyunsaturated fats, even when they are used in very moderate amount (1–1000 mg/L). The reactive oxygen species (ROs) are responsible for oxidative stress, DNA damage, damage to cell membranes and other parts of the cell. which may leads to many diseases such as Parkinson's disease, Alzheimer's disease, cardiovascular diseases, multiple sclerosis, cognitive impairment, cancer, cardiac failure and many others. In recent years, in food industry, there is an interest building in searching for natural and low-cost antioxidants based on plant sources to substitute the synthetic ones such as butylated hydroxyl toluene, butylated hydroxyl anisole, propyl gallate, and tert-butyl hydroquinone because of their negative health consequences (Botterweck et al. 2000). The majority of natural antioxidants are phenolic and terpenolic compounds with the most important groups being flavonoids, tocopherols, and phenolic acids. The hydroxyl group of phenolics is directly attached to the carbon atom of the aromatic ring and the hydrogen atom donated to free radicals, thereby preventing the oxidation of other compounds (Skancke and Skancke 1988). Several studies have revealed that essential oil and their components are natural sources of antioxidants with different modes of action such as free radical scavenging, prevention of chain initiation, reducing agents, termination of peroxides, quenching of singlet oxygen, and binding of metal ion catalysts. Most of the fats found in food are in the form of triglyceride that is the main target for oxidation. Spontaneous reactions of atmospheric oxygen with lipids are known as autoxidation that is an important and common process for oxidation deterioration. Oxidation is one of the most important causes of food deterioration that causes undesirable changes in food value, organoleptic criteria, food nutritional quality and the production of potentially toxic molecules in food. Food undergone extensive oxidation has major defects and has no consumer acceptability. Oxidation during food processing and/or storage may be seen through change in colour, changes in appearance, texture, bad flavors, whereas the variation of its principal components, such as biomolecules like lipids, carbohydrate, protein are not always marked. Their oxidation occurs by a radical-chain reaction mediated by peroxyl radicals that parallels the autoxidation of hydrocarbons. (Fig. 8.1).

This mechanism involves three steps; initiation, propagation and termination. By donating hydrogen radicals (H·) to free radicals, antioxidants prevent oxidative damage and stop the propagation reaction which is called a chain reaction. In the first stage of the chain reaction that is initiation stage, a hydrogen atom (H·) is distracted from a neighbouring carbon to a double bond in an unsaturated fatty acid substrate (RH), forming the alkyl R· radical (free radicals) to initiate the autoxidation chains. The so formed alkyl radical then react with molecular oxygen at a diffusion-controlled rate and produces peroxyl (ROO·) radicals that parallels the autoxidation of hydrocarbons. In the propagation stage these radicals may be stabilised by attacking another susceptible molecule of the substrate (RH) to form a lipid hydroperoxide (ROOH), the oxidized substrate and a new radical (R·) (Amorati et al. 2013). During the last termination stage, the chain reaction continues for many cycles until two radical species quench or react with each other to form non-radical species or stable products (Howard 1974).

Meat and Dairy products are mostly susceptible to oxidation. In meat, lipid oxidation starts at the time of slaughtering, when metabolic processes cease and the unsaturated fatty-acids react with molecular oxygen. This is followed by secondary chain reactions that provokes the production of oxidation rancidity products which affect flavor profile, texture, and color of meat products (Amaral et al. 2018). During the production of butter from milk, the fat content of milk is concentrated about 20

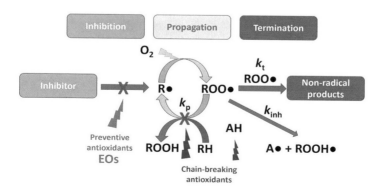

Fig. 8.1 Mechanism of hydrocarbon autoxidation and antioxidant properties of Essential oils. (Adopted and modified from Amorati et al. 2013)

times which enhances the shelf life of butter, but is also subject to an increased risk of lipid oxidation. Another example of oxidation reaction is in cheese here light-induced oxidation impact on the development of rancid off-flavors, mainly because many of these products are packaged in transparent materials. Mayonnaise, one of the most consumed food emulsions, is a lipid-rich emulsion since it is comprised of about 70–80% oil which makes it more sensitive to deterioration by auto-oxidation. Along with general factors influencing lipid oxidation, additional internal and external factors such as packaging material and type of emulsifier used also affects the rate of oxidation (Ghorbani et al. 2016). The incorporation of natural agents such as plant extracts/essential oils can be used for food preservation for several aspects such as:

- **Safety**- slowing/stopping growth of food poisoning micro-organisms
- **Health**- slowing the deterioration of nutrients
- **Quality**- Maintaining texture, taste and aroma
- **Shelf life**- reducing waste and increasing convenience

8.2.1 *In-Vitro Antioxidant Assays of Essential Oils*

Various studies have shown that essential oils represent a huge source of compounds exhibiting strong antioxidant activity. In the food and biological systems antioxidant assays are classified into two ways: Evaluation of lipid peroxidation and measurement of free radical scavenging ability which can be further evaluated by various methods as illustrated in Fig. 8.2.

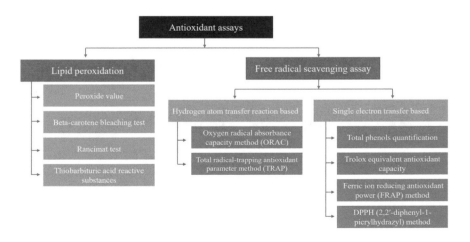

Fig. 8.2 Schematic flow chart of antioxidant assays

8.2.2 Lipid Peroxidation

Lipid peroxidation is a complex process and in assessing lipid peroxidation, several lipid substrates can be used such as oils and fats, linoleic acid, fatty acid methyl esters and low-density lipoproteins. Peroxides are the primary products of oxidation and by evaluating the peroxide level, effectiveness of essential oil against lipid peroxidation can be examined. The antioxidant effect of essential oils can also be evaluated by monitoring the conjugated diene formation at the early stage of lipid peroxidation which can be analysed spectrometrically. (Wei and Shibamoto 2010). Another important method is the β-carotene bleaching method which is commonly referred as coupled oxidation of β-carotene and linoleic acid that estimates the relative ability of antioxidant compounds in essential oils to scavenge the radical of linoleic acid peroxide that oxidizes β-carotene in the emulsion phase. And in the absence of antioxidant β-carotene undergoes a rapid decolourization since the free linoleic acid radical attacks the β-carotene, which results in lose of the double bonds and consequently disappearance of its orange colour (Miguel 2010). Malondialdehyde is formed as a result of secondary product of oxidation which is formed after decomposition of lipid hydroperoxide and it forms a pink chromophore with thiobarbituric acid (TBA). This coloured complex, which absorbs at 532 nm, results in the condensation of 2 M TBA and 1 M malondialdehyde in an acidic environment. However this method is not very specific. The formic acid measurement or the rancimat method is an automated test that measures the conductivity of low molecular weight fatty acids (formic acid) produced during the auto-oxidation of lipids at 100 °C or above it (Miguel 2010). There are many reports on the ability of essential oils to inhibit lipid oxidation by different methods. (Table 8.1).

8.2.3 Free Radical Scavenging Ability

For measuring the free radical scavenging ability, the methods are grouped in two ways according to the chemical reactions involved in the process: hydrogen atom transfer reaction-based methods and single electron transfer reaction-based methods. In the hydrogen atom transfer reaction-based methods, the antioxidant is able to quench free radicals by hydrogen donation while in single electron transfer based-methods the ability of an antioxidant to transfer one electron to reduce any compound, including metals, carbonyl groups and radicals is detected. Oxygen radical absorbance capacity method (ORAC) and the total radical-trapping antioxidant parameter method (TRAP) are the examples of methods based on the transfer of hydrogen. ORAC is a popular method used to have an estimate of the content of antioxidants in food. In both the methods there is a thermal radical generator which is usually 2,2′ azobis(2-amidinopropane) dihydrochloride (AAPH) which gives a steady flux of peroxyl radicals in air saturated solution, another one is a substrate which monitor the reaction progress (UV or fluorescence) and finally the

Table 8.1 Inhibition of lipid oxidation capacity of essential oils using different assays

Essential Oil	Major compounds	Assay	Standard	References
Rosmarinus officinalis	1,8-cineole, α-pinene, β-pinene	β-carotene bleaching assay	BHT	Wang et al. (2008)
Petroselinum crispum	myristicin, apiol, β-pinene, α-pinene	β-carotene bleaching assay	α-Tocopherol, BHT	Zhang et al. (2006)
Melissa officinalis	β-cubebene, α-cadinene, β-caryophyllene,α-cadinolcadinol, caryophyllene oxide	β-carotene bleaching assay	BHT	Radulescu et al. (2021)
Lippia citriodora	limonene, neral, geranial, cital	β-carotene bleaching assay	BHT	Farahmandfar et al. (2018)
Bunium persicum	Caryophyllene, γ-terpinene, cuminyl acetate	β-carotene bleaching assay	BHT	Shahsavari et al. (2008)
Thymus vulgaris	terpinen-4-ol, γ-terpinene, cis-sabinene hydrate, linalool, p-cymene	Rancimat test	BHT, Ascorbic acid	Viuda-Martos et al. (2010)
Oringanum vulgare	carvacrol	Rancimat test	BHT, Ascorbic acid	Viuda-Martos et al. (2010)
Syzygium aromaticum	Eugenol, β-caryophyllene	Rancimat test	BHT, Ascorbic acid	Viuda-Martos et al. (2010)
Ocimum basilicum	linalool, 1,8-cineole, estragole, methyl cinnamate, eugenol	Rancimat test	–	Politeo et al. (2006)
Thymus caespititius	α-terpeniol, p-cymene, γ-terpinene	TBARS method	α-Tocopherol, BHT, BHA	Miguel et al. (2004)
Coriandmm sativum	Linalool, α-pinene, geranyl acetate, γ-terpinene	TBARS method	α-Tocopherol, BHT, BHA	Baratta et al. (1998)
Dodecadenia grandiflora	furanodiene, germacrene D	TBARS method	BHT, quercetin	Joshi et al. (2010)
Persea duthiei	α-pinene, β-pinene, limonene, (E)-nerolidol	TBARS method	BHT, quercetin	Joshi et al. (2010)
Matricaria chamomilla	(Z)-anethole, linalool, limonene, (Z)-β-ocimene, methyleugenol	TBARS method	BHT	
Nigella sativa	p-cymene, thymoquinone, α-thujene	Peroxide value	PG, BHT, BHA	Singh et al. (2014)
Mentha spicata	carvon, cis-carveol, limonene, 1,8-cineole	Peroxide value	BHT	Hussain et al. (2010)
Artemisiadracunculus	(E)-β-farnesene, guaiazulene, bisabolol oxide A, α-farnesene, α-bisabolol	Peroxide value	BHT	Ayoughi et al. (2011)

antioxidant that will inhibit or delay the substrate oxidation by competing with the substrate for the radicals (Huang et al. 2005).

Methods based on the electron transfer include an oxidant (substrate) that abstracts an electron from the antioxidant, and changes the colour of substrate. The degree of colour change is proportional to the antioxidant concentrations. The reaction ends when the colour change stops. This group of the methods includes: (a) total phenols quantification; (b) the Trolox equivalent antioxidant capacity (TEAC) method; (c) the ferric ion reducing antioxidant power (FRAP) method; (d) the Cu(II) reduction ability method; and (e) the 2,2-diphenyl-1-picrylhydrazyl (DPPH) method (MacDonand et al. 2006). Quantification of total phenolic content is done by using Folin–Ciocalteu method which is based on the number of phenolic groups or other potential oxidizable groups present in the sample (Becker et al. 2004). Folin–Ciocalteu is chemically heteropolyphosphotungstates–molybdates. The electron-transfer reaction occurs between antioxidants and molybdenum, which is reduced in the complex, resulting in formation to blue species, which can be measured spectrophotometrically. In several studies it was found that there is a positive correlation between the antioxidant activity of essential oils and the total phenolic content. (Spiridon et al. 2011). In TEAC method, the antioxidant or any reducing agent X, reacts with the coloured and persistent radical ABTS+(2,2'-azino-bis(3-ethylbenzthiazoline-6-sulphonic acid) which has a strong absorption band in the range 600–750 nm Discoloration is compared with that produced by Trolox. (Roginsky and Lissi 2005). In FRAP method, a ferric salt, Fe (III) (TPTZ)2Cl3 (TPTZ = 2,4,6-tripiridil-striazina) is used as an oxidant agent. TEAC and FRAP assays are quite similar, differing only in the pH of the assay: TEAC occurs in neutral pH, while the FRAP method needs an acidic environment (pH 3.6). Essential oils of several plants have reported to possess FRAP activity such as Clove, Coriander, Basil, Mint, Black pepper, Laurel, Marjoram, Everlast, Nutmeg, Fennel, Cinnamon, Sage etc (Politeo et al. 2006). Chelation of transition metals is one of the important mechanisms of the antioxidative action. Transition metal ions can stimulate lipid peroxidation by decomposing lipid hydroperoxides into other components which are able to abstract hydrogen, initiating the chain of reaction of lipid peroxidation (Viuda et al. 2010). Ferrozine is generally used for the determination of chelating activity, which forms complex with Fe2+. In the presence of other chelating agents, the complex formation is disrupted, resulting in the disappearance of red colour of ferrozine-Fe2+ complex. Rate of reduction of red color allows the estimation of chelating activity spectrophotomatically. Fe2+ has the ability to move single electron which permits the formation of many radical reactions, the main aim to inhibit the formation of reactive oxygen species (ROs) is linked with the redox active metal catalysis which involves the chelation of the metal ions. The most commonly used assay is the DPPH assay which is based on the antioxidant that could reduce the stable DPPH which is violet to yellow. DPPH is a nitrogen-centred free radical that could readily accepts an electron or hydrogen radical to form a stable diamagnetic molecule. On reaction of the radical with suitable reducing agents, the electrons become paired off to form the corresponding hydrazine leading to the discoloration of the solution. The antioxidant compounds scavenge the DPPH radical by donating hydrogen atom leading to the formation of anion-radical state which is 1,1-diphenyl-2-picrylhydrazine (DPPH-H) and discoloration of violet colour of DPPH.

Violet DPPH (517nm) (Free Radical)   Yellow DPPH(517nm)

Violet DPPH (517nm) (Free Radical)   Yellow DPPH(517nm)

The other main reactive oxygen and nitrogen species are superoxide anion, peroxyl radical. hydroxyl radical, hydrogen peroxide, singlet oxygen, peroxynitrite, and nitric oxide radical. There are a number of studies stating the free radical scavenging activities and other antioxidant activities of essential oils of several plant families. Some are enlisted in Table 8.2. Many antioxidant rich essential oils are used to produce protective active barriers to be applied directly in food products surface as edible coatings or in films for food packaging purposes. The oils act as barriers that represent a new approach to solving the detrimental impacts of oxygen on food. Some examples are essential oil of *Oregano, Ginger, Rosmery, Satureja* etc (Lourenco et al. 2019).

Table 8.2 Examples of Essential oil possessing antioxidant activity

Plant	Family	Major essential oil constituents	Antioxidant assay	Reference
Lavendular angustifolia	Lamiaceae	linalool, 1,8-cineole, linalyl acetate, caryophyllene	DPPH radical scavenging, ABTS radical scavenging	Yang et al. (2010)
Syzygium aromaticum	Myrtle	eugenol, eugenyl acetate	DPPH radical scavenging, FRAP, TBARS, Rancimat	Politeo et al. (2006)
Limnophilla indica	Plantaginaceae	epi-cyclocolorenone, α-gurjunene, 5-hydroxy-cis-calamenene, β-caryophyllene, α-gurjunene	DPPH radical scavenging, Reducing power, Metal chelating of Fe^{2+}	Kumar et al. (2019)
Citrus limon	Rutaceae	limonene, tricyclene, γ-terpinene	DPPH radical scavenging, ABTS radical scavenging	Yang et al. (2010)
Zanthoxylum armatum	Rutaceae	α-pinene, α-cadinol germacrene-D, E-caryophyllene, 2-undecanone	DPPH radical scavenging, Reducing power, Metal chelating of Fe^{2+}	Dhami et al. (2019)
Ocimum basilicum	Lamiaceae	linalool, estragole, methyl eugenol, methyl cinammate	DPPH radical scavenging, FRAP, TBARS, Rancimat	Politeo et al. (2006)
Premna mucronata	Lamiaceae	ethyl hexanol 1-octen-3-ol, linalool methyl salicylate (E)- caryophyllene	DPPH radical scavenging, Reducing power, Metal chelating of Fe^{2+}	Palariya et al. (2019)
Coriandrum sativum	Apiaceae	linalool, camphor	DPPH radical scavenging, FRAP, TBARS, Rancimat	Politeo et al. (2006)
Salvia reflexa	Lamiaceae	palmitic acid, phytol (E)-caryophyllene caryophyllene oxide	DPPH radical scavenging, Reducing power, Metal chelating of Fe^{2+}	Goswami et al. (2019)
Rosmarius officinalis	Lamiaceae	α-pinene, 1,8-cineole, camphor	DPPH radical scavenging, ABTS radical scavenging	Yang et al. (2010)

(continued)

Table 8.2 (continued)

Plant	Family	Major essential oil constituents	Antioxidant assay	Reference
Coleus barbatus	Lamiaceae	bornyl acetate, n-decanal, sesquisabinene, β-bisabolene, δ-cadinene	DPPH radical scavenging, Reducing power, Metal chelating of Fe2+, H2O2 radical scavenging, Nitric oxide radical scavenging, Superoxide radical scavenging	Kanyal et al. (2021)
Helichrysum italicum	Compositeae	α-pinene, α-cidrene, geraniol	DPPH radical scavenging, FRAP, TBARS, Rancimat	Politeo et al. (2006)
Globba sessiliflora	Zingiberaceae	myrcene, β-caryophyllene, selin-11-en-4α-ol, β-longipinene, manool, germacrene D and β-eudesmol	DPPH radical scavenging, Reducing power, Metal chelating of Fe2+	Kumar et al. (2012)
Caryopteris foetida	verbenaceae	δ-cadinene, β-caryophyllene, (E)-β-farnesene, γ-cadinene, spathulenol, τ-muurolol	DPPH radical scavenging, Reducing power, Metal chelating of Fe2+	Joshi et al. (2021a)
Cinnamomum zeylanicum	Lauraceae	trans-cinnamaldehyde	DPPH radical scavenging, FRAP, TBARS, Rancimat	Politeo et al. (2006)
Nepeta cataria	Lamiaceae	cis-nepetalactone,	DPPH radical scavenging, Reducing power, Metal chelating of Fe2+	Joshi et al. (2021b)
Cotinus coggygria	Anacardiacea	myrcene, pinene, α-terpineol, cymene, sabinene	DPPH radical scavenging, Metal chelating of Fe2+	Thapa et al. (2020)
Foeniculum vulgare	Apiaceae	limonene, estragole, trans-anethole, fenchone, eucalyptol	DPPH free radical scavenging, ferric reducing power assay, thiobarbituric acid reactive species assay, ferrous ion-chelating	Shahat et al. (2011)
Myristica fragrans	Myristicaceae	α-pinene, sabinene, myristicene, β-pinene	DPPH radical scavenging, FRAP, TBARS, Rancimat	Politeo et al. (2006)

(continued)

Table 8.2 (continued)

Plant	Family	Major essential oil constituents	Antioxidant assay	Reference
Origanum vulgare	Lamiaceae	carvacrol, thymol, p-cymene, 3-carene, caryophyllene,	DPPH radical scavenging, Reducing power	Han et al. (2017)
Salvia officinalis	Lamiaceae	thujone, 1,8-cineole	DPPH radical scavenging, FRAP, TBARS, Rancimat	Politeo et al. (2006)
Foeniculum vulgare	Apiaceae	fenchone, trans-anethole	DPPH radical scavenging, FRAP, TBARS, Rancimat	Politeo et al. (2006)
Ocimum basilicum	Lamiaceae	linalool, methyl chavicol, eucalyptol, eugenol, trans-α-bergamotene	DPPH radical scavenging, β-carotene bleaching assay, TBHQ inhibition	Farouk et al. (2016)

8.3 Other Applications of Essential Oils in Food Industry

In addition to the wide use of essential oil as as antioxidant agent, they are used in food products in many other ways like flavouring, fragnancing and a key element of active packaging.

Using essential oils in savoury cooking Thyme, oregano, rosemary, marjoram and asafetida essential oil is used in curries, meatballs, pickles, and savoury meals to add umami flavour.

Using essential oils in organic food processing Thymus, cinnamon and oregano essential oils has antioxidant and antibacterial properties with the objective of its use for the meat industry(minced beef, cooked ham, or dry-cured sausage), Lavender or bergamot essential oil, is used as a flavoring agent in chocolate and chocolate coating (candy melts) (Muriel-Galet et al. 2015).

Using essential oils as food additives Turmeric, citrus and chinese cinnamon essential oils have been used as additives in biodegradable films and coatings for active food packaging. They can provide the films and coatings with antioxidant and antimicrobial properties, depending both on their composition and on the interactions with the polymer matrix.

Using essential oils in baking industry Clove and peppermint essential oils are extremely pungent and used as a flavoring agent in baking industry (cakes, baked goods and candies).

Using essential oils in active food packaging Essential oils can be encapsulated in polymers of edible and biodegradable coatings that provide a slow release to the food surface of packages e.g., fruit, meat, and fish. This approach increases the

Table 8.3 Essential oils and their uses as flavouring agents

Plant essential oil	Flavouring agent
Basil	Flavoring for sauces and condiments
Lemongrass	Flavoring for beverages and sweets
Eucalyptus	Flavoring for beverages, sweets, ice cream
Geranium	Flavoring for sweets, chewing gum
Pepper mint	For flavor liqueurs, ice cream, chewing gum and chocolate
Green mint	Flavor to drinks, sweets, ice cream
citrus	Flavor to glazes, toppings, sauces

stability of volatile components, protecting them from interacting with the food matrix, and increases the antimicrobial activity.

Using essential oils in beverages Lemon-lime sodas use essential oils from lemon, lime, neroli, and orange as the main flavorings, Vanilla essential oils are used as flavouring agent in soft drinks. Stevia (Stevia rebaudiana), whose leaves contain a range of sweet-tasting essential oils that can be used to sweeten beverages. EOs of many plant species are used as flavouring agent in food and beverages (Table 8.3) (Khayreddine 2018).

8.3.1 Aroma

Aroma of essential oil is due to volatile and odorous organic molecules. These molecules are lipophilic and have a low molecular weight (between 100 and 250), and they are made up of saturated and unsaturated hydrocarbons, aldehydes, esters, ketones, alcohol, oxides, phenols, and terpenes, which may produce characteristic odors (Wildwood 1996). Essential oil of many aromatic plants from their different plant parts like flowers leaves provides a pleasant odour to the food products along with flavour. Some examples are Geranium (Pelargonium graveolens L Herit), Lavender (Lavandula officinalis Chaix.), Roman chamomile (Anthemis nobilis Linn.), Rosemary (Rosmarinus officinalis Linn.) etc (Koulivand et al. 2013; Price 1993; Lawless 1995; Svoboda and Deans 1992).

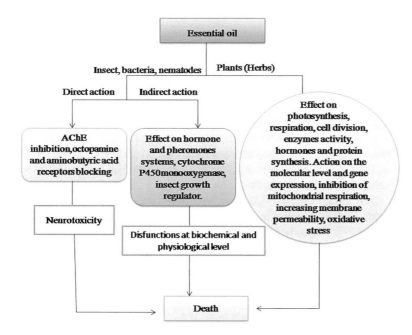

Fig. 8.3 General pathways of essential oils pesticidal actions

8.4 Essential Oil-Based Pesticides for Food Plant Production

Food products such as stored grains, fruits, and other cellulosic materials vulnerable to infestation by various pests, mostly the arthropods, may be preserved when the harmful activity of these insects and anthropods is inhibited. Essential oils are attracting widespread interest for their pesticidal activities like antibacterial, antifungal, antiviral, insecticidal, insect repellent, and deterrent activities (Pisoschi et al. 2018). Because of their hydrophobic/lipophilic nature, they interfere with basic metabolic, biochemical and physiological and behavioural functions of insects. Commonly, essential oils can be inhaled, ingested or skin absorbed by insects than effectively transmitted through the lipid bilayer of cell membranes, disrupting in ion transport, cellular material leakage, alteration in proton motive force-mediated electron flow, and eventually leading to death of pest. The varying bioefficacy and detoxifying activities of EOs sparked the concept for using it as a pesticide. They can be used as fumigants, granular formulations, or direct sprays, and have a variety of effects on insects, ranging from deadly toxicity to repellence and deterrence and the general pathways of essential oil pesticide action are represented in Fig. 8.3. (Mossa 2016). Literature survey has reveasled the efficacy of essential oils of several plant species such as Thymus, Mentha, Artemisia, Limnophila, Salvia, Rosmarinus and many others against many arthropod species like silverleaf whitefly, aphids, bihar hairy caterpillars, cabbage looper, diamondback moth, leaf rollers etc, which plays a significant role in harming the stored grain and horticultural crops (Park and Tak 2016).

8.5   Conclusion

Lipid oxidation has its own significance in the food industry, due to the formation of undesirable off-flavours and potentially toxic compounds. For the prevention antioxidants are being used. Synthetic antioxidants currently used in the food industry (BHA, PGA, BHT) have been found to possess several bad health effects because of their carcinogenic nature. In that sense, there is an interest building in search of natural antioxidants, especially plant-derived antioxidant compounds. Various studies have shown that essential oils are rich source of compounds exhibiting strong antioxidant activity, and that they can be used as natural antioxidants in the food, pharmaceutical, agrochemical and cosmetic industries. Many antioxidant enriched oil are used as edible films and edible coating for food preservation. Beside the antioxidant action essential oils are also used in food industry for the safety and preservation purposes in different ways like flavouring, odouring, repellent and key ingredient of active packaging of several food products.

References

Amaral AB, da Silva MV, Lannes SCDS (2018) Lipid oxidation in meat: mechanisms and protective factors - a review. Food Sci Technol (Campinas) 38:1–15

Amorati R, Foti MC, Valgimigli L (2013) Antioxidant activity of essential oils. J Agric Food Chem 61:10835–10847

Ayoughi F, Marzegar M, Sahari MA, Naghdibadi H (2011) Chemical compositions of essential oils of *Artemisia dracunculus* L. and endemic *Matricaria chamomilla* L. and an evaluation of their antioxidative effects. J Agric Sci Technol 13(1):79–88

Bakkali F, Averbeck S, Averbeck D, Idaomar (2008) Biological effects of essential oils–a review. Food Chem Toxicol 46(2):446–475

Baratta MT, Dorman HD, Deans SG, Biondi DM, Ruberto G (1998) Chemical composition, antimicrobial and antioxidative activity of laurel, sage, rosemary, oregano and coriander essential oils. J Essent Oil Res 10(6):618–627

Becker EM, Nissen LR, Skibsted LH (2004) Antioxidant evaluation protocols: food quality or health effects. Eur Food Res Technol 219(6):561–571

Botterweck AA, Verhagen H, Goldbohm RA, Kleinjans J, Van den Brandt PA (2000) Intake of butylated hydroxyanisole and butylated hydroxytoluene and stomach cancer risk: results from analyses in the Netherlands cohort study. Food Chem Toxicol 38(7):599–605

Burt S (2004) Essential oils: their antibacterial properties and potential applications in foods—a review. Int J Food Microbiol 94(3):223–253

Dhami A, Singh A, Palariya D, Kumar R, Prakash O, Rawat DS, Pant AK (2019) α-pinene rich bark essential oils of *Zanthoxylum armatum* DC. from three different altitudes of Uttarakhand, India and their antioxidant, in vitro anti-inflammatory and antibacterial activity. J Essent Oil-Bear Plants 22(3):660–674

Dhifi W, Bellili S, Jazi S, Bahloul N, Mnif W (2016) Essential oils' chemical characterization and investigation of some biological activities: a critical review. Medicines 3(4):25

Falleh H, Jemaa MB, Saada M, Ksouri R (2020) Essential oils: a promising eco-friendly food preservative. Food Chem 330:127268

Farahmandfar R, Asnaashari M, Pourshayegan M, Maghsoudi S, Moniri H (2018) Evaluation of antioxidant properties of lemon verbena (*Lippia citriodora*) essential oil and its capacity in sunflower oil stabilization during storage time. Food Sci Nutr 6(4):983–990

Farouk A, Fikry R, Mohsen M (2016) Chemical composition and antioxidant activity of *Ocimum basilicum* L. essential oil cultivated in Madinah Monawara, Saudi Arabia and its comparison to the Egyptian chemotype. J Essent Oil-Bear Plants 19(5):1119–1128

Ghorbani Gorji S, Smyth HE, Sharma M, Fitzgerald M (2016) Lipid oxidation in mayonnaise and the role of natural antioxidants: a review. Trends Food Sci Technol 56:88–102

Goswami S, Kanyal J, Prakash O, Kumar R, Rawat DS, Srivastava RM, Pant AK (2019) Chemical composition, antioxidant, antifungal and antifeedant activity of the *Salvia reflexa* Hornem. Essential oil. Asian J Appl Sci 12(4):185–191

Han F, Ma GQ, Yang M, Yan L, Xiong W, Shu JC, Xu HL (2017) Chemical composition and antioxidant activities of essential oils from different parts of the oregano. J Zhejiang Univ Sci B 18(1):79–84

Howard JA (1974) Autoxidation and antioxidants. Basic principles and new developments. Rubber Chem Technol 47:976–990

Huang D, Ou B, Prior RL (2005) The chemistry behind antioxidant capacity assays. J Agric Food Chem 53(6):1841–1856

Hussain AI, Anwar F, Shahid M, Ashraf M, Przybylski R (2010) Chemical composition, and antioxidant and antimicrobial activities of essential oil of spearmint (*Mentha spicata* L.) from Pakistan. J Essent Oil Res 22(1):78–84

Joshi A, Pant AK, Prakash O, Kumar R, Stocki M, Isidorov VA (2021a) Chemical composition, antimicrobial, and antioxidant activities of the essential oils from stem, leaves, and seeds of *Caryopteris foetida* (D. don) Thell. Indian. J Nat Prod Res 12(2):214–224

Joshi M, Himani KR, Prakash O, Pant AK, Rawat DS (2021b) Chemical composition and biological activities of *Nepeta hindostana* (Roth) Haines, *Nepeta graciliflora* Benth. and *Nepeta cataria* L. from India. J Med Herb 12(2):35–46

Joshi SC, Verma AR, Mathela CS (2010) Antioxidant and antibacterial activities of the leaf essential oils of Himalayan Lauraceae species. Food Chem Toxicol 48(1):37–40

Kanyal J, Prakash O, Kumar R, Rawat DS, Srivastava RM, Singh RP, Pant AK (2021) Study on comparative chemical composition and biological activities in the essential oils from different parts of *Coleus barbatus* (Andrews) Bent. ex G. Don. J Essent Oil-Bear Plants 24(4):808–825

Khayreddine B (2018) Essential oils, an alternative to synthetic food additives and thermal treatments. MedCrave Group LLC, pp 1–44

Koulivand PH, Ghadiri MK, Gorji A (2013) Lavender and the nervous system. Evid Based Complement Alternat Med. https://doi.org/10.1155/2013/681304

Kumar R, Kumar R, Prakash O, Srivastava RM, Pant AK (2019) Chemical composition in vitro antioxidant, anti-inflammatory and antifeedant properties in the essential oil of Asian marsh weed *Limnophila indica* L.(Druce). J Pharmacogn Phytochem 8(1):1689–1694

Kumar R, Prakash O, Pant AK, Isidorov VA, Mathela CS (2012) Chemical composition, antioxidant and myorelaxant activity of essential oils of *Globba sessiliflora* Sims. J Essent Oil Res 24(4):385–391

Laranjo M, Fernandez-Leon AM, Potes ME, Agulheiro-Santos AC, Elias M (2017) Use of essential oils in food preservation. In: Méndez-Vilas (ed) Antimicrobial Research: Novel bioknowledge and educational programs (Microbiology Book Series #6). Formatex Research Center, Badajoz, pp 177–188

Lawless J (1995) The illustrated encyclopedia of essential oils: the complete guide to the use of oils in aromatherapy & herbalism. Element Books Ltd, Rockport

Lourenço SC, Moldão-Martins M, Alves VD (2019) Antioxidants of natural plant origins: from sources to food industry applications. Molecules 24(22):4132

MacDonald-Wicks LK, Wood LG, Garg ML (2006) Methodology for the determination of biological antioxidant capacity in vitro: a review. J Sci Food Agric 86(13):2046–2056

Manion CR, Widder RM (2017) Essentials of essential oils. Am J Health Syst Pharm 74(9):153–162

Miguel G, Simoes M, Figueiredo AC, Barroso JG, Pedro LG, Carvalho L (2004) Composition and antioxidant activities of the essential oils of *Thymus caespititius, Thymus camphoratus* and *Thymus mastichina*. Food Chem 86(2):183–188

Miguel MG (2010) Antioxidant activity of medicinal and aromatic plants. A review. Flavour Frag J 25(5):291–312

Mossa AT (2016) Green pesticides: essential oils as biopesticides in insect-pest management. J Environ Sci Technol 9:354–378

Muriel-Galet V, Cerisuelo JP, Lopez-Carballo G, Lara M, Gavara R, Hernandez-Munoz P (2015) Development of antimicrobial films for microbiological control of packaged salad. Int J Food Microbiol 157(2):195–201

Palariya D, Singh A, Dhami A, Pant AK, Kumar R, Prakash O (2019) Phytochemical analysis and screening of antioxidant, antibacterial and anti-inflammatory activity of essential oil of *Premna mucronata* Roxb. leaves. Trends Phytochem Res 3(4):275–286

Park YL, Tak JH (2016) Essential oils for arthropod pest management in agricultural production systems. In: Essential oils in food preservation, flavor and safety. Academic, pp 61–70

Pisoschi AM, Pop A, Georgescu C, Turcuş V, Olah NK, Mathe E (2018) An overview of natural antimicrobials role in food. Eur J Med Chem 143:922–935

Politeo O, Jukić M, Miloš M (2006) Chemical composition and antioxidant activity of essential oils of twelve spice plants. Croat Chem 790(4):545–552

Prakash B, Kiran S (2016) Essential oils: a traditionally realized natural resource for food preservation. Curr Sci 110(10):1890–1892

Price S (1993) The aromatherapy workbook. Thorsons, London

Rădulescu M, Jianu C, Lukinich-Gruia AT, Mioc M, Mioc A, Şoica C, Stana LG (2021) Chemical composition, in vitro and in silico antioxidant potential of *Melissa officinalis* subsp. *officinalis* essential oil. Antioxidants 10(7):1081

Roginsky V, Lissi EA (2005) Review of methods to determine chain-breaking antioxidant activity in food. Food Chem 92(2):235–254

Shahat AA, Ibrahim AY, Hendawy SF, Omer EA, Hammouda FM, Abdel-Rahman FH, Saleh MA (2011) Chemical composition, antimicrobial and antioxidant activities of essential oils from organically cultivated fennel cultivars. Molecules 16(2):1366–1377

Shahsavari N, Barzegar M, Sahari MA, Naghdibadi H (2008) Antioxidant activity and chemical characterization of essential oil of *Bunium persicum*. Plant Foods Hum Nutr 63(4):183–188

Singh S, Das SS, Singh G, Schuff C, de Lampasona MP, Catalan CA (2014) Composition, in vitro antioxidant and antimicrobial activities of essential oil and oleoresins obtained from black cumin seeds (*Nigella sativa* L.). Biomed Res Int 2014:Article ID 918209

Skancke A, Skancke PN (1988) General and theoretical aspects of quinones. Quinonoid Compd 1:1–28

Spiridon I, Bodirlau R, Teaca CA (2011) Total phenolic content and antioxidant activity of plants used in traditional Romanian herbal medicine. Centr Eur J Biol 6(3):388–396

Svoboda KP, Deans SG (1992) A study of the variability of rosemary and sage and their volatile oils in British market: their antioxidative properties. Flavour Fragr J 7:81–87

Thapa P, Prakash O, Rawat A, Kumar R, Srivastava RM, Rawat DS, Pant AK (2020) Essential oil composition, antioxidant, anti-inflammatory, insect Antifeedant and sprout suppressant activity in essential oil from aerial parts of *Cotinus coggygria* Scop. J Essent Oil-Bear Plants 23(1):65–76

Tongnuanchan P, Benjakul S (2014) Essential oils: extraction, bioactivities, and their uses for food preservation. J Food Sci 79(7):1231–1249

Viuda-Martos M, Ruiz Navajas Y, Sánchez Zapata E, Fernández-López J, Pérez-Álvarez JA (2010) Antioxidant activity of essential oils of five spice plants widely used in a Mediterranean diet. Flavour Frag J 25(1):13–19

Wang W, Wu N, Zu YG, Fu YJ (2008) Antioxidative activity of *Rosmarinus officinalis* L. essential oil compared to its main components. Food Chem 108(3):1019–1022

Wei A, Shibamoto T (2010) Antioxidant/lipoxygenase inhibitory activities and chemical compositions of selected essential oils. J Agric Food Chem 58:7218–7225

Wildwood C (1996) The encyclopedia of aromatherapy. Healing Arts Press, Rochester

Yang SA, Jeon SK, Lee EJ, Shim CH, Lee IS (2010) Comparative study of the chemical composition and antioxidant activity of six essential oils and their components. Nat Prod Res 24(2):140–151

Zhang H, Chen F, Wang X, Yao HY (2006) Evaluation of antioxidant activity of parsley (*Petroselinum crispum*) essential oil and identification of its antioxidant constituents. Food Res Int 39(8):833–839

Chapter 9
Positive and Negative Impacts of the Use of Essential Oils in Food

Hartati Soetjipto and November Rianto Aminu

9.1 Introduction

It is well-known that essential oil demonstrated a wide range of biological and pharmacological activities including the antibacterial, antioxidant, antiviral, insecticidal, antifungal, etc (Ipek et al. 2005; Abu-Shanab et al. 2005; De martino et al. 2009). In addition, this compound is also often used in the food industry specially to improve the flavor and the organoleptic properties of different types of foods, and in a variety of household products (De Martino et al. 2009; Evandri et al. 2005). Kelen and Tepe (2008) also reported that essential oil can be used for cancer treatment, food preservations, aromatherapy, and in the perfumery industries.

Essential oils are volatile and aromatic oil extracted from the parts of plant such as roots, stem bark, wood, leaves, flowers, peel of the fruit, fruits, seeds and the whole plant (Deans and Ritchie 1987; Hammer et al. 1999; Sanchez et al. 2010). Whole parts of the plants are usually used for the extraction of plants. This oil plays an important role in plants as protection, communication, chemical protections that these secondary metabolites present, also is decisive in plants resistance against pathogen and herbivores (Wink 1988; de Oliveira et al. 2018). Essential oils have been used since time immemorial as part of folks religious rituals and medicine. Along with the passage of time its use has been developed not only in folk religious ceremonies and medicine but has been penetrated to the food, cosmetic and other industries as a flavoring, additives and fragrance (Singh et al. 2002; Soetjipto 2018.). These oils also contain a lot of secondary metabolite compounds that show various

H. Soetjipto (✉) · N. R. Aminu (✉)
Chemistry Department Faculty of Science and Mathematics, Universitas Kristen Satya Wacana, Salatiga, Indonesia
e-mail: hartati.sucipto@uksw.edu; november.aminu@uksw.edu

© The Author(s), under exclusive license to Springer Nature Switzerland AG 2022
M. Santana de Oliveira (ed.), *Essential Oils*, https://doi.org/10.1007/978-3-030-99476-1_9

bioactivities such as inhibiting bacterial growth, yeast and molds (Chorianopoulos et al. 2008; Burt and Reinders 2003; De Martino et al. 2009; Nazzaro et al. 2013).

Essential oil is not a single compound but are composed of many components with varying levels. The concentration of components is quite different, and major components can constitute up to 85% of the essential oils, while other components can be found only in traces (Cabarkapa et al. 2016). The main constituents of essential oils are terpenoids, especially mono terpenes (C_{10}) and sesquiterpenes (C_{15}), as well as diterpenes (C_{20}). This compound are natural products that are concentrated and have a strong sharp aroma produced by plants as secondary metabolites. As secondary metabolites, essential oils are only found in relatively small amounts, around 1%, but are very important for plant defense because essential oils have antimicrobial properties (Tajkarimi et al. 2010).

There are several methods of extracting essential oils, including solvent extraction, distillation (hydro distillation, hydro diffusion, steam distillation) and effleurage. The disadvantage of these traditional methods more expensive because require a lot of energy and solvents. Essential oils are usually obtained using physical extraction methods, hydro and steam distillation of plant materials, for some materials extraction was obtained by cold pressing (Guenther 2006). To meet the needs in the field of pharmacology and food, essential oils are usually obtained using the steam distillation method. Meanwhile, for other fields such as the fragrance, cosmetics, and perfume industries, extraction methods with lipophilic solvents are used, even the supercritical Carbon dioxide method is preferred and going more attractive (Donelian et al. 2009).

Extraction of essential oils using the microwave assisted hydro distillation (MAH) method has been proven to save extraction time and solvents. By using MAH, the time required to obtain the essential oil is only about 20 min, whereas with Clevenger Hydro distillation it takes time 180 min. The MAH method offers significant advantages over conventional hydro distillation and can therefore replace it on a pilot and industrial scale (Elyemni et al. 2019).

Essential oils are widely used by people every day as a food flavor enhancer, for air freshener, aroma therapy, etc. About 3000 Essential oils are known, 300 of which are commercially important, mainly used in the flavors and fragrances market (Burt et al. 2003). Now the use of essential oils is growing after it is known that their bioactivity is very diverse. For example, it is an antioxidant, antiseptic, medicinal, sedative, anti-inflammatory, spasmolytic and anesthetic (Bakkali et al. 2008; Thaweboon and Thaweboon 2009; Chorianopoulos et al. 2008; Burt and Reinders 2003; De Martino et al. 2009). Some essential oils that are widely traded in the world market include tea tree oil, rosemary, lavender, orange oil, lemon oil, corn mint oil, and eucalyptus oil (Ridder 2021). Meanwhile, several Indonesian essential oils that entered the world market include vetiver, sandalwood, aloes, cinnamon, patchouli, cloves, basil, jasmine, nutmeg, lemon, celery and grapes (Konsulat Jenderal Republik Indonesia 2021).

Various benefits of essential oils have been known in the food sector, in addition to their aroma which can improve the flavor and the organoleptic properties of food, as well as their ability as antimicrobial (antibacterial, antifungal, antiviral) causes essential oils to act as preservatives at the same time. However, not all essential oils can be used in food, the sharp aroma is not suitable for all kinds of food, it is often become an obstacle when applied to food. In this chapter we discuss the role of essential oil especially the positive and negative impacts of using essential oils in food.

9.2   Chemical Compounds of Essential Oil

Essential Oils are composed of a complex mixture of terpenes, terpenoids, phenyl-propanoids, and various compounds of low molecular weight, which can contain 20–60 compounds at different concentration. (Bakkali et al. 2008; Nikmaram et al. 2018; Wińska et al. 2019; De Matos et al. 2019). Terpenes are hydrocarbons formed from several isoprene units (2- methyl- 1,3-butadiene) C_5H_8, bonded to each other in a special regular pattern where the head of one isoprene molecule will bind to the tail of the other isoprene molecule and so on until it forms certain molecules (Fig. 9.1).

Terpenes are synthesized in plant tissues through the mevalonic acid pathway, sometimes accompanied by oxidation reactions, cyclization or rearrangement to form terpenoids. Several functional groups are often found in terpenoid molecules such as alcohols, ethers, aldehydes, ketones, esters (Rassem et al. 2016). It is known that there are two groups of terpenes which are the main constituents of essential oils, monoterpenes (C_{10}) and sesquiterpenes (C_{15}). Examples of molecules include monoterpenes such as pinene, myrcene, thujene, terpinene, etc., while examples for sesquiterpenes include germacrene, caryophyllene, zingiberene.

In addition to the terpene group, aromatic compounds and other groups such as aldehydes, ketones and esters were also found as an essential oil component (Baldim et al. 2019), for example aldehyde, alcohols, phenols, and methoxy derivatives of essential oils are cinnamaldehyde, cinnamic alcohol, eugenol, elemicin, estragole, and anethole (Wang et al. 2009). Beside terpenes and terpenoids, simple hydrocarbons such as alkane, alkene, benzenoids are also known as non-terpenoid

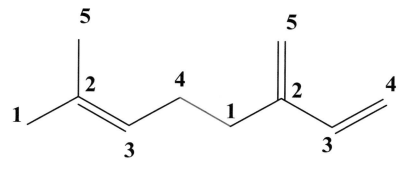

Fig. 9.1   2 isoprene units bonded to form a monoterpene molecule

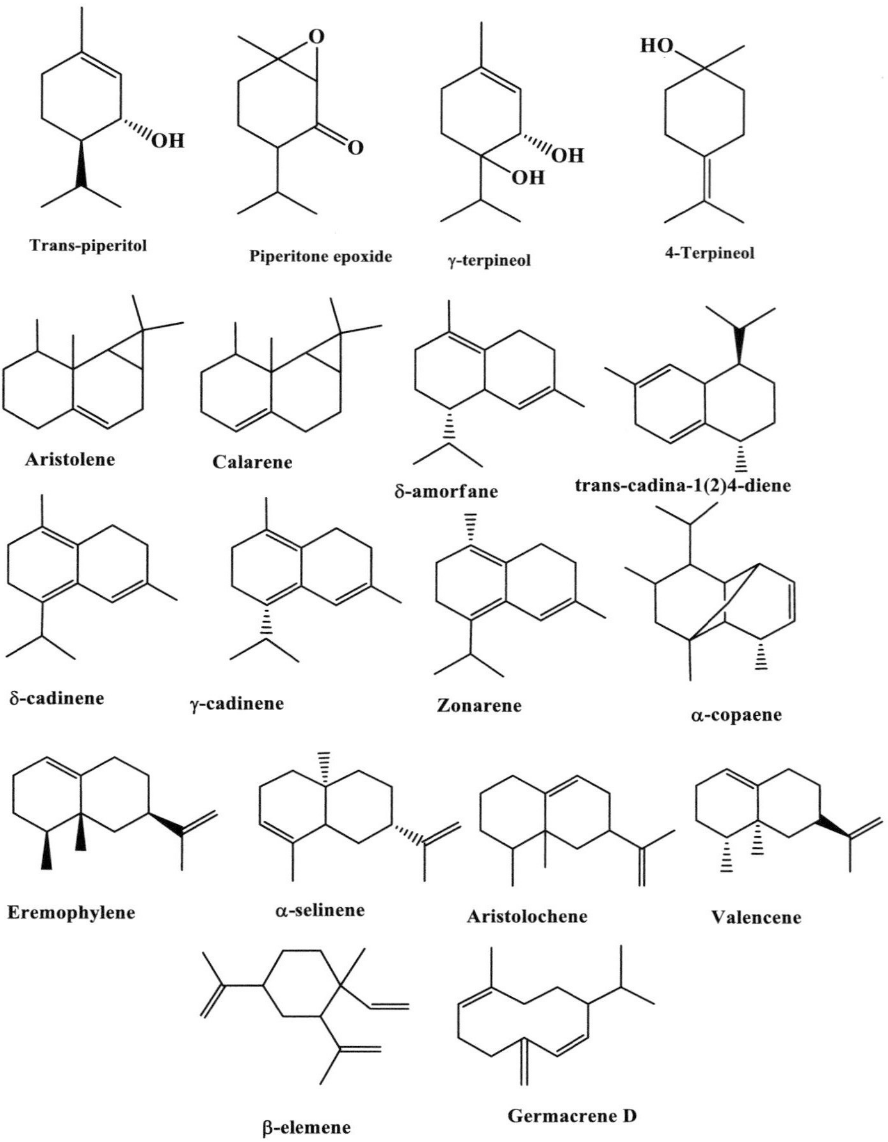

Fig. 9.2 Some molecules of essential oils

hydrocarbons (Trombetta et al. 2005). Some examples of the molecular structure of essential oils can be seen in Fig. 9.2.

Essential oils have characteristics that are characterized by several dominant compounds with levels of 20–70%, while the rest is a mixture of many compounds present in trace amount (Bakkali et al. 2008). For example, fennel seeds (*Foeniculum vulgare* Mill) essential oil is very popular as a flavor enhancer in food and beverages. This oil is composed of 30 compounds, mainly Estragole (35, 51%), trans-Anethole (29.67%), fenkon (1-1,2,3-trimethyl bicyclic) 2.2.1-2-heptanol (22.70%) and the rest is a mixture of compounds with levels less than 3%. Trans-caryophyllene (24.19%) of toothache plant (*Spillanthes paniculata* Wall) essential oil, Limonene (43.4%) of *Citrus lemon* (L) Burm essential oil (Soetjipto 2018). Essential oils contain several important compounds known as bioactive components such as phenols, mono and sesquiterpenes, alcohols, ethers, aldehydes, ketones and carbohydrates (Speranza and Corbo 2010; Tabassum and Vidyasagar 2013)

9.3 Biological Effects

All essential oils have various biological effects such as antimicrobial (antibacterial, anti-fungal, anti-viral), antioxidant, insecticide, anti-cancer, anti-inflammatory, sedative, antiseptic, anesthetic, etc. The various bioactivities of essential oils make essential oils widely used in various industrial fields, such as the food industry, cosmetics, health and other fields (Aureli et al. 1992). In the food sector, antioxidant, antibacterial and antifungal activities are very important because these three properties can be used to complement the desired food properties.

The existence of various properties of biological effects possessed by essential oils, it is necessary to realize that essential oils are composed of many components so that it is also possible for a synergistic effect to occur between all existing components or between a few dominant molecules. For example, limonene, one of the components of citrus essential oil, is more effective as an antibacterial in the form of its essential oil than when used as a single compound, meaning that there is a synergistic effect that plays a role (Van Vuuren and Viljoen 2007).

Among many bioactive compounds, monoterpenes are reported as the most abundant components, mostly 90% in occurrence availability (Bakkali et al. 2008; Caputo et al. 2018). The essential oils of black pepper (*Piper nigrum* L. (Piperaceae)), clove (*Syzygium aromaticum* (L.) Merr. & Perry (Myrtaceae)), geranium (*Pelargonium graveolens* L'Herit (Geraniaceae)), nutmeg (*Myristica fragrans* Houtt. (Myristicaceae)), oregano (*Origanum vulgare* ssp. hirtum (Link) Letsw. (Lamiaceae)) and thyme (*Thymus vulgaris* L. (Lamiaceae)) demonstrated strong inhibitory effects against all tested microorganisms while the main components demonstrated different levels of inhibition (Dorman and Deans 2000; Viuda-Martos et al. 2008).

Related to its antimicrobial properties, essential oils can be used to prevent the growth of spoilage microbes, as food preservatives as well as extending the shelf life (Fratianni et al. 2010). The tendency to avoid chemicals as preservatives in the food sector, as well as the antimicrobial activity possessed by essential oils provide opportunities for the use of essential oils as preservatives in the food sector.

9.3.1 Antibacterial

Antibacterial activity is an activity possessed by essential oils to kill or inhibit the growth of bacteria. Generally, Gram-positive *Staphylococcus aureus* (*S. aureus*) and Gram-negative *Escherichia coli* (*E. coli*) bacteria are used as test bacteria to see the antibacterial effect of the essential oil. *Staphylococcus aureus* (*S. aureus*) is mainly responsible for food poisoning, toxic shock syndrome, endocarditis, and osteomyelitis (Hennekinne et al. 2012). De Martino et al. (2009) reported that *E. coli* and B. *cereus* 4384 were the most sensitive microorganisms, for their research using seven Lamiaceae essential oils (*Hyssopus officinalis* L., *Lavandula angustifolia* Mill., *Melissa officinalis* L., *Ocimum basilicum* L., *Origanum vulgare* L., *Salvia officinalis* L., *Thymus vulgaris* L.), then followed by B. *cereus* 4313, E. *faecalis* and *S. aureus*. Otherwise the most resistant strain was *P. aeruginosa*, showing a low sensitivity. The essential oils of *Origanum vulgare* L and *Thymus vulgaris* L demonstrated the powerful antibacterial activity against the pathogenic bacteria.

Li et al. (2019) showed that some bacteria are responsible for food spoilage and even foodborne diseases. In their research, it was reported that Gram-positive bacteria were more susceptible to essential oils than Gram-negative bacteria because of the condition of the bacterial cell membrane system. The cell membrane of Gram-negative bacteria contains a hydrophilic lipopolysaccharide layer (Shakeri et al. 2014).

Some of the components that make up essential oils which are classified as aldehydes, phenols, alcohol terpenoids have been shown to have high antibacterial effects (Dhifi et al. 2016). Alpha pinene and 1,8-cineol are also components of essential oils that have antibacterial properties (Soković et al. 2010). Linalool is an example of a terpenoid alcohol that is able to damage bacterial cell membranes so that it is reported to be able to inhibit the growth of bacterial cells (Di Pasqua et al. 2007). However, thymol and carvacrol, which are components of several types of *Thymus* and *Oreganum*, can inhibit some pathogenic bacterial strains such as *Escherichia coli, Salmonella enteritidis, Salmonella choleraesuis,* and *Salmonella typhimurium* demonstrated the stronger antibacterial activity than linalool (Penalver et al. 2005). The essential oils can serve as a powerful tool to reduce the bacterial activity (Pisoschi et al. 2018).

There are several methods commonly used to determine antibacterial activity. One method that is often used is the agar diffusion method using paper discs. Paper discs that have been given a certain level of essential oil solution are placed on the media so that they are then incubated for 24 h. The appearance of a transparent zone around the paper disc is measured in diameter and is considered an area of resistance. The antibacterial strength that appears in the form of a transparent zone is expressed as the Minimum Inhibitory Concentration or the minimum dose that indicates inhibition of bacterial growth. Antibacterial activity is considered low if the diameter of the inhibition zone is less than 7 mm, Medium if the DDH is between 7–8 mm and Strong if more than 8 mm (Elgayyar et al. 2001). For example, the strength of antibacterial activity of lime (*Citrus aurantifolia* Swingle) essential oil against *B. subtillis* ATCC 6051 Gram (+) and *P. aeruginosa* FNCC0063 Gram (−) bacteria (1000 µg and 2000 µg) respectively in Medium scale. On the contrary to *Citrus limon* (L) Burm essential oil at a dose of 2000 µg it demonstrated a Strong antibacterial activity against *B. subtilis* ATCC6051 (Gram +) and at 3000 µg also has a Strong antibacterial against *E. coli* 0091IFO (Gram -) (Soetjipto 2018). Other criteria used by Djabou et al. 2013, diameters of inhibition zone (DIZ) were appreciated as follows: Not sensitive (diameter 8.0 mm), moderately sensitive (8.0 < diameter < 14.0 mm), sensitive (14.0 < diameter < 20.0 mm), and extremely sensitive (diameter 20.0 mm).

9.3.2 Antifungal

Antifungal is an activity shared by most essential oils. This activity is very necessary to prevent or reduce food spoilage caused by fungi. There are many examples of fungi that have been reported to be present in many food spoils, such as *Penicillium, Aspergillus, Fusarium, Cladosprium* and many others. Damage due to fungi will cause food to become soft, watery, change taste, color and aroma and it is possible to form toxins (mycotoxins) so that food becomes toxic if consumed. Aspergillus flavus *and A. parasiticus* are the most important species of fungi responsible for aflatoxin contamination of crops prior to harvest or during storage (Yu et al. 2004). Essential oils are reported to be able to inhibit the growth of fungi and the formation of mycotoxin toxins which are very dangerous for food (Maurya et al. 2021).

The antifungal activity of essential oils provides an opportunity for the use of essential oils as preservatives in the food sector to reduce the use of chemicals as synthetic preservatives in food. The ability of this activity is closely related to the content of the constituent components of essential oils such as the content of Trans-anethole, zingiberene, menthol and thymol which is the dominant component of essential oils of fennel, ginger, mint and thyme respectively. These components have anti-fungal activities at 50, 80, 50 and 50% (oil/DMSO; v/v), respectively (Fernanda et al. 2012). Razzaghi-Abyaneh et al. (2008), reported that carvacrol and thymol are the effective compounds against food-borne fungi.

According to Freiesleben and Jager (2014), antifungal compounds will inhibit the growth and activity of fungi by disorganized the structure and function of fungal cell membranes and organelles, and/or causing inhibition of the protein or nuclear material synthesis. Some examples of essential oils that have antifungal effects, such as essential oil of *Thymus vulgaris*, are able to inhibit the growth and formation of aflatoxins from *Aspergillus parasiticus* and A. flavus (Razzaghi-Abyaneh et al. 2009; Kumar et al. 2008). Fennel (Foeniculum vulgare) essential oil with the main content of trans-anethol showed a strong antifungal effect (Patra et al. 2002), also mint essential oil (*Mentha piperita*) with the main component's menthol and 1,8-cineol showed a strong antifungal effect (Griffin et al. 2000). Dwivedy et al.

(2017) studied the potentiality of essential oil of Mentha cardiac as a green alternative to protect fungal contamination in stored dry fruits. The content of nerol (21.89%), citronellal (27.53%), and citronellol (25%) in Cymbopogon Nardus essential oil has been demonstrated to have antifungal effects against food-contaminating *Candida albicans* and *S. aureus* (Cunha et al. 2020).

9.3.3 Antivirus

Several essential oils showed an antiviral effect against the test virus. Viruses require living cells to develop and survive, so that in general for the food sector, viruses will attack the fresh plants/animals or parts such as fresh ready-to-eat food. Therefore, the use of essential oils as food preservatives is almost never used. Some essential oils show antiviral effect against several pathogenic viruses such as herpes simplex virus (HSV-1 and HSV-2), dengue virus type 2, influenza virus, poliovirus, Junin virus and coxsackievirus B1 (Tariq et al. 2019). Wani et al. (2021) reported that the essential oils of cinnamon, bergamot, lemongrass, thyme, lavender have potential as antivirals for influenza type A viruses. Capsid disintegration and virus expansion are the main mechanisms of essential oils causing antiviral effects. This activity will prevent the virus from infecting host cells by adsorption through capsid. Essential oils also inhibit hemagglutinin which is an important membrane protein for many viruses. This membrane is in charge of allowing the virus to enter the host cell (Wani et al. 2021)

In the recent years non-thermal disinfection technologies have widely been used in fresh produce disinfection (Seymour et al. 2002). Birmpa et al. (2018), reported that they have been applied for the decontamination of a wide variety of food products. In advance, essential oils derived from the natural sources have showed broad-spectrum activity, including antifungal, antibacterial and antiviral activities. They also suggest that the combination of non-thermal disinfection technologies with Essential Oils could find potential applications for decontamination of bacteria and viruses in the food industry.

9.3.4 Antioxidant

Antioxidant compounds are very important compounds added to food to prevent food damage due to oxidation. Gutiérrez et al. (2006) reported that antioxidant compounds are very important for the human body. Because basically oxygen is a potentially toxic element that can turn into harmful compounds through metabolism to form a more reactive oxygen such as hydrogen peroxide, hydroxyl free radicals, superoxide and the singlet oxygen, all known as active oxygen. Oxidative stress plays an important role in causing degenerative diseases that are very detrimental to humans such as atherosclerosis, cardiovascular disease, diabetes, cancer (Gutteridge

1993 in Torres-Martínez et al. 2018). Essential Oils are known as rich sources of potential antioxidants that can be investigated to prevent oxidative damage (Yanishlieva-Maslarova 2001). For example, the level of nitric oxide and H_2O_2 production can be decreased by essential oil, whereas no synthase demonstrated a potential to treat oxidative damage (Karimian et al. 2012).

Some essential oils are reported to have antioxidant effects so that essential oils have a potential to be used as natural preservatives in food products. As a natural agent for food preservation, some of essential oil had been qualified as natural antioxidants and proposed as potential substitutes for synthetic antioxidants (Ruberto and Baratta 2000 in Emami et al. (2007)). The ability of essential oils as natural antioxidants is related to the presence of phenol and terpene compounds inside. Both of these compounds are known to have the free radical scavenging activity (Edris 2007). Several investigations have studied the antioxidant activity of monoterpenes and diterpenes or essential oils in vitro. Gamma-terpinene was reported showed a very effective antioxidant (Grassmann 2005). Torres-Martínez et al. (2018) reported that the essential oil of aerial part of *Satureja macrostema* (Moc. and Sessé ex Benth.) showed antioxidant effect. The six components that make up the essential oil include Thymol, Caryophyllene, Limonene, Linalool, Pulegone and menthone which Thymol shows the highest antioxidant effect compared to other components. The essential oils of rosemary (*Rosmarinus officinalis* L.) demonstrated strong inhibition of Lipid Peroxidation in systems of induction (Bozin et al. 2007). Citrus oil also demonstrated an effective antioxidant by DPPH assay (Choi et al. 2000). Bozin et al. 2006 also reported that basil, oregano, and thyme essential oils are rich in monoterpene phenols especially, thymol and carvacrol. The three essential oils mentioned have been studied and reported to have antioxidant activity.

The antioxidative capacity of essential oil is believed to be responsible for the health promoting properties of fruits and vegetables. Three main ways of antioxidant action of carotenoids i.e., quenching of singlet oxygen, hydrogen transfer, or electron transfer (Grassmann 2005).

9.4 Essential Oil and Positive Impact in Food

Laranjo et al. 2017 reported that according to Sauceda (2011). It is estimated that more than 20% of all food produced in the world is spoiled by microorganisms. Food spoilage caused by microbes usually becomes watery, changes taste and aroma. Therefore, it is necessary to add antimicrobial preservatives to foods that use them for a long time (storage period). Based on its work, it is distinguished between food preservatives and food safety. Antimicrobials are used in food to control natural spoilage processes (food preservation) and to control microbial growth (food safety) (Tajkarimi et al. 2010).

Spices have been used traditionally as flavor enhancers, and there have an antimicrobial effect help to extend the shelf life of the food product. One of the natural compounds that have an antimicrobial effect is essential oil. Why are essential oils so useful in the food sector? Essential oils have antimicrobial abilities (antibacterial, anti-fungal and antiviral) so that they can inhibit/kill microbes that grow which will cause food spoilage/damage. These oils are not only able to inhibit food spoilage bacteria/fungi (as a food preservation) but also other pathogenic microbes, so that at the same time it can prevent the transmission of disease through food (as a Food safety).

The abundance of information about the ability of essential oils as antimicrobial and antioxidant makes essential oils increasingly well-known and increasingly playing an important role in various fields, including in the food sector as food preservatives (Hashemi et al. 2017; Pandey et al. 2017). Essential oil also used in the various kinds of cereals, antimicrobial packing of the food items, edible thin film, nano emulsion, preservation of the fruits and vegetables, soft drinks, as the flavoring agents in the carbonated drinks, as the major ingredients in soda/citrus concentrates, seafood preservations, fish, etc. (Mahato et al. 2019).

From what has been studied, it is proven that essential oils can be used as natural food preservatives and pathogen control method in food materials such as meat products, fish, vegetables, rice, fruits and dairy products (Mihai and Popa 2013). Study results show that there are several factors greatly affect the work of essential oils as food preservatives, for example type, effects on organoleptic properties, composition, concentration, biological properties of the antimicrobial, the target microorganism and processing and storage conditions of the targeted food product (Gutierrez et al. 2009).

9.4.1 Food Flavor Agent

Some of the positive effects of using essential oils in food include being able to improve the taste of food (Food Flavor agent), for example the addition of rosemary or marjoram essential oils at a concentration of 200 mg/kg in beef patties improving the flavor of the patties (Mohamed and Mansour 2012). The use of spices as a flavor enhancer in food has been known traditionally for a long time, because it has a unique aroma that causes the food to have a distinctive taste that can be recognized differently from other foods. The aroma that arises from spices comes from the essential oils they contain, so that spices are also an important source of essential oils, for example Indian cuisine curry has a distinctive aroma that comes from curry leaves (*Murraya koenigii*). Curry leaves contain essential oils with a distinctive aroma as flavoring agent for the food in India, Sri Lanka, Africa (Bonde et al. 2007; Jain et al. 2017). The major chemical of essential oil composition of Murraya koenigii (Curry leaf) obtained by steam distillation is composed of Caryophyllene (37.98%), Napthalene (16.30%), Azulene (9.69%), Cyclopentadecanone, 2-hydroxy (8.46%) and α-Pinene (6.51%) (Jamil et al. 2016).

Another example is Oregano (*Origanum vulgare* L), it was known as wild marjoram, an important culinary herb, also a common spice used in Italian foods such as pizza, salads. The addition of oregano in bologna sausages also acceptable to consumers (Viuda-Martos et al. 2010 in Laranjo et al. 2017). The distinctive aroma of oregano is closely related to the essential oil content in it, especially pinene, limonene, thymol and carvacrol. Leyva-López et al. (2017) reported that oregano essential oil is mainly composed of carvacrol, thymol, γ-terpinene, *p*-cymene, linalool, beta-citronellol, 1,8-cineol and beta caryophyllene.

Lemongrass (*Cymbopogon citratus*) is also one of the tropical herbs that is widely used as a flavor enhancer in Southeast Asian foods such as Thai, Indonesian, Filipino, Sri Lankan and Indian, because of its distinctive aroma. The strong lemon scent that emerges from the lemongrass sets it apart from the common citronella type (*C. nardus*). The distinctive aroma of lemon is closely related to the content of essential oils in it, especially citral and myrcene (Barbosa et al. 2008), in addition to geraniol, citronellal, and limonene (Majewska et al. 2019; Mansour et al. 2015; Farias et al. 2019; Kasali et al. 2001). Some foods that often use the taste of lemongrass are soups, processed meats and drinks. Concerning their dietary intake, essential oils are generally considered as safe (GRAS) for their intended use by the U.S. Food and Drug Administration (FDA) (Smith et al. 2005).

9.4.2 Food Preservation

Application of common spices and aromatic plants in cooking is not only related to their impact on sensory and textural properties of foods but also for their antibacterial and health benefits effect (Sobrino-Lopez and Martín-Belloso 2008). The presence of essential oils in food is also able to inhibit the growth of contaminant bacteria and fungi whose presence causes food spoilage. The essential oils of thyme (*Thymus vulgaris* L.), oregano (*Origanum vulgare* L. ssp. Hirtum) and Lemon (*Citrus limon* L.) were effective to inhibit spore germination (Vitoratos et al. 2013). As a food preservative, essential oils are very effective because their high phenolic and terpenoid content makes essential oils have high antimicrobial and antioxidant abilities. As we know the presence of microbes and oxidation reactions are the main causes of food spoilage.

The main ability of essential oils which include antibacterial, antifungal and antioxidant is the main reason essential oils deserve to be named as natural preservatives. Why do we need natural preservatives? Because now many unwanted diseases arise as a result of the use of synthetic preservatives, usually associated with free radicals (Viuda-Martos et al. 2008; Sauceda 2011 in Bhavaniramya et al. 2019). Several diseases such as cancer, stroke, heart disease, neurodegenerative and other degenerative diseases are mostly associated with free radicals, which can be prevented by using free radical scavenging compounds, or antioxidant compounds (Hale et al. 2008). Therefore, essential oils which are known to be rich in phenolic and terpenoid compounds have antimicrobial and antioxidant effects which are very useful for application to edible products (De Oliveira et al. 2011). For example, ten essential oils that have been used as food preservatives can be seen in Table 9.1.

Table 9.1 Essential oils as food preservatives

Essential oil	Main constituents	Antioxidant and antimicrobial effect against	Aroma notes	References
Citrus aurantifolia Swingle (lime)	Limonene, ß-pinene, terpineol, α- terpinolene, gamma-terpinene and citral	*B. subtillis, P. aeruginosa, E. coli*	Fresh	Soetjipto (2018), Lin et al. (2019), and Spadaro et al. (2012)
Citrus limon (L) Burn	Limonene, ß-myrcene, geranyl acetate, ß-pinene, terpineol, α- terpinolene	*B. subtillis, P. aeruginosa, E. coli, Listeria monocytogenes*	Fresh	Soetjipto (2018), Ben Hsouna et al. (2017), and Settani et al. (2012)
F. vulgare Mill (Fennel)	Estragole, trans anethol, fenkon, limonene, α-pinene.	*B. subtillis, E. coli*	Sweet	Soetjipto (2018), Raal et al. (2012), Ruberto and Baratta (2000), Miguel et al. (2010), and Lo Cantore et al. (2004)
Allium sativum (Garlich)	Diallyl di sulfide, di allyl tri sulfide, allyl methyl trisulfide, allyl tetra sulfide	*St. aureus, Salmonella typhimurium, L. monocytogene, E. coli, Campylobacter jejuni*	Pungent, spice	Perricone et al. (2015), Mnayer et al. (2014), Banerjee et al. (2003), Kim et al. (2004), and Corzo-Martínez et al. (2007)
Cinnamomum zeylanicum	Cinnamaldehyde, eugenol, copaene, β-caryophyllene, linalool, eucalyptol, and eugenol	*E. coli, P. aeruginosa, Ent. faecalis, S aureus, S. epidermidis, methicillin-resistant, Klebsiella pneumoniae, Salmonella sp., Vibrio parahaemolyticus, L. innocua, B. cereus, A. niger*	Sweet, wood, spice	Perricone et al. (2015), Alizadeh Behbahani et al. (2020), Gogoi et al. (2021), and Moarefian et al. (2013)

(continued)

Table 9.1 (continued)

Essential oil	Main constituents	Antioxidant and antimicrobial effect against	Aroma notes	References
Syzygium aromaticum Clove	Eugenol, eugenyl acetate, caryophyllene, α-pinene, β-pinene, α-limonene	*B. brevis, B. subtilis, C. botulinum, E. faecalis, Candida spp., A. flavus, A. niger, E. coli, K. pneumoniae, P. aeruginosa, St. aureus, Salmonella spp., L. monocytogene, Serratia sp*	Sweet, spice, wood	Perricone et al. (2015), Selles et al. (2020), and Mohamed et al. (2013)
Ocimum basilicum Basil	Linalool, methylchalvicol or citral, eugenol, methyl eugenol, methyl cinnamate, 1,8-cineole, caryophyllene, camphor, thymol, methyl cinnamate, eugenol, methyl isoeugenol, and elemicine.	*B. brevis, E.coli, A. flavus, A. niger, E. faecalis, E. coli, K. pneumoniae, P. aeruginosa, S. aureus, B. subtilis, C. albicans.*	Fresh, sweet, herb, spice,	Perricone et al. (2015), Joshi (2014), and Chenni et al. (2016)
Zingiber officinale Ginger	β-sesquiphellandrene, zingiberene curcumene, camphene, β-bisabolene	*A. flavus, A. niger, E. faecalis, E. coli, K. pneumoniae, P. aeruginosa, S. aureus, B. subtilis, C. albicans, A. niger, Penicillium spp*	Pungen, spice	Perricone et al. (2015), Sharma et al. (2016), Mahboubi (2019), Kiran et al. (2013), and Bellik (2014)
Origanum vulgare (Oregano)	Sabinyl monoterpenes, terpinen-4-ol, γ-terpinene, carvacrol, thymol	*B. cereus, B. subtilis, C. botulinum, E. faecalis, E. coli, S. aureus, A. niger, L. monocytogenes, K. pneumoniae, P. aeruginosa, Salmonella*	Spice, herb	Perricone et al. (2015), Han and Parker (2017), Teixeira et al. (2013), and Kosakowska et al. (2021)
Peppermint (*Mentha piperita*)	Menthol, menthone, menthylacetate, menthofurane, 1,8-cineole	*B. brevis, S. aureus, V. cholerae, E. faecalis, E. coli, K. pneumoniae, P. aeruginosa, A. flavus, A. niger, S. pyogenes*	Fresh, herb	Perricone et al. (2015), Beigi et al. (2018), Singh et al. (2015), and Rasooli et al. (2008)

There are still many and will continue to increase the use of essential oils as food preservatives.

The use of essential oils as preservatives in edible products is not only added directly to food but it can also be added to food packaging (Amorati et al. 2013). The development of research on food packaging is very important in maintaining food quality, in addition to extending shelf life, maintaining food appearance and protecting food from damage (Fang et al. 2017). Sources of natural preservatives that are applied to food packaging are a new breakthrough in developing packaging products that are easily biodegradable and reduce packaging waste.

The addition of natural preservatives to packaging materials can be added directly or through encapsulation (Rodríguez et al. 2021). The results showed that the essential oil from lemon peel was very effective in its bioactivity against Gram-positive bacteria so that it has the potential to be a candidate for use in packaging preservative systems (Settanni et al. 2012).

Currently essential oils application in active food packaging are strongly linked to their incorporation bio-active compounds of essential oil directly in food or edible/biodegradable food packaging (Akram et al. 2019). The bioactivity of essential oils as antimicrobials and antioxidants has proven to be very efficient in controlling food-borne pathogens (Salehi et al. 2018; Mahomoodally et al. 2019).

Antimicrobial agents have been successfully added to edible composite films and coatings based on polysaccharides or proteins such as starch, cellulose derivatives, chitosan, alginate, fruit puree, whey protein isolated, soy protein, egg albumen, wheat gluten, or sodium caseinate (Valencia-Chamorro et al. 2011). The use of 0.3% essential oil with 1% alginate in fruit packaging gives a good effect in storing fresh pistachios (*Pistacia vera* L) (Hashemi et al. 2021). Edible films can be used as carriers of active compounds such as antimicrobials, antioxidants, texture enhancers or key nutrients, among others (Acevedo-Fani et al. 2015). In active packaging applications, essential oils have been applied in different ways, free and encapsulated, both in non-degradable and biodegradable materials (Rodríguez et al. 2021).

Recently, essential oils are used as an ingredient in the storage of table wine with the aim to increase the shelf life without changing the organoleptic characteristics. Some authors tested a new package, with grapes wrapped with two distinct films, with the addition of a mixture of eugenol, thymol and carvacrol. The result observed that there is a significant reduction in microbial growth as well as lower occurrence of berry decay (Valero et al. 2006; Guillén et al. 2007 in Laranjo et al. 2017).

9.4.3   Food Safety

Some of the pathogenic bacteria in food that need to be observed because of their high pathogenicity are *Bacillus cereus, Salmonella enterica, Escherichia coli, Staphylococcus aureus*, and *Listeria monocytogenes* because they can cause serious diseases in humans such as nausea, vomiting, stomach pain, diarrhea (Bintsis 2017). Among the five pathogenic bacteria, L. monocytogenes is the most dangerous because it can cause death in pregnant women, infants and the elderly (Kraśniewska et al. 2020).

Essential oils are known to have significant antiseptic properties, (anti-bacterial, antifungal and anti-viral), they are also showed antioxidants effect so that the use of essential oils in the food sector as additive compounds is very appropriate, especially in terms of food preservation and protection (Burt 2004; Bhavaniramya et al. 2019). Besides being used as a preservative, essential oils are also being developed in food protection, especially controlling the growth of pathogenic microorganisms (Mathavi et al. 2013; Campos et al. 2016). Food-borne diseases are a growing public health problem in the world. It is estimated that each year in the United States, 31 species of pathogens cause 9.4 million cases of food-borne illnesses (Scallan et al. 2011).

Table 9.1 demonstrates some essential oils that have antimicrobial effects against bacterial/fungal/viral pathogens. Citrus, garlic, clove, mint, ginger, lemongrass, eucalyptus, lavender, oregano, thyme etc. Essential oils exhibit antibacterial, antifungal, antiviral properties against various types of pathogenic microbes (Mith et al. 2014). Some types of essential oils have a main (dominant) component with levels of 20–70% compared to other components which are only in small amounts (Bhavaniramya et al. 2019). The dominant chemical components greatly affect the characteristics and bioactivity of the essential oil. For example, Peacock flower (*C. pulcherrima*) leaf essential oil demonstrated two main components Cubebene 33.87% and Caryophyllene 23.00%, having very strong antibacterial properties against both Gram-positive and Gram-negative bacteria (Constani et al. 2019). Linalool (37.7%) the major compound of *Coriandrum sativum* seed essential oil (Bhuiyan et al. 2009), *Origanum compactum* has carvacrol (30%) and thymols (27%) as the major chemical components (Laghmouchi et al. 2018). However, the presence of many other components with relatively small levels of less than 20% also plays a role in strengthening or increasing the bioactivity. The low molecular weight compounds such as terpenes and terpenoids are present in most of the essential oils are responsible for various activities. The essential oils with the strongest antibacterial properties against foodborne pathogens contain a high percentage of phenolic compounds, such as carvacrol, eugenol and thymol (Laranjo et al. 2017). Cabarkapa et al. (2016) reported that the majority of Essential Oils are classified as Generally Recognized as Safe (GRAS).

9.5   Essential Oil and Negative Impact in Food

Among the 3000 known essential oils, 300 of them have high economic value, are commercially important especially used in the flavors and fragrances market (Burt et al. 2003). Many food products use essential oils as additives, both as a flavor enhancer, as well as a preservative against food spoilage due to microbes and oxidation. Not only as a food preservative, essential oils can also protect food against the development of foodborne infectious diseases caused by pathogenic bacteria. In addition, essential oils are also used as a mixture in basic packaging materials (Hyldgaard et al. 2012).

For the use of essential oils in edible products, there are several things that need to be considered related to the basic properties of essential oils, among others: essential oils are very volatile so that to process them a special method is needed so that the mixing process can be homogeneous and reduce the amount of evaporated oil. Essential oil is also less stable so it is possible to experience unwanted changes during the product processing (Di Pasqua et al. 2005). The nature of essential oils that are difficult to dissolve in water further adds to the difficulty in the mixing process. If the weakness of the properties of essential oils above can be overcome, there is one more very basic characteristic, namely the very sharp aroma. To use essential oils as preservatives or food safety, a minimum dose is required that is able to inhibit the growth of spoilage microbes or has an antioxidant effect, so that at that dose it is possible to appear a strong aroma that is not suitable for food or has a negative effect on the taste of the food (Maurya et al. 2021; Hyldgaard et al. 2012).

The aroma of essential oils is not necessarily in accordance with the processed food or can interfere with the food product so that the food has an undesirable aroma. For this reason, the aroma and taste of food must be acceptable to consumers, meaning that it must pass organoleptic tests.

9.5.1 Volatility

Essential oils as a secondary metabolite of plants have characteristic volatile properties and have a strong and sharp aroma. The sharp aroma that is owned plays a very important role as a means of protecting plants against predators/enemies (insect repellents and against herbivores).

On the other hand, the aroma produced also invites some insects to come and spread the pollens, and seeds. The volatile nature of essential oils is closely related to the constituent components of essential oils which consist of many groups of compounds. For example, terpenes, terpenoids, phenylpropanoids, simple hydrocarbons, coumarin (Niu and Gilbert 2004; Hyldgaard et al. 2012, Trombetta et al. 2005). Swamy et al. 2016 and Matos et al. 2019 who reported that terpenes constitute about 50–95% of the essential oils. Dhifi et al. (2016) also reported that terpenes constitute the majority of chemical classes in essential oils of various plants as an example Cinnamon essential oil (*C. zeylanicum*) contains Cinnamaldehyde, eugenol, copaene, -caryophyllene, linalool, eucalyptol. Oregano essential oil is rich in carvacrol, gamma-terpinene, sabinene and p-cymene. Clove essential oil is rich in eugenol, eugenyl acetate, caryophyllene, β-pinene, α-pinene, limonene. In general, terpenes (mono and sesqui terpenes) and terpenoids are small volatile molecules. The sharp aroma, although in small quantities, is not entirely acceptable to consumers, it can also interfere with organoleptic.

9.5.2 Low Stability

The method for obtaining essential oils is very influential on the quality of essential oils. The incorrect extraction procedure will affect the results obtained and can even damage the essential oil (Adorjan and Buchbauer 2010). Boubechiche et al. (2017) reported that many natural products are unstable at high temperatures and can easily be damaged during hot extraction.

Due to the similarity of the chemical structure of the components that make up essential oils in one group, these components are easily changed from one to another, due to chemical reactions such as oxidation, isomerization, cyclization, or dehydrogenation caused by enzymatic or not (Turek and Stintzing 2013). Several factors are very influential on the stability of essential oils such as heating, light and air, because these factors can cause changes in the components that make up essential oils. Possible chemical reactions include isomerization, oxidation, dehydrogenation, polymerization and thermal re-arrangement (Fig. 9.3).

Changes in the constituent components of essential oils will greatly affect the taste and aroma of essential oils, because the taste and aroma are not only determined by the concentration of one compound but also by the specific odor threshold of many components, which are determined by their structure and volatility (Grosch 2001). For example, the compound p-cymene is often found in old essential oils and is associated with an undesirable aroma in lemon oil (Turek and Stintzing 2011).

As terpenoids tend to be both volatile and thermolabile and may be easily oxidized or hydrolyzed depending on their respective structure (Scott 2005). One example of isomerization and oxidation reactions that occur in trans-anethol molecules to become cis-anethol and anis aldehyde (Fig. 9.4).

Ultraviolet (UV) light and visible (Vis) light are considered to accelerate autoxidation processes by triggering the hydrogen abstraction that results in the formation of alkyl radicals (Choe and Min 2006). Misharina et al. (2003), reported that monoterpenes have been shown to degrade rapidly under the influence of light. There was a change in the reaction of marjoram oil during storage under light so that some unidentified minor components were formed.

Fig. 9.3 Possible conversion reactions in essential oils. (Adapted from Turek and Stintzing 2013)

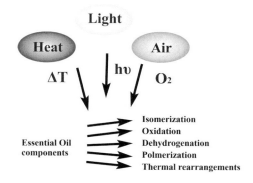

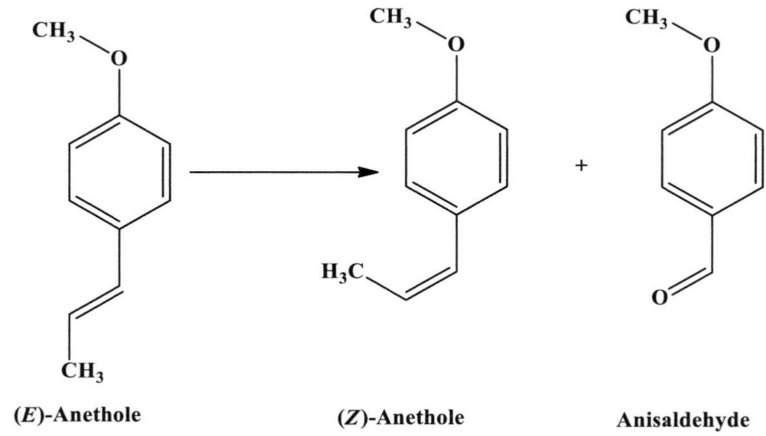

Fig. 9.4 Isomerization and oxidation products of *trans*-anethole detected in fennel oil. (Adapted from Turek and Stintzing 2013)

9.5.3   Low Water Solubility

Essential oils are difficult to dissolve in water; however, the oxidation products of their constituent components are water soluble. The presence of water in the essential oil will cause damage to the essential oil. In its use in the food sector using essential oils as a flavor enhancer or as a preservative in packaging materials, the water content of the food must be maintained to prevent unwanted chemical reactions that will eventually damage the taste of the food.

9.6   Conclusion

The application of essential oils in the food sector is very beneficial, apart from being a preservative, flavor enhancer, as well as controlling microbial growth in foodstuffs and mixed ingredients in food packaging (Food safety). Essential oils can be used as a substitute for chemical preservatives, both antimicrobial and antioxidant, with a "green" label because they are known as Generally Recognized as Safe (GRAS).

There are some negative effects that need to be observed, such as the volatile nature and the strong and sharp aroma, which at certain doses can interfere with the organoleptic properties of the food. Essential oils are less stable because their constituent components easily undergo chemical reactions when exposed to heat, light and air so that they undergo structural changes. In addition, essential oils are also difficult to dissolve in water, so special methods are needed in processing them, the presence of water in essential oils will cause hydrolysis and oxidation reactions.

References

Abu-Shanab B, Adwan GM, Abu-Safiya D, Jarrar N, Adwan K (2005) Antibacterial activities of some plant extracts utilized in popular medicine in Palestine. Turk J Biol 28(2–4):99–102

Acevedo-Fani A, Salvia-Trujillo L, Rojas-Graü MA, Martín-Belloso O (2015) Edible films from essential-oil-loaded nanoemulsions: physicochemical characterization and antimicrobial properties. Food Hydrocoll 47(2015):168–177. https://doi.org/10.1016/j.foodhyd.2015.01.032

Adorjan B, Buchbauer G (2010) Biological properties of essential oils: an updated review. Flavour Fragr J 25:407–426. https://doi.org/10.1002/ffj.2024

Akram MZ, Jalal H, Canogullari S (2019) The use of essential oils in active food packaging: a review of recent studies. Turk J Agric – Food Sci Technol 7(11):1799–1804. https://doi.org/10.24925/turjaf.v7i11.1799-1804.2640

Alizadeh Behbahani B, Falah F, Lavi Arab F, Vasiee M, Tabatabaee Yazdi F (2020) Chemical composition and antioxidant, antimicrobial, and antiproliferative activities of *Cinnamomum zeylanicum* bark essential oil. Evid Based Complement Alternat Med 2020. https://doi.org/10.1155/2020/5190603

Amorati R, Foti MC, Valgimigli L (2013) Antioxidant activity of essential oils. J Agric Food Chem 61(46):10835–10847. https://doi.org/10.1021/jf403496k

Aureli P, Costantini A, Zolea S (1992) Antimicrobial activity of some plant essential oils against. Listeria monocytogenes. J Food Prot 55:334–348

Bakkali F, Averbeck S, Averbeck D, Idaomar M (2008) Biological effects of essential oils--a review. Food Chem Toxicol 46(2):446–475. https://doi.org/10.1016/j.fct.2007.09.106. Epub 2007 Sep 29. PMID: 17996351

Baldim I, Tonani L, von Zeska Kress MR, Oliveira WP (2019) Lippia sidoides essential oil encapsulated in lipid nanosystemas an anti-Candida agent. Ind Crops Prod 127:73–81. https://doi.org/10.1016/j.indcrop.2018.10.064

Banerjee SK, Mukherjee PK, Maulik SK (2003) Garlic as an antioxidant: the good, the bad and the ugly. Phytother Res 17:97–106. https://doi.org/10.1002/ptr.1281

Barbosa LN, Rall VLM, Fernandes AAH, Ushimaru PI, da Silva Probst I, Fernandes A Jr (2008) Essential oils against Foodborne Pathogens and Spoilage Bacteria in Minced Meat. Foodborne Pathogens and Disease 6(6):725–728. https://doi.org/10.1089/fpd.2009.0282

Beigi M, Torki-Harchegani M, Pirbalouti AG (2018) Quantity and chemical composition of essential oil of peppermint (Mentha × piperita L.) leaves under different drying methods. Int J Food Prop 21:1, 267–276. https://doi.org/10.1080/10942912.2018.1453839

Bellik Y (2014) Total antioxidant activity and antimicrobial potency of the essential oil and oleoresin of Zingiber officinale Roscoe. Asian Pac J Trop Dis 4(1):40–44. https://doi.org/10.1016/S2222-1808(14)60311-X

Bhavaniramya S, Vishnupriya S, Al-Aboody MS, Vijayakumar R, Baskaran D et al (2019) Role of essential oils in food safety: antimicrobial and antioxidant applications. Grain Oil Sci Technol 2:49–55. https://doi.org/10.1016/j.gaost.2019.03.001

Bhuiyan MNI, Begum J, Sultana M (2009) Chemical composition of leaf and seed essential oil of Coriandrum sativum L. from Bangladesh. Bangladesh J Pharmacol. https://doi.org/10.3329/bjp.v4i2.2800

Bintsis T (2017) Foodborne pathogens. AIMS Microbiol 3(3):529–563. https://doi.org/10.3934/microbiol.2017.3.529

Birmpa A, Constantinou P, Dedes C (2018) Antibacterial and antiviral effect of essential oils combined with non-thermal disinfection technologies for ready-to-eat Romaine lettuce. J Nutr Food Technol 1(1):24–32. https://doi.org/10.30881/jnfrt.00007

Bonde SD, Nemade LS, Patel MR et al. (2007) Murraya koenigii (curry leaf): ethnobotany, phytochemistry and pharmacology – a review. Int J Pharm Phytopharmacological Res 4(5):45–54

Boubechiche Z, Chihib N-E, Jama C, Hellal A (2017) Comparison of volatile compounds profile and antioxydant activity of Allium sativum essential oils extracted using hydrodistillation, ultrasound-assisted and sono-hydrodistillation processes. Indian J Pharm Educ Res 51(3):S283–S285. https://doi.org/10.5530/ijper.51.3s.30

Bozin B, Mimica-Dukic N, Simin N, Anackov G (2006) Characterization of the volatile composition of essential oils of some lamiaceae spices and the antimicrobial and antioxidant activities of the entire oils. J Agric Food Chem 54(5):1822–1828. https://doi.org/10.1021/jf051922u. PMID: 16506839

Bozin B, Mimica-Dukic N, Samojlik I, Jovin E (2007) Antimicrobial and antioxidant properties of rosemary and sage (Rosmarinus officinalis L. and Salvia officinalis L., Lamiaceae) essential oils. J Agric Food Chem 55(19):7879–7885. https://doi.org/10.1021/jf0715323

Burt S (2004) Essential oils: their antibacterial properties and potential applications in foods-a review. Int J Food Microbiol 94:223–253. https://doi.org/10.1016/j.ijfoodmicro.2004.03.022

Burt SA, Reinders RD (2003) Antibacterial activity of selected plant essential oils against Escherichia coli O157:H7. Lett Appl Microbiol 36:162–167. https://doi.org/10.1046/j.1472-765X.2003.01285.x

Cabarkapa IS, Duragic OM, Kostadinovic LM (2016) Essential oils: mode of antimicrobial activity and potential application in food systems. Agro Food Ind Hi Tech 27(3):61–64

Campos T, Barreto S, Queirós R, Ricardo-Rodrigues S, Félix MR, Laranjo M et al (2016) Conservação de morangos com utilização de óleos essenciais. AGROTEC 18:90–96

Caputo L, Reguilon MD, Miñarro J, De Feo V, Rodriguez-Arias M (2018) Lavandula angustifolia essential oil and linalool counteract social aversion induced by social defeat. Molecules 23:2694. https://doi.org/10.3390/molecules23102694

Chenni M, El Abed D, Rakotomanomana N, Fernandez X, Chemat F (2016) Comparative study of essential oils extracted from Egyptian basil leaves (Ocimum basilicum L.) using hydro-distillation and solvent-free microwave extraction. Molecules 21(1):E113. https://doi.org/10.3390/molecules21010113. PMID: 26797599; PMCID: PMC6273689

Choe E, Min DB (2006) Mechanisms and factors for edible oil oxidation. Compr Rev Food Sci Food Saf 5(4):169–186. https://doi.org/10.1111/j.1541-4337.2006.00009.x

Choi HS, Song HS, Ukeda H, Sawamura M (2000) Radical-scavenging activities of citrus essential oils and their component: detection using 1,1-diphenyl-2-picryl hidrazyl. J Agric Food Chem 48:4156–4161

Chorianopoulos NG, Giaouris ED, Skandamis PN, Haroutounian SA, Nychas GJ (2008) Disinfectant test against monoculture and mixed-culture biofilms composed of technological, spoilage and pathogenic bacteria: bactericidal effect of essential oil and hydrosol of Satureja thymbra and comparison with standard acid-base sanitizers. J Appl Microbiol 104(6):1586–1596. https://doi.org/10.1111/j.1365-2672.2007.03694.x. PMID: 18217930

Constani NRR, Soetjipto H, Hartini S (2019) Antibacterial activity and chemical composition of red peacock flower (Caesalpinia pulcherrima L.) leaf essential oil. Jurnal Kimia Sains dan Aplikasi 22(6):269. https://doi.org/10.14710/jksa.22.6.269-274

Corzo-Martínez M, Corzo N, Villamiel M (2007) Biological properties of onions and garlic. Trends Food Sci Technol 18(12):609–625

Cunha BG, Duque C, Caiaffa KS, Massunari L, Catanoze IA, Dos Santos DM et al (2020) Cytotoxicity and antimicrobial effects of citronella oil (Cymbopogon nardus) and commercial mouthwashes on S. aureus and C. albicans biofilms in prosthetic materials. Arch Oral Biol 109:104577. https://doi.org/10.1016/j.archoralbio.2019.104577

De Martino L, De Feo V, Nazzaro F (2009) Chemical composition and in vitro antimicrobial and mutagenic activities of seven Lamiaceae essential oils. Molecules (Basel, Switzerland) 14(10):4213–4230. https://doi.org/10.3390/molecules14104213

de Matos SP, Teixeira HF, de Lima Á, Veiga-Junior VF, Koester LS (2019) Essential oils and isolated terpenes in nanosystems designed for topical administration: a review. Biomol Ther 9(4):138. https://doi.org/10.3390/biom9040138

de Oliveira MS, Almeida MM, Salazar MLAR, Pires FCS, Bezerra FWF, Cunha VMB, Cordeiro RM, Urbina GRO, Silva Md, Silva APS, HolandaPinto RH, de Carvalho Junior RN (2018)

Potential of Medicinal Use of Essential Oils from Aromatic Plants. In (Ed.), Potential of Essential Oils. IntechOpen. https://doi.org/10.5772/intechopen.78002

De Oliveira TLC, De Araujo Soares R, Ramos EM, Das Graças Cardoso M, Alves E, Piccoli RH (2011) Antimicrobial activity of Satureja montana L. essential oil against Clostridium perfringens type A inoculated in mortadella type sausages formulated with different levels of sodium nitrite. Int J Food Mycrobiol 144:546–555

Deans SG, Ritchie G (1987) Antibacterial properties of plant essential oils. Int J Food Microbiol 5:165–180

Dhifi W, Bellili S, Jazi S, Bahloul N, Mnif W (2016) Essential oils' chemical characterization and investigation of some biological activities: a critical review. Medicines (Basel) 3(4):1–16

Di Pasqua R, De Feo V, Villani F, Mauriello G (2005) In vitro antimicrobial activity of essential oils from Mediterranean Apiaceae, Verbenaceae and Lamiaceae against foodborne pathogens and spoilage bacteria. Ann Microbiol 19:226–230

Di Pasqua R, Betts G, Hoskins N, Edwards M, Ercolini D, Mauriello G (2007) Membrane toxicity of antimicrobial compounds from essential oils. J Agric Food Chem 55:4863–4870. Epub 2007 May 12. PMID: 17497876

Djabou N, Lorenzi V, Guinoiseau E, Andreani S, Giuliani M-C, Desjobert J-M, Bolla J-M, Costa J, Berti L, Luciani A (2013) Phytochemical composition of Corsican Teucrium essential oils and antibacterial activity against foodborne or toxi-infectious pathogens. Food Control 30:354–363

Donelian A, Carlson LH, Lopes TJ, Machado RA (2009) Comparison of extraction of patchouli (Pogostemon cablin) essential oil with supercritical CO2 and by steam distillation. J Supercrit Fluids 48(1):15–20

Dorman HJD, Deans SG (2000) Antimicrobial agents from plants: antibacterial activity of plant volatile oils. J Appl Microbiol 88:308–316

Dwivedy AK, Prakash B, Chanotiya CS, Bisht D, Dubey NK (2017) Chemically characterized Mentha cardiaca L. essential oil as plant-based preservative in view of efficacy against biodeteriorating fungi of dry fruits, aflatoxin secretion, lipid peroxidation and safety profile assessment. Food Chem Toxicol 106:175–184. https://doi.org/10.1016/j.fct.2017.05.043

Edris AE (2007) Pharmaceutical and therapeutic potentials of essential oils and their individual volatile constituents: a review. Phytother Res 21:308–323. https://doi.org/10.1002/ptr.2072

Elgayyar M, Draughon FA, Golden DA, Mount JR (2001) Antimicrobial activity of essential oils from plants against selected pathogenic and saprophytic microorganisms. J Food Prot 64(7):1019–24. https://doi.org/10.4315/0362-028x-64.7.1019. PMID: 11456186

Elyemni M, Bouchra L, Imane N, Taha E, Abdelhak B, Mustapha T, Mahdi C, Noureddine E (2019) Extraction of essential oils of Rosmarinus officinalis L. by two different methods: hydrodistillation and microwave assisted hydrodistillation. Hindawi. Sci World J. https://doi.org/10.1155/2019/3659432

Emami SA, Javadi B, Hassanzadeh MK (2007) Antioxidant activity of the essential oils of different parts of Juniperus communis. subsp. hemisphaerica. and Juniperus oblonga. Pharm Biol 45(10):769–776. https://doi.org/10.1080/13880200701585931

Evandri MG, Battinelli L, Daniele C, Mastrangelo S, Bolle P, Mazzanti G (2005) The antimutagenic activity of Lavandula angustifolia (lavender) essential oil in the bacterial reverse mutation assay. Food Chem Toxicol 43:1381–1387

Fang Z, Zhao Y, Warner RD, Johnson SK (2017) Active and intelligent packaging in meat industry. Trends Food Sci Technol 61:60–71

Farias PKS, Silva JCRL, de Souza CN, da Fonseca FSA, Brandi IV, Martins ER, Azevedo AM, de Almeida AC (2019) Antioxidant activity of essential oils from condiment plants and their effect on lactic cultures and pathogenic bacteria. Cienc Rural 49(2)., Art. no. e2018140

Fernanda C. da Silva, Sara MC, Virgínia M. de S, Deila M. dos SB, Nelson L, Luís RB (2012) Evaluation of antifungal activity of essential oils against potentially mycotoxigenic Aspergillus flavus and Aspergillus parasiticus. Rev Bras Farmacogn 22(5). https://doi.org/10.1590/S0102-695X2012005000052

Fratianni F, De Martino L, Melone A, De Feo V, Coppola R, Nazzaro F (2010) Preservation of chicken breast meat treated with thyme and balm essential oils. J Food Sci 75:M528–M535. https://doi.org/10.1111/j.1750-3841.2010.01791.x

Freisesleben SH, Jager AK (2014) Correlation between plant secondary metabolites and their anti-fungal mechanism—a review. Med Aromat Plants 3:1–6

Gogoi R, Sarma N, Loying R, Pandey SK, Begum T, Lal M (2021) A comparative analysis of bark and leaf essential oil and their chemical composition, antioxidant, anti-inflammatory, antimicrobial activities and genotoxicity of North East Indian Cinnamomum zeylanicum Blume. Nat Prod J 11(1):74–84

Grassmann J (2005) Terpenoids as plant antioxidants. Vitam Horm 72:505–535. https://doi.org/10.1016/S0083-6729(05)72015-X

Griffin GS, Markham LJ, Leach ND (2000) An agar dilution method for the determination of the minimum inhibitory concentration of essential oils. J Essent Oil Res 12:149–255

Grosch W (2001) Evaluation of the key odorants of foods by dilution experiments, aroma models and omission. Chem Senses 26:533–545. https://doi.org/10.1093/chemse/26.5.533

Guenther E (2006) Minyak Atsiri, Jilid I, diterjemahkan oleh Ketaren S. UI Press, Jakarta

Guillén F, Zapata J, Martínez-Romero D, Castillo S, Serrano M, Valero D (2007) Improvement of the overall quality of table grapes stored under modified atmosphere packaging in combination with natural antimicrobial compounds. J Food Sci 72:185–190

Gutiérrez RMP, Luna HH, Garrido SH (2006) Antioxidant activity of Tagetes erecta essential oil. J Chil Chem Soc [online] 51(2):883–886. https://doi.org/10.4067/S0717-97072006000200010, [citado 2022-01-25]

Gutierrez J, Barry-Ryan C, Bourke P (2009) Antimicrobial activity of plant essential oils using food model media: efficacy, synergistic potential and interaction with food components. Food Microbiol 26(2):142–150

Gutteridge JM (1993) Free radicals in disease processes: a compilation of cause and consequence. Free Radic Res Commun 19:141–158. PubMed

Hale AL, Reddivari L, Nzaramba MN et al (2008) Interspecific variability for antioxidant activity and phenolic content among *Solanum* species. Am J Potato Res 85:332. https://doi.org/10.1007/s12230-008-9035-1

Hammer KA, Carson CF, Riley TV (1999) Antimicrobial activity of essential oils and other plant extracts. J Appl Microbiol 86:985–990

Han X, Parker TL (2017) Anti-inflammatory, tissue remodeling, immunomodulatory, and anti-cancer activities of oregano (Origanum vulgare) essential oil in a human skin disease model. Biochim Open 4:73–77

Hashemi SMB, Khorram SB, Sohrabi M (2017) Antioxidant activity of essential oils in foods. In: Hashemi SMB, Khaneghah AM, de Souza Sant'Ana A (eds) Essential oils in food processing. https://doi.org/10.1002/9781119149392.ch8

Hashemi M, Dastjerdi AM, Shakerardekani A et al. (2021) Effect of alginate coating enriched with Shirazi thyme essential oil on quality of the fresh pistachio (Pistacia vera L.). J Food Sci Technol 58:34–43. https://doi.org/10.1007/s13197-020-04510-6

Hennekinne JA, Buyser MLD, Dragacci S (2012) Staphylococcus aureus and its food poisoning toxins: characterization and outbreak investigation. FEMS Microbiol Rev 36:815–836. https://doi.org/10.1111/j.1574-6976.2011.00311.x

Hsouna AB, Halima NB, Smaoui S, Hamdi N (2017) Citrus lemon essential oil: chemical composition, antioxidant and antimicrobial activities with its preservative effect against *Listeria monocytogenes* inoculated in minced beef meat. Lipids Heath Dis 16(1):1–11. https://doi.org/10.1186/s12944-017-0487-5

Hyldgaard M, Mygind T, Meyer RL (2012) Essential oils in food preservation: mode of action, synergies, and interactions with food matrix components. Front Microbiol. 3:12. https://doi.org/10.3389/fmicb.2012.00012. PMID: 22291693; PMCID: PMC3265747

Indonesia's Essential Oil the World's Leading Product (2021) Konsulat Jenderal Republik Indonesia Penang, 467, Jalan Burma, Penang 10350. penang.kjri@kemlu.go.id

Ipek E, Zeytinoglu H, Okay S, Tuylu BA, Kurkcuoglu M, Baser KHC (2005) Genotoxicity and antigenotoxicity of *Origanum* oil and carvacrol evaluated by Ames *Salmonella*/microsomal test. Food Chem 93:551–556

Jain M, Gilhotra R, Singh RP et al (2017) Curry leaf (Murraya Koenigii): a spice with medicinal property. MOJ Biol Med 2(3):236–256. https://doi.org/10.15406/mojbm.2017.02.00050

Jamil R, Nasir NN, Ramli H, Isha R, Ismail NA (2016) Extraction of essential oil from *Murraya koenigii* leaves: potential study for application as natural-based insect repellent. J Eng Appl Sci 11(4). Asian Research Publishing Network (ARPN). http://www.arpnjournals.com/

Joshi RK (2014) Chemical composition and antimicrobial activity of the essential oil of Ocimum basilicum L. (sweet basil) from Western Ghats of North West Karnataka, India. Anc Sci Life 33(3):151–156. https://doi.org/10.4103/0257-7941.144618

Karimian P, Kavoosi G, Saharkhiz MJ (2012) Antioxidant, nitric oxide scavenging and malondial-dehyde scavenging activities of essential oils from different chemotypes of *Zataria multiflora*. Nat Prod Res, Taylor & Francis:1478–6419. https://doi.org/10.1080/14786419.2011.631136

Kasali AA, Oyedeji AO, Ashilokun AO (2001) Volatile leaf oil constituents of Cymbopogon citratus (DC) Stapf. Flavour Fragr J 16(5):377–378

Kelen M, Tepe B (2008) Chemical composition, antioxidant and antimicrobial properties of the essential oils of three salvia species from Turkish flora. Bioresour Technol 99(10):4096–4104

Kim JW, Kim YS, Kyung KH (2004) Inhibitory activity of essentialoils of garlic and onion against bacteria and yeasts. J Food Prot 67: in press

Kiran CR, Chakka AK, Padmakumari Amma KP, Nirmala Menon A, Sree Kumar MM, Venugopalan VV (2013) Essential oil composition of fresh ginger cultivars from North-East India. J Essent Oil Res 25(5):380–387. https://doi.org/10.1080/10412905.2013.796496

Kosakowska O, Węglarz Z, Pióro-Jabrucka E, Przybył JL, Kraśniewska K, Gniewosz M, Bączek K (2021) Antioxidant and antibacterial activity of essential oils and hydroethanolic extracts of Greek oregano (O. vulgare L. subsp. hirtum (Link) Ietswaart) and common oregano (O. vulgare L. subsp. vulgare). Molecules 26(4):988

Kraśniewska K, Kosakowska O, Pobiega K, Gniewosz M (2020) The influence of two-component mixtures from Spanish origanum oil with Spanish marjoram oil or coriander oil on antilisterial activity and sensory quality of a fresh cut vegetable mixture. Foods 9(12):1740. https://doi.org/10.3390/foods9121740

Kumar A, Shukla R, Singh P, Prasad CS, Dubey NK (2008) Assessment of *Thymus vulgaris* L. essential oil as a safe botanical preservative against post-harvest fungal infestation of food commodities. Innov Food Sci Emerg 9:575–580

Laghmouchi Y, Belmehdi O, Skali-Senhaji N, Abrini J (2018) Chemical composition and antibacterial activity of Origanum compactum Benth. essential oils from different areas at northern Morocco. S Afr J Bot 115:120–125. https://doi.org/10.1016/j.sajb.2018.02.002

Laranjo M, Fernández-Léon AM, Potes ME, Agulheiro-Santos AC, Elias M (2017) Use of essential oils in food preservation. In: Méndez-Vilas A (ed) Antimicrobial research: novel bioknowledge and educational programs. Formatex Research Center, Badajoz, pp 177–188

Leyva-López N, Gutiérrez-Grijalva EP, Vazquez-Olivo G, Heredia JB (2017) Essential oils of oregano: biological activity beyond their antimicrobial properties. Molecules 22:989. https://doi.org/10.3390/molecules22060989

Li ZH, Cai M, Liu YS, Sun PL, Luo SL (2019) Antibacterial activity and mechanisms of essential oil from *Citrus medica L. var. sarcodactylis*. Molecules (Basel, Switzerland) 24(8):1577. https://doi.org/10.3390/molecules24081577

Lin LY, Chuang CH, Chen HC, Yang KM (2019) Lime (Citrus aurantifolia (Christm.) Swingle) essential oils: volatile compounds, antioxidant capacity, and hypolipidemic effect. Foods 8(9):398

Lo Cantore P, Iacobellis NS, De Marco A, Capasso F, Senatore F (2004) Antibacterial activity of Coriandrum sativum L. and Foeniculum vulgare Miller var. vulgare (Miller) essential oils. J Agric Food Chem 52(26):7862–7866

Mahato N, Sharma K, Koteswararao R, Sinha M, Baral E, Cho MH (2019) Citrus essential oils: extraction, authentication and application in food preservation. Crit Rev Food Sci Nutr 59(4):611–625

Mahboubi M (2019) *Zingiber officinale* Rosc. essential oil, a review on its composition and bioactivity. Clin Phytosci 5:6. https://doi.org/10.1186/s40816-018-0097-4

Mahomoodally F, Aumeeruddy-Elalfi Z, Katharigatta N, Venugopala HM (2019) Antiglycation, comparative antioxidant potential, phenolic content and yield variation of essential oils from 19 exotic and endemic medicinal plants. Saudi J Biol Sci 26(7):1779–1788. https://doi.org/10.1016/j.sjbs.2018.05.002

Majewska E, Kozłowska M, Gruczyńska-Sękowska E, Kowalska D, Tarnowska K (2019) Lemongrass (Cymbopogon citratus) essential oil: extraction, composition, bioactivity and uses for food preservation – a review. Polish J Food Nutr Sci 69(4):327–341. https://doi.org/10.31883/pjfns/113152

Mansour AF, Fikry RM, Saad MM, Mohamed AM (2015) Chemical composition, antioxidant and antimicrobial activity of (Cymbopogon citratus) essential oil cultivated in Madinah Monawara, Saudi Arabia and its comparison to the Egyptian chemotype. Int J Food Nutr Sci 4(4):29–33

Mathavi V, Sujatha G, Bhavani Ramya S, Karthika Devi B (2013) New trends in food processing, Int J Adv Eng Tech 5:176

Matos I, Machado MP, Schartl M, Coelho MM (2019) Allele-specific expression variation at different ploidy levels in *Squalius alburnoides*. Sci Rep 9:3688. https://doi.org/10.1038/s41598-019-40210-8

Maurya A, Prasad J, Das S, Dwivedy AK (2021) Essential oils and their application in food safety. Front Sustain Food Syst 5:653420. https://doi.org/10.3389/fsufs.2021.653420

Miguel MG, Cruz C, Faleiro L, Simões MT, Figueiredo AC, Barroso JG, Pedro LG (2010) Foeniculum vulgare essential oils: chemical composition, antioxidant and antimicrobial activities. Nat Prod Commun 5(2):319–328, 1934578X1000500231

Mihai A, Popa M (2013) Essential oils utilization in food industry – a literature review. Sci Bull Ser F Biotechnol 17:187–192

Misharina TA, Polshkov AN, Ruchkina EL, Medvedeva IB (2003) Changes in the composition of the essential oil of marjoram during storage. Appl Biochem Microbiol 39(3):311–316. https://doi.org/10.1023/A:1023592030874

Mith H, Duré R, Delcenserie V, Zhiri A, Daube G, Clinquart A (2014) Antimicrobial activities of commercial essential oils and their components against food-borne pathogens and food spoilage bacteria. Food Sci Nutr 2(4):403–416. https://doi.org/10.1002/fsn3.116

Mnayer D, Fabiano-Tixier AS, Petitcolas E, Hamieh T, Nehme N, Ferrant C, Fernandez X, Chemat F (2014) Chemical composition, antibacterial and antioxidant activities of six essentials oils from the Alliaceae family. Molecules 19(12):20034–20053

Moarefian M, Barzegar M, Sattari M (2013) Cinnamomum zeylanicum essential oil as a natural antioxidant and antibactrial in cooked sausage. J Food Biochem 37(1):62–69

Mohamed HMH, Mansour HA (2012) Incorporating essential oils of marjoram and rosemary in the formulation of beef patties manufactured with mechanically deboned poultry meat to improve the lipid stability and sensory attributes. LWT – Food Sci Technol 45(1):79–87

Mohamed AA, Ali SI, El-Baz FK (2013) Antioxidant and antibacterial activities of crude extracts and essential oils of Syzygium cumini leaves. PLoS One 8(4):e60269

Nazzaro F, Fratianni F, Martino D, Coppola R, De Feo V (2013) Effect of essential oils on pathogenic bacteria. Pharmaceuticals 6(12):1451–1474. https://doi.org/10.3390/ph6121451

Nikmaram N, Budaraju S, Barba FJ, Lorenzo JM, Cox RB, Mallikarjunan, et al. (2018) Application of plant extracts to improve the shelf-life, nutritional and health-related properties of ready-to-eat meat products. Meat Sci 145:245–255. https://doi.org/10.1016/j.meatsci.2018.06.031

Niu C, Gilbert ES (2004) Colorimetric method for identifying plant essential oil components that affect biofilm formation and structure. Appl Environ Microbiol 70(12):6951–6956. https://doi.org/10.1128/AEM.70.12.6951-6956.2004

Pandey AK, Kumar P, Singh P, Tripathi NN, Bajpai VK (2017) Essential oils: sources of antimicrobials and food preservatives. Front Microbiol 7:2161. https://doi.org/10.3389/fmicb.2016.02161

Patra M, Shahi SK, Midgely G, Dikshit A (2002) Utilization of essential oil as natural antifungal against nail infective fungi. Flavour Fragr J 17:91–94

Penalver P, Huerta B, Borge C, Astorga R, Romero R, Perea A (2005) Antimicrobial activity of five essential oils against origin strains of the Enterobacteriaceae family. APMIS 113:1–6

Perricone M, Arace E, Corbo MR, Sinigaglia M, Bevilacqua A (2015) Bioactivity of essential oils: a review on their interaction with food components. Front Microbiol 6:76

Pisoschi AM, Pop A, Georgescu C, Turcuş V, Olah NK, Mathe E (2018) An overview of natural antimicrobials role in food. Eur J Med Chem 143:922–935

Raal A, Orav A, Arak E (2012) Essential oil composition of Foeniculum vulgare Mill. fruits from pharmacies in different countries. Nat Prod Res 26(13):1173–1178

Rasooli I, Gachkar L, Yadegarinia D, Bagher Rezaei M, Alipoor Astaneh S (2008) Antibacterial and antioxidative characterisation of essential oils from Mentha piperita and Mentha spicata grown in Iran. Acta Aliment 37(1):41–52

Rassem HHA, Nour AH, Yunus RM (2016) Techniques for extraction of essential oils from plants: a review. Aust J Basic Appl Sci 10(16):117–127

Razzaghi-Abyaneh M, Shams-Ghahfarokhi M, Yoshinari T, Rezaee MB, Jaimand K, Nagasawa H, Sakuda S (2008) Inhibitory effects of Satureja hortensis L. essential oil on growth and aflatoxin production by Aspergillus parasiticus. Int J Food Microbiol 123(3):228–33. https://doi.org/10.1016/j.ijfoodmicro.2008.02.003. Epub 2008 Feb 15. PMID: 18353477

Razzaghi-Abyaneh MM, Shams-Ghahfarokhi MB, Rezaee K, Jaimand S, Alinezhad R, Saberi R, Yoshinari T (2009) Chemical composition and antiaflatoxigenic activity of Carum carvi L., Thymus vulgaris and Citrus aurantifolia essential oils. Food Control 20:1018–1024

Ridder M (2021) Essential oils market worldwide – statistics & facts. Available online: https://www.statista.com/topics/5174/essential-oils/#dossierKeyfigures. Accessed on 14 Nov 2021

Rodríguez M, Núñez EB, Soria LA, García Oliveira P, Prieto Lage M (2021) Essential oils and their application on active packaging systems: a review. Resources 10(7):1–20. https://doi.org/10.3390/resources10010007

Ruberto G, Baratta MT (2000) Antioxidant activity of selected essential oil components in two lipid model systems. Food Chem 69:167–174

Salehi B, Mishra AP, Shukla I, Sharifi-Rad M, Contreras MDM, Segura-Carretero A, Sharifi-Rad J (2018) Thymol, thyme, and other plant sources: health and potential uses. Phytother Res 32(9):1688–1706

Sanchez E, Garcia S, Heredia N (2010) Extract of edible and medicinal plants damage membranes of Vibrio cholerae. Appl Environ Microbiol 76:6888–6894

Sauceda ENR (2011) Uso de agentes antimicrobianos naturales en la conservación de frutas y hortalizas. Ra Ximhai 7(1):153–170

Scallan E, Hoekstra RM, Angulo FJ, Tauxe RV, Widdowson MA, Roy SL, Jones JL, Griffin PM (2011) Foodborne illness acquired in the United States--major pathogens. Emerg Infect Dis 17(1):7–15. https://doi.org/10.3201/eid1701.p11101

Scott RPW (2005) Encyclopedia of analytical science, 2nd edn. https://doi.org/10.1016/B0-12-369397-7/00147-3

Selles SMA, Kouidri M, Belhamiti BT, Amrane AA (2020) Chemical composition, in-vitro antibacterial and antioxidant activities of Syzygium aromaticum essential oil. J Food Meas Charact 13:1–7. https://doi.org/10.1007/s11694-020-00482-5

Settanni L, Palazzolo E, Guarrasi V, Aleo A, Mammina C, Moschetti G, Germanà MA (2012) Inhibition of foodborne pathogen bacteria by essential oils extracted from citrus fruits cultivated in Sicily. Food Control 26(2):326–330

Seymour IJ, Burfoot D, Smith RL, Cox LA, Lockwood A (2002) Ultrasound decontamination of minimally processed fruits and vegetables. Int J Food Sci Technol 37(5):547–557

Shakeri A, Khakdan F, Soheili V, Sahebkar A, Rassam G, Asili J (2014) Chemical composition, antibacterial activity, and cytotoxicity of essential oil from Nepeta ucrainica L. spp. kopetdaghensis. Ind Crop Prod 58:315–321. https://doi.org/10.1016/j.indcrop.2014.04.009

Sharma PK, Singh V, Al M (2016) Chemical composition and antimicrobial activity of fresh rhizome essential oil of Zingiber officinale Roscoe. Pharmacogn J 8(3):185

Singh IP, Kapoor S, Pandey SK, Singh UK, Singh RK (2002) Studies of essential oils. Part 10: anti-bacterial activity of volatile oils of some spices. Phytother Res 16:680–668

Singh R, Shushni MA, Belkheir A (2015) Antibacterial and antioxidant activities of Mentha piperita L. Arab J Chem 8(3):322–328

Smith RL, Cohen SM, Doull J (2005) GRAS flavouring substances 22. Food Technol 59(8):24–62

Sobrino-Lopez A, Martín-Belloso O (2008) Use of nisin and other bacteriocins for preservation of dairy products. Int Dairy J 18:329–343. https://doi.org/10.1016/j.idairyj.2007.11.009

Soetjipto H (2018) Chapter 3: Antibacterial properties of essential oil in some Indonesian herbs. In: El-Shemy HA (ed) Potential of essential oils. IntechOpen, London, pp 41–58

Soković M, Glamočlija J, Marin PD, Brkić D, van Griensven LJLD (2010) Antibacterial effects of the essential oils of commonly consumed medicinal herbs using an in vitro model. Molecules (Basel, Switzerland) 15(11):7532–7546

Spadaro F, Costa R, Circosta C, Occhiuto F (2012) Volatile composition and biological activity of key lime Citrus aurantifolia essential oil. Nat Prod Commun 7(11):1523–1526, 1934578X1200701128

Speranza B, Corbo MR (2010) Essential oils for preserving perishable foods: possibilities and limitations. In: Bevilacqua A, Corbo MR, Sinigaglia M (eds) Application of alternative food preservation technologies to enhance food safety and stability. Bentham Publisher, Sharjah, pp 35–57

Swamy MK, Akhtar MS, Sinniah UR (2016) Antimicrobial properties of plant essential oils against human pathogens and their mode of action: an updated review. Evid Based Complement Alternat Med 2016:3012462. https://doi.org/10.1155/2016/3012462

Tabassum N, Vidyasagar GM (2013) Antifungal investigations on plant essential oils. A review. Int J Pharm Pharm Sci 5:19–28

Tajkarimi MM, Ibrahim SA, Cliver DO (2010) Antimicrobial herb and spice compounds in food. Food Control 21:1199–1218

Tariq S, Wani S, Rasool W, Shafi K, Bhat MA, Prabhakar A, Shalla AH, Rather MA (2019) A comprehensive review of the antibacterial, antifungal and antiviral potential of essential oils and their chemical constituents against drug-resistant microbial pathogens. Microb Pathog 134:103580. https://doi.org/10.1016/j.micpath.2019.103580

Teixeira B, Marques A, Ramos C, Serrano C, Matos O, Neng NR, Nunes ML (2013) Chemical composition and bioactivity of different oregano (Origanum vulgare) extracts and essential oil. J Sci Food Agric 93(11):2707–2714

Thaweboon S, Thaweboon B (2009) In vitro antimicrobial activity of Ocimum americanum L. essential oil against oral microorganisms. Southeast Asian J Trop Med Public Health 40(5):1025–1033

Torres-Martínez R, García-Rodríguez YM, Ríos-Chávez P, Saavedra-Molina A, López-Meza JE, Ochoa-Zarzosa A, Garciglia RS (2018) Antioxidant activity of the essential oil and its major terpenes of Satureja macrostema (Moc. and Sessé ex Benth.) Briq. Pharmacogn Mag 13(4):875–880. https://doi.org/10.4103/pm.pm_316_17

Trombetta D, Castelli F, Sarpietro MG, Venuti V, Cristani M, Daniele C, Saija A, Mazzanti G, Bisignano G (2005) Mechanisms of antibacterial action of three monoterpenes. Antimicrob Agents Chemother 49(6):2474–2478. https://doi.org/10.1128/AAC.49.6.2474-2478.2005

Turek C, Stintzing FC (2011) Evaluation of selected quality parameters to monitor essential oil alteration during storage. J Food Sci 76(9):C1365–C1375. https://doi.org/10.1111/j.1750-3841.2011.02416.x

Turek C, Stintzing FC (2013) Stability of Essential Oils: A Review. Compr Rev Food Sci Food Saf 12:40–53. https://doi.org/10.1111/1541-4337.12006

Valencia-Chamorro SA, Palou L, del Rio MA, Perez-Gago MB (2011) Antimicrobial edible films and coating for fresh and minimally processed fruits and vegetables: a review. Crit Rev Food Sci Nutr 51(9):872–900. https://doi.org/10.1080/10408398.2010.485705

Valero A, Hierro I, González P, Montilla P, Navarro MC (2006) Activity of various essential oils and their main components against L3 larvae of *Anisakis simplex* s.l. In: Govil JN, Singh VK, Arunachalam P (eds) Recent progress in medicinal plants, drug development from molecules. Studium Press LLC, Houston, pp 247–265

Van Vuuren SF, Viljoen AM (2007) Antimicrobial activity of limonene enantiomers and 1,8-cineole alone and incombination. Flavour Fragr J 22:540–544. https://doi.org/10.1002/ffj.1843

Vitoratos A, Bilalis D, Karkanis A, Efthimiadou A (2013) Antifungal Activity of Plant Essential Oils Against Botrytis cinerea, Penicillium italicum and Penicillium digitatum. Notulae Botanicae Horti Agrobotanici Cluj-Napoca 41(1):86–92. https://doi.org/10.15835/nbha4118931

Viuda-Martos M, Ruiz-Navajas Y, Fernández-López J, Pérez-Álvarez JA (2008) Antifungal activities of thyme, clove and oregano essential oils. J Food Saf 27(1):91–101

Viuda-Martos M, El Gendy AE-NGS, Sendra E, Fernández-López J, Abd El Razik KA, Omer EA, Pérez-Alvarez JA (2010) Chemical Composition and Antioxidant and Anti-Listeria Activities of Essential Oils Obtained from Some Egyptian Plants. J Agric Food Chem 58:9063–9070. https://doi.org/10.1021/jf101620c

Wang R, Wang R, Yang B (2009) Extraction of essential oils from five cinnamon leaves and identification of their volatile compound compositions. Innov Food Sci Emerg Technol 10:289–292. https://doi.org/10.1016/j.ifset.2008.12.002

Wani AR, Yadav K, Khursheed A, Rather MA (2021) An updated and comprehensive review of the antiviral potential of essential oils and their chemical constituents with special focus on their mechanism of action against various influenza and coronaviruses. Microb Pathog 152:104620. https://doi.org/10.1016/j.micpath.2020.104620. Epub 2020 Nov 16. PMID: 33212200. ISSN 0882-4010

Wink M (1988) Plant breeding: importance of plant secondary metabolites for protection against pathogens and herbivores. Theor Appl Genet 75(2):225–233

Wińska K, Maczka W, Łyczko J, Grabarczyk M, Czubaszek A, Szumny A (2019) Essential oils as antimicrobial agents—myth or real alternative? Molecules 24:2130. https://doi.org/10.3390/molecules24112130

Yanishlieva-Maslarova N (2001) Sources of natural antioxidants: vegetables, fruits, herbs, spices and teas. In: Yanishlieva N, Pokorny J, Gordor M (eds) Antioxidant in food: practical applications. Woodhead Publishing Ltd, Cambridge, pp 201–249

Yu J, Whitelaw CA, Nierman WC, Bhatnagar D, Cleveland TE (2004) *Aspergillus flavus* expressed sequence tags for identification of genes with putative roles in aflatoxin contamination of crops. FEMS Microbiol Lett 237:333–340

Part III
Essential Oil, Agricultural System Applications

Chapter 10
Control of Phytopathogens in Agriculture by Essential Oils

Maicon S. N. dos Santos, Carolina E. D. Oro, Bianca M. Dolianitis, João H. C. Wancura, Marcus V. Tres, and Giovani L. Zabot

10.1 Introduction

Agricultural systems are a key point of the advances in food security and the constitution of the pillars of the supply chain. With the disturbing imbalance of these complexes, a potential threat to the production arrangement is able of destabilizing the supply structure, from the availability of inputs to obtaining the final product. Moreover, nutritional quality and composition are severely affected by the disorderly action of undesirable microorganisms (Savary et al. 2019). Currently, a range of challenges has been addressed to the agricultural scenario, such as large-scale climate change, the concerning growth in food demand due to an accelerated population expansion, the high risks of pollution disasters, and insect and diseases invasion emergence (Mittal et al. 2020). Accordingly, a severe proportion of serious plant diseases are caused by harmful microbial agents designated phytopathogens.

Phytopathogens are microorganisms with the ability to secrete high amounts of enzymes and colonize plants and animals, optimizing penetration into plant tissues and chitin-based insects and their colonization (Li et al. 2017). Enzymes serve as severe toxins based on secondary metabolites causing critical symptoms in plants such as wilt, chlorosis, and necrosis in leaves and inhibition of plant growth (Peng

M. S. N. dos Santos · B. M. Dolianitis · M. V. Tres · G. L. Zabot (✉)
Laboratory of Agroindustrial Processes Engineering (LAPE), Federal University of Santa Maria (UFSM), Cachoeira do Sul, RS, Brazil
e-mail: giovani.zabot@ufsm.br

C. E. D. Oro
Department of Food Engineering, Regional Integrated University of Upper Uruguai and Missions (URI), Erechim, RS, Brazil

J. H. C. Wancura
Department of Teaching, Research and Development, Sul-Rio-Grandense Federal Institute (IFSul), Charqueadas, RS, Brazil

et al. 2021b). Recently, the occurrence of phytopathogens in crops has become a major obstacle, once the crop yield and final product quality are critically affected (Mishra and Arora 2018; Peng et al. 2021b). Some studies report that the incidence of phytopathogens in agricultural systems causes losses of up to 15% of the main crops annually (Peng et al. 2021b). Nevertheless, other scientific researches showed that losses can reach up to 40%, which indicates annual costs of more than US$200 billion globally (Pacios-Michelena et al. 2021).

The control and prevention of phytopathogens are predominantly conducted through the application of conventional chemical-based products and cultural management from the initial crop cycles to post-harvest and processing (Iantas et al. 2021). The chemicals inhibit mycelial development and tissue colonization by synthesizing a variety of enzymes with high infection potential (Medina-Romero et al. 2021). Nonetheless, these products are extremely worrying elements, exposed to serious problems in human health, severe environmental pollution, and drastic variation in the performance of metabolic processes of organisms that come into contact with or ingest products with concentrations of these pesticides (Hassaan and El Nemr 2020). Moreover, agrochemicals are non-degradable, which leads to the accumulation of high and extremely toxic concentrations in the environment and excessive levels of contamination on a large scale (Palla et al. 2020). Accordingly, the requirement for alternative and environmentally friendly strategies that have antifungal, antibacterial, and anti-mycotoxicgenic performance constitute a crucial trend in agri-food and agri-food-based industries (Maurya et al. 2021). Contextually, the EOs from a range of plants and biomasses consist of an innovative and sustainable approach, aiming at food security and the minimization of environmental toxicity.

The EOs are complex compounds originating from plant secondary metabolism, in apical structures called glandular trichomes (Alonso-Gato et al. 2021). EOs affect microbial cell growth and diffusion, as they obstruct exchanges between the cytoplasmic membrane and the environment, disrupting the functionalities of cell metabolism and promoting a drastic reduction in enzyme synthesis (Palla et al. 2020).

For centuries, EOs are appreciated based on their medical, pharmaceutical, and agricultural potential (Raveau et al. 2020). Thus, these compounds are derived from the integration of a set of extractive strategies, such as hydrodistillation, steam distillation, dry distillation, etc (Wińska et al. 2019). Scientific studies support the use of EOs for fragrance compounds (Sharmeen et al. 2021), pharmaceutical properties (He et al. 2020), cosmetics (Sarkic and Stappen 2018), food security (Bhavaniramya et al. 2019), and agricultural dynamics (Jouini et al. 2020). According to the agricultural context, EOs present a high potential for nematicidal (Eloh et al. 2020; Kundu et al. 2021), antifungal (Puškárová et al. 2017; Schroder et al. 2017), antibacterial (Cazella et al. 2019; Man et al. 2019), insecticidal (de Souza et al. 2021; Zimmermann et al. 2021), and herbicidal purposes (Benchaa et al. 2019; Jouini et al. 2020; Han et al. 2021).

Conclusively, the purpose of this chapter was to support an inclusive and encouraging review on the application of EOs to control phytopathogens in the agriculture approach. Properly, crucial information on the range of the current EOs investigated and the main properties, processes, mechanisms of action, and strategies involving the optimization of EOs extraction were provided. Furthermore, the phytopathogens control potential prescriptions and the main challenges and future perspectives are approached.

10.2 Potential EOs

EOs are products constituted by a combination of volatile substances and very concentrated, which can be obtained from the processing of raw material of plant (Christaki et al. 2021). These oils are mainly used in the food-based business as flavorings and preservatives in the food and the pharmaceutical industries as therapeutic products. These substances have also shown potential to be used as natural herbicides due to allelopathic substances found in some plants, which can interfere with the germination and development of other plants.

Studies have also been managed on the potential of EOs for pest and fungal control in agricultural production areas, showing that EOs can be an appropriate option to synthetic chemicals since they are natural substances that will cause less damage to the environment (Pawlowski et al. 2021). The main advantages that EOs present are: rapid action and degradation in the environment, low phytotoxicity, and toxicity ranging from low to moderate for mammals, being thus an environmentally friendly product (Pawlowski et al. 2021). The main way of obtaining EOs is through extraction procedures using pressurized solvents, supercritical fluid extraction, steam-dragging, and hydrodistillation methods (de Morais 2009).

10.2.1 Characterization and Main Properties

For the extraction of EOs, various parts of plants can be employed such as leaves, fruits, seeds, flowers, and barks. According to Steffens (2010), most of the essential oils come from cultivated plants (65%), followed by trees (33%), such as cedars, eucalyptus and pines, and wild plants (1%). Rosemary, basil, and citronella are some examples of aromatic herbs that are used to obtain essential oils.

Rosemary (*Rosmarinus officinalis* L.) is an aromatic herb well known for its use in cooking and medicine. The species is a shrub from the *Lamiaceae* family. Its leaves, flowers, and roots are potential by-products for the extraction of EOs, being present in its chemical composition compounds with antioxidant and therapeutic properties (Steffens 2010).

Furthermore, Basil (*Ocimum basilicum*) is a plant from the *Lamiaceae* family, mainly used for culinary and medicinal purposes. The species has leaves rich in linalool. The leaves are used in the extraction of EOs, which have repellent and insecticide properties (Pereira and Moreira 2011). Citronella grass (*Cymbopogon winterianus* J.) is a plant *Poaceae* species, which has characteristics such as upright growth with elongated leaves, similar to lemongrass. The species has a higher amount of oil than Cymbopogon nardus, which is the other variety of citronella grass. Therefore, it has become more widely used in the EOs synthesis. The oil removed from the fresh or dried leaves of this plant presents compounds such as citronellol, which makes plant oil an insect repellent (Steffens 2010). In addition to aromatic herbs, some trees and shrubs are also used to obtain essential oils, such as eucalyptus, laurel, and pitangueira.

The genus Eucalyptus presents approximately 600 species of plants that are trees native to Australia and belonging to the family *Myrtaceae*. The oil produced through eucalyptus has antifungal, repellent, and insecticide properties, and cineole is a dominant component of eucalyptus oil (Vitti and Brito 2003). The species *Laurus nobilis* L., known as laurel, is a tree belonging to the *Lauraceae* family that can reach up to 10 meters in height. The leaves are fundamental to obtain EOs, which have antimicrobial properties (Pinheiro 2014). The pitangueira (*Eugenia uniflora* L.) is a tree native to Brazil that has edible fruits. Its leaves and fruits are used for obtaining essential oils due to the amount of sesquiterpenes present in its composition. EOs obtained from this particular species have antimicrobial properties (Auricchio and Bacchi 2003).

EOs obtained from plants present in their chemical composition mainly compounds such as terpenes, which are compounds naturally produced by plants and which are responsible for their aroma and also for performing some functions to them. Terpenes are made up of isoprene molecules, whereas each one is formed by five carbon atoms, once considered the molecule dimensions, the terpenes will receive different denominations. Accordingly, when they are composed of ten carbon atoms, their denomination will be monoterpenes, which is the example of linalool and limonene. When they are made up of 15 carbon atoms, their denomination will be sesquiterpenes, as alpha-saline (de Morais 2009).

In addition to terpenes, essential oils also have aromatic compounds in their composition. The rho – cymene is the most common aromatic compound found in these oils. Moreover, alcohols, esters, aldehydes, ketones, natural oxides, and phenols are also found. Examples of monoterpenoids and alcohols are phenylethyl, benzyl, geraniol, and menthol. The main constituent of esters is terpinyl acetate, which presents relaxing properties. The main aldehydes are aromatic and saturated ones with anti-inflammatory properties. Citral and geranial are some of the main compounds of aldehydes. In ketones, the compounds menthol and thujone can be found. In natural oxides are cineol, a compound found in essential oils derived from eucalyptus. Also, thymol is a phenol that has antimicrobial properties (Monteiro 2002).

The quantity of these compounds as well as whether or not they are present in EOs involves the variety, plant species, and environmental conditions of the site where the plant developed (Oliveira and Brighenti 2018). The compounds present in higher quantities in essential oils are called main compounds and the compounds present in small quantities are called trace compounds (Vitti and Brito 2003).

Some bioactive compounds involve the bioactive properties of EOs. In bergamot oil, volatiles are the main compounds. In cedar oil, the main compound is cedrol, while in clove oil is eugenol. In lemon oil, limonene is the main compound (Monteiro 2002). Depending on the composition, essential oils can be used as antioxidants and antimicrobials. Antioxidants are largely used in the chemical and pharmaceutical industries to prevent disease and food deterioration. EOs that have in their composition one or more hydroxyl groups are preferable to be used as antioxidants. Oregano, basil, and thyme are examples of plants belonging to the family *Lamiaceae* and have compounds in their composition with a hydroxyl group such as thymol and carvacrol. Other compounds, such as phenylpropanoids (e. g., apinol, eugenol, and methyl-eugenol), also have antioxidant activities (Lima and Cardoso 2007).

According to a study developed by Mat Saad et al. (2021), EOs obtained from some species of the family *Myrtaceae*, such as *Baeckea frutescens* and *Leptospermum javanicum*, have antioxidant resources due to the wide presence of compounds such as 1,8 cineol, p-cymene, and a-pinene. The species Pinus pumila belonging to the *Pinaceae* family also has a high antioxidant potential (Peng et al. 2021a). Antimicrobial activity found in essential oils is characterized by the ability of these oils to inhibit the development of microorganisms. Antimicrobial activity in EOs can be determined by increased permeability and loss of cellular constituents, interference in the phospholipid surface of the cell wall of the microorganism, and the alteration of the variety of enzymatic systems. The types of compounds present in the oils allow them to have antimicrobial properties, as is the case with citronellol. It is a compound found in some plant species, such as Lucia-lima (*Lippia citriodora*), which is a *Verbeneaceae* plant species and has anti-parasitic activity (Sarto and Junior 2014).

Furthermore, Sarto and Junior (2014) highlight the essential oil obtained from rosemary-pepper has antimicrobial action. This is a result from the dominant content of compounds such as thymol and carvacrol present in the species. Oils obtained from lemongrass (*Cymbopogon citratus*) and mint (*Mentha piperita*) have been reported in studies due to the presence of antimicrobial properties. In the case of mint, the activity is significantly due to the large volume of compounds such as menthol (Valeriano et al. 2012). The chromatographic method identifies and quantifies a large scale of compounds present in EOs. The disk diffusion method can be used to check the antimicrobial activity of EOs, where through a solid medium the microorganism is tested against the active compound. DPPH (2,2-difenil-1-picril-hidrazil) method is applied to verify the antioxidant potential through the reading by absorbance mixture consisting of the sample and the solution (De Oliveira 2016).

10.3   Insecticidal Properties and Mechanisms of Action

Insecticides used in pest control can be synthetic insecticides or natural insecticides. Natural insecticides are called "botanical insecticides" and come from plants. The botanical insecticides were widely used in the 30th and 40th centuries but were replaced by synthetic insecticides due to various problems in their effectiveness for pest control. Currently, interest has arisen in these products due to the problems caused by synthetic insecticides, such as environmental damage and risks to human health (Corrêa and Salgado 2011). Another problem caused by synthetic insecticides is the emergence of pests resistant to these products, due to their intense and repetitive use. The loss of productivity in crops is largely affected by the presence of pests, and for controlling them the synthetic insecticides have been increasingly used (Rabaioli and da Silva 2016). The place where insecticides will act to control pests is known as the mechanism of action.

Insecticides are classified according to their mechanisms of action in insecticides that act on sodium channels, acetylcholinesterase enzyme-inhibiting insecticides, insecticides that act on receptors of the neurotransmitter acetylcholine, insecticides that act on receptors of the neurotransmitter gamma-aminobutyric acid, and growth-regulating insecticides (Sant'Anna 2009). Insecticides that act on sodium channels can bind to the sodium routes present in the insects nervous cells complex, keeping channels open which will ultimately cause seizures and discoordination leading to death. Acetylcholinesterase enzyme inhibitors, as the name suggests, inhibit the enzyme acetylcholinesterase, which is responsible for the hydrolysis of a neurotransmitter. With the blocking of this enzyme, the hydrolysis of the neurotransmitter does not occur, causing the insect to die through symptoms such as seizures and paralysis. Insecticides that act on the neurotransmitter gamma-aminobutyric receptors are type II pyrethroids and macrocyclic lactones. Finally, growth-regulating insecticides act by inhibiting the growth of insects and can be classified as chitin synthesis inhibitors.

Botanical insecticides are made up of plants that have specific properties and compounds in their constitution that make them interesting to be used for this purpose. The most important group of compounds used for the natural control of insects is pyrethrins. Rotenoids and alkaloids are also used in pest control, which nicotine and nor-nicotine are the most important alkaloids for insect control. The class of terpenoids has insecticide activities. Several compounds of this group have already been tested for the control of insects, such as myrcene and limonene that have plant protection functions (Viegas 2003).

The botanical insecticides can act on the nervous system of insects, as anti-food causing insects to be unable to feed or exhibit repellent activity. Repellent activity is commonly found in essential oils extracted from plants, which occurs through direct contact of the oil with the insect. Examples of essential oils that have repellent action are obtained from lemongrass and eucalyptus, among others (Corrêa and Salgado 2011).

According to Previero et al. (2010), the best-known plants that have insecticide activity and can be used from homemade recipes for application on pests are:

Garlic (Allium sativum): Plant used in feeding as a seasoning; it has allicin, which is a compound with antiseptic activity. It can be used for grain conservation, avoiding the presence of pests. It is used to fight pests such as cochineals and aphids in various crops;

Lemongrass (*Cymbopogon citratus*): Plant belonging to the *Poaceae* family that has in its composition alkaloids and terpenes, which are compounds that give it the ability to be used as an insect repellent;

Eucalyptus (*Eucaliptus citriodora*): Tree originating in Australia, which can be used for various purposes; it has compounds such as pinocarveol, limonene, and myrtenol. It can be used for grain and seed conservation, as well as an insect repellent;

Mint (*Mentha spicata*): Perennial herb from the *Lamiaceae* family, which has compounds such as terpenes, aldehydes, flavonoids, and carotenes. It has application as insect repellent and action against ants;

Rue (*Ruta graveolens*): Aromatic plant rich in flavonoids, alkaloids, and coumarins, which is used in the control of aphids.

Neem (*Azadirachta indica*): Large tree, which has compounds such as meliantrol and azadirachtin. It has several insect properties, such as insect repellent, grain and seed conservation, and caterpillar control.

The main advantages provided by natural insecticides are rapid action and degradation in the environment, selectivity to insects considered beneficial to crops, and low toxicity to mammals. However, they also have some disadvantages, such as rapid degradation. The product degrades so fast and does not have enough time to act on the target. In addition, the action on non-target species and the limited availability of these products for marketing are also considered to be disadvantages (Moreira et al. 2006).

Several studies have been developed for testing the potential natural insecticides in pest control. de Coitinho et al. (2006) reported the use of clove (*Syzygium aromaticum* L.), white pepper (*Piper nigrum* L.), neem (*Azadirachta indica* A. Juss), eucalyptus (*Eucaliptus globulus* Labill. and *Eucaliptus citriodora* Hook.), andiroba (*Carapa guianensis* Aubl.), rosemary (Lippia gracillis HBK.), cedar (Cedrela fissilis Vell.), and peki (*Caryocar brasiliense* Camb.) on corn stored for controlling of *Sitophilus zeamais* Mots, which is one of the main pests found in corn. All oils tested were efficient in controlling adult insects. According to Nali et al. (2004), the efficiency of natural insecticides was tested. The application of products based on garlic, timbo, and neem extract was performed in the control thrips in vines. Products based on timbo and neem extract showed better efficiency than the garlic extract-based product. A study conducted with farmers in the municipality of Arroio do Meio (Brazil) indicated that the plants of the family Asteraceae demonstrated botanical insecticides effects in their crops (Dietrich et al. 2011).

10.4 Essential Oils Extraction Strategies

EOs from aromatic plants have no miscibility with water, and therefore have a hydrophobic nature and represent up to 5% of the weight of dry matter, highly composed of hydrocarbon terpenes and terpenoids. The oil extraction yield will depend mainly on the plant and the chosen extraction method (El Asbahani et al. 2015). During the process of extracting EOs, some chemical changes can occur, such as oxidation, which results in variations in the EO content. Therefore, good sample preparation and extraction temperature control are essential to ensure the success of the extractive process. On a laboratory scale, hydrodistillation is the most used and most classic extraction method (da Silva et al. 2021). Steam distillation, cold pressing, and organic solvent extraction can also be mentioned as the classic and conventional methods for extracting essential oils. However, conventional methods require a lot of extraction time and it is often difficult to remove all the solvent from the essential oil.

Hydrodistillation consists of Clevenger-type equipment, in which the sample is in contact with boiling water. After the stipulated time has elapsed, the oil can be collected from the equipment (Manouchehri et al. 2018). However, as with all methods that employ heat, compounds can be thermally degraded and the quality of EOs compared to other techniques may be lower (Yousefi et al. 2019). The EOs from *Carex meyeriana* Kunth were obtained by hydrodistillation and the extraction parameters were optimized. The best condition found was 9 h of extraction, a ratio of liquid to plant of 43:1 (mL/g), and proportions of 10 mesh to obtain a yield of 0.13% (w/w) (Cui et al. 2018). Higher yields can be obtained from different plant matrices. For example, essential oil of *Elettaria cardamomum* Maton showed a yield of 3.1% (w/w) in the extraction by hydrodistillation for 4 h (Sereshti et al. 2012).

Similar to hydrodistillation, steam distillation procedure is an effective strategy and the quality of the EO obtained is significant, once the steam is forced by leaves or plant material (Yousefi et al. 2019). Steam distillation is a simple, low-cost method for extracting essential oils. In general, this method consumes more time compared to methods that employ organic solvents. However, the quality of the essential oil obtained is superior (Zhu et al. 2020a). The extraction of *Rosmarinus officinalis* L. branches essential oil by steam distillation showed a yield of 2.35%, which was higher than the yield by hydrodistillation (Conde-Hernández et al. 2017), showing that the technique is efficient in extracting essential oils.

Cold pressing is a procedure mainly used for extracting EOs from citrus matrices. Moreover, the sample is cold-pressed and the collected emulsion is separated from the solvent by centrifugation method. The EO from the bark of *Citrus limon* extracted by cold pressing showed a yield of 1.24% and a concentration of 67.1% of limonene in its composition (Himed et al. 2019). Organic solvent extraction is performed on equipment called Soxhlet and it is one of the oldest extraction methods. However, it is not a recommended method for extracting essential oils, mainly because it uses high temperatures and is time-consuming. The extraction process can reach up to 72 h, and the compounds can be modified by time and temperature. Furthermore, the compounds are extracted with the aid of organic solvents that need to be rotary evaporated, since the extracted compounds are diluted in high amounts of solvent and, therefore, the sample needs to be concentrated before analysis. Another problem found in this method is the non-selectivity and possible existence of interfering compounds in the sample (Yousefi et al. 2019).

Accordingly, to optimize the extraction of EOs, innovative green-bases alternatives have been implemented to extract EOs, such as supercritical fluid extraction, subcritical extraction, ultrasound-assisted extraction, microwave-assisted extraction, and instant controlled pressure drop. Supercritical fluid extraction (SFE) is a key technique to obtain high volumes of EOs from plants as it is fast, employs moderate conditions of temperature and pressure, and does not use toxic solvents (Yousefi et al. 2019). SFE employs supercritical carbon dioxide (SC-CO2) as a solvent for the selective extraction of aromatic compounds and EOs from plants. The best temperature and pressure requirements for highly selective extraction using SC-CO2 are 40–50 °C and 8–9 MPa. Under these conditions, SC-CO2 is strongly compressible for being under the near-critical conditions and is characterized as one of the best conditions for extracting EOs from plants as it improves the solubility between the components of EOs and the plant waxes and, consequently, improves the selectivity and yield of the extraction process (del Valle et al. 2019).

Other extraction conditions are also found in the literature since the best condition depends on the plant matrix under study. SFE-CO2 was used as the essential oil extraction method from the tangerine peel and the optimized extraction parameters were 45 °C, 14 MPa, and extraction time of 147 min to obtain a yield of 1.34% (Xiong and Chen 2020). Furthermore, the SFE technique was shown to be effective for the extraction of terpenoids and other lipophilic bioactive from *Mentha piperita* L. (Pavlić et al. 2021) compared to a range of procedures covered in this chapter. Similar to the SFE technique, the subcritical extraction technique occurs when the system pressure or temperature are in a domain area below the critical point. In this technique, water or CO_2 are usually used as a solvent for the extraction of EOs (El Asbahani et al. 2015).

Ultrasound-assisted extraction is a simple and efficient extraction technique that applies sonification principles. This technique can also be used as sample pretreatment to optimize the extraction process by combining different techniques (Balti et al. 2018). Sereshti et al. (2012) reported the optimized conditions of ultrasound-assisted emulsification microextraction of *Elettaria cardamomum* Maton essential oil of 32.5 °C, ultrasound time of 10.5 min, and 120 μL of solvent. Microwave-assisted extraction (MAE) practices the water as the main extraction solvent, mainly because the water has a different polarity than the essential oil and, therefore, facilitates the separation process. It is a method that has been gaining attention because it is fast and has high extraction efficiency and yield (Franco-Vega et al. 2021). High volumes of EOs from *Ferulago angulata* (Apiaceae) were attained by microwave-assisted hydrodistillation (MAHD) and the extraction conditions were optimized. The best extraction conditions were 72 min, operation power of 980 W, and 2 g of biomass for 100 mL of solvent. The authors reported a yield of 6.50% by MAHD compared to 2.65% obtained by hydrodistillation (Mollaei et al. 2019).

Instant controlled pressure drop (DIC) is a technique that uses high-pressure saturated steam for a short time. This technique results in an expansion of the sample due to the abrupt drop in pressure towards the vacuum after the proceeding. The EO from the bark of *Citrus sinensis* was extracted by DIC and presented, under optimized conditions, a yield higher than 16.5 mg/g dry material in 2 min (Allaf et al. 2013). This method can also be used as a pre-treatment of the sample as it is fast and efficient.

10.5   Control of Phytopathogens with Essential Oils

Essential oils in addition to presenting medicinal properties, insecticides, and herbicides have also shown potential for phytopathogen control due to the antifungal and antibacterial properties present in several plants that are used to obtain EOs. The action of EOs on phytopathogens can be both direct and indirect. Some substances found in essential oils cause extravasation of cellular content, as they can act on the cell wall of fungi, affecting cell membranes and causing them to inhibit or reduce mycelial growth (Ramos 2014). The fungi toxic effect has been reported mainly in fungi found in shoots, seeds of a given crop, and also in soil (de Lira Guerra 2013).

Medicinal plants have the potential to be used in the control of phytopathogens since they have antibacterial and antifungal properties that are used to treat some diseases in humans. This causes some to be used against phytopathogens in plants (Ramos 2014). Among these medicinal plants that have this potential, basil (*Ocimum basilicum*), pomegranate (*Punica granatum*), lemon balm (*Melissa officinalis*), marjoram (*Origanum majorana*), eucalyptus (Eucalyptus), and oregano (*Origanum vulgare*) are mentioned. They have already been used in studies where their essential oil has been tested to control phytopathogens in sorgo and soybean (de Lira Guerra 2013).

In addition to these medicinal plants, several other plants have also shown potential for control of phytopathogens due to the phytotoxic properties they have. Mint (*Mentha spicata*), clove (*Syzygium aromaticum*), cinnamon (*Cinnamomum verum*), garlic (*Allium sativum*), and the rue (*Ruta graveolens*) are cited (dos Venturoso 2009). Pinto (2009) demonstrated the efficiency of the use of long pepper (*Piper Longum*), lemongrass (*Cymbopogon citratus*), arabica coffee (*Coffea arabica*), and african lily (*Agaphantus africanus*) in the control of fungi.

10.5.1   *Phytopathogens Diversity*

Phytopathogens are termed microorganisms capable of causing diseases in plants. These phytopathogenic microorganisms can be viruses, bacteria, fungi, and nematodes, among others. In general, phytopathogens can be classified into obligated parasites, optional saprophytes, optional parasites, and accidental parasites (Michereff 2001).

The forced parasites live in the host tissue as is the example of the fungus that causes rust and mildew. Optional saprophytes live most of their lives as parasites, as an example of these, the fungi causing leaf spots are found. Optional parasites develop as saprophytes but pass part or all of their development as parasites. Accidental parasites are saprophytes that under certain conditions, such as in stress conditions, can act as parasites. Another classification of phytopathogens is related to their nutritional requirements, and can be classified as biotrophic, hemibiotrophic, and necrototrophic. Necrotrophic snares are the ones that cause the highest damage to plants. They are responsible for causing leaf spots and stem and root rot in crops and survive in plant cultural remains. Biotrophic are responsible for causing rust and mildew hemibiotrophic sums, and survive in living tissues (Reis et al. 2011). Phytopathogens can cause leaf, root, soil and seed diseases.

Optional parasitic phytopathogens are found in *Rhizopus* spp., *Penicillium* spp., and *Erwinia* spp., which cause rot in seeds and fruits. *Fusarium solani, Sclerotium rolfsii* and *Thielaviopsis basicola* cause rot in the roots, and *Alternaria* spp., *Cercospora* spp., *Colletotrichum gloeosporioides* and *Xanthomonas* spp. are responsible for causing stains in crops. *Plasmopara viticola, Bremia lactucae* and *Pseudoperonospora cubensis* cause the mildew disease. *Oidium* spp. is responsible for the disease known as mildew and Puccinia spp., *Uromyces* spp. and *Hemileia vastatrix* cause the rust (Michereff 2001). Among existing phytopathogens, fungi are the ones that cause the majority of diseases found in plants.

According to Grigolli (2015), the soybean crop has approximately 40 diseases identified in Brazil, which are caused by phytopathogens such as viruses, bacteria, fungi, and nematodes. Among these diseases, they affect the leaves as the target spot caused by a fungus, mildew also caused by fungus characterized by green or yellowish spots, the mildew that develops in the aerial part of the plants and is also caused by fungi. Some diseases affect pods, seeds, and stems such as anthracnose and white mold caused by fungi. Other diseases affect the root part of the crop as the well-known rot of coal and soybean meal caused by fungi.

In rice, the main diseases caused by fungi are brown spot, stain on grains, staining-narrow, false-coal, coal, leaf coal, sheath burning, sheath stain, foot ill, and stem rot. Diseases caused by nematodes are the white tip and nematode of the galls (Silva-Lobo and Filippi 2017). In corn crops, diseases caused by fungi are anthracnose, stem anthracnose, *Fusarium rot, Stenocarpella rot*, rot caused by *Pythium*, and dry stem rot. It is also found in the corn crop bacterial rots that are mainly caused by bacteria of the genus *Pseudomonas* and *Erwinia*. Some diseases caused by viruses have become thin streak and the common mosaic of corn (Casela et al. 2006).

10.5.2   Main Applications

The main applications of EOs to manage phytopathogens are in the control of diseases caused by fungi because they are prominent phytopathogens in existence. The applications of EOs to combat these phytopathogens have been studied in diseases both in leaf diseases, as well as in diseases found in the grains and seeds of certain crops affected. In a study carried out in a greenhouse, Medice et al. (2007) reported that essential plant oils were used, thyme (*Thymus vulgaris* L.), eucalyptus (*Corymbia citriodora*), neem (*Azadirachta indica*), and citronella (*Cymbopogon nardus*), to test the effects on a disease found in soybean crop known as Asian rust. The four oils tested were able to reduce the severity of rust in soybean plants.

EOs from 13 plants: neem (Azadirachta indica A. Juss), clove (*Caryophyllus aromaticus* L.), peppermint (*Mentha piperita* L.), melaleuca (Melaleuca alternifólia Cheel), lemon (Citrus Limonium L.), copaiba (*Copaifera langsdorfii* Desf.), eucalyptus (*Eucalyptus globulus* Labill.), ginger (*Zingiber officinale* Roscoe), basil (*Ocimum basilicum* L.), coconut (*Cocos nucifera* L.), lemongrass (*Cymbopogon citratus* (DC) Stapf.), cinnamon (*Cinnamomum zeylanicum* Breyn.), and citronella (*Cymbopogon winterianus* Jowitt) have been studied by applying at different concentrations in the in vitro control of the phytopathogen responsible for causing the disease known as anthracnose. It was confirmed the EOs as promising strategies according to the studied phytopathogen, being the oil of neem, lemongrass, copaiba, and melaleuca those showed the best results (Ramos 2014).

The effect of EOs of medicinal and aromatic plants such as basil (*Ocinum basilico* L.), stagerian eucalytus (*Eucalyptus officinale* L.), ginger (*Zingiber officinale* L.), copaiba (Copaifera oficinalis L.), juniper (Juniperus communis L.), palmarosa (*Cynbopogon Martinii*), and atlas cedar (*Cedrus atlântica* Manetti) has been studied in the control of Sclerotium wilt in peanut culture in vitro and in vivo assays, noting the aptitude for palmarosa control against this disease caused by fungi (de Lira Guerra 2013). In a study carried out to control phytopathogens present in fennel seeds (*Foeniculum vulgare*) using anise essential oil (*Pimpinella anisum*), the efficiency of the oil against the incidence of fungi in the seeds of the species was found (Neto et al. 2012). Another study carried out with the application of ginger essential oils (*Zingiber officinale*), Tahiti lemon (*Citrus latifolia*) and hydroalcoholic extracts of pariparoba (*Pothomorphe umbellata*) and penicillin (*Alternanthera* sp.) on soybean seeds showed that ginger oil was the one that caused the highest reduction of fungi present in the seeds (Gonçalves et al. 2009). Through these studies, it is possible to notice that the essential oils of eucalyptus, neem, and citronella have wide application in the treatment of phytopathogens.

10.6   Bioproducts Formulation

In terms of chemical pesticides, the typical formulations utilized in agriculture management contain pyrethroids, organophosphates, organochlorines, carbamates, glyphosate, and triazoles (Isman et al. 2011). The latent hazards for non-target organisms and the environment associated with the excessive application of these synthetic insecticides are crucial obstacles to their regulatory approval (Zikankuba et al. 2019). Also, the development of cross-resistance against synthetic insecticides in insect pests with medical interest is another important inconvenience that efforts the eminent and gradual replacement of conventional insecticides by eco-sustainable molecules (Ramakrishnan et al. 2019; Upadhayay et al. 2020; Zhu et al. 2020b).

Essential oils and their compounds are a promising alternative for phytopathogens control. Their molecules are natural products extracted from aromatic plants, having the capacity to inhibit the activity of many phyto- and human pathogens, including insecticidal and repellent mechanisms (Bakthavatsalam 2016). However, the interesting results obtained in lab–scale (or greenhouse) are not commonly observed in the field, where several drawbacks are reported when these bioactive compounds are applied on phytopathogens control. Degradation and volatilization are the major issues that reduce the efficiency of plant-based products under field conditions, resulting in losses that underestimate the potential of these compounds (Borges et al. 2018). Factors like poor water solubility and oxidation are also reported, playing a critical role in the biological activity, application, and persistence of essential oils in agriculture monitoring (Lucia and Guzmán 2021).

Aiming to avoid such difficulties, researchers suggest formulating the bioactive plant products based on solubilization, encapsulation, and/or protection of the molecules with phytopathogen action, employing distinct carriers such as emulsifying agents, surfactants, solvents, defoamers, polymers, plasticizers, stabilizers, and biodegradable antioxidants (Borges et al. 2018). These compounds are applied to the target to guarantee the stability, adherence, and controlled release of the bioactive compounds presented on essential oils and derivates. Knowles (2008) highlights that the utilization of these types of formulations implies that the active ingredients of the essential oils are released into the environment over time, bringing benefits such as minimizing losses of the active compound during processing and storage, a higher period of activity and reduced toxicity to plants and animals.

The formulation approach may differ according to the procedures (and materials) employed for encapsulation. Recently, a significant volume of scientific researches has been focused on the exploration of low-cost formulations for a commercial application. In relation to main options to encapsulation of bioactive compounds from essential oils and their ingredients, it could be mentioned atomization, lyophilization, emulsification, extrusion, fluidized bed, coacervation, and molecular inclusion, which are briefly discussed herein. Atomization (spray-drying) is a prominent technique applied for microencapsulation of EOs, which is considered a low-cost process for use at an industrial scale with costs 30–50 times lower compared to other methods (Bakry et al. 2016). Spray drying involves the atomization of emulsions into a drying chamber in conditions of high temperature (140–200 °C) that implies in the progressive water evaporation and an instantaneous concentration and availability of oils. Water elimination by spray-drying procedure is a large and crucial practice to ensure the microbiological stability of products (de Barros Fernandes et al. 2014).

Lyophilization (freeze-drying) is a process employed for the dehydration of heat-sensitive materials and aromas, like EOs. Previously to the drying process, the oil is dissolved in water and fastly frozen (between −90 and −40 °C) to preserve its chemical characteristics (Sousa et al. 2020). Then, the process pressure is reduced to a partial vacuum and enough heat is added to the system, allowing the frozen moisture in a determined material to sublimate straight from solid to gas phase, keeping typically 2 wt% of water in the product (Bakry et al. 2016). Finally, the dehydrated material is milled until the desired particle size. With fungicidal activity, Borges et al. (2018) describe that freeze-dried elements present a progressive retention of volatile compounds face to spray-dried biomaterials.

Emulsion (emulsification process) involves a thermodynamically unstable system containing at least two non-miscible liquid phases, where one phase contains colloidal particles dispersed in the other phase (Borges et al. 2018), which is a crucial step in the process of micro- and nanoencapsulation of diverse EOs. It is typically employed for the bioactive EOs encapsulation in aqueous solutions that can be applied directly in the liquid state or dried, via spray- or freeze-drying, to generate powders after emulsification (Bakry et al. 2016). Usually, emulsifiers such as Tween 20, Tween 60, Tween 80, Poloxamer 407, etc.; or surfactants are regularly introduced in the emulsion complex aiming to attain a kinetically stable solution. Lucia and Guzmán (2021) described that the small size of the dispersed droplets containing the oil phase permits circumvents significant drawbacks according to their application in the preparation of consumer products, such as the dispersion destabilization as a consequence of gravitational forces, although the current comprehension about the efficiency of micro- and nanoemulsions containing EOs to manage phytopathogens remains still limited.

Coacervation is a cost-effectiveness alternative and is the oldest and largely recommended method for microencapsulation of oleaginous compounds and EOs (García-Saldaña et al. 2016). The process includes the electrostatic attraction involving two biopolymers of opposite charges with the coacervates formation happening over a limited pH range, where the liquid phase is distinct from the polymer-rich (coacervate) phase (Hernández-Nava et al. 2020). Coacervation process is classified into simple and complex coacervation. In a classic coacervation procedure, the polymer is "salted out" by the significant performance of electrolytes, or desolvated by the inclusion of a water-miscible non-solvent, or still by temperature modifying (Bakry et al. 2016). These conditions promote the interactions between macromolecules. Complex coacervation is a process in which two or more oppositely charged polymers and polysaccharides are involved (Timilsena et al. 2019). Simple coacervation has advantages over complex coacervation, mainly regarding costs and malleable operations.

In an extrusion method for encapsulation of EOs, extracted core and coating material are processed through a nozzle region at high pressure into an ionic solution in constant agitation, where the gel beads formed are obtained after some minutes and dried (Arriola et al. 2016). This method has the benefit of increasing the oils' stability against oxidation and promoting the progress of the shelf-life of the compounds compared to the spray-drying technique (Arriola et al. 2019). However, the technique is not frequently applied compared to spray-drying and has not been fully investigated in the formulation of compounds for agriculture disease control (Borges et al. 2018). This fact is associated mainly due to the costs involved in the process, considerably higher than spray-drying and also due to the large particles (from 150 to 2000 µm) that limit their implementation in various applications (Bakry et al. 2016).

Fluidized bed (air suspension coating or spray coating) is an innovative and effective alternative coating methods to coat solid core particles, with the interesting growth of applications in food and pharmaceutical industries (Rungwasantisuk and Raibhu 2020). However, it is still poorly explored in phytopathogens management. This technique has been employed to give an additional coating after a spray-dried process, increasing the protection of the encapsulated essential oil. According to this strategy, a high volume of solid particles are conducted to air suspension and the encapsulating material is sprayed onto these components, producing a coating, where the encapsulating material is able to be a concentrated solution or dispersion, a hot melt, or an emulsion (del Carmen Razola-Díaz et al. 2021).

The Molecular inclusion complex is an encapsulation method that occurs at a molecular magnitude, where a guest (active) compound is apprehend by a host (polymer) by physicochemical forces (Ozkan et al. 2019). The physical interactions involved in host-guest complexation are typically a combination of hydrogen bonding, van der Waal forces, hydrophobic effects, and electrostatic effects (Steiner et al. 2018). An extremely important molecule is cyclodextrin, an enzymatically modified form of starch molecules comprised by a hydrophilic external region and an internal hydrophobic area, being popular for encapsulation due to their ability to accommodate and stabilize molecules in their cavity (Tian et al. 2021). The molecular constitution of cyclodextrin is widely similar to a hollow cone, where the size of the inner cavity is in the nanoscale (Saifullah et al. 2019). Tian et al. (2021) highlighted that the encapsulating effectiveness is widely limited by the low solubility of the cyclodextrins, although recent studies have been demonstrated the capacity in increasing the stability of the molecules, producing EOs from liquid to powder solution by establishing inclusion arrangements with prolonging of the release of the active compound.

In summary, Table 10.1 presents some works where the techniques described previously are applied to formulate EOs and their bioactive constituents for use in phytopathogen agro-control.

10.7   Challenges and Future Outlooks

The wide uses of EOs as sources of potential bioactive compounds comprehend extensive fields of study and applications over time. Initially, EOs were largely applied for medicinal, pharmaceutical, and food-related purposes. Correspondingly, Fig. 10.1 presents a temporal viewpoint of EOs from the EOs potential discovery until the recent uses of these materials. The timeline specifies the key applications of each period of time. The application of EOs as an antimicrobial source and widespread use in agriculture has been the focus of a range of research in recent years. The current study provided the scenario for the use of EOs as a tool to manage phytopathogens in agriculture. The expansion of this strategy is directly associated with the exploration of more sustainable approaches in the management procedures of agricultural species. Moreover, Fig. 10.2 indicates appreciable future

Table 10.1 Methods to formulating bioactive compounds for phytopathogenic control in agriculture

Method	Bioactive compound	Function	Target organism	Application	Assay	References
Atomization	*Rosmarinus officinalis* oil	Insecticidal	*Tribolium confusum*	–	In vivo	Ahsaei et al. (2020)
Atomization	Garlic oil	Antimicrobial	*Pseudomonas syringae pv. tomato*	Tomato plants	Inoculation	Cortesi et al. (2017)
Atomization	Thyme oil	Antifungal	*Colletotrichum gloeosporioides*	Mango fruit	In vitro	Esquivel-Chávez et al. (2021)
Lyophilization	*Peumus boldus* oil	Antifungal	Fungal mycoflora	Peanut seeds	In situ	Girardi et al. (2018)
Lyophilization	*Illicium verum* oil	Antifungal	*Aspergillus flavus*	Lotus seeds	In vivo	Li et al. (2020)
Emulsification	*Satureja hortensis* oil	Herbicidal	*Amaranthus Retroflexus and Chenopodium album*	Weed seeds	In vivo	Hazrati et al. (2017)
Emulsification	Purslane oil	Larvicidal	*Sitophilus granarius*	Granary weevil	In situ and in vivo	Sabbour and El-Aziz (2016)
Emulsification	Neem/citronella oils mixture	Antifungal	*Rhizoctonia solani Sclerotium rolfsii*	–	In vitro	Ali et al. (2017)
Coacervation	*Satureja hortensis* L. oil	Herbicidal	–	*Lycopersicon esculentum* Mill. and *Amaranthus retroflexus* L	In vitro	Taban et al. (2020)
Coacervation	*Peumus boldus* oil	Antifungal	*Penicillium sp. and Aspergillus sp.*	Peanut seeds	In vivo	Girardi et al. (2016)
Coacervation	Tea tree oil	Antifungal	*Botrytis cinerea*	Cherry tomato	In vitro and in vivo	Yue et al. (2020)
Extrusion	Clove, thyme and cinnamon oil	Antifungal	*Aspergillus niger and Fusarium verticillioides*	–	In vitro	Soliman et al. (2013)
Molecular inclusion	Thyme and betel leaf oils	Antimicrobial	*Colletotrichum gloeosporioides*	Sapota fruit	In vitro and in vivo	Gundewadi et al. (2021)
Molecular inclusion	Clove and oregano oil	Antifungal	*Fusarium oxysporum*	–	In vitro	Estrada-Cano et al. (2017)

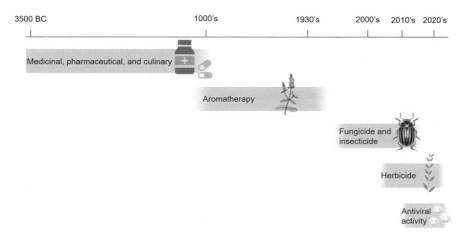

Fig. 10.1 Temporal arrangement of EOs applications, pointing the EOs main significant in the different periods

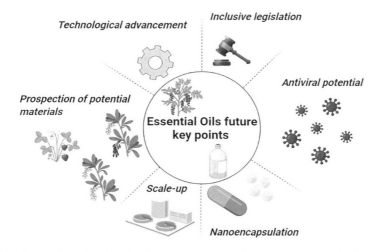

Fig. 10.2 Noticeable and challenging key points corresponding to the recent applications of EOs

perspectives and challenges intimately associated with the future application of EOs approach.

Initially investigated as a potential component for application in fragrances and flavors, the diffusion of the technique has been universally verified in the agri-food complex of industries, since the process involves from the initial to the final post-harvest and processing procedures. The nanoencapsulation of EOs as a pathway of food preservation regarding degradation processes and microbial agents is an excellent precursor of advances in food conservation and product availability efficiency.

This strategy has been disseminated as one of the main advances with an antimicrobial purpose in the fields of nanoscience and technology.

Therefore, due to the abiotic and management influence on the chemical characterization of EOs, sustainable extraction strategies have been considered, such as microwave, supercritical fluid, and ultrasound extraction procedures. The narratives regarding the diversification of environmentally friendly alternatives are essentially based on the significant reduction of agricultural solvents, on extraction performance, and on the final product cost-effectiveness. Nonetheless, disturbances involving the stability of bioactive compounds have been the key to the improvement of nanoencapsulation technology. Accordingly, the step-by-step encapsulation process encloses several stages that involve temperature and pressure alterations, which directly influence the bioactive functions of the biochemical compounds and can cause an accentuated product degradation.

Furthermore, the improvement in the industrial scale of processes involving the extraction of EOs from plant matrices has been discussed recently. The application of pre-treatments previously to the hydrolyzation and distillation protocols is a determining factor as to the expansion of the extraction efficiency potential. The integration of technologies is prominent and one of the main challenges regarding the optimization of obtaining high-quality EOs. The diffusion of the extraction proportions of essential oils does not follow the extraction carried out in lower scalars. Among the main bottlenecks evidenced, the uncertain economic and energy viability, the phenological cycles of the species that have been precursors in the production of EOs, and strict regulation regarding the continuous production of EOs and demand stand out, which directly affects prices marketing, and industrial and technological processing employed.

Moreover, studies have highlighted the prospecting of a range of raw materials and plant biomass that serve as effective matrices in obtaining EOs. However, in-depth investigations are required regarding post-extraction and product supply procedures, such as storage, transportation, and distribution. Environmental and social frontiers are significant obstacles in the end-product supply chain. Appropriately, the use of new plant matrices in supply chain contexts has been discussed as appropriate targets to optimize production and availability according to the demand stipulations.

Additionally, legislation regarding the use of EOs has covered categories of utmost economic importance, such as the food chain and pharmaceutical and alternative medicine industries. Similar to a range of bioactive products, regulation around the application of EOs for agriculture has been at the center of a diversity of global institutions. Nonetheless, comprehensive legislation is distinct in different fields of application. Since the chemical compounds of EOs are unstable, strict authority and monitoring are required for operations using EOs. The Food and Agriculture Organization of the United Nations (FAO) has introduced guidelines regarding the emergence of OEs for multiple uses. Accordingly, the Food and Agricultural Legislation (FAOLEX) provides policies aimed at employing EOs in agriculture. Furthermore, EU regulations date back to the 2000s and are predominantly based on consumer safety and protection. On the other hand, the specificities are heterogeneous and underdeveloped countries significantly lack advertising and reliable information that comprehends health risks, chemical exposure, reaction, and storage. This is one of the main obstacles when it comes to accelerated population growth and the necessity for commercialization and production policies for biochemical products.

Finally, some reports have indicated the requirement for EOs in the coronavirus era. Since its sudden appearance, the medical and pharmaceutical industries have continually searched for compounds that have activity against the SARS-CoV-2 virus. With a range of phytochemical properties, a variety of EOs has antiviral effects, as well as valuable anti-inflammatory, antioxidant, and high potential control compounds for respiratory disorders. The use of EOs compared to the combination of drugs and chemicals has shown the use of these extracts as excellent candidates for the adoption of antiviral agents. EOs are capable of obstructing viral functions in the human organism, preventing the replication of viral cells, and causing the inhibition of viral enzymes. Hence, thorough research should be encouraged as a strategy of optimizing the effectiveness of these products as key elements in a variety of fundamental applications for sustainable and innovative advances.

References

Ahsaei SM, Rodríguez-Rojo S, Salgado M et al (2020) Insecticidal activity of spray dried micro-encapsulated essential oils of Rosmarinus officinalis and Zataria multiflora against Tribolium confusum. Crop Prot 128:104996. https://doi.org/10.1016/j.cropro.2019.104996

Ali EOM, Shakil NA, Rana VS et al (2017) Antifungal activity of nano emulsions of neem and citronella oils against phytopathogenic fungi, Rhizoctonia solani and Sclerotium rolfsii. Ind Crop Prod 108:379–387. https://doi.org/10.1016/j.indcrop.2017.06.061

Allaf T, Tomao V, Ruiz K, Chemat F (2013) Instant controlled pressure drop technology and ultrasound assisted extraction for sequential extraction of essential oil and antioxidants. Ultrason Sonochem 20:239–246. https://doi.org/10.1016/j.ultsonch.2012.05.013

Alonso-Gato M, Astray G, Mejuto JC, Simal-Gandara J (2021) Essential oils as antimicrobials in crop protection. Antibiotics 10:1–12. https://doi.org/10.3390/antibiotics10010034

Arriola NDA, de Medeiros PM, Prudencio ES et al (2016) Encapsulation of aqueous leaf extract of Stevia rebaudiana Bertoni with sodium alginate and its impact on phenolic content. Food Biosci 13:32–40. https://doi.org/10.1016/j.fbio.2015.12.001

Arriola NDA, Chater PI, Wilcox M et al (2019) Encapsulation of Stevia rebaudiana Bertoni aqueous crude extracts by ionic gelation – effects of alginate blends and gelling solutions on the polyphenolic profile. Food Chem 275:123–134. https://doi.org/10.1016/j.foodchem.2018.09.086

Auricchio M, Bacchi E (2003) Eugenia uniflora L. "Brazilian cherry" leaves: pharmacobotanical, chemical and pharmacological properties. Rev Inst Adolfo Lutz 62:55–61

Bakry AM, Abbas S, Ali B et al (2016) Microencapsulation of oils: a comprehensive review of benefits, techniques, and applications. Compr Rev Food Sci Food Saf 15:143–182. https://doi.org/10.1111/1541-4337.12179

Bakthavatsalam N (2016) Semiochemicals. In: Ecofriendly pest management for food security. Elsevier, San Diego, pp 563–611

Balti MA, Hadrich B, Kriaa K (2018) Lab-scale extraction of essential oils from Tunisian lemongrass (Cymbopogon flexuosus). Chem Eng Process Process Intensif 124:164–173. https://doi.org/10.1016/j.cep.2017.12.012

Benchaa S, Hazzit M, Zermane N, Abdelkrim H (2019) Chemical composition and herbicidal activity of essential oils from two Labiatae species from Algeria. J Essent Oil Res 31:335–346. https://doi.org/10.1080/10412905.2019.1567400

Bhavaniramya S, Vishnupriya S, Al-Aboody MS et al (2019) Role of essential oils in food safety: antimicrobial and antioxidant applications. Grain Oil Sci Technol 2:49–55. https://doi.org/10.1016/j.gaost.2019.03.001

Borges DF, Lopes EA, Fialho Moraes AR et al (2018) Formulation of botanicals for the control of plant-pathogens: a review. Crop Prot 110:135–140. https://doi.org/10.1016/j.cropro.2018.04.003

Casela CR, da Ferreira AS, de Pinto NFJA (2006) Doenças na Cultura do Milho, 1st edn. Embrapa, Sete Lagoas

Cazella LN, Glamoclija J, Soković M et al (2019) Antimicrobial activity of essential oil of Baccharis dracunculifolia DC (Asteraceae) aerial parts at flowering period. Front Plant Sci 10:1–9. https://doi.org/10.3389/fpls.2019.00027

Christaki S, Moschakis T, Kyriakoudi A et al (2021) Recent advances in plant essential oils and extracts: delivery systems and potential uses as preservatives and antioxidants in cheese. Trends Food Sci Technol 116:264–278. https://doi.org/10.1016/j.tifs.2021.07.029

Conde-Hernández LA, Espinosa-Victoria JR, Trejo A, Guerrero-Beltrán J (2017) CO2-supercritical extraction, hydrodistillation and steam distillation of essential oil of rosemary (Rosmarinus officinalis). J Food Eng 200:81–86. https://doi.org/10.1016/j.jfoodeng.2016.12.022

Corrêa JCR, Salgado HRN (2011) Atividade inseticida das plantas e aplicações: revisão. Insecticidal activities of plants and applications: a review. Rev Bras Plantas Med 13:500–506. https://doi.org/10.1590/s1516-05722011000400016

Cortesi R, Quattrucci A, Esposito E et al (2017) Natural antimicrobials in spray-dried microparticles based on cellulose derivatives as potential eco-compatible agrochemicals. J Plant Dis Prot 124:269–278. https://doi.org/10.1007/s41348-016-0055-7

Cui H, Pan HW, Wang PH et al (2018) Essential oils from Carex meyeriana Kunth: optimization of hydrodistillation extraction by response surface methodology and evaluation of its antioxidant and antimicrobial activities. Ind Crop Prod 124:669–676. https://doi.org/10.1016/j.indcrop.2018.08.041

da Silva BD, Bernardes PC, Pinheiro PF et al (2021) Chemical composition, extraction sources and action mechanisms of essential oils: natural preservative and limitations of use in meat products. Meat Sci 176:108463. https://doi.org/10.1016/j.meatsci.2021.108463

de Barros Fernandes RV, Marques GR, Borges SV, Botrel DA (2014) Effect of solids content and oil load on the microencapsulation process of rosemary essential oil. Ind Crop Prod 58:173–181. https://doi.org/10.1016/j.indcrop.2014.04.025

de Coitinho RLBC, de Oliveira JV, Junior MGCG, Gomes CA (2006) Residual effect of natural insecticides in the control of Sitophilus zeamais mots. On stored corn. Rev Caatinga 19:183–191

de Lira Guerra Y (2013) Prospecting for essential oils for the control of Sckerotium wilt in peanuts. University Federal Rural of Pernambuco, Recife, Brazil

de Morais LAS (2009) Óleos Essenciais no Controle Fitossanitário. In: Biocontrole de doenças de plantas: uso e perspectivas. Embrapa Meio Ambiente, Jaguariúna, pp 139–152

De Oliveira C (2016) Chemical characterization, antioxidant activity and antimicrobial of Baccharis oreophila Malme essential oil. Technological University of Paraná, Pato Branco, Brazil

de Souza MT, de Souza MT, Bernardi D et al (2021) Insecticidal and oviposition deterrent effects of essential oils of Baccharis spp. and histological assessment against Drosophila suzukii (Diptera: Drosophilidae). Sci Rep 11:1–15. https://doi.org/10.1038/s41598-021-83557-7

del Carmen Razola-Díaz M, Guerra-Hernández EJ, García-Villanova B, Verardo V (2021) Recent developments in extraction and encapsulation techniques of orange essential oil. Food Chem 354:129575. https://doi.org/10.1016/j.foodchem.2021.129575

del Valle JM, Calderón D, Núñez GA (2019) Pressure drop may negatively impact supercritical CO2 extraction of citrus peel essential oils in an industrial-size extraction vessel. J Supercrit Fluids 144:108–121. https://doi.org/10.1016/j.supflu.2018.09.005

Dietrich F, Aparecida A, Strohschoen G et al (2011) Use of botanical pesticides in organic agriculture in Arroio do Meio/RS. R Bras Agrociência 17:251–255

dos Venturoso LR (2009) Plant extracts in the control of the soybean phytopathogenic fungi. University Federal Grande Dourados

El Asbahani A, Miladi K, Badri W et al (2015) Essential oils: from extraction to encapsulation. Int J Pharm 483:220–243. https://doi.org/10.1016/j.ijpharm.2014.12.069

Eloh K, Kpegba K, Sasanelli N et al (2020) Nematicidal activity of some essential plant oils from tropical West Africa. Int J Pest Manag 66:131–141. https://doi.org/10.1080/0967087 4.2019.1576950

Esquivel-Chávez F, Colín-Chávez C, Virgen-Ortiz JJ et al (2021) Control of mango decay using antifungal sachets containing of thyme oil/modified starch/agave fructans microcapsules. Futur Foods 3:100008. https://doi.org/10.1016/j.fufo.2020.100008

Estrada-Cano C, Anaya Castro MA, Castellanos LM et al (2017) Antifungal activity of microcapsulated clove (Eugenia caryophyllata) and Mexican oregano (Lippia berlandieri) essential oils against Fusarium oxysporum. J Microb Biochem Technol 9. https://doi.org/10.4172/1948-5948.1000342

Franco-Vega A, López-Malo A, Palou E, Ramírez-Corona N (2021) Effect of imidazolium ionic liquids as microwave absorption media for the intensification of microwave-assisted extraction of Citrus sinensis peel essential oils. Chem Eng Process – Process Intensif 160. https://doi.org/10.1016/j.cep.2020.108277

García-Saldaña JS, Campas-Baypoli ON, López-Cervantes J et al (2016) Microencapsulation of sulforaphane from broccoli seed extracts by gelatin/gum Arabic and gelatin/pectin complexes. Food Chem 201:94–100. https://doi.org/10.1016/j.foodchem.2016.01.087

Girardi NS, García D, Robledo SN et al (2016) Microencapsulation of Peumus boldus oil by complex coacervation to provide peanut seeds protection against fungal pathogens. Ind Crop Prod 92:93–101. https://doi.org/10.1016/j.indcrop.2016.07.045

Girardi NS, Passone MA, García D et al (2018) Microencapsulation of Peumus boldus essential oil and its impact on peanut seed quality preservation. Ind Crop Prod 114:108–114. https://doi.org/10.1016/j.indcrop.2018.01.036

Gonçalves GG, De MLPV, Salgado LA (2009) Essential oils and plant extracts to control phytopatogen of soybean grain. Hortic Bras 27:102–107

Grigolli JFJ (2015) Manejo de Doenças na Cultura da Soja. In: Tecnologia e Produção: Soja 2014/2015. pp 135–156

Gundewadi G, Rudra SG, Gogoi R et al (2021) Electrospun essential oil encapsulated nanofibers for the management of anthracnose disease in Sapota. Ind Crop Prod 170:113727. https://doi.org/10.1016/j.indcrop.2021.113727

Han C, Shao H, Zhou S et al (2021) Chemical composition and phytotoxicity of essential oil from invasive plant, Ambrosia artemisiifolia L. Ecotoxicol Environ Saf 211:111879. https://doi.org/10.1016/j.ecoenv.2020.111879

Hassaan MA, El Nemr A (2020) Pesticides pollution: classifications, human health impact, extraction and treatment techniques. Egypt J Aquat Res 46:207–220. https://doi.org/10.1016/j.ejar.2020.08.007

Hazrati H, Saharkhiz MJ, Niakousari M, Moein M (2017) Natural herbicide activity of Satureja hortensis L. essential oil nanoemulsion on the seed germination and morphophysiological features of two important weed species. Ecotoxicol Environ Saf 142:423–430. https://doi.org/10.1016/j.ecoenv.2017.04.041

He T, Li X, Wang X et al (2020) Chemical composition and anti-oxidant potential on essential oils of Thymus quinquecostatus Celak. from Loess Plateau in China, regulating Nrf2/Keap1 signaling pathway in zebrafish. Sci Rep 10:1–18. https://doi.org/10.1038/s41598-020-68188-8

Hernández-Nava R, López-Malo A, Palou E et al (2020) Encapsulation of oregano essential oil (Origanum vulgare) by complex coacervation between gelatin and chia mucilage and its properties after spray drying. Food Hydrocoll 109:106077. https://doi.org/10.1016/j.foodhyd.2020.106077

Himed L, Merniz S, Monteagudo-Olivan R et al (2019) Antioxidant activity of the essential oil of citrus limon before and after its encapsulation in amorphous SiO2. Sci Afr 6:e00181. https://doi.org/10.1016/j.sciaf.2019.e00181

Iantas J, Savi DC, da Silva Schibelbein R et al (2021) Endophytes of Brazilian medicinal plants with activity against phytopathogens. Front Microbiol 12:1–18. https://doi.org/10.3389/fmicb.2021.714750

Isman MB, Miresmailli S, Machial C (2011) Commercial opportunities for pesticides based on plant essential oils in agriculture, industry and consumer products. Phytochem Rev 10:197–204. https://doi.org/10.1007/s11101-010-9170-4

Jouini A, Verdeguer M, Pinton S et al (2020) Potential effects of essential oils extracted from Mediterranean aromatic plants on target weeds and soil microorganisms. Plan Theory 9:1–24. https://doi.org/10.3390/plants9101289

Knowles A (2008) Recent developments of safer formulations of agrochemicals. Environmentalist 28:35–44. https://doi.org/10.1007/s10669-007-9045-4

Kundu A, Dutta A, Mandal A et al (2021) A comprehensive in vitro and in silico analysis of nematicidal action of essential oils. Front Plant Sci 11:1–15. https://doi.org/10.3389/fpls.2020.614143

Li J, Gu F, Wu R et al (2017) Phylogenomic evolutionary surveys of subtilase superfamily genes in fungi. Sci Rep 7:1–15. https://doi.org/10.1038/srep45456

Li Y, Wang Y, Kong W et al (2020) Illicium verum essential oil, a potential natural fumigant in preservation of lotus seeds from fungal contamination. Food Chem Toxicol 141:111347. https://doi.org/10.1016/j.fct.2020.111347

Lima RK, Cardoso MG (2007) Lamiaceae family: important essential oils with biological and antioxidant activity. Rev Fitos 3:14–24

Lucia A, Guzmán E (2021) Emulsions containing essential oils, their components or volatile semiochemicals as promising tools for insect pest and pathogen management. Adv Colloid Interf Sci 287:102330. https://doi.org/10.1016/j.cis.2020.102330

Man A, Santacroce L, Jacob R et al (2019) Antimicrobial activity of six essential oils against a group of human pathogens: a comparative study. Pathogens 8:1–11. https://doi.org/10.3390/pathogens8010015

Manouchehri R, Saharkhiz MJ, Karami A, Niakousari M (2018) Extraction of essential oils from damask rose using green and conventional techniques: microwave and ohmic assisted hydrodistillation versus hydrodistillation. Sustain Chem Pharm 8:76–81. https://doi.org/10.1016/j.scp.2018.03.002

Mat Saad H, Rahman SNSA, Navanesan S et al (2021) Evaluation of antioxidant activity and phytochemical composition of *Baeckea frutescens* and *Leptospermum javanicum* essential oils. South African Journal of Botany 141:474–479

Maurya A, Prasad J, Das S, Dwivedy AK (2021) Essential oils and their application in food safety. Front Sustain Food Syst 5. https://doi.org/10.3389/fsufs.2021.653420

Medice R, Alves E, de Assis RT et al (2007) Essential oils used in the control of Asian soybean rust Phakopsora pachyrhizi Syd. & P. Syd. Ciênc Agrotec 31:83–90

Medina-Romero YM, Hernandez-Hernandez AB, Rodriguez-Monroy MA, Canales-Martínez MM (2021) Essential oils of Bursera morelensis and Lippia graveolens for the development of a new biopesticides in postharvest control. Sci Rep 11:1–10. https://doi.org/10.1038/s41598-021-99773-0

Michereff SJ (2001) Fundamentos de Fitopatologia. Universidade Federal Rural de Pernambuco, Recife

Mishra J, Arora NK (2018) Secondary metabolites of fluorescent pseudomonads in biocontrol of phytopathogens for sustainable agriculture. Appl Soil Ecol 125:35–45. https://doi.org/10.1016/j.apsoil.2017.12.004

Mittal D, Kaur G, Singh P et al (2020) Nanoparticle-based sustainable agriculture and food science: recent advances and future outlook. Front Nanotechnol 2. https://doi.org/10.3389/fnano.2020.579954

Mollaei S, Sedighi F, Habibi B et al (2019) Extraction of essential oils of Ferulago angulata with microwave-assisted hydrodistillation. Ind Crop Prod 137:43–51. https://doi.org/10.1016/j.indcrop.2019.05.015

Monteiro ARP (2002) Antimicrobial activity of essential oils. University Fernando Pessoa

Moreira MD, Picanço MC, da Silva EM et al (2006) Uso de inseticidas botânicos no controle de pragas. Control Altern Pragas e Doenças Viçosa EPAMIG/CTZM 1:89–120

Nali LR, Barbosa FR, de Carvalho CAL, dos Santos JBC (2004) Efficiency of natural insecticides and thiamethoxam in the control of thrips in vine. Pestic REcotoxicol e Meio Ambient 14:103–108

Neto ACA, Araújo PC, de Souza WCO et al (2012) Essential oil of anise in incidence and control of pathogens in weed seeds sweet (Foeniculum vulgare Mill.). Verde Agroecol e Desenvolv Sustentável 7:170–176

Oliveira MF, Brighenti AM (2018) Controle de plantas daninhas: métodos físico, mecânico, cultural, biológico e alelopatia. Embrapa. 196p. ISBN 978-85-7035-851-6

Ozkan G, Franco P, De Marco I et al (2019) A review of microencapsulation methods for food antioxidants: principles, advantages, drawbacks and applications. Food Chem 272:494–506. https://doi.org/10.1016/j.foodchem.2018.07.205

Pacios-Michelena S, Aguilar González CN, Alvarez-Perez OB et al (2021) Application of Streptomyces antimicrobial compounds for the control of phytopathogens. Front Sustain Food Syst 5:1–13. https://doi.org/10.3389/fsufs.2021.696518

Palla F, Bruno M, Mercurio F et al (2020) Essential oils as natural biocides in conservation of cultural heritage. Molecules 25:1–11. https://doi.org/10.3390/molecules25030730

Pavlić B, Teslić N, Zengin G et al (2021) Antioxidant and enzyme-inhibitory activity of peppermint extracts and essential oils obtained by conventional and emerging extraction techniques. Food Chem 338. https://doi.org/10.1016/j.foodchem.2020.127724

Pawlowski Â, de Abreu Kuzey C, de Bastos KP et al (2021) Potencial alelopático dos óleos essenciais de capim-limão, citronela e lavanda

Peng X, Feng C, Wang X et al (2021a) Chemical composition and antioxidant activity of essential oils from barks of Pinus pumila using microwave-assisted hydrodistillation after screw extrusion treatment. Ind Crop Prod 166:2–13. https://doi.org/10.1016/j.indcrop.2021.113489

Peng Y, Li SJ, Yan J et al (2021b) Research progress on phytopathogenic fungi and their role as biocontrol agents. Front Microbiol 12. https://doi.org/10.3389/fmicb.2021.670135

Pereira RDCA, Moreira ALM (2011) Manjericão: cultivo e utilização, 1st edn. Embrapa Agroindústria Tropical, Fortaleza

Pinheiro LS (2014) Antifungal activity in vitro of essential oil Laurus nobilis L. (Laurel) on strains Cryptococcus neoformans. University Federal da Paraíba, João Pessoa, Brazil

Pinto JMA (2009) Plant extracts for the control of common bean anthracnose disease. University Federal of Lavras, Lavras, Brazil

Previero CA, Júnior BCL, Florencio LK, dos Santos DL (2010) Recipes from plants with insecticides properties in pest control, 1st ed. Centro Universitário Luterano de Palmas – CEULP/ULBRA

Puškárová A, Bučková M, Kraková L et al (2017) The antibacterial and antifungal activity of six essential oils and their cyto/genotoxicity to human HEL 12469 cells. Sci Rep 7:1–11. https://doi.org/10.1038/s41598-017-08673-9

Rabaioli V, da Silva CP (2016) Prospecting of different species of plants with biopesticides action in the agriculture of Mato Grosso do Sul Volmir. Cienc Biol Agrar Saúde 20:188–195

Ramakrishnan B, Venkateswarlu K, Sethunathan N, Megharaj M (2019) Local applications but global implications: can pesticides drive microorganisms to develop antimicrobial resistance? Sci Total Environ 654:177–189. https://doi.org/10.1016/j.scitotenv.2018.11.041

Ramos K (2014) Essential oils in control of Colletotrichum gloeosporioides. University Camilo Castelo Branco, São Paulo, Brazil

Raveau R, Fontaine J, Lounès-Hadj Sahraoui A (2020) Essential oils as potential alternative biocontrol products against plant pathogens and weeds: a review. Foods 9. https://doi.org/10.3390/foods9030365

Reis EM, Casa RT, Bianchin V (2011) Inoculum reduction measures to manage Armillaria root disease in a severely infected stand of ponderosa pine in South-Central Washington: 35-year results. Summa Phytopathol 37:85–91. https://doi.org/10.1093/wjaf/27.1.25

Rungwasantisuk A, Raibhu S (2020) Application of encapsulating lavender essential oil in gelatin/gum-arabic complex coacervate and varnish screen-printing in making fragrant gift-wrapping paper. Prog Org Coat 149:105924. https://doi.org/10.1016/j.porgcoat.2020.105924

Sabbour MM, El-Aziz SEA (2016) Efficacy of three essential oils and their nano-particles against Sitophilus granarius under laboratory and store conditions. J Entomol Res 40:229. https://doi.org/10.5958/0974-4576.2016.00042.6

Saifullah M, Shishir MRI, Ferdowsi R et al (2019) Micro and nano encapsulation, retention and controlled release of flavor and aroma compounds: a critical review. Trends Food Sci Technol 86:230–251. https://doi.org/10.1016/j.tifs.2019.02.030

Sant'Anna FB (2009) Main mechanisms of insecticide resistance on insects. Pubvet 3:unpaginated

Sarkic A, Stappen I (2018) Essential oils and their single compounds in cosmetics-a critical review. Cosmetics 5:1–21. https://doi.org/10.3390/cosmetics5010011

Sarto MPM, Junior GZ (2014) Antimicrobial activity of essential oils marcella. Uningá Rev 20:98–102. https://doi.org/10.1590/1809-43921989191363

Savary S, Willocquet L, Pethybridge SJ et al (2019) The global burden of pathogens and pests on major food crops. Nat Ecol Evol 3:430–439. https://doi.org/10.1038/s41559-018-0793-y

Schroder T, Gaskin S, Ross K, Whiley H (2017) Antifungal activity of essential oils against fungi isolated from air. Int J Occup Environ Health 23:181–186. https://doi.org/10.1080/10773525.2018.1447320

Sereshti H, Rohanifar A, Bakhtiari S, Samadi S (2012) Bifunctional ultrasound assisted extraction and determination of Elettaria cardamomum Maton essential oil. J Chromatogr A 1238:46–53. https://doi.org/10.1016/j.chroma.2012.03.061

Sharmeen JB, Mahomoodally FM, Zengin G, Maggi F (2021) Essential oils as natural sources of fragrance compounds for cosmetics and cosmeceuticals. Molecules 26. https://doi.org/10.3390/molecules26030666

Silva-Lobo VL, Filippi MCC (2017) Manual de Identificação de Doenças da Cultura do Arroz, 1st edn. Embrapa, Brasília

Soliman EA, El-Moghazy AY, El-Din MSM, Massoud MA (2013) Microencapsulation of essential oils within alginate: formulation and in vitro evaluation of antifungal activity. J Encapsulation Adsorpt Sci 03:48–55. https://doi.org/10.4236/jeas.2013.31006

Sousa S, Maia ML, Correira-Sá L et al (2020) Chemistry and toxicology behind insecticides and herbicides. In: Controlled release of pesticides for sustainable agriculture. Springer International Publishing, Cham, pp 59–109

Steffens AH (2010) Study of the chemical composition of essencial oils obtained by steam distillation in a laboratory and industrial scale. Catholic University of Rio Grande do Sul, Caxias do Sul, Brazil

Steiner BM, McClements DJ, Davidov-Pardo G (2018) Encapsulation systems for lutein: a review. Trends Food Sci Technol 82:71–81. https://doi.org/10.1016/j.tifs.2018.10.003

Taban A, Saharkhiz MJ, Khorram M (2020) Formulation and assessment of nano-encapsulated bioherbicides based on biopolymers and essential oil. Ind Crop Prod 149:112348. https://doi.org/10.1016/j.indcrop.2020.112348

Tian Q, Zhou W, Cai Q et al (2021) Concepts, processing, and recent developments in encapsulating essential oils. Chin J Chem Eng 30:255–271. https://doi.org/10.1016/j.cjche.2020.12.010

Timilsena YP, Akanbi TO, Khalid N et al (2019) Complex coacervation: principles, mechanisms and applications in microencapsulation. Int J Biol Macromol 121:1276–1286. https://doi.org/10.1016/j.ijbiomac.2018.10.144

Upadhayay J, Rana M, Juyal V et al (2020) Impact of pesticide exposure and associated health effects. In: Pesticides in crop production. Wiley, pp 69–88

Valeriano C, Picolli RH, Cardoso MG, Alves E (2012) Antimicrobial activity of essential oils against sessile and planktonic pathogens of food source. Rev Bras de Plantas Med 14:57–67

Viegas C (2003) Terpenes with insecticidal activity: an alternative to chemical control of insects. Quim Nova 26:390–400. https://doi.org/10.1590/s0100-40422003000300017

Vitti AMS, Brito JO (2003) Eucalyptus essential oil. São Paulo

Wińska K, Mączka W, Łyczko J, Grabarczyk M, Czubaszek A, Szumny A (2019) Essential oils as antimicrobial agents—myth or real alternative? Molecules 24(11):1–21

Xiong K, Chen Y (2020) Supercritical carbon dioxide extraction of essential oil from tangerine peel: experimental optimization and kinetics modelling. Chem Eng Res Des 164:412–423. https://doi.org/10.1016/j.cherd.2020.09.032

Yousefi M, Rahimi-Nasrabadi M, Pourmortazavi SM et al (2019) Supercritical fluid extraction of essential oils. TrAC – Trends Anal Chem 118:182–193. https://doi.org/10.1016/j.trac.2019.05.038

Yue Q, Shao X, Wei Y et al (2020) Optimized preparation of tea tree oil complexation and their antifungal activity against Botrytis cinerea. Postharvest Biol Technol 162:111114. https://doi.org/10.1016/j.postharvbio.2019.111114

Zhu JJ, Yang JJ, Wu GJ, Jiang JG (2020a) Comparative antioxidant, anticancer and antimicrobial activities of essential oils from Semen Platycladi by different extraction methods. Ind Crop Prod 146:112206. https://doi.org/10.1016/j.indcrop.2020.112206

Zhu Q, Yang Y, Zhong Y et al (2020b) Synthesis, insecticidal activity, resistance, photodegradation and toxicity of pyrethroids (A review). Chemosphere 254:126779. https://doi.org/10.1016/j.chemosphere.2020.126779

Zikankuba VL, Mwanyika G, Ntwenya JE, James A (2019) Pesticide regulations and their malpractice implications on food and environment safety. Cogent Food Agric 5:1601544. https://doi.org/10.1080/23311932.2019.1601544

Zimmermann RC, de Carvalho Aragão CE, de Araújo PJP et al (2021) Insecticide activity and toxicity of essential oils against two stored-product insects. Crop Prot 144. https://doi.org/10.1016/j.cropro.2021.105575

Chapter 11
Volatile Allelochemicals

Alicia Ludymilla Cardoso de Souza, Chrystiaine Helena Campos de Matos, and Renan Campos e Silva

11.1 Introduction

Allelochemicals are chemical species involved in interactions of individuals of different species (Kost 2008). In plants, allelochemicals are released into the environment through volatilization of living parts of the plant, leaching of plant foliage, decomposition of plant material and root exudation (Scavo et al. 2019). Allelochemicals are subclassified depending on the adaptive effect on the species involved in the interaction. Thus, allomones have a neutral effect on the receiver, but modify their behavior to benefit the emitter; the cairomoniums are favorable to the receiver, but not to the emitter, and the synomones benefit both the emitter and the receiver (Hickman et al. 2021). This classification encompasses volatile compounds, since they are present in several multitrophic interactions, which makes them candidates for research and technology for agriculture (Silva et al. 2018).

Volatiles mediate the interaction of plants with pollinators, herbivores and their natural enemies, other plants and microorganisms and are used to protect against biotic and abiotic stresses and to provide information, and potentially misinformation, to mutualists and competitors, as shown in Fig. 11.1. As knowledge about these interactions increases, the underlying mechanisms become increasingly

A. L. C. de Souza (✉)
Programa de Pós-graduação em Química, Universidade Federal de São Carlos,
São Carlos, São Paulo, Brazil

C. H. C. de Matos (✉)
Programa de Pós Graduação em Agroquímica, Instituto Federal de Educação, Ciência e
Tecnologia Goiano, Rio Verde, Goias, Brazil

R. Campos e Silva (✉)
Programa de Pós-graduação em Química, Universidade Federal do Pará, Belém, Para, Brazil
e-mail: campos.silva@icen.ufpa.br

M. Santana de Oliveira (ed.), *Essential Oils*, https://doi.org/10.1007/978-3-030-99476-1_11

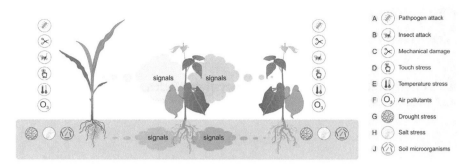

Fig. 11.1 Chemical response of the plant from different stimuli (Ninkovic et al. 2021)

complex. The mechanisms of biosynthesis and perception of volatiles are slowly being discovered.

The growing scientific knowledge can be used to design and apply volatile-based agricultural strategies (Baldwin 2010; Bouwmeester et al. 2019). In addition, diverse and abundant volatiles are also directly involved in the interaction between microorganism-insects, microbe-microbe and plant-microbe, and are also susceptible to environmental conditions, food source and growth stage, for example, exponential versus stationary (Beck and Vannette 2017).

Volatile plant mixtures are dominated by four classes of secondary metabolites: terpenoids, compounds with aromatic rings, the derived fatty acid mentioned above, and volatiles derived from amino acids other than L-phenylalanine, being attributed to terpenoids play the central role in generating the chemical diversity of plant volatiles and appear to be under strong diversified selection (Baldwin 2010). Such compounds, when emitted by a given organism, induce specific responses in the recipient, and may influence their behavior and aptitude, this aspect places these compounds with possible alternatives for the management of agricultural pests (Holighaus and Rohlfs 2016). Among terpenoids, the two homoterpenes 4,8-dimethyl-1.3 (*E*),7-nonatriene and 4,8,12-trimethyl- 1.3 (E), 7 (*E*),11-tridecatetraene are the most frequently reported volatile plants induced by herbivores. They can be synthesized by plants of many species from terpenic nerolidol, geranil and linalol alds without any herbivory mediation (Dicke 1994).

Natural volatile allelochemicals and derivatives are sustainable alternatives, offering many advantages to commercial nematicides (Faria et al. 2020). The importance of plant volatiles, in addition to the general appeal of fragrances and flavors to humans, has made these secondary metabolites a target for metabolic engineering (Tumlinson 2014). As information transmitters, volatiles have provided plants with solutions to the challenges associated with being rooted in soil and real estate (Baldwin 2010). In agriculture, volatile compounds are extremely important, since they excel the plant's defense mechanisms for greater resistance or tolerance to impending stress, reactive species of extinction oxygen, have potent antimicrobial and allelopathic effects, and can be important in regulating plant growth, development and senescence through interactions with plant hormones. Current limits and disadvantages that can hinder the use of volatiles in the open field are analyzed and solutions for better exploration in sustainable agriculture of the future are envisioned (Brilli et al. 2019).

In view of the importance of volatile compounds in communication and the possibility of application in agriculture, either as pesticides or herbicides, or to prepare the soil, thus enabling greater success in crops and increasing productivity, this chapter presents the main aspects related to volatile allelochemicals and their importance in interspecific interactions and their commercial exploitation.

11.2   The Role of Volatiles in Ecological Interactions

Living beings generally interact with the environment from different signals, be they vibrational, visual, tactile, olfactory or chemical signals, also called semiochemicals. In this sense, Chemical Ecology arises having as object of study the chemical compounds that permeate these interactions that can occur with organisms of the same species or with organisms of different species, being thus called intraspecific interactions and interspecific interactions, respectively (Bergström 2007; Bergström 2008).

The semiochemicals (from the Greek semeîon = sign), participants of these interactions, have a great structural diversity and compounds with different forms of action (Zarbin et al. 2009). These, when they participate in intraspecific interactions, that is, the communication between an emitting individual and a signal receiver that are of the same species, are called pheromones and receive classifications for each type of function they perform. When they participate in interspecific interactions, whose emitting individual and signal receiver are of different species, they are called allelochemicals (Thöming 2021).

Allelochemicals can have several functionalities, substantially impacting the growth of neighboring plants, the defensive biochemistry of the plant, and may also alter the production of allelochemicals of neighboring plants (Berens et al. 2017). In addition to repelling herbivores, the compounds produced by plants can act as warning signals to warn neighboring plants of an imminent attack of herbivores or pathogens, and serve as an inter and intra species signal, the chemical compounds trigger specific activities in organisms (Yang et al. 2018).

Allelochemicals have their classification based on cost-benefit interactions between emitter and receptor, being divided into allomones (compounds that favor the emitting species), kairomones (compounds that favor the receiving species) and sinomones (compounds in which they favor both species) (Bergström 2007; Bakthavatsalam 2016). These compounds can mediate ecological interactions of species of different trophic levels, such as insects and plants, but even though many methods using semiochemicals in agriculture (in pest management and control) are recognized, studies targeting volatile allelochemicals are more recent (Thöming 2021).

Some allelochemicals evoke a behavioral response in the individual signal receptor that is adaptively favorable to the receiver, but not to the emitter, the kairomones, previously mentioned. The best known group of allelochemicals that act as kairomones are those involved in the localization of food. Such substances are emitted by predators, parasites, parasitoids, herbivores and fungi during their search for food and/or oviposition sites, the relationship of some species and their volatiles are listed in Table 11.1.

Another group of allelochemicals presents behavioral or physiological response that is favorable for both the emitter and the receptor. The so-called sinomones are compounds that are often involved in the pollination process, as can be seen in Table 11.2 (Kost 2008; Abd El-Ghany 2019).

In addition, the class of allelochemicals that act as allomons includes repellent or toxic compounds, which provide defense against attacks, presenting a behavioral signal that will primarily benefit only the emitter. An example of the action of allomones in plants is the action of granular trichomes that cover the leaves of plants and stems, which release herbivore-inhibiting allomones under stress conditions as a defense process (Abd El-Ghany 2019).

According to Dudareva et al. (2013), volatiles are usually involved in performing different functions in plants, such as defense, protection, the ammunition needed to attract or repel insects (both above and below ground) and reproduction. Xie et al. (2021) in a recent review, they report volatile organic compounds as effective auxiliaries of the ecological, economic and sustainable social development of the

Table 11.1 Examples of allelochemicals that act as kairomones

Collector(s)	Prey / Host	Compound(s)	Origin of Kairomone
Elatophilus hebraicus	Matsucoccus josephi	(2E,6E,8E)-5,7-Dimethyl-2,6,8-decatrien-4-one	Sexual pheromone of M. josephi
Lutzomyia longipalpis	Vulpes vulpes	Benzaldehyde 4-Methyl-2-pentanone	Anal and caudal glands of V. vulpes
Acerophagus coccois, Aenasius vexans	Phenacoccus herreni	O-Caffeoylserine	Body surface of de P. herreni
Depressaria pastinacella	Pastinaca sativa	Octyl butyrate	Essential oil components found exclusively in tissues consumed by D. pastinacella
Malthodes fuscus	Fomes fomentarius, Fomitopsis pinicola	Oct-1-en-3-ol	Volatiles emitted by F. fomentarius e F. pinicola

Font: Kost (2008)

Table 11.2 Examples of allelochemicals involved in the attraction of different species of animal pollinators

Pollinator	Plant	Flower coloring	Main compound(s)
Several insects: Drosophilids, scarabs, small scarals	*Rubus* spp. *Ranunculus* spp. *Chamaedora linearis*	White, purple White, yellow, red, purple, orange, cream	(E)-β-Ocimene
Bombus spp.	*Cimicifuga Simplex Polemonium foliosissimum*	White Purple	α-Pinene β-Pinene
Night moths *Manduca sexta*	*Nicotiana alata*	Lime green, red, white, yellow	1,8-Cineole Linalool
Necrophilic insects *Lucilia* spp. *Calliphora* spp.	*Helicodiceros muscivorus*	Purple	Dimethyl sulfide Dimethyl trisulfide
Eidolon helvum Rousettus aegyptiacus	*Adansonia digitata*	White	Dimethyl disulfide 3-Methyl butanal

Font: Kost (2008)

substantial agricultural industry, through the allelopathy of these compounds released by plants.

In plant-plant interaction, many studies appear describing allelopathic activities and potentials in different plant species (Zhou et al. 2010; Abd-ElGawad et al. 2021). In a recent study, Sothearith et al. (2021) describe the invasive behavior of an ornamental plant, *Lantana camara* L. (Verbenaceae), in which its released allelochemicals suppressed several processes of native plant species and led to death. In this sense, in relation to plant allelopathic effects, research studies and development of bioherbicides, ecological tools for weed infestation control, have been highlighted (Puig et al. 2018; Mushtaq et al. 2020; De Mastro et al. 2021).

Furthermore, activities of diverse interactions between insect-plants occur commonly, where some phytophagous insects that are adapted to recognize plant allelochemicals, use them for various functions, from the recognition of their host, in order to accept or reject it (for places of oviposition and feeding of larvae, for example), as well as the use of these volatiles in sexual communication. What can be emphasized is that some plants, especially *Orixa japonica*, present chemical barriers that instill these interactions and drive away certain hosts (Nishida 2014). Fungal allelochemicals have also demonstrated insect repellent behavior, and another point is effective and ecological for pest management, but the mode of action of these volatiles is little known, requiring further studies related to insect-fungus interactions (Holighaus and Rohlfs 2016).

With the growing interest in knowledge related to allelochemicals, thousands of structures of volatile compounds involved in interspecific interactions are known and reported in the literature, with chemical and structural diversity. However, related to the chemical communication of insects, it is still incomplete the understanding of the complex olfactory mechanism of recognition of these compounds and the way these interactions are performed. The studies appear with several incompatibilities of the results and thus there is still much to be developed so that in short allelochemicals have their practical use in agricultural strategies (Thöming 2021).

Overall, it is clear that of the available allelochemicals, not all of them will be ecologically relevant or active for interactions. Many have a chemical fingerprint, which shows that they are specific to some interactions and that in some way they will not function in isolation in nature, that is, not being the only stimulus, since chemical communication also occurs from the olfactory, tactile, visual and other ways, with no known "rule" for the occurrence of these interactions (Poldy 2020).

11.3 Main Volatile Allelochemicals Present in Nature

The identification of plant allelochemicals is the key to understanding plant-plant allelopathic interactions. Several volatile organic compounds have been investigated and identified from a variety of plant species and present chemically diverse (Kong et al. 2019). Most allelochemicals are secondary metabolites belonging to three main groups: phenolic compounds, terpenoids and alkaloids (Gniazdowska and Bogatek 2005).

The phenolic compounds of plants are the most important allelochemicals and represent a large group of organic compounds. Such species are composed of whose molecular structures contain at least one hydroxyl group and possibly other substituents directly linked to the benzene ring, and can be categorized as simple phenols, flavonoids, stilbenes, coumarins, lignins and tannins according to their different structures (Li et al. 2021).

The allelopathic mechanism of phenolic compounds demonstrates that such compounds interfere with various essential plant enzymes and physiological processes, such as cinnamic and benzoic acids, shown in Fig. 11.2, which interfere in hormonal activity, membrane permeability, photosynthesis, respiration and synthesis of organic compounds (Latif et al. 2017).

The compounds of the terpenoid group include a class of hydrocarbons and their derivatives containing oxygen (e.g., alcohols, aldehydes, carboxylic acids and

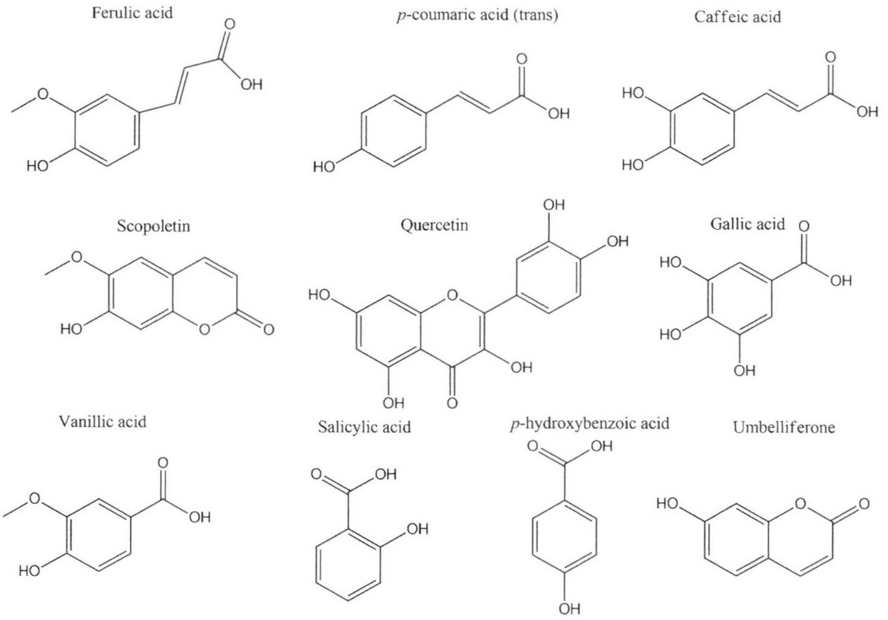

Fig. 11.2 Examples of phenolic compounds with known allelopathic properties (Latif et al. 2017)

esters) (Li et al. 2021). Terpenoids, especially monoterpenes and oxygen-containing sesquiterpenes, have active allelopathic activity in plants and can be slowly dissolved and released, exhibiting effective allelopathic effects (Tholl 2015).

The allelopathic action of monoterpenes, such as 1,4-cineol and 1,8-cineol, has been studied due to its potential as herbicide candidates, presenting action that prolong the time of weed germination and also reduce its development. In addition to monoterpenes, some sesquiterpenes also exhibit this allelopathic effect, and can be seen in Fig. 11.3 (Ninkuu et al. 2021).

Alkaloids form a group of basic heterocyclic compounds containing nitrogen, of plant origin and are named accordingly due to their alkaline chemical nature Alkaloids are widely distributed throughout the plant kingdom and have often reported playing important defensive roles in plant interactions against herbivores (Latif et al. 2017; Souto et al. 2021). Several alkaloids present repellent, larvicidal, and insecticide activity, and their chemical structures can be observed in Fig. 11.4.

Allelochemicals derived from plants corroborate for a management of agricultural pests with greater ecological potential, but for their intact use it is necessary to know in depth the structural characteristics of chemical species and their modes of action on the different receptor individuals and their interaction with other ecosystem organisms (Kong et al. 2019).

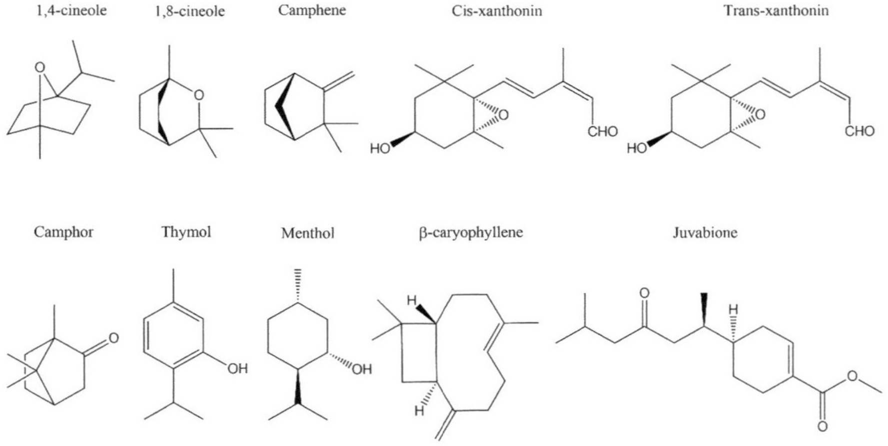

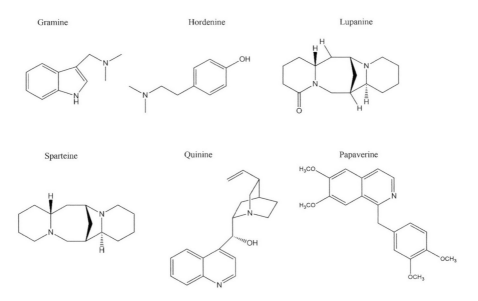

Fig. 11.3 Examples of monoterpenes and sesquiterpenes with known allelopathic properties (Latif et al. 2017)

Fig. 11.4 Examples of alkaloids with known allelopathic properties (Latif et al. 2017)

11.4 Importance of Communication Mediated by Volatiles in Agriculture

Current agriculture focuses efforts on sustainable methods of crop protection for predictable and economical food production. For several decades, conventional pesticides have served the world well; however, for sustainable pest management, there is a need to replace these external applications with environmentally friendly crop protection approaches (Mbaluto et al. 2020). Consequently, there is a great need for new and alternative methods of insect pest control. An interesting source of ecological pesticides are biocide compounds, which occur naturally in plants as allelochemicals (secondary metabolites), helping plants to resist, tolerate or compensate for stress caused by insect pests (Gajger and Dar 2021) and also to interfere with the growth of other plants, i.e. bioherbicides. Thus, ecological pest management is an alternative to conventional herbicides (Kong et al. 2019).

Plant volatiles organic compounds (VOC's) facilitate communication between plants and organisms of other trophic levels, i.e. herbivores and their natural enemies. There is also growing evidence that plant VOC's provide direct defense against various abiotic and biotic stresses. The ability of volatile compounds of plants to act as a sign of attraction and reliable desuasion for herbivores and pathogens and a sign of attraction for beneficial insects presents new perspectives for their commercial use as bait in sustainable agriculture (Maurya 2020).

Thus, VOCs expose great importance for the future development of green agriculture. Allelopathy among allelochemical compounds demonstrate activity in relation to disease resistance and prevention of plant pests, impacting competition (inhibiting the risk of weeds), regulating plant growth, breaking dormancy, affecting the content of reactive oxygen species (ROS) and enzymatic activity, modulating the respiration and photosynthesis of plants and their role as a signal conducting substance, as can be seen in Fig. 11.5.

In agriculture, allelochemicals, including the volatiles, play essential roles in pest control, acting in direct and indirect defense of vegetables and helping in strategies to mitigate attacks by natural predators. Due to complex volatile-mediated interactions, non-target effects should be considered when manipulating volatiles in an agricultural context. The integration of this perspective into agroecosystems can help manage the chemical characteristics of plants more efficiently and responsibly (Silva et al. 2018).

Several volatile compounds used plant-plant interactions are examples of how they influence defenses against herbivores. One study showed that green leaf volatiles ((Z)-3-hexenal, (Z)-3-hexen-1-ol, 2 and (Z)-3-hexenil acetate), emitted by plants damaged by herbivores, induce an increase in defenses of neighboring plants, for example, when corn seedlings exposed to volatiles green leaves exhibited higher levels of jasmonic acid and sesquiterpene emissions than seedlings not exposed to green leaf volatiles (Engelberth et al. 2004). In the same sense, soybean (*Glycine max*) reacted to *rhyssomatus nigerrimus* infestation by emitting high levels of 1-octen-3-ol, 6-methyl-5-hepten-2-one, (E)-β-ocimene, salicylaldehyde, unknown 10, linalool, methyl salicylate, (Z)-8-dodecenyl acetate (ester 5), ketone 2 and geranyl acetone, in an attempt to mitigate the effect of the attack (Espadas-Pinacho et al. 2021).

On another hand, another study showed that tomato plants, when attacked by whitefly (*Bemisia tabaci*), one of the most important invasive pests of crops in the world, released a mixture of volatiles composed mainly of β-cayophyllene, β-mycene and ρ-cimene, which in turn induce neighboring tomato plants to initiate

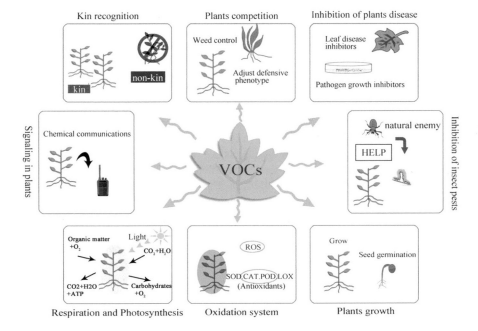

Fig. 11.5 Allelopathic relationship between VOC's produced by plants (Xie et al. 2021)

defenses dependent on salicylic acid and suppress the defenses dependent on jas-monic acid, thus making the neighboring tomato plants more susceptible to white-flies, this shows that *Bemisia tabaci* is able to interfere with this transfer of information, which may be an important reason for this and other whitefly species to have had as much success as invasive pests (Zhang et al. 2019).

Undamaged plants constantly release volatile compounds that can be exploited by con- or hetero-specific neighbors, such interactions are called allelobiotics, as can be seen in Fig. 11.6. This interaction increases the competitive capacity of exposed plants because the allocation of resources for root biomass can contribute to adequacy by facilitating greater absorption of nutrients, especially in habitats characterized by low productivity (Ninkovic et al. 2019). Allelobiotic responses of plants were observed in the volatile interaction between plants of different species, for example, potato exposed to onion released significantly higher amounts of (E)-nerolidol and (3E, 7E)-4,8,12-Trimetiltrideca-1,3,7,11-tetraene (TMTT) repel-ling aphids in laboratory experiments and also entailed a significant reduction in the abundance of aphids in the field (Ninkovic et al. 2013). The same research group investigated the effect of this interaction on the ladybugs of the species *Coccinella septempunctata* and showed that the odor of potatoes exposed to onion swells was significantly more attractive to ladybugs than that of unexposed potatoes, as can be seen in Fig. 11.6. When presented individually, TMTT was attractive to ladybugs, while (*E*) -nerolidol was repellent. The volatile exchange between unattacked plants

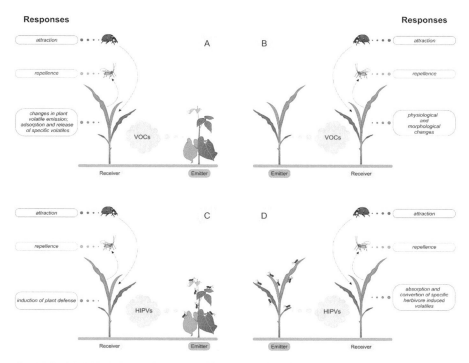

Fig. 11.6 Allelobiotic interaction by volatile compounds (Ninkovic et al. 2016)

and the consequent increase in attractiveness to ladybugs can be a mechanism that contributes to the increase in the abundance of natural enemies in complex plant habitats (Vucetic et al. 2014).

Plant-plant interactions are also important for weed control, because many volatiles have a delayed effect on the growth of other species and aiming to find compost with such potential, several allelochemicals with herbicide activity were isolated from different crops, and allelochemicals with herbicide activity can be categorized into two main groups: phenolics and terpenoids (Bachheti et al. 2020; Schandry and Becker 2020). The natural-ecological effect of preventing the growth of neighboring species is called allelopathy and has been known since antiquity. The use of allelopathic crops in agriculture is currently being carried out, for example, as components of crop rotation, for intercropping, as cover crops or as green fertilizer (Cheng and Cheng 2015). Based on this, several studies have been conducted in order to discover sources of bioherbicides and plant volatiles gain prominence in this branch, for example, volatiles present in the aqueous extract of leaves and branches of *Tinospora tuberculata* composed mainly of 2,3-butanodiol were able to inhibit weed growth in rice fields, thus being able to suppress weeds in rice fields and to develop new herbicides based on the release of phytotoxic compounds by this plant (Aslani et al. 2015).

Other interactions mediated by volatile compounds that are important from the point of view of agriculture are insect-microbe, plant-microbe and microbe-microbe interactions. These interactions are particularly interesting when they occur with the participation of endophytic microorganisms, as demonstrated in in vivo and in vitro assays, where the volatile compound 5-pentil-2-furaldehyde, produced by *Oxyporus latemarginatus*, exhibited strong antifungal activity in the presence of the phytopathogens *Alternaria alternata*, *Colletotrichum gloeosporioides* and *Fusarium oxysporum f. sp. Lycopesici*, known to cause problems in food-producing species (Lee et al. 2009).

The volatile beneficial effects emitted during the interaction between microorganisms was also verified in the postharvest period of strawberries, because the volatiles emitted by *Daldinia eschscholtzii*, mainly elemycin, were responsible for the inhibition of *Colletotrichum acutatum*, which caused anthracnose in postharvest strawberry fruits (Khruengsai et al. 2021). Microorganisms can also interact, directly or indirectly with insects, through improvements in the attractiveness of their host plants, such interaction involves complex mechanisms, since the organisms are not in direct contact. This effect was verified when fava seedlings (*Vicia faba*) colonized by arbuscular mycorrhizal fungi became more attractive by regulating the concentrations of (Z)-3 hexenyl acetate, naphthalene and (R)-germacrene D and the suppression of the production of (E)-caryophyllene and (E)-β-farnesene. This finding shows that suppressed emission of sesquiterpenes in mycorrhizal plants may be a key chemical mechanism of attractiveness of mycorrhizal plants to aphids (Babikova et al. 2014).

In addition, microorganisms present in the soil also establish friendly or antagonistic relationships with plants that can modulate various characteristics of plant growth and health that can be measured by volatile allelochemicals and these interactions can be well utilized in the agricultural sector (Bailly and Weisskopf 2017; Raza and Shen 2020). An example of beneficial interaction is the interaction of the arbuscular mycorrhizal fungus *Funneliformis mosseae* IMA1 with *Vitis vinifera* cv. Sangiovese leaf tissue, which causes the increase in the content of volatiles related to plant defenses under attack by pathogens / herbivores or linked to water stress, such as (*E*)-hexenal, 3-hexenal, geraniol, benzaldehyde and methyl salicylate, was observed in mycorrhizal plants (Velásquez et al. 2020). On the contrary, plants can react by releasing volatile compounds to mitigate negative effects of pathogen attacks, as occurs with lima beans (*Phaseolus lunatus*) that released defense volatiles as nonanal to combat infestation by *Pseudomonas syringae* (Yi et al. 2009).

References

Abd El-Ghany NM (2019) Semiochemicals for controlling insect pests. J Plant Prot Res 59:1–11. https://doi.org/10.24425/jppr.2019.126036

Abd-ElGawad AM, Elgamal AM, EI-Amier YA, Mohamed TA, El Gendy AE-NG, Elshamy AI (2021) Chemical composition, allelopathic, antioxidant, and anti-inflammatory activities of sesquiterpenes rich essential oil of *Cleome amblyocarpa* Barratte & Murb. Plants 10:1294

Aslani F, Juraimi AS, Ahmad-Hamdani MS, Alam MA, Hashemi FSG, Omar D, Hakim MA (2015) Phytotoxic interference of volatile organic compounds and water extracts of *Tinospora tuberculata* Beumee on growth of weeds in rice fields. S Afr J Bot 100:132–140. https://doi. org/10.1016/j.sajb.2015.04.011

Babikova Z, Gilbert L, Bruce T, Dewhirst SY, Pickett JA, Johnson D (2014) Arbuscular mycorrhizal fungi and aphids interact by changing host plant quality and volatile emission. Funct Ecol 28:375–385. https://doi.org/10.1111/1365-2435.12181

Bachheti A, Sharma A, Bachheti RK, Husen A, Pandey DP (2020) Plant allelochemicals and their various applications BT – co-evolution of secondary metabolites. In: Mérillon J-M, Ramawat KG (eds) Co-evolution of secondary metabolites, 1st edn. Springer International Publishing, Cham, pp 441–465

Bailly A, Weisskopf L (2017) Mining the volatilomes of plant-associated microbiota for new biocontrol solutions. Front Microbiol 8:1–12. https://doi.org/10.3389/fmicb.2017.01638

Bakthavatsalam N (2016) Semiochemicals. In: Omkar BT-EPM for FS (ed) Ecofriendly pest management for food security, 1st edn. Academic, San Diego, pp 563–611

Baldwin IT (2010) Plant volatiles. Curr Biol 20:392–397. https://doi.org/10.1016/j.cub.2010.02.052

Beck JJ, Vannette RL (2017) Harnessing insect-microbe chemical communications to control insect pests of agricultural systems. J Agric Food Chem 65:23–28. https://doi.org/10.1021/acs. jafc.6b04298

Berens ML, Berry HM, Mine A, Argueso CT, Tsuda K (2017) Evolution of hormone signaling networks in plant defense. Annu Rev Phytopathol 55:401–425. https://doi.org/10.1146/ annurev-phyto-080516-035544

Bergström G (2007) Chemical ecology = chemistry + ecology! Pure Appl Chem 79:2305–2323. https://doi.org/10.1351/pac200779122305

Bergström LGW (2008) Chemical communication by behaviour-guiding olfactory signals. Chem Commun:3959–3979. https://doi.org/10.1039/B712681F

Bouwmeester H, Schuurink RC, Bleeker PM, Schiestl F (2019) The role of volatiles in plant communication. Plant J 100:892–907. https://doi.org/10.1111/tpj.14496

Brilli F, Loreto F, Baccelli I (2019) Exploiting plant volatile organic compounds (VOCS) in agriculture to improve sustainable defense strategies and productivity of crops. Front Plant Sci 10:1–8. https://doi.org/10.3389/fpls.2019.00264

Cheng F, Cheng Z (2015) Research progress on the use of plant allelopathy in agriculture and the physiological and ecological mechanisms of allelopathy. Front Plant Sci 6:1–16. https://doi. org/10.3389/fpls.2015.01020

De Mastro G, El Mahdi J, Ruta C (2021) Bioherbicidal potential of the essential oils from Mediterranean Lamiaceae for weed control in organic farming. Plants (Basel) 10:818

Dicke M (1994) Local and systemic production of volatile herbivore-induced terpenoids: their role in plant-carnivore mutualism. J Plant Physiol 143:465–472. https://doi.org/10.1016/ S0176-1617(11)81808-0

Dudareva N, Klempien A, Muhlemann JK, Kaplan I (2013) Biosynthesis, function and metabolic engineering of plant volatile organic compounds. New Phytol 198:16–32. https://doi. org/10.1111/nph.12145

Engelberth J, Alborn HT, Schmelz EA, Tumlinson JH (2004) Airborne signals prime plants against insect herbivore attack. Proc Natl Acad Sci U S A 101:1781–1785

Espadas-Pinacho K, López-Guillén G, Gómez-Ruiz J, Cruz-López L (2021) Induced volatiles in the interaction between soybean (Glycine max) and the Mexican soybean weevil (Rhyssomatus nigerrimus). Braz J Biol 81:611–620. https://doi.org/10.1590/1519-6984.227271

Faria JMS, Barbosa P, Teixeira DM, Mota M (2020) A review on the nematicidal activity of volatile allelochemicals against the pinewood nematode. Environ Sci Proc 3:1. https://doi.org/10.3390/ iecf2020-08003

Gajger IT, Dar SA (2021) Plant allelochemicals as sources of insecticides. Insects 12:1–21. https:// doi.org/10.3390/insects12030189

Gniazdowska A, Bogatek R (2005) Allelopathic interactions between plants. Multi site action of allelochemicals. Acta Physiol Plant 27:395–407

Hickman DT, Rasmussen A, Ritz K, Birkett MA, Neve P (2021) Review: allelochemicals as multi-kingdom plant defence compounds: towards an integrated approach. Pest Manag Sci 77:1121–1131. https://doi.org/10.1002/ps.6076

Holighaus G, Rohlfs M (2016) Fungal allelochemicals in insect pest management. Appl Microbiol Biotechnol 100:5681–5689. https://doi.org/10.1007/s00253-016-7573-x

Khruengsai S, Pripdeevech P, Tanapichatsakul C (2021) Antifungal properties of volatile organic compounds produced by Daldinia eschscholtzii MFLUCC 19-0493 isolated from Barleria prionitis leaves against Colletotrichum acutatum and its post- harvest infections on strawberry fruits. https://doi.org/10.7717/peerj.11242

Kong CH, Xuan TD, Khanh TD, Tran HD, Trung NT (2019) Allelochemicals and signaling chemicals in plants. Molecules 24:1–19. https://doi.org/10.3390/molecules24152737

Kost C (2008) Chemical communication. In: Jørgensen SE, Fath BD (eds) Encyclopedia of ecology. Elsevier, Amsterdam, pp 557–575

Latif S, Chiapusio G, Weston LA (2017) Allelopathy and the role of allelochemicals in plant defence. Adv Bot Res 82:19–54. https://doi.org/10.1016/bs.abr.2016.12.001

Lee SO, Kim HY, Choi GJ, Lee HB, Jang KS, Choi YH, Kim J (2009) Mycofumigation with Oxyporus latemarginatus EF069 for control of postharvest apple decay and Rhizoctonia root rot on moth orchid. J Appl Microbiol 106:1213–1219. https://doi.org/10.1111/j.1365-2672.2008.04087.x

Li B, Yin Y, Kang L, Feng L, Liu Y, Du Z, Tian Y, Zhang L (2021) A review: application of allelochemicals in water ecological restoration—algal inhibition. Chemosphere 267. https://doi.org/10.1016/j.chemosphere.2020.128869

Maurya AK (2020) Application of plant volatile organic compounds (VOCs) in agriculture BT – new frontiers in stress management for durable agriculture. In: Rakshit A, Singh HB, Singh AK, Singh US, Fraceto L (eds) . Springer Singapore, Singapore, pp 369–388

Mbaluto CM, Ayelo PM, Duffy AG, Erdei AL, Tallon AK, Xia S, Caballero-Vidal G, Spitaler U, Szelényi MO, Duarte GA, Walker WB, Becher PG (2020) Insect chemical ecology: chemically mediated interactions and novel applications in agriculture. Arthropod Plant Interact 14:671–684. https://doi.org/10.1007/s11829-020-09791-4

Mushtaq W, Ain Q, Siddiqui MB, Alharby H, Hakeem KR (2020) Allelochemicals change macromolecular content of some selected weeds. S Afr J Bot 130:177–184. https://doi.org/10.1016/j.sajb.2019.12.026

Ninkovic V, Dahlin I, Vucetic A, Petrovic-obradovic O, Glinwood R, Webster B (2013) Volatile exchange between undamaged plants – a new mechanism affecting insect orientation in intercropping. PLoS One 8:e69431. https://doi.org/10.1371/journal.pone.0069431

Ninkovic V, Markovic D, Dahlin I (2016) Decoding neighbour volatiles in preparation for future competition and implications for tritrophic interactions. Perspect Plant Ecol Evol Syst 23:11–17. https://doi.org/10.1016/j.ppees.2016.09.005

Ninkovic V, Rensing M, Dahlin I, Markovic D, Dahlin I (2019) Who is my neighbor? Volatile cues in plant interactions. Plant Signal Behav 14:e1634993. https://doi.org/10.1080/1559232 4.2019.1634993

Ninkovic V, Markovic D, Rensing M (2021) Plant volatiles as cues and signals in plant communication. Plant Cell Environ 44:1030–1043. https://doi.org/10.1111/pce.13910

Ninkuu V, Zhang L, Yan J, Fu Z, Yang T, Zeng H (2021) Biochemistry of terpenes and recent advances in plant protection. Int J Mol Sci 22. https://doi.org/10.3390/ijms22115710

Nishida R (2014) Chemical ecology of insect–plant interactions: ecological significance of plant secondary metabolites. Biosci Biotechnol Biochem 78:1–13. https://doi.org/10.1080/0916845 1.2014.877836

Poldy J (2020) Volatile cues influence host-choice in arthropod pests. Animals 10:1984. https://doi.org/10.3390/ani10111984

Puig CG, Gonçalves RF, Valentão P, Andrade PB, Reigosa MJ, Pedrol N (2018) The consistency between phytotoxic effects and the dynamics of allelochemicals release from Eucalyptus globulus leaves used as bioherbicide green manure. J Chem Ecol 44:658–670. https://doi.org/10.1007/s10886-018-0983-8

Raza W, Shen Q (2020) Volatile organic compounds mediated plant-microbe interactions in soil. In: Sharma V, Salwan R, Al-Ani LKT (eds) Molecular aspects of plant beneficial microbes in agriculture, 1st edn. Academic Press, London, pp 209–219

Scavo A, Abbate C, Mauromicale G (2019) Plant allelochemicals: agronomic, nutritional and ecological relevance in the soil system. Plant Soil 442:23–48. https://doi.org/10.1007/s11104-019-04190-y

Schandry N, Becker C (2020) Allelopathic plants: models for studying plant–interkingdom interactions. Trends Plant Sci 25:176–185. https://doi.org/10.1016/j.tplants.2019.11.004

Silva RF, Rabeschini GBP, Peinado GLR, Cosmo LG, Rezende LHG, Murayama RK, Pareja M (2018) The ecology of plant chemistry and multi-species interactions in diversified agroecosystems. Front Plant Sci 871:1–7. https://doi.org/10.3389/fpls.2018.01713

Sothearith Y, Appiah KS, Mardani H, Motobayashi T, Yoko S, Eang Hourt K, Sugiyama A, Fujii Y (2021) Determination of the allelopathic potential of Cambodia's medicinal plants using the dish pack method. Sustainability 13:1294

Souto AL, Sylvestre M, Tölke ED, Tavares JF, Barbosa-Filho JM, Cebrián-Torrejón G (2021) Plant-derived pesticides as an alternative to pest management and sustainable agricultural production: prospects, applications and challenges. Molecules 26:4835. https://doi.org/10.3390/molecules26164835

Tholl D (2015) Biosynthesis and biological functions of terpenoids in plants. Adv Biochem Eng Biotechnol 148:63–106. https://doi.org/10.1007/10_2014_295

Thöming G (2021) Behavior matters—future need for insect studies on odor-mediated host plant recognition with the aim of making use of allelochemicals for plant protection. J Agric Food Chem 69:10469–10479. https://doi.org/10.1021/acs.jafc.1c03593

Tumlinson JH (2014) The importance of volatile organic compounds in ecosystem functioning. J Chem Ecol 40:212–213. https://doi.org/10.1007/s10886-014-0399-z

Velásquez A, Valenzuela M, Carvajal M, Fiaschi G, Avio L, Giovannetti M, D'Onofrio C, Seeger M (2020) The arbuscular mycorrhizal fungus Funneliformis mosseae induces changes and increases the concentration of volatile organic compounds in Vitis vinifera cv. Sangiovese leaf tissue. Plant Physiol Biochem 155:437–443. https://doi.org/10.1016/j.plaphy.2020.06.048

Vucetic A, Dahlin I, Petrovic-Obradovic O, Glinwood R, Webster B, Ninkovic V (2014) Volatile interaction between undamaged plants affects tritrophic interactions through changed plant volatile emission. Plant Signal Behav 9:e29517. https://doi.org/10.4161/psb.29517

Xie Y, Tian L, Han X, Yang Y (2021) Research advances in allelopathy of volatile organic compounds (VOCs) of plants. Horticulturae 7:1–16. https://doi.org/10.3390/horticulturae7090278

Yang XF, Li LL, Xu Y, Kong CH (2018) Kin recognition in rice (*Oryza sativa*) lines. New Phytol 220:567–578. https://doi.org/10.1111/nph.15296

Yi HS, Heil M, Adame-Álvarez RM, Ballhorn DJ, Ryu CM (2009) Airborne induction and priming of plant defenses against a bacterial pathogen. Plant Physiol 151:2152–2161. https://doi.org/10.1104/pp.109.144782

Zarbin PHG, Rodrigues MACM, Lima ER (2009) Feromônios de insetos: tecnologia e desafios para uma agricultura competitiva no Brasil. Quim Nova 32:722–731. https://doi.org/10.1590/s0100-40422009000300016

Zhang P, Wei J, Zhao C, Zhang Y, Li C, Liu S, Dicke M (2019) Airborne host – plant manipulation by whiteflies via an inducible blend of plant volatiles. PNAS 116:7387–7396. https://doi.org/10.1073/pnas.1818599116

Zhou Z, Daviet J-C, Marin B, Macian F, Salle J-Y, Zhou N, Zhu Y (2010) Vital and functional outcomes of the first-ever hemispheric stroke, epidemiological comparative study between Kunming (China) and Limoges (France). Ann Phys Rehabil Med 53:547–558. https://doi.org/10.1016/j.rehab.2010.09.001

Chapter 12
Phytotoxic Activity of Essential Oils

Ahmed A. Almarie

12.1   Plant Phytotoxicity

Some plants have defensive mechanisms when subjected to external influences such as the organisms within the surrounding environment including neighboring plants. Such a defense mechanism is known as allelopathy or plant phytotoxicity (Soltys et al. 2013). Phytotoxic plants release compounds which responsible to show the phytotoxic effect from different parts to the environment by leaching, volatilization, exudate from living plant tissue, or by the decomposition of plant residues Fig. 12.1. Hence, it was responsible for inhibiting the germination and the growth of neighboring organisms (Bitas et al. 2013).

Phytotoxic compounds are produced in plants as secondary metabolites. These compounds are involved in many ecological advantages such as protecting predators including neighbouring plants. Phytotoxic symptoms on targeted or receiver plants are shown in different ways, such as the reduction in both the length and mass of radicle and roots, extension shoot and coleoptile, swelling or necrosis of root tips, destruction of the cell wall, curling of the root axis, lack of root hairs, decrease in the number of seminal roots, reduced in plant dry weight accumulation, leaf discoloration and lower in reproductive capacity Fig. 12.2 (Khalaj et al. 2013).

The term Weeds or either Weed refers to plants that grow where they are not wanted or welcomed. If weed growth is left uncontrolled, it can cause significant reductions in the quality and quantity of agricultural production as a result of competition with economic crops on the basic growth requirements such as water, carbon dioxide, oxygen, sunlight, nutrients, and space. Thereby weeds become (farmer's enemy number 1). Weeds are also associated relatively with the location where the plant grows. For an instance, some of these plants may be useful in the

A. A. Almarie (✉)
College of Agriculture – University of Anbar, Ramadi, Iraq
e-mail: ag.ahmed.abdalwahed@uoanbar.edu.iq

Fig. 12.1 Methods of
phytotoxic compounds
released from the donor
plant into the environment.
(Almarie 2020)

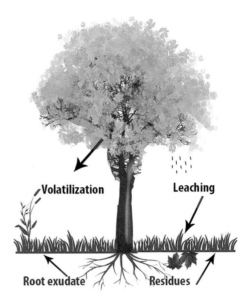

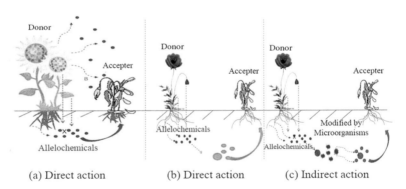

(a) Direct action   (b) Direct action   (c) Indirect action

Fig. 12.2 Direct and indirect phytotoxic mechanisms of donor plant to the targeted plants. (Soltys et al. 2013)

gardens or characterized to have medicinal properties. However, if these plants grow in the fields or orchards and consequently affect the normal agricultural production negatively, then it is termed as weeds or weed plants (Appleby 2005).

12.2   Weed Plants

12.2.1   *Weed Control Methods*

Human efforts of controlling weeds began with the use of cultural practices such as tillage, planting, crop rotation, fertilizer application, irrigation, etc., that are adapted to create favorable conditions for the crop. If properly used, the practices can help in suppressing weeds. On the other hand, culture methods alone, cannot control weeds; they can only help to reduce the weed population. Culture methods, therefore, can be effectively used in combination with other methods. Every method of weed control has its advantages and disadvantages. No single method is successful under all weed situations. Most often, a combination of these methods gives effective and economic control than a single method. These methods of controlling weeds were later developed in the form of mechanical weed control such as hand pulling, hand hoeing, and planting in rows to facilitate machinery use, but again these methods did not attain the desired benefits (Zimdahl 2013).

Later, a new mechanism of weed control was developed through the use of chemical inputs. Chemical weed control began on a small scale. Since the nineteenth century, a combination of salt and ash powder was used to control weed plants that grow on either side of the railway. The use of synthetic herbicides, however, began in the 1940s with the development of some organic herbicides, specifically the 2,4-D. This herbicide is considered a growth regulator used in high doses to control broadleaf weeds (Appleby 2005). Then, chemical weed control was widely used as a form of weed control and achieved a dominant role in the crop management, more efficient, economical and low costs as compared to other methods and contributed strongly to the increase in the agricultural yields and reduce losses due to weeds (Cobb and Reade 2010). As a result of using chemical weed control, the traditional method of weed control such as cultivation and hand weeding has been great being decreased (Gianessi 2013). A new method to control weeds was created by producing different types of synthetic herbicides according to the mode of action of these compounds against weed plants. By the 1990s, the number of compounds that have been used in herbicides in many different formulas reached more than 180 compounds (Powles and Yu Bitas). The total value of the global agrochemical market was between 31 and 35 billion US$ and of the products, herbicides accounted for 48% followed by fungicides at 22% (Zhang 2018). Nowadays, chemical weed control becomes an integral part of the complex world of technical inputs required for modern agricultural production and has been accepted as a standard tool of trade by farmers throughout the world (Zimdahl 2018).

12.2.2 Negative Impact of Synthetic Herbicides

Using synthetic herbicides even at recommended rates can lead to a negative impact on the ecosystems, especially the harmful effects that come from their residues. On the other hand, the efficiency of these compounds will be slowly decreased due to the increase in the resistance of the weed plants as a result of the continuous use of these compounds to control the same weed species. Hence, using these synthetic herbicides continuously becomes a double-edged sword.

A report from WHO mentioned that no segment of the population is completely protected from the risks of pesticide exposure and the high-risk groups are not only of the people of the developing countries but all the countries over the world (Aktar et al. 2009). The hazards of synthetic pesticides are summarized by their impact through food commodities, surface water, groundwater, and soil contamination and the effects on soil fertility (beneficial soil microorganisms) and non-target vegetation (Potts et al. 2010). For example, Glyphosate; a common non-selective systemic herbicide promoted by the manufacturers as a safer herbicide reported tracing of its residues in both humans and animal urine. It was then suggested that the use of glyphosate may have to be re-evaluated to reduce human exposure to the dangers of synthetic herbicides (Niemann et al. 2015). So, less harmful products and at the same time be effective on weeds are urgently needed to avoid the risks posed by synthetic herbicides Fig. 12.3.

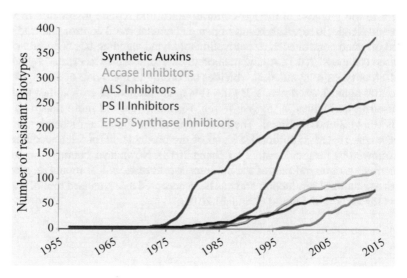

Fig. 12.3 Number of resistant plant species for several herbicides according to their modes of action. (Heap 2014)

12.3   Phytotoxic Activity of Essential Oils

Essential oils are a complex mixture of secondary metabolites mainly terpenoids. Essential oils are biosynthesized via different isoprenoid pathways such as the methylerythritol phosphate (MEP) pathway and mevalonic acid (MVA) pathway. The essential oils isolated from different plant species exhibited significant phytotoxic activity against various plant species. The phytotoxic activity of essential oils was found to inhibit germination and seedling growth of targeted plant species. The essential oils have been described as potent biological agents such as phytotoxic (Abd-ElGawad et al. 2021).

The phytotoxicity in essential oils is due to certain compounds within the components of vegetable oils. Otherwise, most studies have been conducted to screen the phytotoxicity of whole essential oils as active ingredients of bioherbicides.

12.3.1   Essential Oil as Natural Weed Killers

Essential oil is a concentrated volatile liquid consisting of different types of secondary plant metabolites but mainly composed of terpenoids and phenolics. Technically, essential oils are defined as odiferous bodies by oily nature obtained from plants in different ways, such as cold and hot pressing, distillation, and extraction using organic solvents (Baser and Buchbauer 2015).

Essential oils produced from specific types of plants can be used for different purposes. Most of the essential oil usage is influenced by donor or producer plants and their surroundings such as scent to attract certain animals and insects, aiding in pollination, protection or as repellent agents, energy reserve, wound healing, and prevent water evaporation. Essential oils can be obtained from different parts of plants such as the leaves, flowers, fruit, seeds, roots, rhizomes, bark, and wood (Fornari et al. 2012).

Biosynthetically, essential oil components are composed of two groups. The first group is the terpenoids, which is considered the main group; mostly, of the monoterpenoids, sesquiterpenoids. The second group is non-terpenoids, which may contain aromatic compounds such as phenylpropanoids, short-chain aliphatic structures, nitrogenated and sulfuric substances (Baser and Buchbauer 2015).

Essential oils can be isolated from plants by several processes such as expressed oils, steam distillation, solvent extraction, fractional distillation and percolation, and carbon dioxide extraction. The process of steam distillation is the most widely accepted method for extracting essential oils on a large scale.

Recently, the effectiveness of essential oils has been investigated on some species of weed, demonstrating the ability to inhibit germination and the development of seedlings. The reasons that encouraged the use of essential oils as alternative compounds to conventional herbicides are due to a less harmful effect on the environment and are almost as effective as synthetic herbicides. Furthermore, there are no contradictions and obstacles to be used as bioherbicides in all aspects of agriculture, specifically in organic farming as compared to the use of synthetic pesticides, which has attracted a lot of interest in the safety and health of the consumers (Dayan et al. 2011).

12.3.2 Natural Weed Killer's Mode of Action

The term "mode of action" refers to the sequence of weed killers from absorption into plants until plant death. Understanding weed killer's mode of action is helpful to know the control process. In general, weed killer with the same mode of action produces similar injury symptoms, because the outward appearance of an injury is a function of herbicide effect on the plant at the cellular level. Therefore, it is much easier to diagnose symptoms belonging to different weed killer's modes of action within the same modes of action. The study of the injury symptoms of the targeted plant tissues resulting from the application of weed killer helps to determine how it interacts with the biological and physical systems of the targeted plant. Injury symptoms in targeted weed species depend on the type of weed killer, the rate of application, stage of growth, type of exposure, and the plant species receptor involved. All weed killers work by disrupting one or more than one of the natural mechanisms of the targeted plant tissues such as the stomatal system through the influence of the guard cells, photosynthesis by the distraction of chlorophyll pigment, and targeting cell membrane and other cellular systems.

Briefly, weed killers are divided according to their mode of action into two groups; systemic or contact weed killers. Systemic weed killers are absorbed and transported through the plant's tissues which causes the killing of the entire plant. While contact weed killers affect only the contacting part of the plant. Moreover, weed killers which affect both narrow and broadleaf weeds are called non-selective. Otherwise, weed killers affect one of these groups called selective weed killers Fig. 12.4.

Regarding the phytotoxic effect of essential oils mechanisms on targeted plant tissues that have been identified by one or more than one process of below:

Fig. 12.4 Weed killer's mode of action

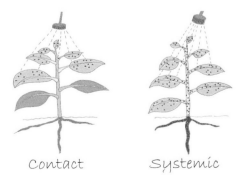

Contact   Systemic

1. Changes in membrane permeability and inhibition of plant nutrient uptake.
2. Inhibition of cell division, elongation, and submicroscopic structure.
3. Effects on plant photosynthesis and respiration.
4. Effects on various enzymatic functions and activities.
5. Effects on the synthesis of plant endogenous hormones.
6. Effects on protein synthesis.

Bioherbicidal mechanisms of the essential oils as post-contact formulations weed killers are strictly fast-acting. They generally disrupt the cuticular layer of the foliage which results in the rapid desiccations or burn-down of young tissues (Cheema et al. 2012).

Membrane disruption can be considered as one of the underlying mechanisms of essential oil's phytotoxic effects, which result in cell death and growth inhibition. Bioactive compounds in essential oils such as terpenoids are less specific and attack a multitude of proteins by building hydrogen, hydrophobic and ionic bonds and as a result of this, modulating their 3D structures and in consequence their bioactivities (Wink 2015).

Monoterpenes are considered lipophilic compounds; hence, there is, therefore, the possibility of plant cell membrane expansion as a result of the accumulation of monoterpenes, thereby destroying membrane structure (Azimova et al. 2011; Poonpaiboonpipat et al. 2013).

Moreover, the monoterpenes compounds in essential oils uncoupled the oxidative photophosphorylation (transform ADP to form ATP using the energy of sunlight) As a result, monoterpenes cause a reduction in cellular respiration leading to

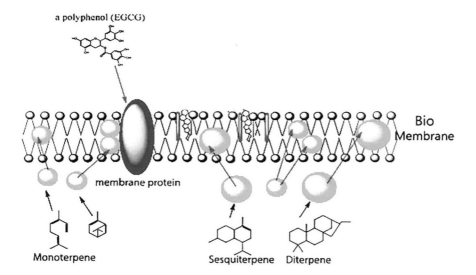

Fig. 12.5 Interaction of terpenoids with the plant cell membrane. (Wink 2015)

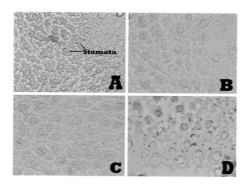

Fig. 12.6 Light compounds microscope of the lower leaf surface of narrow-leaf weed spraying with natural weed killer; Lemongrass essential oil as the active ingredient (Almarie 2017)
(**a**) Control
(**b**) Treated: Stomata still open as a result broken their mechanism
(**c**) Treated: Beginning of disappearance cell wall at 2 days after treatment
(**d**) Treated: Smashing plant cells and overlapping their contents at days after Treatment

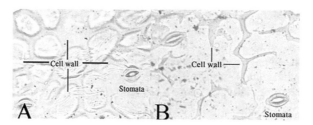

Fig. 12.7 Light compounds microscope of the lower leaf surface of broadleaf weed spraying with natural weed killer (*Lemongrass* essential oil as an active ingredient). (Almarie 2017)
(**a**) Control
(**b**) Treated: Cell wall destroyed

a perturbation in the ATP production. Thus, disorders in physiological processes in plants are induced Fig. 12.5 (Khalaj et al. 2013).

Regarding the mode action of essential oils tested on some weed plant's tissues, injury symptoms begin to appear after absorbing the essential oils by leaf membranes of the targeted weed leaves. At the early phytotoxicity of the essential oil application, the stomata still open resulting from disabling the mechanism of the guard cells which are responsible for opening and closing the stomata. As a result of losing control of the mechanism of the stomata, transpiration, and gases O_2 and CO_2 as well as the water exchange process uncontrolled. Then, essential oil's compounds began to accumulate in the cell membrane and conjugated with active components especially the membrane proteins which led to the expansion and rupture of the leaf membranes which allowed the transmission components of cells between each other randomly. The rupturing of tissues is accompanied by inhibition of all

Fig. 12.8 Injury symptoms of the *D. australe* weed plant affecting lemongrass essential oil. (1) Before spraying (2) 1 day after spraying (3) 2 days after spraying (4) 3 days after spraying. (Almarie 2017)

biological processes such as photosynthesis, water, and nutrient translocation and respiration, thus resulting in wilting and death (Figs. 12.6 and 12.7) (Almarie et al. 2016).

Bioherbicidal mechanisms of essential oils were studied by another researcher and found that essential oils used in organic agriculture as post-contact formulations weed killer are strictly fast-acting. They generally disrupt the cuticular layer of the foliage which results in the rapid desiccations or burn-down of young tissues (Dayan et al. 2009).

The results obtained from screening weed tissues affected by the application of the essential oils when compared with the healthy tissues of the same weed species, showed similar herbicidal mechanics caused by the contact synthetic herbicides. Therefore, the effects of essential oils on both major types of weeds (grassy and broad leaves) make them a non-selective weed killer Fig. 12.8.

12.4 Essential Oil Plants with Common Phytotoxic Activity

12.4.1 Trees

Trees are characterized by a high content of essential oils according to their biomass as compared with herbal plants. The trees' essential oil great purposes against internal and external effects such as infection from microorganisms and injuries from climbing animals as well as used as a defense mechanism against neighboring plants. Trees Essential oils were screened to identify their phytotoxic activity against different species of plants especially weeds from many researchers are listed in the Table 12.1 below:

12.4.2 Herbal

Herbal plants, including medicinal herbs, produce hundreds of chemical compounds for numerous functions such as defense against insects, fungi, diseases, and herbivorous animals. Various phytotoxic activities have been identified in different herbals especially essential oils herbal plants. Herbal plants belong to certain families that are characterized by phytotoxic activity as compared with other plant families which are Cupressaceae, Lamiaceae, and Rutaceae (Ismail et al. 2012).

However, the phytotoxic activity of essential oils of these plants ranged from low to high depending on the type and content of phytotoxic compounds in essential oils and even the climate or environmental conditions of plant growth. Terpenes and terpenoids are found in a variety of herbal plants which consider the most bioactive compounds and the main content of essential oils. The herbal plants showing a broad spectrum of phytotoxic activity on another plant species regarding their essential oils are listed in the Table 12.2 below:

12.5 Future of the Essential Oils as Commercial Natural Weed Killers

Natural weed killers have been involved as a new approach to improving agricultural systems to help resolve current environmental problems as a result of the negative impact of synthetic herbicides. Commercial natural weed killers' products have appeared in markets that combine the benefits of bioherbicides according to less detrimental to the wider environment care. Natural weed killers are urgently needed by organic and conventional agricultural to reduce residues of synthetic herbicides. Today, natural weed killers posted in the market as safe and successful products to control weeds. The natural products are mostly oil components (terpenes and phenolics). However, natural weed killers did not reach the desired limit for many reasons such as the expense of production and intellectual property issues. Nevertheless, there are natural herbicidal phytotoxins for which the problems in commercialization could be overcome with new production technologies, are presented in the Table 12.3.

Table 12.1 Common trees with high phytotoxic activity

Tree Name	Affected plant species	References
Eucalyptus *Eucalyptus globulus*	*Amaranthus blitoide* *C, dactylon*	Rassaeifar et al. (2013)
Eucalyptus *Eucalyptus salubris*	*Solanum elaeagnifolium*	Zhang et al. (2012)
Eucalyptus *Eucalyptus citriodora*	*A. viridis*	Vaid (2015)
Eucalyptus *Eucalyptus citriodora*	*S. arvensis* *Sonchus oleraceus* *Xanthium strumarium* *A. fatua*	Benchaa et al. (2018)
Eucalyptus *Eucalyptus cinerea*	*S. arvensis* *Erica vesicaria* *Scorpiurus muricatus*	Grichi et al. (2016)
Pine *Pinus nigra*	*P. canariensis* *Trifolium campestre* *S. arvensis*	Amri et al. (2017)
Pine *Pinus pinea*	*S. arvensis* *Trifolium campestre* *L. rigidium* *P. canariensis*	Ulukanli et al. (2014)
Eucalyptus *Eucalyptus globulus*	*Chenopodium album, Raphanus raphanistrum,* *Melilotus indicus, and Sisymbrium irio*	Almarie (2021)
Eucalyptus *Eucalyptus lehmanii*	*Sinapis arvensis* *Diplotaxis harra* *Trifolium campestre Desmazeria rigida* *Phalaris canariensis*	Grichi et al. (2016)
Pine *Pinus brutia* *Pinus pinea*	*L. sativa* *Lepidium sativum* *P. oleracea*	Ulukanli et al. (2014)
Boldo *Peumus boldus*	*Amaranths hybrids* *P. oleracea*	Verdeguer et al. (2011)
Mediterranean cypress *Cupressus sempervirens*	*L. rigidum* *Phalaris canariensis* *Trifolium campestre* *Sinapis arvensis*	Amri et al. (2012)
Monterey cypress *Cupressus macrocarpa*	*Digitaria australe* *A. hybridus*	Almarie et al. (2016)
Black tea-tree *Melaleuca bracteata*	*Panicum virgatum* *D. longiflora* *Stachytarpheta indica* *Aster subulatus*	
Manuka *Leptospermum scoparium*	*Digitaria sanguinalis*	Dayan et al. (2011)
Japanese prickly-ash *Zanthoxylum piperitum*	*Amaranthus tricolor*	Chotsaeng et al. (2017)

Table 12.2 Common herbal plants with high phytotoxic activity

Herbal plant	Affected plant species	References
Artemisia *Artemisia scoparia*	*Achyranthes aspera, Cassia occidentalis Parthenium hysterophorus Echinochloa crus-galli, Ageratum conyzoides*	Kaur et al. (2010)
Catnip *Nepeta meyeri*	*Amaranthus Retroflexus Portulaca olerace Bromus danthoniae, Agropyron cristatum Lactuca serriola Bromus tectorum Bromus intermedius Chenopodium album Cynodon dactylon Convolvulus arvensis*	Mutlu et al. (2010)
	Bromus danthoniae Lactuca serriola	Kekeç et al. (2013)
Catnip *Nepeta cataria*	*Hordeum spontaneum Taraxacum officinale Avena fatua*	Saharkhiz et al. (2016)
Catmint *Anisomeles indica*	*Bidens pilosa C. occidentalis, A. viridis E. crus-galli*	Batish et al. (2012)
Gum rockrose *Cistus ladanifer*	*A. hybridus Conyza canadensis Parietaria judaica*	Verdeguer et al. (2012)
Lemongrass *Cymbopogon citratus*	*E. crus-galli*	Poonpaiboonpipat et al. (2013)
Lemongrass *C. citratus*	*P. virgatum Chloris barbata, Euphorbia hirta Stachytarpheta indica*	Almarie et al. (2016)
Savory herb *Satureja khuzestanica Satureja rechingeri*	*Secale cereale*	Taban et al. (2013)
Mexican devil *Eupatorium adenophorum*	*P.s minor*	Ahluwalia et al. (2014)
Geranium *Pelargonium graveolens*	*Silybum marianum*	Saad and Abdelgaleil (2014)
Wormwood *Artemisia judaica*	*S. marianum*	
Caraway *Carum carvi*	*Phalaris canariensis*	Marichali et al. (2014)

(continued)

Table 12.2 (continued)

Herbal plant	Affected plant species	References
Thyme *Thymus daenensis*	*A. retroflexus* *Avena fatua* *Datura stramonium* *Lepidium sativum*	Kashkooli and Saharkhiz (2014)
Mexican mint *Plectranthus amboinicus*	*L. sativa* *Sorghum bicolor*	Pinheiro et al. (2015)
Marigold *Tagetes minuta*	*Chenopodium murale* *P. minor & A. viridis*	Arora et al. (2015)
Rasp-leaf pelargonium *Pelargonium radula*	*Digitaria australe* *A. hybridus*	Almarie et al. (2016)
Clove *Syzygium aromaticum*	*Cassia occidentalis* *Bidens pilosa*	Vaid et al. (2010)

Table 12.3 Commercial natural Weed killer found in marketplaces

Trade name	Active ingredient	References
BurnOut	Clove oil & citric acid	Islam et al. (2018)
Bioganic Safety	Clove oil & thyme oil	Dayan and Duke (2010)
Weed Zap	Clove oil & cinnamon oil	Islam et al. (2018)
Eco SMART	(Eugenol) Main component in clove oil	Dayan et al. (2009)
Matran II	Clove oil	Islam et al. (2018)
Eco-Exempt® HC	Clove oil	Duke et al. (2018)
GreenMatc EX TM	Lemongrass	Gared (2019)
Green Match EXTM	Lemongrass	Islam et al. (2018)
Avenger®	Citrus oil	Duke et al. (2018)
Weed Blitz® Organic	Pine oil	
Organic InteceptorTM	Pine oil	Dayan and Duke (2010)

12.6 Conclusion

Overview about the phytotoxic activity of essential oils as a natural weed killer against a wide range of weed species according to the latest investigations conducted in the current decade is briefed. Essential oils can be useful to control weeds which should be considered as a new approach in agricultural sustainability to reduce weed losses and keep the environment safe from the risk of synthetic herbicides. The current review also turns out that essential oils components such as terpenoids and flavonoids showed the highest phytotoxicity which is considered the dominant compound found in essential oils. The phytotoxic effect of essential oils become a new approach which been used in agriculture Systems. Essential oils can act as eco-friendly weed killers and have great value in sustainable agriculture. Nowadays, there is remarkable progress through the use of essential oils on the market that is derived from different plants as natural weed killers. With increasing emphasis on organic agriculture and environmental protection, increasing attention has been paid to research, and the physiological and ecological mechanisms of essential oils as natural weed killers are gradually being elucidated. Moreover, progress has been made in research on the associated molecular mechanisms. The phytotoxic activity of essential oils requires further research for widespread application in agricultural production worldwide.

References

Abd-ElGawad AM, El Gendy AENG, Assaeed AM, Al-Rowaily SL, Alharthi AS, Mohamed TA et al (2021) Phytotoxic effects of plant essential oils: a systematic review and structure-activity relationship based on chemometric analyses. Plan Theory 10(1):36. https://doi.org/10.3390/plants10010036

Ahluwalia V et al (2014) Chemical analysis of essential oils of *Eupatorium adenophorum* and their antimicrobial, antioxidant and phytotoxic properties. J Pest Sci 87(2):341–349. https://doi.org/10.1007/s10340-013-0542-6

Aktar W, Sengupta D, Chowdhury A (2009) Impact of pesticides use in agriculture: their benefits and hazards. Interdiscip Toxicol 2(1):1–12. https://doi.org/10.2478/v10102-009-0001-7

Almarie AA, Mamat AS, Wahab Z, Rukunudin IH (2016) Chemical composition and phytotoxicity of essential oils isolated from Malaysian plants. Allelopath J 37(1):55–69

Almarie AAA (2017) Allelopathic potential of essential oils isolated from local plants on common weeds found in Malaysian croplands. Unpunished Ph.D. thesis submitted to UniMAP Malaysia

Almarie AAA (2020) Roles of terpenoids in essential oils and its potential as natural weed killers: recent developments. In: Essential oils-bioactive compounds, new perspectives and applications. IntechOpen. https://doi.org/10.5772/intechopen.9132

Almarie A (2021) Bioherbicidal potential of eucalyptus and clove oil and their combinations on four weedy species. Iraqi J Sci:1494–1502. https://doi.org/10.24996/ijs.2021.62.5.13

Amri I, Mancini E, De Martino L, Marandino A, Lamia H, Mohsen H, De Feo V (2012) Chemical composition and biological activities of the essential oils from three Melaleuca species grown in Tunisia. Int J Mol Sci 13(12):16580–16591. https://doi.org/10.3390/ijms131216580

Amri I, Hanana M, Jamoussi B, Hamrouni L (2017) Essential oils of *Pinus nigra* JF Arnold subsp. laricio Maire: chemical composition and study of their herbicidal potential. Arab J Chem 10:S3877–S3882. https://doi.org/10.1016/j.arabjc.2014.05.026

Appleby AP (2005) A history of weed control in the United States and Canada-a sequel. Weed Sci 53(6):762–768. https://doi.org/10.1614/WS-04-210.1

Arora K, Batish DR, Singh HP, Kohli RK (2015) Allelopathic potential of the essential oil of wild marigold (Tagetes minuta L.) against some invasive weeds. J Environ Agric Sci 3:56–60. http://jeas.agropublishers.com/wp-content/uploads/2017/09/JEAS-3-9.pdf

Azimova SS, Glushenkova AI, Vinogradova VI (2011) Lipids, lipophilic components and essential oils from plant sources. Springer Science & Business Media. https://doi.org/10.1007/2F978-0-85729-323-7

Batish DR, Singh HP, Kaur M, Kohli RK, Singh S (2012) Chemical characterization and phytotoxicity of volatile essential oil from leaves of *Anisomeles indica* (Lamiaceae). Biochem Syst Ecol 41:104–109. https://doi.org/10.1016/j.bse.2011.12.017

Baser KHC, Buchbauer G (2015) Handbook of essential oils: science, technology, and applications. CRC Press. https://www.crcpress.com/Handbook-of-Essential-Oils-Science-Technology-and-Applications-Second/Baser-Buchbauer/p/book/9781466590465

Benchaa S, Hazzit M, Abdelkrim H (2018) Allelopathic effect of *Eucalyptus citriodora* essential oil and its potential use as bioherbicide. Chem Biodivers. https://doi.org/10.1002/cbdv.201800202

Bitas V, Kim HS, Bennett JW, Kang S (2013) Sniffing on microbes: diverse roles of microbial volatile organic compounds in plant health. Mol Plant Micro Interact 26(8):835–843. https://doi.org/10.1094/MPMI-10-12-0249-CR

Cobb AH, Reade JP (2010) Herbicide discovery and development. Herbicides Plant Physiol, 2nd edn, pp 27–49. https://doi.org/10.1104/pp.114.241992

Cheema ZA, Farooq M, Wahid A (2012) Allelopathy: current trends and future applications. Springer Science & Business Media. https://doi.org/10.1007/978-3-642-30595-5

Chotsaeng N, Laosinwattana C, Charoenying P (2017) Herbicidal activities of some allelochemicals and their synergistic behaviors toward *Amaranthus tricolor* L. Molecules 22(11):1841. https://doi.org/10.3390/molecules22111841

Dayan FE, Cantrell CL, Duke SO (2009) Natural products in crop protection. Bioorg Med Chem 17(12):4022–4034. https://doi.org/10.1016/j.bmc.2009.01.046

Dayan FE, Duke SO (2010) Natural products for weed management in organic farming in the USA. Outlooks Pest Manag 21(4):156–160. https://doi.org/10.1564/21aug02

Dayan FE, Howell JL, Marais JP, Ferreira D, Koivunen M (2011) Manuka oil, a natural herbicide with preemergence activity. Weed Sci 59(4):464–469. https://doi.org/10.1614/WS-D-11-00043.1

Duke SO, Owens DK, Dayan FE (2018) Natural product-based chemical herbicides. In: Weed control. CRC Press, pp 153–165

Fornari T, Vicente G, Vázquez E, García-Risco MR, Reglero G (2012) Isolation of essential oil from different plants and herbs by supercritical fluid extraction. J Chromatogr A 1250:34–48. https://doi.org/10.1016/j.chroma.2012.04.051

Gared Shaffer (2019) SDSU Extension. https://extension.sdstate.edu/organic-herbicides

Gianessi LP (2013) The increasing importance of herbicides in worldwide crop production. Pest Manag Sci 69(10):1099–1105. https://doi.org/10.1002/ps.3598

Grichi A, Nasr Z, Khouja ML (2016) Phytotoxic effects of essential oil from *eucalyptus lehmanii* against weeds and its possible use as a bioherbicide. Bulletin Environ Pharmacol Life Sci 5:17–23

Heap I (2014) Herbicide resistant weeds. In: Integrated pest management. Springer, pp 281–301. https://doi.org/10.1007/978-94-007-7796-5_12

Islam AM, Yeasmin S, Qasem JRS, Juraimi AS, Anwar MP (2018) Allelopathy of medicinal plants: current status and future prospects in weed management. Agric Sci 9(12):1569–1588. https://doi.org/10.4236/as.2018.91211

Ismail A, Lamia H, Mohsen H, Bassem J (2012) Herbicidal potential of essential oils from three Mediterranean trees on different weeds. Curr Bioact Compd 8(1):3–12. https://doi.org/10.2174/157340712799828197

Kashkooli AB, Saharkhiz MJ (2014) Essential oil compositions and natural herbicide activity of four Denaei Thyme (Thymus daenensis Celak.) Ecotypes. J Essen Oil-Bear Plants 17(5):859–874. https://doi.org/10.1080/0972060X.2014.884946

Kaur S, Singh HP, Mittal S, Batish DR, Kohli RK (2010) Phytotoxic effects of volatile oil from Artemisia scoparia against weeds and its possible use as a bioherbicide. Ind Crops Prod 32(1):54–61. https://doi.org/10.1016/j.indcrop.2010.03.007

Kekeç G, Mutlu S, Alpsoy L, Sakçali MS, Atici Ö (2013) Genotoxic effects of catmint (Nepeta meyeri Benth.) essential oils on some weed and crop plants. Toxicol Ind Health 29(6):504–513. https://doi.org/10.1177/0748233712440135

Khalaj MA, Amiri M, Azimi MH (2013) Allelopathy: physiological and sustainable agriculture important aspects. Int J Agron Plant Prod 4(5):950–962. https://doi.org/10.2478/eje-2019-0014

Marichali AKAHSA, Hosni K, Dallali S, Ouerghemmi S, Ltaief HBH, Benzarti S, Sebei H (2014) Allelopathic effects of *Carum carvi* L. essential oil on germination and seedling growth of wheat, maize, flax and canary grass. Allelopath J 34(1):81

Mutlu S, Atici O, Esim N (2010) Bioherbicidal effects of essential oils of *Nepeta meyeri* Benth. on weed spp. Allopathy journal 26(2):291–300. https://acgpubs.org/doc/2018080720472751-RNP-EO_1304-027.pdf

Niemann L, Sieke C, Pfeil R, Solecki R (2015) A critical review of glyphosate findings in human urine samples and comparison with the exposure of operators and consumers. J Verbr Lebensm 10(1):3–12. https://doi.org/10.1007/s00003-014-0927-3

Pinheiro PF, Costa AV, Alves TDA, Galter IN, Pinheiro CA, Pereira AF, Fontes MMP (2015) Phytotoxicity and cytotoxicity of essential oil from leaves of *Plectranthus amboinicus*, carvacrol, and thymol in plant bioassays. J Agric Food Chem 63(41):8981–8990. https://doi.org/10.1021/acs.jafc.5b03049

Poonpaiboonpipat T, Pangnakorn U, Suvunnamek U, Teerarak M, Charoenying P, Laosinwattana C (2013) Phytotoxic effects of essential oil from *Cymbopogon citratus* and its physiological mechanisms on barnyardgrass (Echinochloa crus-galli). Ind Crop Prod 41:403–407. https://doi.org/10.1016/j.indcrop.2012.04.057

Potts SG, Biesmeijer JC, Kremen C, Neumann P, Schweiger O, Kunin WE (2010) Global pollinator declines: trends, impacts and drivers. Trends Ecol Evol 25(6):345–353. https://doi.org/10.1016/j.tree.2010.01.007

Rassaeifar M, Hosseini N, Asl NHH, Zandi P, Aghdam AM (2013) Allelopathic effect of *eucalyptus globulus* essential oil on seed germination and seedling establishment of *Amaranthus blitoides* and *Cyndon dactylon*. Trakia J Sci 11(1):73–81. http://tru.uni-sz.bg/tsj/vol11N1_013/M.Rasaeifar.pdf

Saad LMMG, Abdelgaleil SAM (2014) Allelopathic potential of essential oils isolated from aromatic plants on *Silybum marianum*. Glob Adv Res J Agric Sci 3:289–297. http://garj.org/full-articles/allelopathic-potential-of-essential-oils-isolated-from-aromatic-plants-on-silybum-marianum-1.pdf?view=inline

Saharkhiz MJ, Zadnour P, Kakouei F (2016) Essential oil analysis and phytotoxic activity of catnip (Nepeta cataria L.). Am J Essen Oils Nat Prods 4(1):40–45. http://www.essencejournal.com/vol4/issue1/pdf/3-2-12.1.pdf

Soltys D, Krasuska U, Bogatek R, Gniazdowska A (2013) Allelochemicals as bioherbicides—present and perspectives. In: Herbicides—current research and case studies in use. In Tech, pp 517–542. https://doi.org/10.5772/56185

Taban A, Saharkhiz MJ, Hadian J (2013) Allelopathic potential of essential oils from four *Satureja spp*. Bio Agric Hortic 29(4):244–257. https://doi.org/10.1080/01448765.2013.830275

Vaid S, Batish DR, Singh HP, Kohli RK (2010) Phytotoxic effect of eugenol towards two weedy species. Bioscan 5(3):339–341

Vaid S (2015) Phytotoxicity of citronellol against *Amaranthus viridis* L. Int J Eng Appl Sci 2:94–96. https://www.neliti.com/publications/257763/phytotoxicity-of-citronellol-against-amaranthus-viridis-l

Ulukanli Z, Karaborklu S, Bozok F, Burhan ATES, Erdogan S, Cenet M, karaaslan, M. G. (2014) Chemical composition, antimicrobial, insecticidal, phytotoxic and antioxidant activities of Mediterranean *Pinus brutia* and *Pinus pinea* resin essential oils. Chin J Nat Medic 12(12):901–910. https://doi.org/10.1016/S1875-5364(14)60133-3

Verdeguer M, García-Rellán D, Boira H, Pérez E, Gandolfo S, Blázquez MA (2011) Herbicidal activity of *Peumus boldus* and *Drimys winterii* essential oils from Chile. Molecules 16(1):403–411. https://doi.org/10.3390/molecules16010403

Verdeguer M, Blázquez MA, Boira H (2012) Chemical composition and herbicidal activity of the essential oil from a *Cistus ladanifer* L. population from Spain. Nat Prod Res 26(17):1602–1609. https://doi.org/10.1080/14786419.2011.592835

Wink M (2015) Biochemistry of plant secondary metabolism. Annu. plant reviews, vol 40. Wiley. https://www.nhbs.com/biochemistry-of-plant-secondary-metabolism-book

Zhang J, An M, Hanwen W, Liu DL, Stanton R (2012) Chemical composition of essential oils of four Eucalyptus species and their phytotoxicity on silverleaf nightshade (Solanum elaeagnifolium Cav.) in Australia. Plant Growth Regul 68(2):231–237. https://doi.org/10.1007/s10725-012-9711-5

Zhang W (2018) Global pesticide use: profile, trend, cost/benefit and more. Proc Int Acad Ecol Environ Sci 8(1):1. http://www.iaees.org/publications/journals/piaees/articles/2018-8(1)/global-pesticide-use-Profile-trend-cost-benefit.pdf

Zimdahl RL (2013) Fundamentals of weed science. Academic Press. https://www.elsevier.com/books/fundamentals-of-weed-science/Zimdahl/978-0-12-811143-7

Zimdahl RL (2018) Fundamentals of weed science. Academic Press. https://www.sciencedirect.com/book/9780128111437/fundamentals-of-weed-science

Part IV
Essential Oils as Antiparasitic Agents

Chapter 13
Antileishmanial Activity of Essential Oils

José Weverton Almeida-Bezerra, Victor Juno Alencar Fonseca,
Johnatan Wellisson da Silva Mendes, Roberta Dávila Pereira de Lima,
Antonia Thassya Lucas dos Santos, Saulo Almeida de Menezes,
Benedito Yago Machado Portela, Lilian Cortez Sombra Vandesmet,
Felicidade Caroline Rodrigues, José Jailson Lima Bezerra,
Viviane Bezerra da Silva, Rafael Pereira da Cruz, Allyson Francisco dos
Santos, Cícero Jorge Verçosa, Jamile Maria Pereira Bastos Lira de
Vasconcelos, Maria Eliana Vieira Figueroa, Clêidio da Paz Cabral,
Gabriel Messias da Silva Nascimento, Maria Ivaneide Rocha,
Marcio Pereira do Nascimento, Priscilla Augusta de Sousa Fernandes,
Francisco Sydney Henrique da Silva,
and Maria Flaviana Bezerra Morais-Braga

13.1 Introduction

Besides being an important component of the plant defense system against pathogenic attacks and environmental stress, the secondary metabolism of plants provides a useful range of natural products (Piasecka et al. 2015). Due to their biological

J. W. Almeida-Bezerra (✉) · V. J. A. Fonseca · J. W. da Silva Mendes · A. T. L. dos Santos ·
R. P. da Cruz · G. M. da Silva Nascimento · M. I. Rocha · M. P. do Nascimento ·
P. A. de Sousa Fernandes · M. F. B. Morais-Braga
Regional University of Cariri – URCA, Crato, CE, Brazil
e-mail: weverton.almeida@urca.br

R. D. P. de Lima
Federal University of Cariri – UFCA, Crato, CE, Brazil

S. A. de Menezes
Federal University of Rio Grande do Sul – UFRGS, Porto Alegre, RS, Brazil

B. Y. M. Portela
Federal University of Ceara – UFC, Fortaleza, CE, Brazil

L. C. S. Vandesmet
Catholic University Center of Quixadá, Quixadá, CE, Brazil

F. C. Rodrigues · J. J. L. Bezerra · V. B. da Silva
Federal University of Pernambuco – UFPE, Recife, PE, Brazil

activities, the secondary metabolites of plants have been increasingly used as medicinal substances and food additives for therapeutic, aromatic and culinary purposes. The characteristics and concentration of secondary molecules and the biosynthesis by a plant are defined by the identity of the species and genetic, ontogenic, morphogenetic, physiological, developmental, and environmental factors. This suggests that various taxonomic groups of plants have adaptive physiological responses to deal with stress and defensive stimuli (Yang et al. 2018; Isah 2019).

Terpenes and terpenoids (the oxygenated derivatives of terpenes) are chemical compounds that represent the majority of molecules in the composition of essential oils (EOs) (Matos et al. 2019). This class of molecules is characterized by a different number of isoprene (C_5H_8) units (Blowman et al. 2018). Depending on the number of these units, terpenes can be categorized into hemiterpenes, monoterpenes, sesquiterpenes, diterpenes, triterpenes, among others (Rubulotta and Quadrelli 2019; Sharma et al. 2021). They can also be divided into groups such as acyclic, monocyclic, and bicyclic (Blowman et al. 2018). The terpenoid is a type of terpene that has oxygen attached to its structure (Sharma et al. 2021).

Essential oils, which are one of the substance types formed by terpenes, are widely used and studied for their pharmacological, biological, and permeation enhancing properties. However, several terpenes and EOs are sensitive to environmental conditions and may undergo volatilization and chemical degradation (Matos et al. 2019). Essential oils are natural products with a complex composition and are used in different ways, namely, through inhalation, topical application onto the skin, and oral consumption. There are, therefore, three main routes of ingestion or application: the skin system, the olfactory system, and the gastrointestinal system. Understanding these routes is important to clarify the mechanisms of action of EOs (Koyama and Heinbockel 2020).

The biological and pharmacological activities of EOs investigated so far include antibacterial (Ács et al. 2018), antifungal (Mutlu-Ingok et al. 2020), antiviral (Brochot et al. 2017), antileishmanial (Oliveira et al. 2020), antioxidant (Menezes Filho et al. 2020), cytotoxic (Contini et al. 2020), and anti-inflammatory (Saldanha et al. 2019) activities.

Leishmaniasis is a collection of diseases caused by parasitic protozoa of more than 20 species of *Leishmania*. The disease has three main forms: the tegumentary (most common form), the visceral (most severe form), and the mucocutaneous (most disabling form). Humans are contaminated by these parasites by the bite of infected female phlebotomine sandflies (WHO 2021). The clinical manifestations of leishmaniasis are quite mutable and can range from localized skin lesions to dissipation of life-threatening visceral disease (Meira and Gedamu 2019). Currently,

A. F. dos Santos · C. J. Verçosa · J. M. P. B. L. de Vasconcelos · M. E. V. Figueroa ·
C. da Paz Cabral
Pernambuco Department of Education and Sports, Recife, PE, Brazil

F. S. H. da Silva
State University of Ceará – UECE, Fortaleza, CE, Brazil

more than 1 billion people worldwide are in endemic areas of leishmaniasis and are at risk of infection (WHO 2021).

The first-line drugs for the treatment of leishmaniasis are antimonials. In resistant cases, pentavalents, amphotericin B deoxycholate, liposomal amphotericin B, and paromomycin are used as secondary options. However, these drugs have their use limited because of side effects, high costs, induction of resistance in parasites, and administration in hospitalized patients (Albuquerque et al. 2020). Therefore, research for new compounds is needed. In this sense, EOs have been increasingly investigated for their effectiveness against species of the genus *Leishmania*, to serve as an alternative for the treatment of leishmaniasis (Mahmoudvand et al. 2016; Sharifi-Rad et al. 2018; Rottini et al. 2019; Macêdo et al. 2020; Ferreira et al. 2020; Vandesmet et al. 2020; Gomez et al. 2021).

Therefore, this review seeks to understand the action of EOs against *Leishmania* species, parasites that cause vector-borne diseases known as leishmaniasis and which represent a serious public health problem.

13.2 Methodology

13.2.1 Database Search

Articles were searched through consultations in the Scopus© database (https://www.scopus.com/). As keywords, the descriptors "Essential oil AND *Leishmania*" were used, only in the English language.

13.2.2 Inclusion and Exclusion Criteria

Only scientific articles that addressed specific information about the potential of EOs extracted from different plant species against *Leishmania* spp. and published in the last 10 years (2011–2021) were selected. Regarding the exclusion criteria, review articles, e-books, book chapters, editorials, course completion works, dissertations, theses, abstracts published in congress proceedings, and articles on the potential of extracts, isolated chemical compounds, EOs commercialized without identification of the species, non-active EOs, and fixed oils against *Leishmania* spp. were discarded.

13.2.3 Data Screening and Information Categorization

Initially, 186 scientific articles were identified and selected in the Scopus© database. After applying the exclusion criteria, 72 documents that did not fit the theme of this review were discarded (Fig. 13.1). Finally, 114 articles containing data on the potential of EOs against *Leishmania* spp. were included (Fig. 13.1). The information collected in the articles was categorized into: (1) "Essential oils against *Leishmania* spp."; (2) "Terpenes"; (3) "Mechanisms of action"; (4) "Other compounds present in essential oils"; and (5) "Other applications". Further details about the species, active concentration of essential oils, evolutionary form of *Leishmania* spp., major constituents, and mechanism of action were also organized and presented in a table.

13.3 Results

Of 186 articles, 114 met the inclusion criteria and were selected for data extraction (Table 13.1). Of the 114 studies, 111 are *in vitro* (97.4%), 2 *in vivo/in vitro* (1.7%), and 1 *in vivo* (0.9%) assays of EOs with leishmanicidal activity. The *Leishmania* species most used in the assays were: *L. amazonensis*, used in 54 (47.4%) of the studies, *L. infantum*, in 33 (28.9%) of the studies, and *L. major*, in 21 (18.4%) of the studies. Table 13.1 presents the EOs of plant species from 74 genera belonging to 26 families, among which the most frequent were Lamiaceae with 14 genera (18.9%), Asteraceae with 9 genera (12.1%), and Myrtaceae with 8 genera (10.8%).

Of the 114 studies included in the review, 100 (87.7%) performed the chemical characterization of the EOs and 14 (12.3%) did not. Carvacrol was the major constituent most present in the EOs, being reported in 8 studies (7%), followed by thymol, cited in 7 studies (6.1%), and α-pinene and 1,8-cineole cited in 5 studies (4.3%) each.

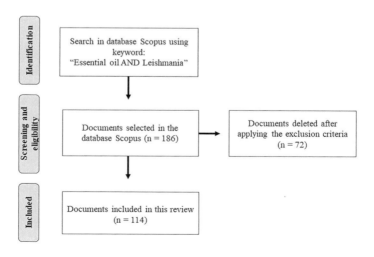

Fig. 13.1 Flowchart of selection of scientific documents included in this review

Table 13.1 Antileishmanial activity of aromatic species

Family/Species	Evolutionary form	Species	IC$_{50}$	Majority constituents	Reference
Própolis tunisiana	Promastigote	L. major	5.29 µg/mL	α-pinene (36.7%)	Jihene et al. (2020)
Própolis tunisiana	Promastigote	L. infantum	3.67 µg/mL	α-pinene (36.7%)	Jihene et al. (2020)
Própolis tunisiana	Amastigote	L. major	7.38 µg/mL	α-pinene (36.7%)	Jihene et al. (2020)
Própolis tunisiana	Amastigote	L. infantum	4.96 µg/mL	α-pinene (36.7%)	Jihene et al. (2020)
Amaranthaceae					
Dysphania ambrosioides (L.) Mosyakin & Clemants	Amastigote	L. amazonensis	4.9 µg/mL para L. amazonensis	–	Machín et al. (2019)
Dysphania ambrosioides (L.) Mosyakin & Clemants	Amastigote	L. amazonensis	4.7 µg/mL	Carvacrol (62%)	Monzote et al. (2011)
Dysphania ambrosioides (L.) Mosyakin & Clemants	Promastigote	L. amazonensis	2.9 µg/mL	Carvacrol (62%)	Monzote et al. (2011)
Dysphania ambrosioides (L.) Mosyakin & Clemants	Amastigote	L. amazonensis	4.6 µg/mL	–	Monzote et al. (2014d)
Dysphania ambrosioides (L.) Mosyakin & Clemants	Promastigote	L. amazonensis	3.7 µg/mL	–	Monzote et al. (2014d)
Dysphania ambrosioides (L.) Mosyakin & Clemants	Promastigote	L. tropica	1.83 µg/mL	4-careno (56.59%); o-cimeno (41.46%)	Ali et al. (2021)
Anacardiaceae					
Myracrodruon urundeuva (Engl.) Fr. All.	Promastigote	L. amazonensis	205 µg/mL	β-myrcene (42.46%); α-myrcene (37.23%)	Carvalho et al. (2017)
Myracrodruon urundeuva (Engl.) Fr. All.	Amastigote	L. amazonensis	44.5 µg/mL	β-myrcene (42.46%); α-myrcene (37.23%)	Carvalho et al. (2017)
Pistacia vera L.	Amastigote	L. tropica	21.3 µg/mL	Limonene (26.21%)	Mahmoudvand et al. (2015b)
Pistacia lentiscus L.	Promastigote	L. infantum	11.28 µg/mL	Myrcene (33.46%)	Bouyahya et al. (2019)

(continued)

Table 13.1 (continued)

Family/Species	Evolutionary form	Species	IC$_{50}$	Majority constituents	Reference
Pistacia lentiscus L.	Promastigote	*L. infantum*	8 μg/mL	α-pinene (20.46%)	Bouyahya et al. (2019)
Annonaceae					
Annona crassiflora Mart.	Promastigote	*L. infantum*	25.97 μg/mL	α-amorphene (43.6%)	Oliani et al. (2013)
Annona coriacea Mart	Promastigote	*L. major*	305.20 μg/mL	Bicyclogermacrene (36%)	Siqueira et al. (2011)
Annona coriacea Mart	Promastigote	*L. infantum*	39.93 μg/mL	Bicyclogermacrene (36%)	Siqueira et al. (2011)
Annona coriacea Mart	Promastigote	*L. brasiliensis*	261.20 μg/mL	Bicyclogermacrene (36%)	Siqueira et al. (2011)
Annona coriacea Mart	Promastigote	*L. amazonensis*	160.20 μg/mL	Bicyclogermacrene (36%)	Siqueira et al. (2011)
Bocageopsis multiflora (Mart.) R.E.Fr.	Promastigote	*L. amazonensis*	14.6 μg/mL	Spathulenol (16.2%)	Oliveira et al. (2014)
Guatteria australis A.St.-Hil.	Promastigote	*L. infantum*	30.71 μg/mL	Germacrene B (50.66%)	Siqueira et al. (2015)
Coriandrum sativum L.	Promastigote	*L. donovani*	26.58 μg/mL	E)-2-undecenal; (E)-2-decenal; (E)-2- Dodecenal	Donega et al. (2014)
Apiaceae					
Ferula galbaniflua Boiss. & Buhse	Promastigote	*L. amazonensis*	95.70 μg/mL	Methyl-8-pimaren-18-oate (41.82%)	Andrade et al. (2016)
Ferula communis L.	Promastigote	*L. major*	0.11 μg/mL	–	Essid et al. (2015)
Ferula communis L.	Promastigote	*L. infantum*	0.05 μg/mL	–	Essid et al. (2015)
Pseudotrachydium kotschyi (Boiss.) Pimenov & Kljuykov	Amastigote	*L. major*	–	Z-α-trans- Bergamotol (23.25%)	Ashrafi et al. (2020b)
Arecaceae					
Scheelea phalerata Mart. ex Spreng	Promastigote	*L. amazonensis*	165.5 μg/mL	Phytol (36.7%)	Oliveira et al. (2020)

Asteraceae

Artemisia absinthium L.	Promastigote	L. major	1.49 µg/mL	Chamazuleno (39.2%)	Mathlouthi et al. (2018)
Artemisia absinthium L.	Amastigote	L. amazonensis	13.4 µg/mL	Acetato de trans-sabinil (36.7%)	Monzote et al. (2014a)
Artemisia absinthium L.	Promastigote	L. amazonensis	14.4 µg/mL	Acetato de trans-sabinil (36.7%)	Monzote et al. (2014a)
Artemisia absinthium L. (E2)	Promastigote	L. infantum	<100 µg/mL	Cis-Epoxyocimene (59.9%)	Bailen et al. (2013)
Artemisia absinthium L. (SNC)	Promastigote	L. infantum	<100 µg/mL	–	Bailen et al. (2013)
Artemisia annua L.	Amastigote	L. donovani	7.3 µg/mL	Camphor (52.06%)	Islamuddin et al. (2014a)
Artemisia annua L.	Promastigote	L. donovani	14.63 µg/mL	Camphor (52.06%)	Islamuddin et al. (2014a)
Artemisia campestris L.	Promastigote	L. major	2.20 µg/mL	β-pineno (32%)	Mathlouthi et al. (2018)
Artemisia campestris L.	Promastigote	L. infantum	44 µg/mL	β-pinene (32.95%)	Aloui et al. (2016)
Artemisia dracunculus L.	Promastigote	L. tropica	111 µg/mL	p-allyanisole (67.62%)	Ghanbariasad et al. (2021b)
Artemisia dracunculus L.	Promastigote	L. major	114 µg/mL	p-allyanisole (67.62%)	Ghanbariasad et al. (2021b)
Artemisia herba alba Asso	Promastigote	L. major	1.20 µg/mL	α-thujone (29.3%)	Mathlouthi et al. (2018)
Artemisia herba alba Asso	Promastigote	L. infantum	68 µg/mL	Camphor (36.82%)	Aloui et al. (2016)
Artemisia ludoviciana Nutt.	Promastigote	L. infantum	<64 mg/mL	Camphor (40.6%); 1,8-cineole (25.5%)	Baldemir et al. (2018)
Eremanthus erythropappus (DC) McLeisch	Promastigote	L. amazonensis	9.53 µg/mL	α-bisabolol (85.98%)	Gomes et al. (2020)

(continued)

Table 13.1 (continued)

Family/Species	Evolutionary form	Species	IC$_{50}$	Majority constituents	Reference
Matricaria chamomilla L.	Promastigote	*L. amazonensis*	3.33 µg/mL	–	Jorjani et al. (2017)
Matricaria chamomilla L.	Amastigote	*L. amazonensis*	14.56 µg/mL	–	Jorjani et al. (2017)
Matricaria chamomilla L.	Promastigote	*L. amazonensis*	60.16 µg/mL	β-farnesene (52.73%)	Andrade et al. (2016)
Matricaria recutita L.	Promastigote	*L. amazonensis*	10.4 µg/mL	–	Hajaji et al. (2018)
Matricaria recutita L.	Promastigote	*L. infantum*	10.8 µg/mL	–	Hajaji et al. (2018)
Melampodium divaricatum (Rich.) DC.	Amastigote	*L. amazonensis*	10.7 µg/mL	E-caryophyllene (56.0%)	Moreira et al. (2019)
Melampodium divaricatum (Rich.) DC.	Promastigote	*L. amazonensis*	24.2 µg/mL	E-caryophyllene (56.0%)	Moreira et al. (2019)
Pluchea carolinensis (Jacq.) G. Don.	Promastigote	*L. amazonensis*	24.7 µg/mL	Selin-11-en-4α-ol (51.0%)	García et al. (2017)
Pluchea carolinensis (Jacq.) G. Don.	Amastigote	*L. amazonensis*	6.2 µg/mL	Selin-11-en-4α-ol (51.0%)	García et al. (2017)
Pulicaria vulgaris Gaertn.	Promastigote	*L. major*	25.64 µg/mL	Thymol (50.22%)	Sharifi-Rad et al. (2018)
Pulicaria vulgaris Gaertn.	Promastigote	*L. infantum*	18.54 µg/mL	Thymol (50.22%)	Sharifi-Rad et al. (2018)
Tagetes lucida Cav.	Promastigote	*L. amazonensis*	118.8 µg/mL	Methyl chavicol (97%)	Monzote et al. (2020b)
Tagetes lucida Cav.	Promastigote	*L. tarentolae*	61.4 µg/mL	Methyl chavicol (97%)	Monzote et al. (2020b)

Vanillosmopsis arborea Baker	Promastigote	*L. amazonensis*	7.35 µg/mL	α-bisabolol (97.9%)	Colares et al. (2013)
Vanillosmopsis arborea Baker	Amastigote	*L. amazonensis*	12.58 µg/mL	α-bisabolol (97.9%)	Colares et al. (2013)
Vernonia brasiliana (L.) Druce	Promastigote	*L. infantum*	39.01 µg/mL	β-cariofileno (21.47%)	Mondego-Oliveira et al. (2021)
Vernonia polyanthes Less	Promastigote	*L. infantum*	19.4 µg/mL para	Myrcene (34.3%)	Moreira et al. (2017).
Bixaceae					
Bixa orellana L.	Amastigote	*L. amazonensis*	8.5 µg/mL	–	Machín et al. (2019)
Bixa orellana L.	Amastigote	*L. amazonensis*	8.5 µg/mL	Ishwarane (18.6%)	Monzote et al. (2014c)
Bixa orellana L. (Nanocomplexo)	Amastigote	*L. amazonensis*	15.4 µg/mL	–	Machín et al. (2019)
Burseraceae					
Bursera graveolens Triana & Planch.	Amastigote	*L. amazonensis*	36.7 µg/mL	Limonene (26.5%)	Monzote et al. (2012).
Protium heptaphyllum (Aubl.) Marchand		*L. amazonensis*	9.02 µg/mL	–	Cabral et al. (2021)
Protium ovatum Engl.	Promastigote	*L. amazonensis*	2.28 µg/mL	–	Estevam et al. (2017)
Canellaceae					
Cinnamodendron dinisii Schwacke	Promastigote	*L. amazonensis*	54.05 µg/mL	α-pinene (35.41%)	Andrade et al. (2016)

(continued)

Table 13.1 (continued)

Family/Species	Evolutionary form	Species	IC$_{50}$	Majority constituents	Reference
Euphorbiaceae					
Croton nepetifolius Baill.	Promastigote	L. amazonensis	9.87 µg/mL	Methyl eugenol (33.89%)	Morais et al. (2019).
Croton rhamnifolioides Pax & K. Hoffm.	Promastigote	L. braziliensis	127.43 µg/mL	–	Alcântara et al. (2021)
Croton rhamnifolioides Pax & K. Hoffm.	Promastigote	L. infantum	111.84 µg/mL	–	Alcântara et al. (2021)
Croton linearis Jacq.	Promastigote	L. amazonensis	20.0 µg/mL	–	Díaz et al. (2018)
Croton linearis Jacq	Amastigote	L. amazonensis	13.8 µg/mL	–	Díaz et al. (2018)
Fabaceae					
Copaifera sp.	Promastigote	L. amazonensis	18 µg/mL	–	Moraes et al. (2018).
Copaifera sp.	Promastigote	L. infantum	16 µg/mL	–	Moraes et al. (2018).
Copaifera guianensis Desf.	Promastigote	L. amazonensis	590 µg/mL	–	Moraes et al. (2018).
Copaifera guianensis Desf.	Promastigote	L. infantum	366 µg/mL	–	Moraes et al. (2018).
Copaifera reticulata Ducke	Amastigote	L. infantum	0.52 µg/mL	β-linalool (73.21%)	Rottini et al. (2019)
Copaifera reticulata Ducke	Promastigote	L. infantum	7.88 µg/mL	β-linalool (73.21%)	Rottini et al. (2019)
Geraniaceae					
Pelargonium graveolens L'Hér.	Promastigote	L. major	0.28 µg/mL	Citronellol (24.75%)	Essid et al. (2015)
Pelargonium graveolens L'Hér.	Promastigote	L. infantum	0.11 µg/mL	Citronellol (24.75%)	Essid et al. (2015)

Lamiaceae

Elsholtzia ciliata (Thunb.) Hyl.	Promastigote	L. mexicana	8.49 nl/mL	Geranial (23.4%)	Le et al. (2017)
Lavandula luisieri (Lavandula stoechas var. luisieri)	Promastigote	L. infantum	63 µg/mL	Necrodane derivatives (36%)	Machado et al. (2019)
Lavandula luisieri (Lavandula stoechas var. luisieri)	Promastigote	L. tropica	38 µg/mL	Necrodane derivatives (36%)	Machado et al. (2019)
Lavandula luisieri	Promastigote	L. major	31 µg/mL	Necrodane derivatives (36%)	Machado et al. (2019)
Lavandula stoechas L.	Promastigote	L. major	0.9 µg/mL	Fenchone (31.81%); camphor (29.60%)	Bouyahya et al. (2017b)
Mentha australis R.Br.	Promastigote	L. donovani	3.7 µg/mL	β-linalool (22.9%)	Ibrahim et al. (2017)
Melissa officinalis L.	Promastigote	L. braziliensis	<125 µg/mL	Geranial (35.69%); Z citral (25.51%)	Costa et al. (2016)
Mentha pulegium L.	Promastigote	L. major	1.3 µg/mL	Menthone (21.1%); pulegone (40.9%)	Bouyahya et al. (2017c)
Nepeta curvidens Boiss. & Balansa	Amastigote	L. major	71.02 µg/mL	–	Ashrafi et al. (2020a)
Origanum compactum Benth.	Promastigote	L. major	0.13 µg/mL	Carvacrol (43.5%)	Bouyahya et al. (2017a)
Origanum compactum Benth.	Promastigote	L. infantum	0.02 µg/mL	Carvacrol (43.5%)	Bouyahya et al. (2017a)
Origanum compactum Benth.	Promastigote	L. tropica	0.22 µg/mL	Carvacrol (43.5%)	Bouyahya et al. (2017a)
Ocimum canum Sims	Promastigote	L. amazonensis	17.4 µg/mL	Thymol (42.15%); p-cymene (21.17%)	Silva et al. (2018)
Ocimum canum Sims	Amastigote	L. amazonensis	13.1 µg/mL	Thymol (42.15%); p-cymene (21.17%)	Silva et al. (2018)

(continued)

Table 13.1 (continued)

Family/Species	Evolutionary form	Species	IC$_{50}$	Majority constituents	Reference
Ocimum gratissimum L.	Promastigote	*L. mexicana*	4.85 nl/mL	Eugenol (86.5%)	Le et al. (2017)
Origanum onites L.	Promastigote	*L. donovani*	17.8 µg/mL	Carvacrol (70.6%)	Tasdemir et al. (2019)
Plectranthus amboinicus (Lour.) Spreng	Promastigote	*L. amazonensis*	58.2 µg/mL	Carvacrol (71%)	Monzote et al. (2020c)
Rosmarinus officinalis L.	Promastigote	*L. infantum*	1.2 µg/mL	1,8-Cineole (23.6%)	Bouyahya et al. (2017c)
Satureja khuzestanica Jamzad	Amastigote	*L. major*	–	–	Kheirandish et al. (2011)
Tetradenia riparia (Hochst.) Codd	Promastigote	*L. amazonensis*	15.67 ng/mL	–	Cardoso et al. (2015)
Tetradenia riparia (Hochst.) Codd	Amastigote	*L. amazonensis*	15.67 ng/mL	–	Cardoso et al. (2015)
Tetradenia riparia (Hochst.) Codd	Promastigote	*L. amazonensis*	0.03 µg/mL	–	Demarchi et al. (2016)
Tetradenia riparia (Hochst.) Codd	Amastigote	*L. amazonensis*	0.5 µg/mL	–	Demarchi et al. (2016)
Teucrium polium Decne.	Promastigote	*L. major*	0.15 µg/mL	Carvacrol (56.06%)	Essid et al. (2015)
Teucrium polium Decne.	Promastigote	*L. infantum*	0.09 µg/mL	Carvacrol (56.06%)	Essid et al. (2015)
Teucrium polium Decne.	Promastigote	*L. donovani*	2.3 µg/mL	–	Ibrahim et al. (2017)
Thymus capitellatus Hoffmanns. & Link	Promastigote	*L. infantum*	37 µg/mL	1,8-cineol (58.6%)	Machado et al. (2014)
Thymus capitellatus Hoffmanns. & Link	Promastigote	*L. tropica*	35 µg/mL	1,8-cineol (58.6%)	Machado et al. (2014)
Thymus capitellatus Hoffmanns. & Link	Promastigote	*L. major*	62 µg/mL	1,8-cineol (58.6%)	Machado et al. (2014)
Thymus hirtus sp. *algeriensis*	Promastigote	*L. major* e	0.43 µg/mL	–	Ahmed et al. (2011)

Thymus hirtus sp. *algeriensis*	Promastigote	*L. infantum*	0.25 μg/mL	–	Ahmed et al. (2011)
Zataria multiflora Boiss.	Promastigote	*L. tropica*	3.2 μL/mL	Thymol (41.81%); carvacrol (28.85%)	Dezaki et al. (2016)
Zataria multiflora Boiss.	Amastigote	*L. tropica*	8.3 μL/mL	Thymol (41.81%); carvacrol (28.85%)	Dezaki et al. (2016)
Lauraceae					
Cinnamomum cassia (L.) J. Presl	Promastigote	*L. mexicana*	2.92 nl/mL	Trans-Cinnamaldehyde (83.6%)	Le et al. (2017)
Cinnamomum verum J.Presl	Promastigote	*L. mexicana*	21 μg/mL	Cinnamaldehyde (73.3%)	Andrade-Ochoa et al. (2021)
Cinnamomum zeylanicum Blume	Promastigote	*L. tropica*	7.56 μg/mL	Cinnamaldehyde (62.04%)	Ghanbariasad et al. (2021a)
Cinnamomum zeylanicum Blume	Promastigote	*L. major*	16.53 μg/mL	Cinnamaldehyde (62.04%)	Ghanbariasad et al. (2021c)
Cryptocarya aschersoniana Mez	Promastigote	*L. amazonensis*	4.46 μg/mL	Limonene (42.3%)	Andrade et al. (2018a)
Endlicheria bracteolata (Meisn.)	Amastigote	*L. amazonensis*	3.54 μg/mL	Guaiol (46.4%)	Sales et al. (2018)
Endlicheria bracteolata (Meisn.)	Promastigote	*L. amazonensis*	7.94 μg/mL	Guaiol (46.4%)	Sales et al. (2018)
Nectandra gardneri Meisn.	Amastigote	*L. infantum*	2.7 μg/mL	Intermediol (58.2%)	Bosquiroli et al. (2017)
Nectandra gardneri Meisn.	Amastigote	*L. amazonensis*	2.1 μg/mL	Intermediol (58.2%)	Bosquiroli et al. (2017)
Nectandra hihua (Ruiz & Pav.) Rohwer	Amastigote	*L. infantum*	0.2 μg/mL	Bicyclogermacrene (28.1%)	Bosquiroli et al. (2017)
Nectandra hihua (Ruiz & Pav.) Rohwer	Amastigote	*L. amazonensis*	0.2 μg/mL	Bicyclogermacrene (28.1%)	Bosquiroli et al. (2017)
Nectandra megapotamica (Spreng.) Mez	Promastigote	*L. amazonensis*	6.66 μg/mL	–	Almeida et al. (2020)
Ocotea dispersa (Nees & Mart.) Mez	Promastigote	*L. amazonensis*	4.67 μg/mL	α-eudesmol (20.9%)	Alcoba et al. (2018)

(continued)

Table 13.1 (continued)

Family/Species	Evolutionary form	Species	IC$_{50}$	Majority constituents	Reference
Ocotea odorifera (Vell.) Rohwer	Promastigote	*L. amazonensis*	11.67 μg/mL	Safrole (36.3%)	Alcoba et al. (2018)
Meliaceae					
Guarea macrophylla Vahl	Promastigote	*L. amazonensis*	11.8 μg/mL	–	Oliveira et al. (2019)
Myrtaceae					
Campomanesia xanthocarpa (Mart.) O.Berg	Promastigote	*L. amazonensis*	70 μg/mL	–	Ferreira et al. (2020)
Campomanesia xanthocarpa (Mart.) O.Berg	Amastigote	*L. amazonensis*	6 μg/mL	–	
Eugenia gracillima Kiaersk.	Promastigote	*L. braziliensis*	74.64 μg/mL	–	Sampaio et al. (2021)
Eugenia gracillima Kiaersk.	Promastigote	*L. infantum*	80.4 μg/mL	–	Sampaio et al. (2021)
Eugenia piauhiensis Vellaff.	Amastigote	*L. amazonensis*	4.59 μg/mL	γ-Elemene (23.5%)	Nunes et al. (2021)
Eugenia piauhiensis Vellaff.	Promastigote	*L. amazonensis*	6.43 μg/mL	γ-Elemene (23.5%)	Nunes et al. (2021)
Eugenia pitanga (O.Berg) Nied.	Promastigote	*L. amazonensis*	6.10 μg/mL	–	Kauffmann et al. (2017)
Myrcia ovata Cambess.	Promastigote	*L. amazonensis*	8.69 μg/mL	Geranial (52.6%); Neral (37.1%)	Gomes et al. (2020)
Myrciaria plinioides D.Legrand	Promastigote	*L. amazonensis*	14.16 μg/mL	Spathulenol (21.12%)	Kauffmann et al. (2019)
Myrciaria plinioides D.Legrand	Promastigote	*L. infantum*	101.50 μg/mL	Spathulenol (21.12%)	Kauffmann et al. (2019)

Species	Stage	Parasite	Value	Compound	Reference
Myrtus communis L.	Promastigote	*L. tropica*	8.4 µg/mL	α-pinene (24.7%)	Mahmoudvand et al. (2015a)
Psidium myrsinites DC.	Promastigote	*L. braziliensis*	52.2 µg/mL	–	Vandesmet et al. (2020)
Syzygium aromaticum (L.) Merr. & L.M.Perry	Promastigote	*L. major*	654.76 µg/mL	Eugenol (65.41%)	Moemenbellah-Fard et al. (2020)
Syzygium aromaticum (L.) Merr. & L.M.Perry	Promastigote	*L. tropica*	180.24 µg/mL	Eugenol (65.41%)	Moemenbellah-Fard et al. (2020)
Syzygium aromaticum (L.) Merr. & L.M.Perry	Promastigote	*L. amazonensis*	60.0 µg/mL	Eugenol (59.75%); eugenyl Acetate (29.24%)	Islamuddin et al. (2014b)
Syzygium aromaticum (L.) Merr. & L.M.Perry	Amastigote	*L. amazonensis*	43.9 µg/mL	Eugenol (59.75%); eugenyl Acetate (29.24%)	Rodrigues et al. (2015)
Syzygium cumini (L.) Skeels	Promastigote	*L. amazonensis*	60 mg/L	α-pinene (31.85%); (Z)-b-ocimene (28.98%)	Dias et al. (2013)
Piperaceae					
Piper aduncum L.	Promastigote	*L. amazonensis*	25.9 µg/mL	Bicyclogermacrene (20.9%)	Bernuci et al. (2016)
Piper aduncum L.	Amastigote	*L. amazonensis*	36.2 µg/mL	Bicyclogermacrene (20.9%)	Bernuci et al. (2016)
Piper aduncum L.	Promastigote	*L. braziliensis*	77.9 µg/mL	–	Ceole et al. (2017)
Piper aduncum var. *ossanum*	Promastigote	*L. amazonensis*	19.3 µg/mL	Piperitone (20.07%)	Gutiérrez et al. (2016)
Piper aduncum var. *ossanum*	Promastigote	*L. infantum*	32.5 µg/mL	Piperitone (20.07%)	Gutiérrez et al. (2016)
Piper angustifolium Ruiz & Pav.	Amastigote	*L. infantum*	1.43 µg/mL	Spathulenol (23.8%)	Bosquiroli et al. (2015)

(continued)

Table 13.1 (continued)

Family/Species	Evolutionary form	Species	IC$_{50}$	Majority constituents	Reference
Piper claussenianum (Miq.) C.DC.	Promastigote	*L. amazonensis*	21.3 µg/mL	(E)-nerolidol (83.29%)	Marques et al. (2011)
Piper cernuum Vell.	Amastigote	*L. amazonensis*	–	β-elemene (30.0%)	Capello et al. (2015)
Piper demeraranum (Miq.) C.DC.	Promastigote	*L. amazonensis*	86 µg/mL	β-elemene (33.1%)	Carmo et al. (2012)
Piper demeraranum (Miq.) C.DC.	Amastigote	*L. amazonensis*	78 µg/mL	β-elemene (33.1%)	Carmo et al. (2012)
Piper demeraranum (Miq.) C.DC.	Promastigote	*L. guyanensis*	22.7 µg/mL	β-elemene (33.1%)	Carmo et al. (2012)
Piper demeraranum (Miq.) C.DC.	Amastigote	*L. guyanensis*	22.7 µg/mL	β-elemene (33.1%)	Carmo et al. (2012)
Piper diospyrifolium Kunth	Promastigote	*L. amazonensis*	13.5 µg/mL	–	Bernuci et al. (2016)
Piper diospyrifolium Kunth	Amastigote	*L. amazonensis*	76.1 µg/mL	–	Bernuci et al. (2016)
Piper duckei C.DC.	Promastigote	*L. amazonensis*	46 µg/mL	Trans-caryophyllene (27.1%)	Carmo et al. (2012)
Piper duckei C.DC.	Amastigote	*L. amazonensis*	42.4 µg/mL	Trans-caryophyllene (27.1%)	Carmo et al. (2012)
Piper duckei C.DC.	Promastigote	*L. guyanensis*	15.2 µg/mL	Trans-caryophyllene (27.1%)	Carmo et al. (2012)
Piper hispidum Sw.	Amastigote	*L. amazonensis*	3.4 µg/mL	–	Houël et al. (2015)
Piper tuberculatum Jacq	Promastigote	*L. brasiliensis*	143.59 µg/mL	β-pinene (27.74%)	Sanchez-Suarez et al. (2013)
Piper tuberculatum Jacq	Promastigote	*L. infantum*	133.97 µg/mL	β-pinene (27.74%)	Sanchez-Suarez et al. (2013)

Plant species	Stage	Leishmania	Value	Major compound	Reference
Piper var. *brachypodom* (Benth.) C. DC.	Promastigote	*L. infantum*	23.68 µg/mL	trans-ß-caryophyllene (20.2%)	Leal et al. (2013)
Piper var. *brachypodom* (Benth.) C. DC.	Amastigote	*L. infantum*	62.82 µg/mL	Trans-ß-caryophyllene (20.2%)	Leal et al. (2013)
Piper marginatum Jacq.	Amastigote	*L. amazonensis*	0.58 µg/mL	3,4-methylenedioxypropiophenone (22.9%)	Macêdo et al. (2020)
Piper marginatum Jacq.	Promastigote	*L. amazonensis*	7.9 µg/mL	3,4-methylenedioxypropiophenone (22.9%)	Macêdo et al. (2020)
Poaceae					
Cymbopogon citratus (DC.) Stapf	Promastigote	*L. infantum*	25 µg/mL	Geranial (45.7%); Neral (32.5%)	Machado et al. (2012a)
Cymbopogon citratus (DC.) Stapf	Promastigote	*L. tropica*	52 µg/mL	Geranial (45.7%); Neral (32.5%)	Machado et al. (2012a)
Cymbopogon citratus (DC.) Stapf	Promastigote	*L. major*	38 µg/mL	Geranial (45.7%); Neral (32.5%)	Machado et al. (2012a)
Ranunculaceae					
Nigella sativa L.	Promastigote	*L. infantum*	62.1 µg/mL	Thymoquinone (42.4%)	Mahmoudvand et al. (2015a)
Nigella sativa L.	Promastigote	*L. tropica*	53.3 µg/mL	Thymoquinone (42.4%)	Mahmoudvand et al. (2015a)
Nigella sativa L.	Promastigote	*L. tropica*	–	–	Abamor and Allahverdiyev 2016
Nigella sativa L.	Amastigote	*L. tropica*	–	–	Abamor and Allahverdiyev 2016
Rosaceae					
Agrimonia pilosa Ledeb	Promastigote	*L. donovani*	<100 µg/mL	–	Dhami et al. (2021)
Agrimonia pilosa Ledeb	Amastigote	*L. donovani*	<100 µg/mL	–	Dhami et al. (2021)

(continued)

Table 13.1 (continued)

Family/Species	Evolutionary form	Species	IC$_{50}$	Majority constituents	Reference
Rubiaceae					
Mitracarpus frigidus (Willd. ex Roem. & Schult.) K.Schum.	Promastigote	*L. major*	47.2 µg/mL	Linalool (29.29%)	Fabri et al. (2012)
Mitracarpus frigidus (Willd. ex Roem. & Schult.) K.Schum.	Promastigote	*L. amazonensis*	89.7 µg/mL	Linalool (29.29%)	Fabri et al. (2012)
Rutaceae					
Citrus limon L.	Amastigote	*L. major*	4.2 µg/mL	Neryl acetate (29.5%)	Maaroufi et al. (2021)
Citrus sinensis (L.) Osbeck	Promastigote	*L. tropica*	151.13 µg/mL	Limonene (71.264%)	Ghanbariasad et al. (2021a)
Citrus sinensis (L.) Osbeck	Promastigote	*L. major*	108.31 µg/mL	Limonene (71.264%)	Ghanbariasad et al. (2021a)
Haplophyllum tuberculatum A. Juss.	Promastigote	*L. mexicana*	6.48 µg/mL	–	Hamdi et al. (2018)
Ruta chalepensis L.	Promastigote	*L. major*	1.13 µg/mL	2-undecanone (84.28%)	Ahmed et al. (2011)
Ruta chalepensis L.	Promastigote	*L. infantum*	1.13 µg/mL	2-undecanone (84.28%)	Ahmed et al. (2011)
Salicaceae					
Casearia sylvestris SW.	Amastigote	*L. amazonensis*	14.0 µg/mL	E-caryophyllene (22.2%)	Moreira et al. (2019)
Casearia sylvestris SW.	Promastigote	*L. amazonensis*	29.8 µg/mL	E-caryophyllene (22.2%)	Moreira et al. (2019)
Verbenaceae					
Aloysia gratissima (Gillies & Hook.) Tronc.	Promastigote	*L. amazonensis*	25 µg/mL	–	Garcia et al. (2018)
Aloysia gratissima (Gillies & Hook.) Tronc.	Amastigote	*L. amazonensis*	0.16 µg/mL	–	Garcia et al. (2018)

Species	Form	Leishmania	Concentration	Major compound	Reference
Lantana camara L.	Promastigote	*L. braziliensis*	72.31 μg/mL	(E)-caryophyllene (23.75%)	Barros et al. (2016)
Lantana camara L.	Promastigote	*L. infantum*	0.25 μg/mL para *L. amazonensis*	Germacrene D (24.90%)	Machado et al. (2012b)
Lantana camara L.	Promastigote	*L. amazonensis*	18 μg/mL para *L. infantum*	Germacrene D (24.90%)	Machado et al. (2012b)
Lippia berlandieri Schauer	Promastigote	*L. mexicana*	59 μg/mL	Thymol (58.3%); p-Cymene (24.6%)	Andrade-Ochoa et al. (2021)
Lippia gracilis Schauer	Promastigote	*L. infantum*	86.32 μg/mL	Thymol (61.84%)	Melo et al. (2013)
Lippia sidoides Cham.	Amastigote	*L. amazonensis*	34.4 μg/mL	Thymol (78.37%)	Medeiros et al. (2011)
Lippia sidoides Cham.	Promastigote	*L. amazonensis*	44.3 μg/mL	Thymol (78.37%)	Medeiros et al. (2011)
Lippia sidoides Cham.	Promastigote	*L. infantum*	54.8 μg/mL	Carvacrol (43.7%)	Farias-Junior et al. (2012)
Lippia origanoides Kunth	Promastigote	*L. brasiliensis*	0.39 μg/mL	–	Neira et al. (2018)
Zingiberaceae					
Alpinia speciosa K. Schum	Promastigote	*L. brasiliensis*	67.18 μg/mL	1,8-cineole (28.46%)	Pereira et al. (2018)
Curcuma longa L.	Amastigote	*L. amazonensis*	63.3 μg/mL	Turmerone (55.43%)	Teles et al. (2019)
Zingiber zerumbet (L.) Sm.	Promastigote	*L. mexicana*	3.34 nl/mL	Zerumbone (60.3%)	Le et al. (2017)
Amomum aromaticum Roxb.	Promastigote	*L. mexicana*	9.25 nl/mL	Eucalyptol (55.2%)	Le et al. (2017)
Zygophyllaceae					
Bulnesia sarmientoi Lorentz ex Griseb.	Promastigote	*L. amazonensis*	85.56 μg/mL	Guaiol (48.29%)	Andrade et al. (2016)

Seventeen articles (14.9%) performed tests to verify the possible mechanism of action of the EOs, while 97 (85.1%) did not. In this sense, among the studies that investigated the mechanism of action, the ones by Demarchi et al. (2015) and Demarchi et al. (2016) with the EO of *Tetradenia riparia* stood out with the best result in terms of IC_{50} (0.03 µg/ml). The leishmanicidal potential of *T. riparia* EO against *L. amazonensis* was explained by the oil's ability to modify the ultrastructure of promastigotes, suggesting an autophagic process with chromatin condensation; presence of blebbings and nuclear fragmentation; decreased macrophage infection rate by amastigotes; and, finally, inhibition of granulocyte and macrophage colony-stimulating factor, interleukin-4 (IL-4), IL-10 and tumor necrosis factor. Other EOs are also noteworthy for their ability to inhibit parasites at low concentrations, such as those from *Origanum compactum* (IC_{50} = 0.02 µg/mL), *Ferula communis* (IC_{50} = 0.05 µg/mL), and *Teucrium polium* (IC_{50} = 0.09 µg/mL) against *L. infantum* isolates.

13.4 Discussion

13.4.1 Essential Oils Against Leishmania spp.

The genus *Leishmania* is a group of flagellated parasites comprising more than 20 different species distributed in the subgenus *Leishmania* or *Viannia*, whose main vectors are phlebotomine sandflies of the genus *Lutzomyia* and *Phlebotomus* (Espinosa et al. 2018). Members of the genus *Leishmania* differentiate from proliferative promastigotes in the insect vector gut into infective metacyclic promastigotes in the foregut of the insect. The parasites are inoculated by the vector as flagellated promastigotes into the mammalian host, where they infect macrophages, differentiating into amastigote forms (Rocha et al. 2005).

Leishmania parasites can be divided according to their clinical forms and manifestations, geographic distribution, and reservoir. *Leishmania (L.) amazonensis*, *L. mexicana L, L. (L.) tropica*, and *L. (V.) guyanensis* are more prevalent in South America and are characterized by causing multiple or individual ulcerative lesions, a condition called cutaneous leishmaniasis. In addition to *L. (V.) braziliensis*, which can cause mucocutaneous changes, *L. infantum* and *L. donovani* cause the most serious conditions called visceral leishmaniasis, which include, but are not limited to persistent fever, splenomegaly, and weight loss (Burza et al. 2018).

The main anti-*Leishmania* therapeutic methods involve the use of pentavalent antimonials, amphotericin B, paromomycin, pentamidine, and miltefosine; however, there is great resistance to treatment adherence due to their high toxicity and side effects, in addition to the financial impact in more poor regions (Roatt et al. 2020). There is also concern about the development of resistant strains and variable response to treatment depending on the parasite species. In Brazil, strains of *L. infantum* resistant to miltefosine have been isolated in patients whose treatment was unsuccessful. According to Roatt et al. (2020), this finding suggests a natural resistance to this drug because ince it had not yet been used in the country (Carnielli et al. 2019).

The exploration of the plant kingdom is one of the only options for the development of therapeutic agents with high safety and cost-benefit profile for various health problems, as highlighted by Bekhit et al. (2018). The investigation of new compounds that can be used in the treatment of leishmaniasis begins with ethnobotanical studies, which provide information about the medicinal properties of various plant species based on the knowledge disseminated in traditional communities.

Ethnobotanical studies and the investigation of the therapeutic potential of plants make it possible to track new bioactive molecules with the potential to become new drugs in the future. Passero et al. (2021) list 216 species distributed in 76 genera that present contributions to the experimental treatment of leishmaniasis, opening a wide range of options for investigations in the field. A review published by Rocha et al. (2005) found about 239 chemically defined natural molecules reported in the literature which were evaluated for anti-*Leishmania* activity, including alkaloids, terpenes, various lactones, flavonoids, diterpenes, steroids, lipids, carbohydrates, proteins, coumarins, phenylpropanoids, and depsides. Recently, a review published by Fampa et al. (2021) highlighted about 30 volatile compounds that were also evaluated for their anti-*Leishmania* activity.

13.4.2   Terpenes

According to the data obtained, analyses show that, among the different compounds that constitute EOs, terpenes are the most abundant, present both as sesquiterpenes and monoterpenes. The anti-*Leishmania* activity of compounds present in EOs can be attributed to their lipophilic character. Several studies indicate that these substances act by breaking the microbial cytoplasmic membrane, making it permeable, affecting polarization and compromising biological barriers and the enzyme matrix (Cristani et al. 2007).

The EO of *Myrciaria plinioides* leaves was effective against *L. amazonensis* promastigotes and presented an IC_{50} value of 14.16 ± 7.40 μg/mL; however, the activity against *L. infantum* promastigotes was less pronounced, with an IC_{50} value of 101.50 ± 5.78 μg/ml (Kauffmann et al. 2019). The anti-*Leishmania* activity was attributed to the presence of the sesquiterpenes spathulenol (**1**) and caryophyllene oxide (**2**), which represent 36.32% of the total components that can cause alterations in the mitochondrial membrane potential, in addition to modification of the redox index, inhibition of cellular isoprenoid biosynthesis, and changes in the plasma membrane (Santos et al. 2008; Rodrigues et al. 2013; Monzote et al. 2014c).

The EO of *Lantana camara* was able to cause 100% inhibition of proliferation of *L. amazonensis* at concentrations above 3 μg/mL, and about 90% inhibition in *L. chagasi* at the concentration of 250 μg/mL (Machado et al. 2012b). The presence of germacrene-D (**3**) in the composition of the EO was considered to be responsible for the inhibitory effect on the growth of promastigote cultures. This hypothesis is based on the activity of amphotericin B, which is able to act as an antifungal and antiparasitic agent, as suggested by tests in germacrene-D (**3**). It is noteworthy that Biavatti et al. (2001) observed a toxic effect of the EO in tests using brine shrimp and mammalian cells *in vitro*. However, the authors mentioned that this effect was not related to the presence of germacrene-D (**3**), as it did not present a toxic effect in the same models.

Another terpene with anti-*Leishmania* activity widely cited in the literature is pinene (**4,10**). More than 40 components were found through gas chromatography analysis in the EO of propolis, with 36.17% of α-pinene (**4**). In *in vitro* tests, pinene (**4,10**) was effective against the promastigotes and amastigotes of *L. major* and *L. infantum*, with IC_{50} of 5.29 μg/mL and 3.67 μg/mL for promastigotes, and 7.38 μg/mL and 4.96 μg/ml for amastigotes of *L. major* and *L. infantum*, respectively. Furthermore, the EO exhibited synergistic activity with amphotericin B, inhibiting the growth of *Leishmania* by more than 98%. Although the activity was attributed to its major compound, the authors did not rule out a synergy of pinene (**4,10**) with the less expressive components present in the EO (Jihene et al. 2020).

In tests performed by Dias et al. (2013), the EO of *Syzygium cumini* showed good activity against the promastigote forms of *L. amazonensis*. At all concentrations and time points analyzed, significantly higher mortality was observed in the treatment than in the control groups, leading to the conclusion that *S. cumini* EO has leishmanicidal rather than leishmanistatic activity. The greatest efficacy was seen within 24 hours of exposure, with an IC_{50} of 36 mg/L. Although the author did not perform specific tests to determine the mechanisms of action through which the EO acts, leishmanicidal activity was attributed to the lipophilic characteristic of the EO, mentioned above. With 31.85% of α-pinene (**4**) in its composition, its action can be compared to that of the EO of *Cinnamodendron dinisii*, which has 35.41% α-pinene (**4**) (Andrade et al. 2016). Although *C. dinisii* EO has a higher concentration of pinene (**4,10**) in its composition, its activity was lower than that of *S. cumini* EO. It is possible that the minor compounds in these species interfere in the action of pinene (**4,10**).

Bouyahya et al. (2019) tested the EO of leaves and fruits of *Pistacia lentiscus*, obtaining an IC_{50} of 11.28 and 8 μg/mL, respectively, against *L. infantum*, 17.52 and 21.42 μg/mL against *L. major*, and 23.5 and 26.2 μg/mL against *L. tropica*. Both EOs presented better results than the standard drug glucantime and although they were obtained from *P. lentiscus*, both presented major compounds at different concentrations. In the EO of the leaves, was in higher concentration, 33.46%, while α-pinene (**4**) represented only 19.20%. The EO of the fruits presented 20.46% of α-pinene (**4**), and the second compound with the highest concentration was limonene (**5**), corresponding to 18.26%. This shows that the composition of EO can change according to the part of the plant from which it is extracted.

It is known that besides varying according to the part of the plant from which it is extracted, the composition of the EO can be altered by environmental factors such as climate, time of collection, and geographic location (Do Carmo et al. 2012; Essid et al. 2015; Bouyahya et al. 2019). Variability is also present in plants of same genus but different species. This is the case of *Artimisia* plants studied by Mathlouthi et al. (2018). In their tests, they showed a remarkable anti-*Leishmania* activity, with an IC_{50} of 2.20 μg/mL and 1.20 μg/mL for *Artemisia campestres* and *Artemisia herba-alba*, respectively, both against the promastigote forms of *L. major*. *Artemisia herba-alba* had β-thujone (**8**) (29.4%) and 1,8-cineole (**9**) (14.8%), with only a small fraction of β-pinene (**10**) (2.3%), while *A. campestres* had β-pinene (**10**) (32%) and limonene (**5**) (17.3%), but β-thujone (**8**) was absent.

Although many EOs have shown better results than the isolated compounds, several factors may be involved in these processes. In a study by Do Carmo et al. (2012), the EO of *Piper duckei* showed a lower result than its major compound, trans-caryophyllene, against *L. amazonensis* promastigotes. The IC_{50} was 46 μg/mL for the EO, and 96 μg/mL for the isolated compound. The authors reported that, during the experiments, it was possible to observe that the purity of trans-caryophyllene is an important factor for the activity against *L. amazonensis*. The oxidation of trans-caryophyllene to its corresponding oxides affects the results; depending on the level of oxidation, activity may not be observed. Another plant of the *Piper* genus, *Piper cernuum* Vell, also had caryophyllene (**11**) in its composition (16%). In *in vitro* tests with macrophages infected with *L. amazonensis*, the isolated compound reached greater efficiency in reducing parasite infection in macrophages at concentrations of 2 and 10 μg/mL, leading to infection rates of 105 ± 16 and 101 ± 7, respectively, both lower than values obtained with amphotericin B (34 ± 5 at 0.1 μg/mL), but superior to those obtained with the EO (131 ± 15 at 2 μg/mL and 115 ± 13 at 10 μg/mL). According to Capello et al. (2015), the effect of the EO may be associated with bioactive sesquiterpenes present in its composition.

In a research carried out by Essid et al. (2015), compounds of the EOs extracted from *F. communis*, *T. polium*, and *Pelargonium graveolens* exhibited strong inhibitory activity against the growth of promastigote forms of *L. major* and *L. infantum*, with IC_{50} values <1 μg/mL. Their main constituents were β-caryophyllene (**11**), carvacrol (**12**) and citronellol (**13**) respectively. In tests with the isolated compounds, β-caryophyllene (**11**) was the most active, with an IC_{50} of 1.06 ± 0.37 μg/mL for *L. infantum* and 1.33 ± 0.52 μg/mL for *L. major*. Carvacrol (**12**) had an IC_{50} of 7.35 ± 1.78 g/mL for *L. infantum* and 9.15 ± 0.12 g/mL for *L. major*. Very low activity was recorded for citronellol (**13**). It is interesting to note that the isolated compounds showed lower activity than the EO.

According to Carvalho et al. (2017), EOs are more effective than their individual chemical constituents. Their bioactivity depends on the additive and synergistic action of the components. The EO of *Cymbopogon citratus* and its major constituents citral (**14**) (neral (**15**) 40% + geranial (**16**) 60%) and myrcene (**6,7**) were tested against *L. infantum* by Machado et al. (2012a), resulting in IC_{50} values of 25 μg/mL for the EO, 42 μg/mL for citral (**14**), and 164 μg/mL for myrcene (**6,7**), thus showing the best result for the EO. In a work carried out by Moreira et al. (2017), the EO of *Vernonia polyanthes* Less presented an IC_{50} of 19.4 μg/mL against *L. infantum*, lower than the IC_{50} of zerumbone (**17**) (9 μg/mL), a monoterpene present in the EO.

On the other hand, in the work by Leal et al. (2013), the EOs of *Piper brachypodom* and *Piper var. brachypodom* presented trans-ß-caryophyllene (**11**) as the major component (20.2%). The results showed that the EOs were more active against *L. infantum* promastigotes (IC_{50} 23.43 and 23.68 µg/mL, respectively). However, none of these EOs was active against the intracellular forms of this protozoan. Trans-ß-caryophyllene (**11**) had an IC_{50} of 24.02 µg/mL against *L. infantum* promastigotes, a result slightly lower than that obtained for the EOs, but it was active against amastigote forms, with an IC_{50} of 53.39 µg/mL. The author stated that it is much more difficult for components to reach intracellular forms because they need to penetrate barriers and reach the place where the parasite is alive, as opposed to free forms in whose case the product can act directly on the parasite. Considering the similarity of the results, it is possible to say that the action of *Piper* EOs is due to its major constituent, and that the constituents with lower expression possibly acted negatively, preventing the action of the EOs in the intracellular forms of *L. infantum*.

According to Cristani et al. (2007), the activity of monoterpernes such as carvacrol (**12**) and thymol (**18**) results from the disturbance of the lipid fraction of the plasma membrane of microorganisms, as bacteria. Other studies point to the same type of interaction in parasites and claim that terpenes are responsible for the hydrophobic characteristic of EOs, allowing their diffusion across the cell membrane of parasites such as *Leishmania* and affecting intracellular metabolic pathways and organelles (Andrade et al. 2016).

In a work carried out by de Medeiros et al. (2011), the incubation of *L. amazonensis* promastigotes with *Lippia sidoides* EO and its main constituent thymol (**18**) efficiently inhibited the growth of the parasite. IC_{50}/48 h values were 44.38 and 19.47 µg/mL for EO and thymol (**18**), respectively. The treatment of intracellular amastigotes with the EO at concentrations of 25, 50 and 100 µg/mL caused a significant decrease in the survival rate of the parasites, with an IC_{50} value of 34.4 µg/mL. The authors also pointed out that, while thymol (**18**) had low selectivity against promastigotes and showed toxicity to mammalian macrophages, the EO showed low toxicity to mammalian cells, a fact attributed to the protective effect of other constituents.

Study conducted by Farias-Junior et al. (2012) brought the first analysis of the anti-*Leishmania* properties of *L. sidoides* EO, in which carvacrol (**12**) instead of thymol (**18**), was the main constituent. It was demonstrated that the carvacrol-rich (**12**) EO had an IC_{50} lower than that of the EO whose main constituent was thymol (**18**) against *L. chagasi* promastigotes. Although it is logical to attribute such activity to carvacrol (**12**), the EO also had 6% of thymol (**18**), and thus there is a possibility of a synergistic effect between thymol (**18**) and carvacrol (**12**) to explain the greater anti-*Leishmania* effect observed in this EO.

Essid et al. (2015) suggest that the inhibitory activity of carvacrol (**12**) is enhanced in the presence of its isomer thymol (**18**) and its precursors γ-terpene and *p*-cymene (**19**), as demonstrated by Lambert et al. (2001). In their studies, the EOs of *F. communis*, *T. polium*, and *P. graveolens* reduced by more than 90% the number of parasites in a dose-dependent manner, in the case of *L. infantum* and *L. major*, presenting anti-*Leishmania* activity greater than amphotericin B. The authors highlight that the mechanism of action of the EOs may involve changes in the mitochondrial membrane.

The relationship between carvacrol (12) and p-cymene (19) was also suggested by Bouyahya et al. (2017a). In their study, the EO of *O. compactum* extracted from different plant phases (vegetative, flowering and post flowering) showed effective action against three *Leishmania* species in a dose-dependent manner, being the EO obtained in the flowering phase the most active against the three parasites tested. The author also speculated that the involved mechanisms of action may include induction of apoptosis, disruption of the electron transport chain, and inhibition of DNA topoisomerase (Castro et al. 1992).

Monzote et al. (2011) brought another perspective to the action of carvacrol (12). Treatment of *L. amazonensis*-infected murine macrophages with the EO of *Chenopodium ambrosioides* L. proved to inhibit parasite growth. The authors attribute this activity to ascaridol (20) and also mention that the toxicity exhibited by the sample could have been caused by the different compounds present in the EO or by the interaction between them. This hypothesis was formulated from the study of Monzote et al. (2009) that showed that ascaridol (20) forms a highly reactive carbon-centered free radical. The authors suggested that, through its phenolic hydroxyl group, carvacrol (12) serves to attenuate the cytotoxic activity of ascaridol (20) by eliminating the free radical (Dapkevicius et al. 2002; Guimarães et al. 2010).

The EOs of *Lippia gracilis* Schauer genotypes 106 and 110 were analyzed and tested against *L. chagasi* promastigotes, resulting in IC_{50} values of 86.32 μg/mL^{-1} and 77.26 μg/mL^{-1}, respectively (de Melo et al. 2013). The authors also showed that thymol (18) and carvacrol (12), the main compounds of the EOs, which also had exhibitory activity, and the latter (IC_{50} of 2.3 μg/mL^{-1}) had similar performance to amphotericin B (0.51 μg/mL^{-1}).

Both compounds were also found in the EO of *Z. multiflora*, which showed a significant anti-*Leishmania* effect on the promastigote forms of *L. tropica*. Furthermore, it was shown that the promastigote forms of *L. tropica* without treatment were able to infect 84.1% of macrophages, while promastigotes treated with *Z. multiflora* EO had potency to infect only 11.3% (Dezaki et al. 2016).

Thymoquinone (21), the major compound (43.4%) of the EO of *Nigella sativa* L. (Ranunculaceae), showed an inhibitory capacity for parasitic growth of *L. tropica* promastigotes, with IC_{50}/72 h of 1.16 mg/mL, and *L. infantum*, with IC_{50}/72 h of 1.47 mg/mL, while the EO presented IC_{50}/72 h values of 9.3 mg/mL for *L. tropica* and 11.7 mg/mL for *L. infantum* (Mahmoudvand et al. 2015a). An assay was also carried out to evaluate the inhibition of the infection in macrophages: the promastigotes of *L. tropica* were able to infect only 13 and 27.3%, and those of *L. infantum* infected only 16.3 and 33.6% of the murine macrophages when treated with thymoquinone (21) and the EO of *N. sativa*, respectively. However, despite the results showing the high anti-*Leishmania* potential of thymoquinone (21), this coumpound was more cytotoxic compared to EO (Mahmoudvand et al. 2015a).

Forty-four compounds were detected through GC–MS in the EO of *Pluchea carolinensis*; selin-11-en-4α-ol (22) (51%) was the major compound (García et al. 2017). In this study, *in vitro* assays for antiparasitic evaluation of the EO showed the ability to inhibit 100% of the growth of promastigote and amastigote forms of *L. amazonensis* at concentrations of 100 and 200 µg/mL, with a lower IC_{50} on amastigote (6.2 ± 0.1 µg/ mL) than promastigote (24.7 ± 7.1 µg/mL) forms. In *in vivo* models of cutaneous leishmaniasis in BALB/c mice, no mortality or weight loss was observed in the treated groups. The administration of the EO of *P. carolinensis* demonstrated to control the size of the lesions and parasite load of animals infected with *L. amazonensis*. The authors of the work suggest that the results found *in vitro* and *in vivo* on the anti-*Leishmania* effect of EO may be due to the major compound selin-11-en-4α-ol (22), but indicate the need to reiterate analyses with the isolated compound to elucidate its mechanism of action.

Because the intracellular forms of *Leishmania* species complete part of their cell cycle inside macrophages, it is important to establish the selectivity index (SI) of the EO and its components (Moreira et al. 2019). More toxic compounds must be more selective for protozoa than host cells. SI values greater than 1 are considered more selective for activity against parasites, and values lower than 1 are considered more selective for activity against cells.

In their studies, Moreira et al. (2019) established the SI ratio for the EO of *Casearia sylvestris* SW. and its major compound (22.2%) *E*-caryophyllene (23), with values of 2.9 and 5.8, respectively. This was an interesting result, as both were moderately toxic against BALB/c mouse macrophages. The EO presented an IC_{50} of 29.8 µg/mL on *L. amazonensis* promastigotes, better than the result for *E*-caryophyllene (23) (49.9 µg/mL). On amastigote forms, *E*-caryophyllene (23) had a better result (10.7 µg/mL) than the EO (14 µg/mL) (Fig. 13.2).

13.4.3   Mechanisms of Action

13.4.3.1   Morphological Changes

Chemical analyses revealed 97.9% of α-bisabolol (24) in the constitution of the EO of *Vanillosmopsis arborea* (Colares et al. 2013). The compound and the EO showed efficiency in inhibiting the growth of *L. amazonensis* promastigotes with $IC_{50}/24$ h of 4.95 *µg*/mL and 7.35 *µg*/mL, respectively. The parasites showed alterations such as severe cell damage with loss of morphology, discontinuity of the nuclear membrane, increased mitochondrial volume and kinetoplast, and presence of vesicles with an electrondense display with lipid inclusion in the plasma membrane. In addition, the SI, especially for intracellular amastigotes, showed that the compound (9383) was less toxic than the EO (11,526) (Colares et al. 2013). The apoptotic mechanism can be seen in Fig. 13.3.

The above results corroborate the findings of Hajaji et al. (2018), in which α-bisabolol (22) isolated from the EO of *Matricaria recutita* L. showed SI values of 5.5 and 6.7 for *L. amazonensis* and *L. infantum* amastigotes, respectively, and IC_{50} of 16.0 ± 1.2 and 9.5 ± 0.1 µg/mL on *L. amazonensis* and *L. infantum* promastigotes, respectively. The researchers demonstrated the ability of the compound to

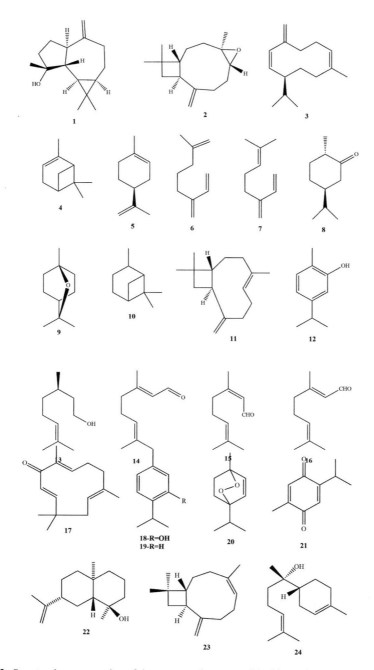

Fig. 13.2 Structural representation of the compounds presented in this section

Fig 13.2 (continued)

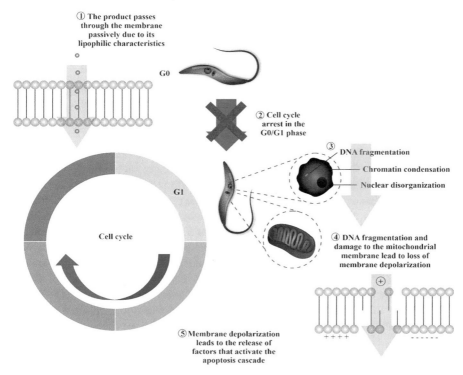

Fig. 13.3 The apoptotic mechanism. (1) The products pass through the parasite membrane passively, due to their lipophilic characteristics; (2) Then, several authors have observed that volatile compounds act by inhibiting the cell cycle in the G0/G1 phase; (3) Phenotypic alterations, such as DNA fragmentation, chromatin condensation and nuclear disorganization, have also been reported, whose images can be found in the studies mentioned in this section; (4) Depolarization of the mitochondrial membrane, which plays a crucial role as a therapeutic target in protists such as *Leishmania*, is the main mechanism promoted by these compounds. (5) Together, these pathways share characteristics responsible for the release of factors that activate the apoptosis cascade, which despite being a programmed process, is the main form of cell death induced by chemical agents

affect plasma membrane permeability without causing necrotic effects, and to activate a programmed cell death process by cellular enhancement of phosphatidylserine externalization and membrane damage, with an apoptosis percentage of 21.66 (IC_{50}) and 40% (IC_{90}) for *L. amazonensis* and 17 (IC_{50}) and 20% (IC_{90}) for *L. infantum* after 24 h of treatment.

The EO of *Cryptocarya aschersoniana* was rich in limonene (**5**) (42%) and had remarkable activity against *L. amazonensis* promastigotes (IC_{50} = 4.46 µg/mL) in the study by Andrade et al. (2018b). However, it was highly toxic to mouse macrophages, with a CC_{50} of 7.71 µg/mL. According to the authors, compounds with CC_{50} below 10 µg/mL are highly toxic, above 10 and below 100 µg/mL are moderately toxic, and above 100 and below 1000 µg/mL are non-toxic. This type of classification allows evaluating the cytotoxicity of a compound and understanding the mechanisms of action of different substances in their interactions with tissues. The authors recognized that, as this was an *in vitro* test, it did not replicate the actual architecture of the living tissue in which the underlying cells could repair the damage suffered (Andrade et al. 2018b).

The EO from *Vernonia brasiliana* (L.) Druce was rich in terpenes, with the major component being β-caryophyllene (11) (Mondêgo-Oliveira et al. 2021). The EO showed activity against *L. infantum* promastigotes, with IC_{50} of 39.01 μg/mL and SI of 1.61, being more toxic to parasites than to DH82 cells. Although the IC_{50} of the standard drug miltefosine was higher (2.54 μg/mL), it was more toxic to DH82 cells, with an SI of 0.55. When tested in combined therapy, there was an antagonistic effect. According to the author, this shows that although both products are bioactive against *Leishmania*, this does not mean that the products will act synergistically. The mechanisms of action of *V. brasiliana* EO were tested and, after 72 hours in contact with *L. infantum* promastigotes at IC_{50} of 39.01 μg/mL, important structural changes were observed, with decreased mitochondrial membrane potential and increased reactive species of oxygen (ROS) production, inducing a late apoptosis.

Although little research has been carried out to identify the mechanisms of action by which EOs and their constituents act, a general analysis of the findings suggests disturbances in the plasma membrane of *Leishmania* causing significant morphological alterations that can induce apoptosis. In the work by Machado et al. (2012a), *C. citratus* EO induced the death of *L. infantum* promastigotes in which depolarization of the mitochondrial potential was observed, involving cell-cycle arrest at the G0/G1 phase and nuclear disorganization, with chromatin condensation. In a study by Aloui et al. (2016), *A. campestres* presented β-pinene (10) (32.95%) and was active against *L. infantum* promastigotes (IC_{50} = 44 μg/mL). Furthermore, the EO increased the proportion of cells in the subG0/G1 phase, indicating DNA degradation in promastigotes, suggesting alterations of the apoptotic type.

The EO of *Myrcia ovata* caused growth inhibition of *L. amazonensis*, with a considerable difference at 20 and 30 mg/mL compared to the untreated control (Amorim Gomes et al. 2020). Both concentrations caused 100% inhibition with IC_{50}/96 h of 8.69 mg/mL. The authors observed that after incubation for 3 days with 10 mg/mL of EO, the parasites showed accumulation of lipid bodies, nucleolus disorganization, and the appearance of structures suggestive of autophagosome; and after 4 days of treatment with 5 mg/mL, the parasites showed mitochondrial enlargement (Amorim Gomes et al. 2020). This effect was attributed to the main constituents of the EO geranial (16) and neral (15). The effect of citral (14), which is a mixture of geranial (16) and neral (15) isomers, already tested on *L. amazonensis*, caused ultrastructural changes that included mitochondrial damage and presence of two or more flagella in the parasites, among other effects (Santin et al. 2009).

Neral (15) (cis-citral) and geranial (16) (trans-citral) together represented about 81% of the EO of *C. citratus*, which was able to kill 65% of *L. infantum* and *L. major* promastigotes and 80% of *L. tropica* promastigotes at a concentration of 50 μg/ml (Machado et al. 2012a). In turn, at the same concentration, citral (14) killed about 45% of *L. infantum* and *L. tropica* promastigotes, and about 60% of *L. major* promastigotes. Furthermore, none of them showed cytotoxicity in bovine aortic endothelial cells and macrophage lineage in the MTT test (Machado et al. 2012a).

The investigations of Sen et al. (2010) showed that promastigotes treated with EO and citral (14) showed prominent ultrastructural effects such as the appearance of aberrant-shaped cells with cell body septation, cytoplasmic disorganization, increased cytoplasmic clearance and loss of intracellular content, presence of autophagosomal structures, characterized by intense cytoplasmic vacuolization, in addition to irregular surface with blebs formation and rupture of the membrane. Another factor highlighted is the presence of membrane vesicles in the flagellar pocket, characteristic of an exocytosis process, and it is possible that they resulted from the secretion of abnormal lipids, which accumulate as a consequence of the effect of citral (14). *Cymbopogon citratus* EO and citral (14) further promoted sustained mitochondrial membrane depolarization, which is a typical feature of metazoan apoptosis and has been observed to play a key role in drug-induced death in protists such as *Leishmania*. The authors also noted the presence of myelin-like figures as multilamellar bodies, where the nuclear chromatin was organized similarly to the nucleus of apoptotic cells, with disruption of the nuclear membrane. The authors' main hypothesis is that EO and citral (14) may have a passive entry and accumulate in the cell membranes of the parasite, leading to an increase in membrane permeability and formation of structures known as autophagosomes (Rodrigues et al. 2002) that are probably involved in an intense process of remodeling of intracellular organelles irreversibly damaged by the EO and citral (14).

Islamuddin et al. (2014a) showed that camphor (25) (52.06%) was the major component in the chemical composition of the EO of *Artemisia annua* leaves. The EO exhibited IC_{50} of 14.63 ± 1.49 µg/mL and 7.3 ± 1.85 µg/mL against *L. donovani* promastigotes and amastigotes, respectively. In their evaluations, the authors reported changes in cell morphology, shrinkage in promastigotes that became round in shape, with ruptured flagella and no motility. The apoptosis mechanism was also recognized by the externalization of phosphatidylserine in the cell membrane, evidenced by increased annexin V binding. The authors also observed DNA fragmentation in apoptotic cells, showing an increased proportion of cells in the subG0/G1 phase when treated with *A. annua* EO. Also at the intracellular level, treatment with EO was able to cause depolarization of the parasite's mitochondrial membrane, leading to permeabilization of the inner mitochondrial membrane and consequent release of apoptotic factors.

Monzote et al. (2014b) demonstrated that the EO of *Bixa orellana* presented activity against the intracellular amastigote form of *L. amazonensis*, with IC_{50} of 8.1 µg/mL and SI of 7, and cytotoxic concentration sevenfold higher for the host cells than for the parasites. The EO also showed the ability to control the progression of established cutaneous leishmaniasis in BALB/c mice, with significant differences in lesion size and parasite load between animals treated with EO compared to controls, with no deaths observed after 14 days of application intraperitoneal of the EO. According to the authors, the geranylgeraniol (26) present in the composition of the EO (9.1%) may be associated with such activity, since it has been reported that this compound promotes alterations in the mitochondrial structure, including swelling and formation of circular cristae (Vannier-Santos and Castro 2009). In addition, the compound has also been observed to cause kinetoplast DNA disorganization (Vannier-Santos and Castro 2009) as well as increased superoxide anion production, leading to apoptosis (Lopes et al. 2012).

In the study of the antiparasitic action of the EO of *Lavandula luisieri*, Machado et al. (2019) observed an effect on cell viability in promastigotes of *L. infantum*, with $IC_{50}/24$ h equal to 63 µg/mL, *L. tropica*, with $IC_{50}/24$ h equal to 38 µg/mL, and *L. major*, with $IC_{50}/48$ h equal to 31 µg/mL. In the MTT test, no toxicity was observed at the doses tested ($CC_{50} > 200$ µg/mL; SI > 3.17). The authors suggest that the action of the EO is linked to oxygenated monoterpenes (75.7%) in its chemical composition, and necrodane derivatives as major compounds (36%). The effects of the EO were verified from image analysis in Scanning Electron Microscope (SEM) and Transmission Electron Microscopy (TEM), in which round and aberrant shapes, cell body septation, disorganization of cytoplasmic organelles, and many autophagosomal structures featured by intense cytoplasmic vacuolization were observed in *L. infantum* promastigotes. The EO was able to induce mitochondria swelling and mitochondrial membrane disorganization indicated by the presence of complex invaginations and formation of concentric membranous structures. These data can be explained by the ability to induce depolarization of the mitochondrial potential, which can promote apoptosis (Arnoult et al. 2002). The arrest of cells in the G0/G1 phase was also detected, with a reduction in the number of cells in the S and G2/M phases; the authors suggested that this may have occurred due to an decrease in mitochondrial membrane potential and since this reduces the energy available.

The analysis of the EO of *Eremanthus erythropappus* conducted by Amorim Gomes et al. (2020) revealed the presence of 13 constituents, corresponding to 94.22% of its composition, with 85.98% of α-bisabolol (**22**). The authors verified a percentage of inhibition of *L. amazonensis* promastigotes of 35% under concentrations of 5 and 10 mg/mL of *E. erythropappus* EO, and almost 100% inhibition using concentrations higher than 20 and 30 mg/mL after 96 h of treatment, with $IC_{50}/96$ h of 9.53 mg/ml. The ultrastructural analysis showed that after 3 days of incubation with 10 mg/mL of EO, the parasites showed accumulation of lipid bodies, demonstrating a possible mechanism of action of the compound.

De Medeiros et al. (2011) also pointed out that the treatment with *L. sidoides* EO induced remarkable changes in the morphology of the parasites, particularly the accumulation of large lipid droplets in the vicinity of the plasma membrane. At high EO concentrations, membrane disruption, increased lipid electron density, and loss of cytoplasmic content, alterations compatible with loss of cell viability and cell death by necrosis (Menna-Barreto et al. 2009), were also observed. Furthermore, characteristics such as parasite swelling, presence of wrinkled or ruptured membranes, and loss of cytoplasmic material in promastigotes were present, supporting the deleterious effects of EO on the plasma membrane so widely disseminated in the literature. The hypothesis of the authors is that the constituents of the EO penetrate into the cell and impair the ergosterol biosynthesis pathway, and they may also react directly with the membrane through their reactive hydroxyl portion. Thus, the extensive membrane damage may be due to a combined effect of the two events (Nafiah et al. 2011).

Subsequently, Monzote et al. (2014a) demonstrated that NADH- and succinate-dependent reduction of cytochrome-C was inhibited in mitochondrial fractions of *L. amazonensis* and liver mitochondria from BALB/c mice in the presence of *C. ambrosioides* EO and its pure major compounds, carvacrol (**12**) and thymol (**18**).

Their findings suggested that such reduction was not specifically sensitive to EO in *Leishmania* mitochondria, however, the existence of other more sensitive and more selective targets, such as mitochondrial membrane potential, was not ruled out. The authors could not establish whether the loss of mitochondrial membrane potential was a primary effect of EO (directly influencing mitochondrial functions) or arose subsequent to other cellular effects triggering apoptosis via mitochondria. Furthermore, they suggested that other parasite damages caused by EO such as free radical-triggered DNA or protein-alterations, or parasite-specific transporters such as the P2 amino-purine transporter (De Koning 2001), DNA triggered by free radicals or protein alterations, or parasite-specific transporters, such as the P2 amino-purine transporter (De Koning 2001), could contribute to specific killing of *Leishmania*.

Tasdemir et al. (2019) performed tests with both carvacrol (**12**) and thymol (**18**), the main constituents of the EO of *Origanum onites*, reporting for the first time their effect on *L. donovani* amastigotes. The authors suggested that the EO permeates the cell membrane and kills parasites by affecting the cytoplasmic metabolic pathways or organelles, and not by compromising the integrity of the parasite's membrane, as presented by several studies in this section. They reached this conclusion based on a flow cytometry study performed by Santoro et al. (2007) and also highlighted the importance of the presence of the hydroxyl group in the bioactivity of phenolic compounds such as carvacrol (**12**) and thymol (**18**) (Dorman and Deans 2000; Ultee et al. 2002).

13.4.3.2 Immunological Changes

The evaluation of the EO of *Pseudotrachydium kotschyi* revealed the presence of Z-α-trans-bergamotol (**27**) (23.25%), durylaldehyde (**28**) (16.07%), and α-bergamotene (**29**) (10.48%) (Ashrafi et al. 2020a). It was observed that the EO had anti-*Leishmania* potential at a concentration of 5000 µg/mL and suggested that these compounds are involved in the biological activities of the oil, for it was observed that EO was able to protect macrophages against infection by promastigotes. Their data indicated that EO exerts anti-*Leishmania* activity by affecting the levels of TNF-α and TGF-β1 in macrophages. These cytokines were determined in *Leishmania*-infected macrophages after treatment with EO. The immunological mechanism can be seen in the Fig. 13.4.

The EO of *Artemisia absinthium* inhibited the *in vitro* growth of *L. amazonensis* promastigotes and amastigotes, with IC_{50} of 14.4 ± 3.6 µg/mL and 13.4 ± 2.4 µg/mL, respectively (Monzote et al. 2014c). The activity *in vivo* was evaluated in a

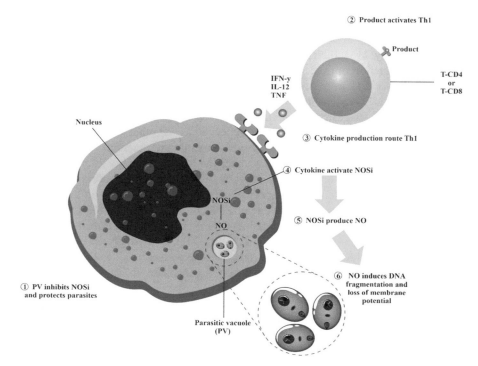

Fig. 13.4 Modulation of the immune response by essential oils and their compounds. (1) Parasitophorous vacuoles (PV) contain a plasma membrane and may represent a specific adaptation to minimize the toxic effects of reactive nitrogen intermediates generated by the host cell. (2) The modulation of the response occurs in cells with a T-helper type 1 (Th1) cytokine profile. (3) This profile is associated with the production of cytokines such as IFN-γ, IL-12 and TNF. (4) These cytokines lead to the activation of anti-*Leishmania* activities, mainly through the activation of the inducible nitric oxide synthase (iNOS) enzyme. (5) The production of nitric oxide (NO) induces an oxidative explosion in infected cells. (6) This oxidative explosion is associated with the process of loss of the parasite's mitochondrial membrane potential, however, it can also cause extensive nuclear DNA fragmentation in axenic and intracellular amastigotes

model of cutaneous leishmaniasis in BALB/c mice, where control of lesion size and parasite burden was observed. Furthermore, no evidence of mortality in the treated groups or weight loss greater than 10% were observed during the study. The authors suggested that *Artemisia* EO may improve Th1 immune responses and microbicide activation of macrophages.

Nunes et al. (2021) found 69.76% of hydrocarbon sesquiterpenes in the EO of *Eugenia piauhiensis* Vellaff. (Myrtaceae), with 23.5% being γ-elemene **(30)** and 11.94% (*E*)-β-caryophyllene **(23)**. The EO and the isolated compound γ-elemene **(30)** presented greater activity against amastigote (EC$_{50}$ = 4.59 ± 0.07 μg/mL and 8.06 ± 0.12 μg/mL, respectively) than promastigote (IC$_{50}$ = 6.43 ± 0.18 μg/mL and 9.82 ± 0.15 μg/mL, respectively) forms of *L. amazonensis*. The authors suggested that this difference could be indicative of immunomodulatory activity and macrophage activation, as experiments of macrophage infection models *in vitro* revealed increased levels of TNF-α, IL-12, NO, and ROS in the supernatant of *L. amazonensis*-infected macrophages, suggesting an activation of the Th1 (not Th2) profile, a mechanism that has been the objective of anti-*Leishmania* drugs.

In the work by Carvalho et al. (2017), the EO of *Myracrodruon urundeuva*, rich in myrcene (6,7) (α-myrcene (6) 37.23% and β-myrcene (7) 42.46%), caused morphological changes such as cells with rounded or completely spherical shapes, with the presence of cell debris, typical of cell lysis. Furthermore, the results obtained against both forms of *L. amazonensis* (IC_{50} = 205 µg/mL for promastigotes; 104.5 µg/mL for axenic amastigotes; 44.5 µg/mL for intracellular amastigotes) suggest an increase in the phagocytic capacity of macrophages. According to the authors, this increase can be triggered by immunomodulatory mechanisms. One way to assess this activity is by determining the NO content. NO production is stimulated by protective cytokines, such as IFN-γ, and is extremely reactive, causing damage to the parasite's proteins and DNA. However, their tests with the EO of *M. urundeuva* did not promote NO production, suggesting that phagocytosis was not stimulated by immunomodulatory mechanisms.

In line with the immunomodulator role of NO, Jihene et al. (2020) showed that *Leishmania*-infected macrophages produced 36.8% more NO than uninfected ones. Furthermore, uninfected macrophages treated with 14.76, 7.38 and 3.69 µg/mL of propolis EO produced 50.4%, 38.1% and 25% respectively more NO than control cells. Macrophages infected and treated with EO showed a significant increase in NO levels, reaching 230% at the highest concentration.

The EO of *Nectranda hihua*, composed mainly of sesquiterpenes (89%), especially bicyclogermacrene (31) (28.1%), showed activity against intracellular *L. infantum* amastigotes (IC_{50} = 0.2 ± 1.1 mg/mL). The SI values were 249.4 and 149.0 for murine fibroblasts and macrophages, respectively, reflecting the oil's highly selective action on amastigote forms. The EO of *Nectranda gardneri* was active in intracellular amastigotes of *L. infantum* and *L. amazonensis* (IC_{50} = 2.7 ± 1.3 and 2.1 ± 1.06 mg/mL, respectively), with low cytotoxicity. This EO was also composed mainly of sesquiterpenes (85.4%), with intermediol (32) being the main component (58.2%) (Bosquiroli et al. 2017). The authors observed that the EO of the two species induced a significant increase in NO production by *L. amazonensis*-infected cells, however, in the case of *L. infantum*, only the EO from *N. gardneri* was active, suggesting that the anti-*Leishmania* activity of the EO may be associated with this important mechanism (Olekhnovitch and Bousso 2015).

Bosquiroli et al. (2015) demonstrated the inhibition of proliferation of intracellular amastigotes 24 h after the EO of *Piper angustifolium* was added to infected cells. The infection rate decreased in a range of 88.1 to 100% from the lowest to the highest concentration, with an IC_{50} of 1.43 µg/mL for *L. infantum* and low cytotoxicity for mammalian cells compared to amphotericin B, although the latter is more active. A significant increase in NO release was found after treatment with the EO at concentrations of 6.25 and 12.5 µg/mL; however, at concentrations of 25 and 50 µg/mL, the EO did not induce a significant increase in NO release, showing an atypical result that may be due to the presence of certain compounds in the EO.

The EO of *Curcuma longa* expressed anti-*Leishmania* action against promastigote and amastigote forms of *L. amazonensis* (Teles et al. 2019). The concentration of 125 μg/mL generated a decrease of 80.73% of promastigote and 40.75% of amastigote forms in infected cells. In terms of possible mechanisms of action, the authors evaluated the production of nitrite, an indirect measure to quantify NO. They found that the EO inhibited the production of NO in macrophages. Thus, the authors suggested the existence of other possible mechanisms involved in the activity of *C. longa* EO against intracellular amastigotes yet to be investigated.

13.4.3.3 Antioxidants

Since the loss of membrane balance can lead to the entry of ions into the cells, causing polarization changes; verifying the antioxidant capacities of EOs may serve to detetct this activity. According to Bouyahya et al. (2017b), antioxidant tests serve to express mechanisms of action involving polarization and chemical behavior in the presence of the product being tested.

Ahmed et al. (2011) found the compound camphor (**25**) (13.82%) in the composition of the EO of *Thymus hirtus sp. Algeriensis* and verified its anti-*Leishmania* activity. They found an IC_{50} of 0.43 μg/mL for *L. major* promastigotes and 0.25 μg/mL for *L. infantum* promastigotes. The composition and anti-*Leishmania* activity of the EO of *Ruta chalepensis* was investigated in the same study, highlighting the presence of 84.28% of 2-undecanone, and inhibitory action only against *L. infantum* promastigotes. Their tests to assess antioxidant potential through DPPH free radical scavenging showed a low antioxidant power for the EO, suggesting that anti-*Leishmania* activity was not correlated with antioxidant activity of the EO.

High concentrations of camphor (**25**) (36.82%) and compounds such as α-thujone (**33**) (7.65%) and β-thujone (**8**) (7.21%) were found in the EO of *A. herba-alba* (Aloui et al. 2016). The EO was tested against promastigote forms of *L. infantum*, revealing inhibitory power with an IC_{50} of 68 μg/mL. Antioxidant capacity by DPPH radical scavenging, with an IC_{50} of 9.1 mg/mL, and intense reducing capacity by the of ferric reducing antioxidant power (FRAP) assay, with a result of 27.48 mM Fe2+, were also observed. The effect of the EO on the cell membrane assessed through measurement of lactate dehydrogenase showed no induction of cytolysis even after prolonged incubation time (72 h). Flow cytometric analysis of *L. infantum* promastigotes detected DNA degradation by the increase in the proportion of cells in the sub-G0/G1 phase between the applied doses, followed by a decrease in the number of cells in the S and G_2/M phases. Annexin V/7-ADD staining showed that treatment with the EO caused the parasites to express apoptotic profiles without inducing necrosis.

13.4.3.4   Enzymatic Activity

According to the study by Marques et al. (2011), the EO of *Piper claussenianum* leaves was rich in sesquiterpenes, with nerolidol (**34**) being the major component (81%), and caused 62.17% inhibition in the levels of arginase activity. Pretreatment of *L. amazonensis* promastigotes with the EO reduced the percentage of macrophage infection by 42.7%, and the treatment of already infected macrophages promoted a reduction of 31.25% of the infected cells. Cytotoxicity of the EO in macrophage and fibroblast cell lines was absent at concentrations ranging from 40 to 0.56 mg/mL. The authors also performed treatment with the EO of *P. claussenianum* and INF-γ together, which provided an increase in NO production of 20.5% in cells infected with *Leishmania*. Such production was considered by the authors as a useful strategy for infection control by inhibiting arginase activity levels in the parasite.

In parasites of the genus *Leishmania*, arginase activity is essential for the growth of the protozoans (Vincendeau et al. 2003; Roberts et al. 2004) in addition to being associated with cytotoxic processes and immunological mechanisms due to the role in NO synthesis (Kanyo et al. al. 1996; Da Silva et al. 2002). Thus, arginase activity is a potential target of anti-*Leishmania* pharmacological compounds.

Oxygenated monoterpenes, especially 1,8-cineole (**9**) (23.6%) and camphor (**25**) (18.7%), were predominant in the EO of *Rosmarinus officinalis* L. (Bouyahya et al. 2017c). In chemical analyses of the EO of *Melaleuca leucadendra* L. (Myrtaceae), there was 61% of 1,8-cineole (**9**) (Monzote et al. 2020b). In their assays with *L. amazonensis*, the authors demonstrated that 1,8-cineole (**9**) had an IC_{50} value of 68.3 ± 3.4 µg/mL and no cytotoxicity against macrophages at 200 µg/mL. Despite this, the authors did not associate the antiprotozoal activity to the compound, suggesting that the activity of the EO may result from complex interactions between its constituents, and that even components in smaller amounts can play a critical role.

In a computational analysis of the structure and binding of 1,8-cineole (**9**) isolated from *Croton nepetifolius* EO in relation to the enzyme *L. infantum* trypanothione reductase (LiTR), in the structural representation of LiTR coupled to 1,8-cineole (**9**), favorable interactions of different types were formed, as Van der Waals, hydrophobic and hydrogen bonds, with participation of 7 residues (Gly197; Tyr221; Arg222), and the ligand established H bonding interaction with Gly196 within a radius of 3.68 Å (Morais et al. 2019). Turkano et al. (2018) demonstrated the RT inhibition of the compound 2-(diethylamino)ethyl 4-((3-(4-nitrophenyl)-3-oxopropyl)amino)benzoate with the participation of the residues Tyr221, Gly197, Asn254, Arg222, and Arg228, which are essential for LiTR inactivation, suggesting a possible mechanism of action against the *Leishmania* species tested.

13.4.4   Other Compounds Present in Essential Oils

Although most results point to terpenes as the main constituents present in EOs, other compounds, such as phenylpropanoids, have shown strong anti-*Leishmania* activity. One of the main representatives of this class is eugenol (**35**).

In a research carried out by Moemenbellah-Fard et al. (2020), 33 components were identified in the EO of *Syzygium aromaticum*, and among the main ones, eugenol (**35**) (65.41%), trans-caryophyllene (12.06%), eugenol acetate (9.85%), and caryophyllene oxide (**2**) (3.0%) stood out. The EO and eugenol (**35**) were tested as for their antiparasitic activity against *L. major* promastigotes, reaching IC_{50} values of 654 µg/mL and 517 µg/mL, respectively, and against *L. tropica* promastigotes, with IC_{50} of 180 µg/mL and 233 µg/mL, respectively.

In the studies by Islamuddin et al. (2013), the EO of *S. aromaticum* revealed a concentration of 59.75% of eugenol (**35**) and 29.24% of eugenyl acetate (**36**). The authors found an anti-*Leishmania* effect against intracellular promastigote and amastigote forms of *L. donovani*, with IC_{50} of 21 mg/mL and 15.24 mg/mL, respectively. In this study, it was indicated that EO-induced cell death occurred due to loss of membrane integrity, with evidence indicating late apoptosis. The authors also reported that EO-treated promastigotes exhibited a hypodiploid peak in subG0/G1, and the parasites presented reduced DNA content, thus confirming the occurrence of DNA fragmentation and induction of apoptosis. It is noteworthy that the mechanisms of action presented were similar to those presented by terpene-rich EOs.

Analysis of the EO of *Ocimum gratissimum* identified the presence of 86.5% of eugenol (**35**) (Le et al. 2017). This EO had its anti-*Leishmania* activity against *L. mexicana* tested using concentrations of 25 and 50 nL/mL, with IC_{50} of 4.85 nL/mL. Cytotoxicity tests showed survival of more than 80% of the analyzed mammalian cells after 72 hours, at the maximum concentration used (Le et al. 2017).

Methyl-eugenol (**37**) was reported as the major compound (33.89%) of the EO of *C. nepetifolius,* followed by *E*-caryophyllene (**23**) (21.23%) and 1,8-cineole (**9**) (10.44%). According to Morais et al. (2019), these compounds were likely responsible for the anti-*Leishmania* activity of the EO at concentrations of 100, 50, 25, 12.5, and 6.25 µg/mL against *L. amazonensis* (IC_{50} = 9.87 ± 2.21 µg/mL) and *L. braziliensis* (IC_{50} = 9.08 ± 2.59 µg/ml). In addition, at the concentration of 100 µg/mL, the EO presented toxicity against macrophages statistically similar to amphotericin B. It is important to note that, although the largest fractions of these EOs are phenylpropanoids, there are terpenes in considerable concentrations present in their composition.

This is the case of the EO of leaves of *Scheelea phalerata* Mart. ex Spreng (Arecaceae). The EO had phytol (**38**) as a major compound in percentages of 36.7% and 26.1% in plants collected in the dry and rainy seasons, respectively; the EO extracted in the rainy season also presented 18.7% of palmitic acid (**39**), as found in the work of Oliveira et al. (2020). Nevertheless, only the EO extracted in the rainy season had an effect against *L. amazonensis* promastigotes (IC_{50} = 165.05 ± 33.26 µg/mL). The authors suggested the role of compounds produced in this season in the inhibitory effect on parasites, emphasizing a synergistic action between the main components of the EO, phytol (**38**) and palmitic acid (**39**), since the EO extracted during the dry season showed a higher concentration of phytol (**38**) but no anti-*Leishmania* activity. Another hypothesis addressed in the study was based on the possibility that other compounds present in the EO are capable of altering the activity of phytol (**38**) by the formation of compounds, promoting the inactivation of the molecule.

The compound methyl chavicol, also called estragole (**40**), was found in the EO of *Tagetes lucida* Cav., constituting approximately 97% of the oil. The EO was tested against *L. tarentolae* promastigotes, resulting in an IC_{50} of 61.4 ± 2.4 µg/mL, and against *L. amazonensis* promastigotes, with an IC_{50} of 118.8 ± 1.2 µg/mL. Estragole (**40**) proved to be more effective than the EO, with an IC_{50} of 28.5 ± 1.0 µg/mL and 25.5 ± 3.3 µg/mL for *L. tarentolae* and *L. amazonensis*, respectively (Monzote et al. 2020a). The authors observed that the EO promoted inhibition of oxygen consumption in *L. tarentolae* at the maximum tested concentration of 100 µg/mL; however parasites treated with estragole (**40**) remained with normal oxygen consumption, suggesting that the EO targets the mitochondria of protozoa. Furthermore, estragole (**40**) was able to cause mitochondrial rupture. The authors suggested that the molecule acts as a mitochondrial uncoupler, although it is only a weak inhibitor of mitochondrial electron transfer in *Leishmania*.

13.4.5   Other Applications

Other forms of application for EOs have been explored, as in the case of EO eluted in nanoemulsions and nanogel. These mixtures can be used topically, improving the pharmacodynamic profiles of the product. In the study by Ghanbariasad et al. (2021a), the EO from *Citrus sinensis*, whose major compound was limonene (**5**) (71.26%), was used against *L. tropica* and *L. major* promastigotes, and IC_{50} values of 151.13 µg/mL and 108.31 µg/mL, respectively, were observed. Then, the nanogel based on *C. sinensis* nanoemulsion was prepared to improve its stability. According to the author, the advantage of converting nanoemulsions into nanogels is the increase in viscosity, which promotes the accumulation of the solution and improves the hydration of the application site. The nanometric dispersion of the EO and the better hydration lead to better penetration of the EO in to the locality. It is suggested that this type of application could also prevent the entry of environmental pathogens into the lesion, reducing the chance of secondary infection. In tests, the viability against *L. major* and *L. tropica* was reduced to less than 10% when used at a concentration of 9.15 mg, which was a better result than that obtained with EO alone in topical application (Ghanbariasad et al. 2021a).

13.4.6   Perceptions, Conclusions and Perspectives

Although EOs are presented as important candidates in the search for new anti-*Leishmania* drugs, we observed that some steps are still needed, especially considering that most studies did not perform the *in vivo* analyses necessary to identify the main characteristics of the compounds (bioavailability, pharmacokinetics, pharmacodynamics etc.) in new pharmacological approaches. The investigation of compounds in *in vivo* assays is essential to leverage new therapeutic hypotheses, since many compounds are discarded for not showing results *in vivo* or *in vitro*, as discussed by Brito et al. (2013). However, the authors emphasize that the mechanisms of action and interaction of drugs in humans are often discovered after their indication and use.

Another important highlight is that the evaluations presented in this section used the promastigote form to screen the most prominent compounds, probably due to handling, cost and duration of the tests. However, it is important to mention that studies conducted with amastigotes cultivated in macrophages are considered the best choice for evaluating the potential of the compounds in initial evaluation models, although, in experimental stages of sandflies, for example, there is no difference between promastigotes and amastigotes as to the development of the infection, as observed by Fampa et al. (2021) in *L. donovani*. This condition is important, considering that the morbidity and mortality associated with *Leishmania* is caused by this evolutionary form (Brito et al. 2013).

This question is evident in the studies by Tasdemir et al. (2019) who found discrepancies between the efficacy of thymol **(18)** and carvacrol **(12)** *in vitro* and *in vivo*, with reduced effects in animals. The authors attributed this result to non-ideal pharmacokinetics and physicochemistry, such as very fast absorption, low solubility, low bioavailability and elimination rate, considered the main obstacles in the development of drugs from the EO and its volatile components (Wang et al. 2009; Nagoor Meeran et al. 2017).

Despite the importance of the initial investigation of compounds, it is important to mention that some authors leave clues about the steps to follow after their studies, through the elucidation of some mechanisms of action. They cited, for example, the release of NO or the observation of the ultrastructural effects of compounds on the parasites. Although there are cost and equipment limitations, it is important to set a path for future investigations of active substances, minimizing secondary studies aimed at screening mechanisms, which are important due to the phenotypic and genotypic differences presented by the *Leishmania* species used in the bioassays.

Another alternative is presented by Andrade-Ochoa et al. (2021) who, based on the varied chemical structures and biological activities exhibited by the compounds, suggested the use of *in silico* methodologies to identify different therapeutic targets for EO constituents. Analyses performed by Ogungbe and Setzer (2013) provided evidence of the interaction of different structural types of terpenoids with certain targets in *Leishmania* that may support new phytochemical investigations and synthetic modifications in compounds or the synthesis of new antiparasitic structures.

It is possible to conclude that the anti-*Leishmania* activity of EOs stems from to the lipophilic character of their constituents, such as terpenes and phenylpropanoids, which can passively cross the membranes and disturb the osmotic balance of the cells. This may partly explain why many of the EOs have a certain degree of toxicity for mammalian cells. Given the few studies that have tested the mechanisms of action of EOs, research aimed at elucidating these bioactivities is necessary.

References

Abamor ES, Allahverdiyev AM (2016) A nanotechnology based new approach for chemotherapy of cutaneous Leishmaniasis: TIO2@AG nanoparticles – Nigella sativa oil combinations. Exp Parasitol 166:150–163. https://doi.org/10.1016/j.exppara.2016.04.008

Ács K, Balázs VL, Kocsis B, Bencsik T, Böszörményi A, Horváth G (2018) Antibacterial activity evaluation of selected essential oils in liquid and vapor phase on respiratory tract pathogens. BMC Complement Altern Med 18:1–9. https://doi.org/10.1186/s12906-018-2291-9

Ahmed SBH, Sghaier RM, Guesmi F, Kaabi B, Mejri M, Attia H, Laouini D, Smaali I (2011) Evaluation of antileishmanial, cytotoxic and antioxidant activities of essential oils extracted from plants issued from the leishmaniasis-endemic region of Sned (Tunisia). Nat Prod Res 25:1195–1201. https://doi.org/10.1080/14786419.2010.534097

Albuquerque RDDG, Oliveira AP, Ferreira C, Passos CLA, Fialho E, Soares DC, Amaral VF, Bezerra GB, Esteves RS, Santos MG, Albert ALM, Rocha L (2020) Anti-*Leishmania amazonensis* activity of the terpenoid fraction from *Eugenia pruniformis* leaves. An Acad Bras Ciênc 92:1–14. https://doi.org/10.1590/0001-3765202020201181

Alcântara IS, Martins AOBPB, Oliveira MRC, Coronel C, Gomez MCV, Rolon M, Wanderley AG, Quintans Júnior LJ, Araújo AAS, Freitas PR, Coutinho HDM, Menezes IRA (2021) Cytotoxic potential and antiparasitic activity of the *Croton rhamnifolioides* Pax leaves. & K. Hoffm essential oil and its inclusion complex (EOCr/β-CD). Polym Bull 2021:1–14. https://doi.org/10.1007/s00289-021-03556-6

Alcoba AET, Melo DCDE, Andrade PMDE, Dias HJ, Pagotti MC, Magalhães LG, Júnior WGF, Crotti AEM, Miranda MLD (2018) Chemical composition and *in vitro* antileishmanial and cytotoxic activities of the essential oils of *Ocotea dispersa* (Nees) Mez and *Ocotea odorifera* (Vell) Rohwer (Lauraceae). Nat Prod Res 32:2865–2868. https://doi.org/10.1080/14786419.2017.1385007

Ali N, Nabi M, Shoaib M, Shah I, Ahmed G, Shakirullah Z, Ali Shah SW, Ghias M, Khan S, Ali W (2021) GC/MS analysis, anti-leishmanial and relaxant activity of essential oil of *Chenopodium ambrosioides* (L.) from Malakand region. Pak J Pharm Sci 34:577–583. https://doi.org/10.36721/PJPS.2021.34.2.REG.577-583.1

Almeida KC, Silva BB, Alves CC, Vieira TM, Crotti AE, Souza JM, Martins CHG, Ribeiro AB, Squarisi IS, Tavares DC, Bernabé LS, Magalhães LG, Miranda MLD (2020) Biological properties and chemical composition of essential oil from *Nectandra megapotamica* (Spreng.) Mez. leaves (Lauraceae). Nat Prod Res 34:3149–3153. https://doi.org/10.1080/14786419.2019.1608539

Aloui Z, Messaoud C, Haoues M (2016) Asteraceae *Artemisia campestris* and *Artemisia herba-alba* essential oils trigger apoptosis and cell cycle arrest in *Leishmania infantum* promastigotes. Evid Based Complement Altern Med 2016:1–15. https://doi.org/10.1155/2016/9147096

Amorim Gomes G, Martins-Cardoso K, dos Santos FR (2020) Antileishmanial activity of the essential oils of *Myrcia ovata* Cambess. and *Eremanthus erythropappus* (DC) McLeisch leads to parasite mitochondrial damage. Nat Prod Res 0:1–5. https://doi.org/10.1080/14786419.2020.1827402

Andrade MA, Azevedo CDS, Motta FN, Santos ML, Silva CL, Santana JM, Bastos IM (2016) Essential oils: *in vitro* activity against *Leishmania amazonensis*, cytotoxicity and chemical composition. BMC Complement Altern Med 16:1–8. https://doi.org/10.1186/s12906-016-1401-9

Andrade PM, De Melo DC, Alcoba AET, Ferreira Junior WG, Pagotti MC, Magalhaes LG, Miranda ML (2018a) Chemical composition and evaluation of antileishmanial and cytotoxic activities of the essential oil from leaves of *Cryptocarya aschersoniana* Mez.(Lauraceae Juss.). An Acad Bras Cienc 90:2671–2678. https://doi.org/10.1590/0001-3765201820170332

Andrade MA, Azevedo CS, Motta FN, Santos ML, Silva CL, Santana JM, Bastos IMD (2018b) Essential oils: in vitro activity against *Leishmania amazonensis*, cytotoxicity and chemical composition. BMC Complement Altern Med 16:1–8. https://doi.org/10.1186/s12906-016-1401-9

Andrade-Ochoa S, Chacón-Vargas KF, Sánchez-Torres LE, Rivera-Chavira BE, Nogueda-Torres B, Nevárez-Moorillón GV (2021) Differential antimicrobial effect of essential oils and their main components: insights based on the cell membrane and external structure. Membranes 11:1–17. https://doi.org/10.3390/membranes11060405

Arnoult D, Akarid K, Grodet A, Petit PX, Estaquier J, Ameisen JC (2002) On the evolution of programmed cell death: apoptosis of the unicellular eukaryote *Leishmania major* involves cysteine proteinase activation and mitochondrion permeabilization. Cell Death Differ 9:65–81. https://doi.org/10.1038/sj.cdd.4400951

Ashrafi B, Beyranvand F, Ashouri F, Rashidipour M, Marzban A, Kheirandish F, Veiskarami S, Ramak P, Shahrokhi S (2020a) Characterization of phytochemical composition and bioactivity assessment of *Pseudotrachydium kotschyi* essential oils. Med Chem Res 29:1676–1688. https://doi.org/10.1007/s00044-020-02594-5

Ashrafi B, Rashidipour M, Gholami E, Sattari E, Marzban A, Kheirandish F, Khaksarian M, Taherikalani M (2020b) Soroush S Investigation of the phytochemicals and bioactivity potential of essential oil from *Nepeta curvidens* Boiss. & Balansa. S Afr J Bot 135:109–116. https://doi.org/10.1016/j.sajb.2020.08.015

Bailen M, Julio LF, Diaz CE, Sanz J, Martínez-Díaz RA, Cabrera R, Burillo J, Gonzalez-Coloma A (2013) Chemical composition and biological effects of essential oils from *Artemisia absinthium* L. cultivated under different environmental conditions. Ind Crop Prod 49:102–107. https://doi.org/10.1016/j.indcrop.2013.04.055

Baldemir A, Karaman Ü, İlgün S, Kaçmaz G, Demirci B (2018) Antiparasitic efficacy of *Artemisia ludoviciana* nutt. (Asteraceae) essential oil for *Acanthamoeba castellanii, Leishmania infantum* and *trichomonas vaginalis*. Indian J Pharm Educ Res 52:416–425. https://doi.org/10.5530/ijper.52.3.48

Barros LM, Duarte AE, Morais-Braga MFB, Waczuk EP, Vega C, Leite NF, Menezes IRA, Coutinho HDM, Rocha JBT, Kamdem JP (2016) Chemical characterization and trypanocidal, leishmanicidal and cytotoxicity potential of *Lantana camara* L. (Verbenaceae) essential oil. Molecules 21:1–16. https://doi.org/10.3390/molecules21020209

Bekhit AA, El-Agroudy E, Helmy A, Ibrahim TM, Shavandi A, Bekhit AEDA (2018) *Leishmania* treatment and prevention: natural and synthesized drugs. Eur J Med Chem 160:229–244. https://doi.org/10.1016/j.ejmech.2018.10.022

Bernuci KZ, Iwanaga CC, Fernandez-Andrade CMM, Lorenzetti FB, Torres-Santos EC, Faiões VDS, Gonçalves JE, Amaral W, Deschamps C, Scodro RBL, Cardoso RF, Baldin VP, Cortez DAG (2016) Evaluation of chemical composition and antileishmanial and antituberculosis activities of essential oils of *Piper* species. Molecules 21:1–17. https://doi.org/10.3390/molecules21121698

Biavatti MW, Vieira PC, Silva MFGF, Fernandes JB, Albuquerque S, Magalhães CMI, Pagnocca FC (2001) Chemistry and bioactivity of *Raulinoa echinata* Cowan, an endemic Brazilian rutaceae species. Phytomedicine 8:121–124. https://doi.org/10.1078/0944-7113-00016

Blowman K, Magalhães M, Lemos MFL, Cabral C, Pires IM (2018) Anticancer properties of essential oils and other natural products. Evid Based Complement Altern Med 2018:1–13. https://doi.org/10.1155/2018/3149362

Bosquiroli LSS, Demarque DP, Rizk YS, Cunha MC, Marques MCS, Matos MDFC, Arruda CC (2015) In vitro anti-*leishmania infantum* activity of essential oil from *Piper angustifolium*. Rev Bras Farmacogn 25:124–128. https://doi.org/10.1016/j.bjp.2015.03.008

Bosquiroli LSS, dos Ferreira ACS, Farias KS, da Costa EC, Matos MDFC, Kadri MCT, Rizk YS, Alves FM, Perdomo RT, Carollo CA, Arruda CCP (2017) *In vitro* antileishmania activity of sesquiterpene-rich essential oils from nectandra species. Pharm Biol 55:2285–2291. https://doi.org/10.1080/13880209.2017.1407803

Bouyahya A, Assemian ICC, Mouzount H, Bourais I, Et-Touys A, Fellah H, Benjouad A, Dakka N, Bakri Y (2019) Could volatile compounds from leaves and fruits of *Pistacia lentiscus* constitute a novel source of anticancer, antioxidant, antiparasitic and antibacterial drugs? Ind Crop Prod 128:62–69. https://doi.org/10.1016/j.indcrop.2018.11.001

Bouyahya A, Dakka N, Talbaoui A, Et-Touys A, El-Boury H, Abrini J, Bakri Y (2017a) Correlation between phenological changes, chemical composition and biological activities of the essential oil from Moroccan endemic oregano (*Origanum compactum* Benth). Ind Crop Prod 108:729–737. https://doi.org/10.1016/j.indcrop.2017.07.033

Bouyahya A, Et-Touys A, Abrini J, Talbaoui A, Fellah H, Bakri Y, Dakka N (2017b) *Lavandula stoechas* essential oil from Morocco as novel source of antileishmanial, antibacterial and antioxidant activities. Biocatal Agric Biotechnol 12:179–184. https://doi.org/10.1016/j.bcab.2017.10.003

Bouyahya A, Et-Touys A, Bakri Y, Talbaui A, Fellah H, Abrini J, Dakka N (2017c) Chemical composition of *Mentha pulegium* and *Rosmarinus officinalis* essential oils and their antileishmanial, antibacterial and antioxidant activities. Microb Pathog 111:41–49. https://doi.org/10.1016/j.micpath.2017.08.015

Brito AMG, Dos Santos D, Rodrigues SA, Brito RG, Xavier-Filho L (2013) Plants with anti-*Leishmania* activity: integrative review from 2000 to 2011. Pharmacogn Rev 7:34–41. https://doi.org/10.4103/0973-7847.112840

Brochot A, Guilbot A, Haddioui L, Roques C (2017) Antibacterial, antifungal, and antiviral effects of three essential oil blends. Microbiology 6:e00459. https://doi.org/10.1002/mbo3.459

Burza S, Croft SL, Boelaert M (2018) Leishmaniasis. Lancet 392:951–970. https://doi.org/10.1016/S0140-6736(18)31204-2

Cabral RSC, Fernandes CC, Dias ALB, Batista HRF, Magalhães LG, Pagotti MC, Miranda MLD (2021) Essential oils from *Protium heptaphyllum* fresh young and adult leaves (Burseraceae): chemical composition, in vitro leishmanicidal and cytotoxic effects. J Essent Oil Res 33:276–282. https://doi.org/10.1080/10412905.2020.1848651

Capello TM, Martins EGA, De Farias CF, Figueiredo CR, Matsuo AL, Passero LFD, Oliveira-Silva D, Sartorelli P, Lago JHG (2015) Chemical composition and *in vitro* cytotoxic and antileishmanial activities of extract and essential oil from leaves of *Piper cernuum*. Nat Prod Commun 10:285–288. https://doi.org/10.1177/1934578x1501000217

Cardoso BM, Mello TFP, Lopes SN, Demarchi IG, Lera DSL, Pedroso RB, Cortez DA, Gazim ZC, Aristides SMA, Silveira TGV, Lonardoni MVC (2015) Antileishmanial activity of the essential oil from *Tetradenia riparia* obtained in different seasons. Mem Inst Oswaldo Cruz 110:1024–1034. https://doi.org/10.1590/0074-02760150290

Carmo DFM, Amaral ACF, Machado GMC, Leon LL, Silva JRA (2012) Chemical and biological analyses of the essential oils and main constituents of *Piper* species. Molecules 17:1819–1829. https://doi.org/10.3390/molecules17021819

Carnielli JBT, Monti-Rocha R, Costa DL, Sesana AM, Pansini LNN, Segatto M, Mottram JC, Costa CHN (2019) Natural resistance of *Leishmania infantum* to miltefosine contributes to the low efficacy in the treatment of visceral leishmaniasis in Brazil. Am J Trop Med Hyg 101:789–794. https://doi.org/10.4269/ajtmh.18-0949

Carvalho CES, Sobrinho-Junior EPC, Brito LM, Nicolau LAD, Carvalho TP, Moura AKS, Rodrigues KAF, Carneiro SMP, Arcanjo DDR, Citó AMGL, Carvalho FAA (2017) Anti-*Leishmania* activity of essential oil of *Myracrodruon urundeuva* (Engl.) Fr. All.: composition, cytotoxity and possible mechanisms of action. Exp Parasitol 175:59–67. https://doi.org/10.1016/j.exppara.2017.02.012

Castro C, Jimenez M, Gonzalez-De La Parra M (1992) Inhibitory effect of Piquerol A on the growth of epimastigotes of *Trypanosoma cruzi*. Planta Med 58:281–282. https://doi.org/10.1055/s-2006-961457

Ceole LF, Cardoso MDG, Soares MJ (2017) Nerolidol, the main constituent of *Piper aduncum* essential oil, has anti-*Leishmania braziliensis* activity. Parasitology 144:1179–1190. https://doi.org/10.1017/S0031182017000452

Colares AV, Almeida-Souza F, Taniwaki NN, Souza CDSF, Costa JGM, Calabrese KDS, Abreu-Silva AL (2013) *In vitro* antileishmanial activity of essential oil of *Vanillosmopsis arborea* (Asteraceae) baker. Evid Based Complement Altern Med 2013. https://doi.org/10.1155/2013/727042

Contini A, Di Bello D, Azzarà A, Giovanelli S, D'Urso G, Piaggi S, Pinto B, Pistelli L, Scarpato R, Testi S (2020) Assessing the cytotoxic/genotoxic activity and estrogenic/antiestrogenic potential of essential oils from seven aromatic plants. Food Chem Toxicol 138:111205. https://doi.org/10.1016/j.fct.2020.111205

Costa AR, Pereira PS, Barros LM, Duarte AE, Vega Gomez MC, Rolón M, Vidal CAS, Maia AJ, Morais-Braga MFB, Coutinho HD (2016) The cytotoxicity activity and evaluation of antiprotozoa *Melissa officcinalis* L. (Cidro-Melisa). Rev Cuba Plantas Med 21:1–13

Cristani M, D'Arrigo M, Mandalari G, Castelli F, Sarpietro MG, Micieli D, Venuti V, Bisignano G, Saija A, Trombetta D (2007) Interaction of four monoterpenes contained in essential oils with model membranes: implications for their antibacterial activity. J Agric Food Chem 55:6300–6308. https://doi.org/10.1021/jf070094x

Dapkevicius A, Van Beek TA, Lelyveld GP, Van-Veldhuizen A, de Groot A, Linssen JP, Venskutonis R (2002) Isolation and structure elucidation of radical scavengers from *Thymus vulgaris* leaves. J Nat Prod 65:892–896. https://doi.org/10.1021/np010636j

Demarchi IG, Terron MS, Thomazella MV, Mota CA, Gazim ZC, Cortez DAG, Aristides SMA, Silveira TGV, Lonardoni MVC (2016) Antileishmanial and immunomodulatory effects of the essential oil from *Tetradenia riparia* (Hochstetter) Codd. Parasite Immunol 38:64–77. https://doi.org/10.1111/pim.12297

Demarchi IG, Thomazella MV, Terron MS, Lopes L, Gazim ZC, Cortez DAG, Donatti L, Aristides SMA, Silveira TGV, Lonardoni MVC (2015) Antileishmanial activity of essential oil and 6,7-dehydroroyleanone isolated from *Tetradenia riparia*. Exp Parasitol 157:128–137. https://doi.org/10.1016/j.exppara.2015.06.014

Dezaki ES, Mahmoudvand H, Sharififar F, Fallahi S, Monzote L, Ezatkhah F (2016) Chemical composition along with anti-leishmanial and cytotoxic activity of *Zataria multiflora*. Pharm Biol 54:752–758. https://doi.org/10.3109/13880209.2015.1079223

Dhami DS, Pandey SC, Shah GC, Bisht M, Samant M (2021) *In vitro* antileishmanial activity of the essential oil from *Agrimonia pilosa*. Natl Acad Sci Lett 44:195–198. https://doi.org/10.1007/s40009-020-00992-2

Dias CN, Rodrigues KAF, Carvalho FAA, Carneiro SM, Maia JG, Andrade EH, Moraes DF (2013) Molluscicidal and leishmanicidal activity of the leaf essential oil of *Syzygium cumini* (L.) skeels from Brazil. Chem Biodivers 10:1133–1141. https://doi.org/10.1002/cbdv.201200292

Díaz JG, Arranz JCE, Batista DGJ, Fidalgo LM, Acosta JLV, Macedo MB, Cos P (2018) Antileishmanial potentialities of *Croton linearis* leaf essential oil. Nat Prod Commun 13:629–634. https://doi.org/10.1177/1934578X1801300527

Donega MA, Mello SC, Moraes RM, Jain SK, Tekwani BL, Cantrell CL (2014) Pharmacological activities of cilantro's aliphatic aldehydes against *Leishmania donovani*. Planta Med 80:1706–1711. https://doi.org/10.1055/s-0034-1383183

Dorman HJD, Deans SG (2000) Antimicrobial agents from plants: antibacterial activity of plant volatile oils. J Appl Microbiol 88:308–316. https://doi.org/10.1046/j.1365-2672.2000.00969.x

Espinosa OA, Serrano MG, Camargo EP, Teixeira MMG, Shaw JJ (2018) An appraisal of the taxonomy and nomenclature of trypanosomatids presently classified as *Leishmania* and *Endotrypanum*. Parasitology 145:430–442. https://doi.org/10.1017/S0031182016002092

Essid R, Rahali FZ, Msaada K, Sghair I, Hammami M, Bouratbine A, Aoun K, Limam F (2015) Antileishmanial and cytotoxic potential of essential oils from medicinal plants in Northern Tunisia. Ind Crop Prod 77:795–802. https://doi.org/10.1016/j.indcrop.2015.09.049

Estevam EB, Deus IPD, Silva VP, Silva EAD, Alves CC, Alves JM, Cazal CM, Magalhães LG, Pagotti MC, Esperandim VR, Souza AF, Miranda ML (2017) *In vitro* antiparasitic activity and chemical composition of the essential oil from *Protium ovatum* leaves (Burceraceae). An Acad Bras Ciênc 89:3005–3013. https://doi.org/10.1590/0001-3765201720170310

Fabri RL, Coimbra ES, Almeida AC, Siqueira EP, Alves TMA, Zani CL, Scio E (2012) Essential oil of *Mitracarpus frigidus* as a potent source of bioactive compounds. An Acad Bras Ciênc 84:1073–1080. https://doi.org/10.1590/S0001-37652012000400021

Fampa P, Florencio M, Santana RC, Rosa D, Soares DC, Guedes HLM, Silva AC, Siqueira AC, Pinto-da-Silva LH (2021) Anti-*Leishmania* effects of volatile oils and their isolates. Rev Bras Farmacogn. https://doi.org/10.1007/s43450-021-00146-5

Farias-Junior AP, Rios MC, Moura TA, Almeida R, Alves P, Blank A, Fernandes RPM, Scher R (2012) Leishmanicidal activity of carvacrol-rich essential oil from *Lippia sidoides* cham. Biol Res 45:399–402. https://doi.org/10.4067/S0716-97602012000400012

Ferreira FBP, Ramos-Milaré ÁCFH, Gonçalves JE, Lazarin-Bidóia D, Nakamura CV, Sugauara RR, Fernandez CMM, Gazim ZC, Demarchi IG, Silveira TGV, Lonardoni MVC (2020) *Campomanesia xanthocarpa* (Mart.) O. Berg essential oil induces antileishmanial activity and remodeling of the cytoplasm organelles. Nat Prod Res 2020:1–5. https://doi.org/10.1080/14786419.2020.1827401

Garcia MCF, Soares DC, Santana RC, Saraiva EM, Siani AC, Ramos MFS, Danelli MDGM, Souto-Padron TC, Pinto-Da-Silva LH (2018) The *in vitro* antileishmanial activity of essential oil from *Aloysia gratissima* and guaiol its major sesquiterpene against *Leishmania amazonensis*. Parasitology 145:219–1227. https://doi.org/10.1017/S0031182017002335

García M, Scull R, Satyal P, Setzer WN, Monzote L (2017) Chemical characterization antileishmanial activity and cytotoxicity effects of the essential oil from leaves of *Pluchea carolinensis* (Jacq.) G. Don. (Asteraceae). Phytother Res 31:1419–1426. https://doi.org/10.1002/ptr.5869

Ghanbariasad A, Amoozegar F, Rahmani M, Zarenezhad E, Osanloo M (2021a) Impregnated nanofibrous mat with nanogel of *Citrus sinensis* essential oil as a new type of dressing in cutaneous leishmaniasis. Biointerface Res Appl Chem 11:11066–11076. https://doi.org/10.33263/BRIAC114.1106611076

Ghanbariasad A, Azadi S, Agholi M, Osanloo M (2021b) The nanoemulsion-based nanogel of *Artemisia dracunculus* essential oil with proper activity against *Leishmania tropica* and *Leishmania major*. Nanomed Res J 6:89–95. https://doi.org/10.22034/NMRJ.2021.01.010

Ghanbariasad A, Valizadeh A, Ghadimi SN, Fereidouni Z, Osanloo M (2021c) Nanoformulating *Cinnamomum zeylanicum* essential oil with an extreme effect on *Leishmania tropica* and *Leishmania major*. J Drug Deliv Sci Technol 63:102436. https://doi.org/10.1016/j.jddst.2021.102436

Gomes GA, Martins-Cardoso K, Santos FR, Florencio M, Rosa D, Zuma AA, Santiago GMP, Motta MC, Carvalho MG, Fampa P (2020) Antileishmanial activity of the essential oils of *Myrcia ovata* Cambess. and *Eremanthus erythropappus* (DC) McLeisch leads to parasite mitochondrial damage. Nat Prod Res 8:1–5. https://doi.org/10.1080/14786419.2020.1827402

Gomez MCV, Rolón M, Coronel C, Carneiro JNP, Santos ATL, Almeida-Bezerra JW, Menezes SA, Silva LE, Coutinho HDM, Amaral W, Ribeiro-Filho J, Morais-Braga MFB (2021) Antiparasitic effect of essential oils obtained from two species of *Piper* L. native to the Atlantic forest. Biocatal Agric Biotechnol 32:101958. https://doi.org/10.1016/j.bcab.2021.101958

Guimarães AG, Oliveira GF, Melo MS, Cavalcanti SCH, Antoniolli AR, Bonjardim LR, Silva FA, Santos JPA, Rocha RF, Moreira JCF, Araújo AAS, Gelain DP, Quintans-Júnior LJ (2010)

Bioassay-guided evaluation of antioxidant and antinociceptive activities of carvacrol. Basic Clin Pharmacol Toxicol 107:949–957. https://doi.org/10.1111/j.1742-7843.2010.00609.x

Gutiérrez Y, Montes R, Scull R, Sánchez A, Cos P, Monzote L, Setzer WN (2016) Chemodiversity associated with cytotoxicity and antimicrobial activity of *Piper aduncum* var. ossanum. Chem Biodivers 13:1715–1719. https://doi.org/10.1002/cbdv.201600133

Hajaji S, Sifaoui I, López-Arencibia A, Reyes-Batlle M, Jiménez IA, Bazzocchi IL, Valladares B (2018) Leishmanicidal activity of α-bisabolol from Tunisian chamomile essential oil. Parasitol Res 117:2855–2867. https://doi.org/10.1007/s00436-018-5975-7

Hamdi A, Bero J, Beaufay C, Flamini G, Marzouk Z, Vander HY, Quetin-Leclercq J (2018) *In vitro* antileishmanial and cytotoxicity activities of essential oils from *Haplophyllum tuberculatum* A. Juss leaves stems and aerial parts. BMC Complement Altern Med 18:1–10. https://doi.org/10.1186/s12906-018-2128-6

Houël E, Gonzalez G, Bessière JM, Odonne G, Eparvier V, Deharo E, Stien D (2015) Therapeutic switching: from antidermatophytic essential oils to new leishmanicidal products. Mem Inst Oswaldo Cruz 110:106–113. https://doi.org/10.1590/0074-02760140332

Ibrahim SRM, Abdallah HM, Mohamed GA, Farag MA, Alshali KZ, Alsherif EA, Ross SA (2017) Volatile oil profile of some lamiaceous plants growing in Saudi Arabia and their biological activities. Z Naturforsch C 72:35–41. https://doi.org/10.1515/znc-2015-0234

Isah T (2019) Stress and defense responses in plant secondary metabolites production. Biol Res 52:39. https://doi.org/10.1186/s40659-019-0246-3

Islamuddin M, Chouhan G, Want MY, Tyagi M, Abdin MZ, Sahal D, Afrin F (2014a) Leishmanicidal activities of *Artemisia annua* leaf essential oil against Visceral Leishmaniasis. Front Microbiol 5:1–16. https://doi.org/10.3389/fmicb.2014.00626

Islamuddin M, Sahal D, Afrin F (2014b) Apoptosis-like death in *Leishmania donovani* promastigotes induced by eugenol-rich oil of *Syzygium aromaticum*. J Med Microbiol 63:74–85. https://doi.org/10.1099/jmm.0.064709-0

Islamuddin M, Sahal D, Afrin F (2013) Apoptosis-like death in *Leishmania donovani* promastigotes induced by eugenol-rich oil of *Syzygium aromaticum*. J Med Microbiol 63:74–85. https://doi.org/10.1099/jmm.0.064709-0

Jihene A, Rym E, Ines KJ, Majdi H, Olfa T, Abderrabba M (2020) Antileishmanial potential of propolis essential oil and its synergistic combination with amphotericin B. Nat Prod Commun 15:1–8. https://doi.org/10.1177/1934578X19899566

Jorjani O, Raeisi M, Hezarjaribi HZ, Soltani M, Soosaraei M (2017) Studying the chemical composition *in vitro* activity of *Cinnamomum zeylanicum* and *Eugenia caryophyllata* essential oils on *Leishmania major*. J Pharm Sci Res 9:1300–1304

Kauffmann C, Ethur EM, Arossi K, Hoehne L, Freitas EM, Machado GMC, Cavalheiro MMC, Flach A, Costa LAMA, Gnoatto SCB (2017) Chemical composition and evaluation preliminary of antileishmanial activity in vitro of essential oil from leaves of *Eugenia pitanga* a native species of Southern of Brazil. J Essent Oil Bear Plants 20:559–569. https://doi.org/10.1080/0972060X.2017.1281767

Kauffmann C, Giacomin AC, Arossi K, Pacheco LA, Hoehne L, Freitas EM, Machado GMC, Cavalheiro MMC, Gnoatto SCB, Ethur EM (2019) Antileishmanial in vitro activity of essential oil from *Myrciaria plinioides* a native species from Southern Brazil. Braz J Pharm Sci 55:1–8. https://doi.org/10.1590/s2175-97902019000217584

Kheirandish F, Delfan B, Farhadi S, Ezatpour B, Khamesipour A, Kazemi B, Ebrahimzade F, Rashidipour M (2011) The effect of *Satureja khuzestanica* essential oil on the lesions induced by *Leishmania major* in BALB/c mice. Afr J Pharm Pharmacol 5:648–653. https://doi.org/10.5897/AJPP11.130

Koning HP (2001) Uptake of pentamidine in *Trypanosoma brucei brucei* is mediated by three distinct transporters: implications for cross-resistance with arsenicals. Mol Pharmacol 59:586–592. https://doi.org/10.1124/mol.59.3.586

Koyama S, Heinbockel T (2020) The effects of essential oils and terpenes in relation to their routes of intake and application. Int J Mol Sci 21:1558. https://doi.org/10.3390/ijms21051558

Lambert RJW, Skandamis PN, Coote PJ, Nychas GJE (2001) A study of the minimum inhibitory concentration and mode of action of oregano essential oil thymol and carvacrol. J Appl Microbiol 91:453–462. https://doi.org/10.1046/j.1365-2672.2001.01428.x

Le TB, Beaufay C, Nghiem DT, Mingeot-Leclercq MP, Quetin-Leclercq J (2017) In vitro antileishmanial activity of essential oils extracted from Vietnamese plants. Molecules 22:1–12. https://doi.org/10.3390/molecules22071071

Leal SM, Pino N, Stashenko EE, Martínez JR, Escobar P (2013) Antiprotozoal activity of essential oils derived from *Piper* spp. grown in Colombia. J Essent Oil Res 25:512–519. https://doi.org/1 0.1080/10412905.2013.820669

Lopes MV, Desoti VC, Caleare ADO, Ueda-Nakamura T, Silva SO, Nakamura CV (2012, 2012) Mitochondria superoxide anion production contributes to geranylgeraniol-induced death in *Leishmania amazonensis*. Evid Based Complement Altern Med. https://doi.org/10.1155/2012/298320

Maaroufi Z, Cojean S, Loiseau PM, Yahyaoui M, Agnely F, Abderraba M, Mekhloufi G (2021) In vitro antileishmanial potentialities of essential oils from *Citrus* limon and *Pistacia lentiscus* harvested in Tunisia. Parasitol Res 120:1455–1469. https://doi.org/10.1007/s00436-020-06952-5

Macêdo CG, Fonseca MYN, Caldeira AD, Castro SP, Pacienza-Lima W, Borsodi MPG, Sartoratto A, Silva MN, Salgado CG, Rossi-Bergmann B, Castro KCF (2020) Leishmanicidal activity of *Piper marginatum* Jacq. from Santarém-PA against *Leishmania amazonensis*. Exp Parasitol 210:107847. https://doi.org/10.1016/j.exppara.2020.107847

Machado M, Martins N, Salgueiro L, Cavaleiro C, Sousa MC (2019) *Lavandula luisieri* and *Lavandula viridis* essential oils as upcoming anti-protozoal agents: a key focus on leishmaniasis. Appl Sci 9. https://doi.org/10.3390/app9153056

Machado M, Pires P, Dinis AM, Santos-Rosa M, Alves V, Salgueiro L, Cavaleiro C (2012a) Monoterpenic aldehydes as potential anti-*Leishmania* agents: activity of *Cymbopogon citratus* and citral on *L. infantum* *L. tropica* and *L. major*. Exp Parasitol 130:223–231. https://doi.org/10.1016/j.exppara.2011.12.012

Machado RRP, Valente W, Lesche B, Coimbra ES, Souza NB, Abramo C, Soares GLG, Kaplan MAC (2012b) Essential oil from leaves of *Lantana camara*: a potential source of medicine against leishmaniasis. Rev Bras Farmacogn 22:1011–1017. https://doi.org/10.1590/S0102-695X2012005000057

Machado M, Dinis AM, Santos-Rosa M, Alves V, Salgueiro L, Cavaleiro C, Sousa MC (2014) Activity of *thymus capitellatus* volatile extract 18-cineole and borneol against *Leishmania* species. Vet Parasitol 200:39–49. https://doi.org/10.1016/j.vetpar.2013.11.016

Machín L, Tamargo B, Piñón A, Atíes RC, Scull R, Setzer WN, Monzote L (2019) *Bixa orellana* L. (Bixaceae) and *Dysphania ambrosioides* (L.) Mosyakin & Clemants (Amaranthaceae) essential oils formulated in nanocochleates against *Leishmania amazonensis*. Molecules 24:4222. https://doi.org/10.3390/molecules24234222

Mahmoudvand H, Tavakoli R, Sharififar F, Minaie K, Ezatpour B, Jahanbakhsh S, Sharifi I (2015a) Leishmanicidal and cytotoxic activities of *Nigella sativa* and its active principle thymoquinone. Pharm Biol 53:1052–1057. https://doi.org/10.3109/13880209.2014.957784

Mahmoudvand H, Ezzatkhah F, Sharififar F, Sharifi I, Dezaki ES (2015b) Antileishmanial and cytotoxic effects of essential oil and methanolic extract of *Myrtus communis* L. Korean J Parasitol 53:21–27. https://doi.org/10.3347/kjp.2015.53.1.21

Mahmoudvand H, Dezaki ES, Ezatpour B, Sharifi I, Kheirandish F, Rashidipour M (2016) In vitro and in vivo antileishmanial activities of *Pistacia vera* essential oil. Planta Med 82:279–284. https://doi.org/10.1055/s-0035-1558209

Marques AM, Barreto ALSB, Curvelo JAR, Romanos MTV, Soares RMDA, Kaplan MAC (2011) Antileishmanial activity of nerolidol-rich essential oil from *Piper clausseanum*. Rev Bras Farmacogn 21:908–914. https://doi.org/10.1590/S0102-695X2011005000157

Mathlouthi A, Belkessam M, Sdiri M, Diouani MF, Souli A, El-Bok S, Ben-Attia M (2018) Chemical composition and anti-leishmania major activity of essential oils from *Artemesia* spp.

grown in Central Tunisia. J Essent Oil Bearing Plants 21:1186–1198. https://doi.org/10.108 0/0972060X.2018.1526128

Matos SP, Teixeira HF, Lima ÁA, Veiga-Junior VF, Koester LS (2019) Essential oils and isolated terpenes in nanosystems designed for topical administration: a review. Biomol Ther 9:138. https://doi.org/10.3390/biom9040138

Medeiros MGF, Silva AC, Citó AM, Borges AR, Lima SG, Lopes JAD, Figueiredo RCBQ (2011) *In vitro* antileishmanial activity and cytotoxicity of essential oil from *Lippia sidoides*. Cham Parasitol Int 60:237–241. https://doi.org/10.1016/j.parint.2011.03.004

Meira CS, Gedamu L (2019) Protective or detrimental? Understanding the role of host immunity in leishmaniasis. Microorganisms 7:695. https://doi.org/10.3390/microorganisms7120695

Melo JO, Bitencourt TA, Fachin AL, Cruz EMO, Jesus HCR, Alves PB, Arrigoni-Blank MF, Franca SC, Beleboni RO, Fernandes RPM, Blank AF, Scher R (2013) Antidermatophytic and antileishmanial activities of essential oils from *Lippia gracilis* Schauer genotypes. Acta Trop 128:110–115. https://doi.org/10.1016/j.actatropica.2013.06.024

Menezes Filho ACP, Oliveira Filho JG, Castro CFS (2020) Avaliações antioxidante e antifúngica dos óleos essenciais de *Hymenaea stigonocarpa* Mart. ex Hayne e *Hymenaea courbaril* L. J Biotechnol Biodivers 8:104–114. https://doi.org/10.20873/jbb.uft.cemaf.v8n2.menezes

Menna-Barreto RFS, Salomão K, Dantas AP, Santa-Rita RM, Soares MJ, Barbosa HS, Castro SL (2009) Different cell death pathways induced by drugs in *Trypanosoma cruzi*: an ultrastructural study. Micron 40:157–168. https://doi.org/10.1016/j.micron.2008.08.003

Moemenbellah-Fard MD, Abdollahi A, Ghanbariasad A, Osanloo M (2020) Antibacterial and leishmanicidal activities of *Syzygium aromaticum* essential oil versus its major ingredient eugenol. Flavour Fragr J 35:534–540. https://doi.org/10.1002/ffj.3595

Mondêgo-Oliveira R, Sousa JCS, Moragas-Tellis CJ, Souza PVR, Chagas MDSS, Behrens MD, Hardoim DJ, Taniwaki NN, Chometon TQ, Bertho AL, Calabrese KS, Almeida-Souza F, Abreu-Silva AL (2021) *Vernonia brasiliana* (L.) Druce induces ultrastructural changes and apoptosis-like death of *Leishmania infantum* promastigotes. Biomed Pharmacother 133:111025. https://doi.org/10.1016/j.biopha.2020.111025

Monzote L, García M, Pastor J, Gil L, Scull R, Maes L, Cos P, Gille L (2014a) Essential oil from *Chenopodium ambrosioides* and main components: activity against Leishmania their mitochondria and other microorganisms. Exp Parasitol 136:20–26. https://doi.org/10.1016/j.exppara.2013.10.007

Monzote L, García M, Scull R, Cuellar A, Setzer WN (2014b) Antileishmanial activity of the essential oil from *Bixa orellana*. Phyther Res 28:753–758. https://doi.org/10.1002/ptr.5055

Monzote L, Gutiérrez Y, Machin L, Staniek K, Scull R, Satyal P, Gille L, Setzer WN (2020a) Antileishmanial activity and influence on mitochondria of the essential oil from *Tagetes lucida* cav. And its main component. Sci Pharm 88:1–8. https://doi.org/10.3390/scipharm88030031

Monzote L, Nance MR, García M, Scull R, Setzer WN (2011) Comparative chemical cytotoxicity and antileishmanial properties of essential oils from *Chenopodium ambrosioides*. Nat Prod Commun 6:281–286. https://doi.org/10.1177/1934578x1100600232

Monzote L, Pinón A, Scull R, Setzer WN (2014d) Chemistry and leishmanicidal activity of the essential oil from *Artemisia absinthium* from Cuba. Nat Prod Commun 9:1799–1804. https://doi.org/10.1177/1934578x1400901236

Monzote L, Scherbakov AM, Scull R, Satyal P, Cos P, Shchekotikhin AE, Gille L, Setzer WN (2020b) Essential oil from *Melaleuca leucadendra*: antimicrobial antikinetoplastid antiprolif-erative and cytotoxic assessment. Molecules 25. https://doi.org/10.3390/molecules25235514

Monzote L, Stamberg W, Staniek K, Gille L (2009) Toxic effects of carvacrol caryophyllene oxide and ascaridole from essential oil of *Chenopodium ambrosioides* on mitochondria. Toxicol Appl Pharmacol 240:337–347. https://doi.org/10.1016/j.taap.2009.08.001

Monzote L, Hill GM, Cuellar A, Scull R, Setzer WN (2012) Chemical composition and anti-proliferative properties of *Bursera graveolens* essential oil. Nat Prod Commun 7(11):1531–1534. https://doi.org/10.1177/1934578X1200701130

Monzote L, Pastor J, Scull R, Gille L (2014c) Antileishmanial activity of essential oil from *Chenopodium ambrosioides* and its main components against experimental cutaneous leishmaniasis in BALB/c mice. Phytomedicine 21:1048–1052. https://doi.org/10.1016/j.phymed.2014.03.002

Monzote L, Scherbakov AM, Scull R, Gutiérrez YI, Satyal P, Cos P, Shchekotikhin AE, Gille L, Setzer WN (2020c) Pharmacological assessment of the carvacrol chemotype essential oil from *Plectranthus amboinicus* growing in Cuba. Nat Prod Commun 15:1–12. https://doi.org/10.117 7/1934578X20962233

Moraes ARDP, Tavares GD, Rocha FJS, Paula E, Giorgio S (2018) Effects of nanoemulsions prepared with essential oils of copaiba- and andiroba against *Leishmania infantum* and *Leishmania amazonensis* infections. Exp Parasitol 187:12–21. https://doi.org/10.1016/j.exppara.2018.03.005

Morais SM, Cossolosso DS, Silva AAS, Moraes M, Teixeira M, Campello C, Bonilla O, Paula V, Vila-Nova N (2019) Essential oils from *Croton* species: chemical composition *in vitro* and *in silico* antileishmanial evaluation antioxidant and cytotoxicity activities. J Braz Chem Soc 30:2404–2412. https://doi.org/10.21577/0103-5053.20190155

Moreira RRD, Martins GZ, Varandas R, Cogo J, Perego CH, Roncoli G, Sousa MDC, Nakamura CV, Salgueiro L, Cavaleiro C (2017) Composition and leishmanicidal activity of the essential oil of *Vernonia polyanthes* Less (Asteraceae). Nat Prod Res 31:2905–2908. https://doi.org/1 0.1080/14786419.2017.1299723

Moreira RRD, Santos AG, Carvalho FA, Perego CH, Crevelin EJ, Crotti AEM, Cogo J, Cardoso MLC, Nakamura CV (2019) Antileishmanial activity of *Melampodium divaricatum* and *Casearia sylvestris* essential oils on *Leishmania amazonensis*. Rev Inst Med Trop Sao Paulo 61:e33. https://doi.org/10.1590/s1678-9946201961033

Mutlu-Ingok A, Devecioglu D, Dikmetas DN, Karbancioglu-Guler F, Capanoglu E (2020) Antibacterial antifungal antimycotoxigenic and antioxidant activities of essential oils: an updated review. Molecules 25:4711. https://doi.org/10.3390/molecules25204711

Nafiah MA, Mukhtar MR, Omar H, Ahmad K, Morita H, Litaudon M, Awang K, Hadi AH (2011) N-cyanomethylnorboldine: a new aporphine isolated from Alseodaphne perakensis (Lauraceae). Molecules 16:3402–3409. https://doi.org/10.3390/molecules16043402

Nagoor Meeran MF, Javed H, Taee HA, Azimullah S, Ojha SK (2017) Pharmacological properties and molecular mechanisms of thymol: prospects for its therapeutic potential and pharmaceutical development. Front Pharmacol 8:1–34. https://doi.org/10.3389/fphar.2017.00380

Neira LF, Mantilla JC, Stashenko E, Escobar P (2018) Toxicidad genotoxicidad y actividad anti-*Leishmania* de aceites esenciales obtenidos de cuatro (4) quimiotipos del género *Lippia*. B Latinoam Caribe Pl 17:68–83

Nunes TAL, Costa LH, Sousa JMS, Souza VMR, Rodrigues RRL, Val MDCA, Pereira ACTC, Ferreira GP, Silva MV, Costa JMAR, Véras LMC, Diniz RC, Rodrigues KAF (2021) *Eugenia piauhiensis* Vellaff. essential oil and γ-elemene its major constituent exhibit antileishmanial activity promoting cell membrane damage and *in vitro* immunomodulation. Chem Biol Interact 339:109429. https://doi.org/10.1016/j.cbi.2021.109429

Ogungbe IV, Setzer WN (2013) In-silico Leishmania target selectivity of antiparasitic terpenoids. Molecules 18:7761–7847. https://doi.org/10.3390/molecules18077761

Olekhnovitch R, Bousso P (2015) Induction propagation and activity of host nitric oxide: lessons from leishmania infection. Trends Parasitol 31:653–664. https://doi.org/10.1016/j.pt.2015.08.001

Oliani J, Siqueira CAT, Sartoratto A, Queiroga CL, Moreno PRH, Reimão JQR, Tempone AG, Diaz IEC, Fischer DCH (2013) Chemical composition and *in vitro* antiprotozoal activity of the volatile oil from leaves of *Annona crassiflora* Mart. (Annonaceae). Pharmacologyonline 3:8–15

Oliveira DM, Furtado FB, Gomes AAS, Belut BR, Nascimento EA, Morais SAL, Martins CHG, Santos VCO, Silva CV, Teixeira TL, Cunha LCS, Oliveira A, Aquino FJT (2020) Chemical constituents and antileishmanial and antibacterial activities of essential oils from *Scheelea phalerata*. ACS Omega 5:1363–1370. https://doi.org/10.1021/acsomega.9b01962

Oliveira EA, Martins EGA, Soares MG, Chagas-Paula DA, Passero LFD, Sartorelli P, Baldim JL, Lago JHGA (2019) Comparative study on chemical composition antileishmanial and cytotoxic activities of the essential oils from leaves of *Guarea macrophylla* (Meliaceae) from two different regions of São Paulo State Brazil using multivariate statistical analysis. Braz Chem Soc 30:1395–1405. https://doi.org/10.21577/0103-5053.20190035

Oliveira ESC, Amaral ACF, Lima ES, Silva JRA (2014) Chemical composition and biological activities of *Bocageopsis multiflora* essential oil. J Essent Oil Res 26:161–165. https://doi.org/1 0.1080/10412905.2013.840809

Passero LFD, Brunelli ES, Sauini T, Pavani TFA, Jesus JA, Rodrigues E (2021) The potential of traditional knowledge to develop effective medicines for the treatment of leishmaniasis. Front Pharmacol 12:1408. https://doi.org/10.3389/fphar.2021.690432

Pereira PS, Maia AJ, Duarte AE, Oliveira-Tintino CDM, Tintino SR, Barros LM, Vega-Gomez MC, Rolón M, Coronel C, Coutinho HDM, Silva TG (2018) Cytotoxic and anti-kinetoplastid potential of the essential oil of *Alpinia speciosa* K. Schum. Food Chem Toxicol 119:387–391. https://doi.org/10.1016/j.fct.2018.01.024

Piasecka A, Jedrzejczak-Rey N, Bednarek P (2015) Secondary metabolites in plant innate immunity: conserved function of divergent chemicals. New Phytol 206:948–964. https://doi. org/10.1111/nph.13325

Roatt BM, de Oliveira Cardoso JM, De Brito RCF, Coura-Vital W, de Oliveira Aguiar-Soares RD, Reis AB (2020) Recent advances and new strategies on leishmaniasis treatment. Appl Microbiol Biotechnol 104:8965–8977. https://doi.org/10.1007/s00253-020-10856-w

Roberts SC, Tancer MJ, Polinsky MR, Gibson KM, Heby O, Ullman B (2004) Arginase plays a pivotal role in polyamine precursor metabolism in *Leishmania*: characterization of gene deletion mutants. J Biol Chem 279:23668–23678. https://doi.org/10.1074/jbc.M402042200

Rocha LG, Almeida JRGS, Macêdo RO, Barbosa-Filho JM (2005) A review of natural products with antileishmanial activity. Phytomedicine 12:514–535. https://doi.org/10.1016/j. phymed.2003.10.006

Rodrigues JCF, Rodriguez C, Urbina JA (2002) Ultrastructural and biochemical alterations induced by promastigote and amastigote forms of *Leishmania amazonensis*. Antimicrob Agents Chemother 46:487–499. https://doi.org/10.1128/AAC.46.2.487

Rodrigues KA, Amorim LV, de Oliveira JM, Dias CN, Moraes DF, Andrade EH, Maia JG, Carneiro SM, Carvalho FA (2013) *Eugenia uniflora* L. essential oil as a potential anti-*Leishmania* agent: effects on *Leishmania amazonensis* and possible mechanisms of action. Evid Based Complement Altern Med:2013. https://doi.org/10.1155/2013/279726

Rodrigues KA, Amorim LV, Dias CN, Moraes DF, Carneiro SM, Carvalho FA (2015) *Syzygium cumini* (L.) Skeels essential oil and its major constituent α-pinene exhibit anti-*Leishmania* activity through immunomodulation *in vitro*. J Ethnopharmacol 16:32–40. https://doi.org/10.1016/j. jep.2014.11.024

Rottini MM, Amaral ACF, Ferreira JLP, Oliveira ESC, Silva JRA, Taniwaki NN, Dos Santos AR, Almeida-Souza F, de Souza CDSF, Calabrese KDS (2019) *Endlicheria bracteolata* (Meisn.) essential oil as a weapon against *Leishmania amazonensis*: *in vitro* assay. Molecules 24:1–13. https://doi.org/10.3390/molecules24142525

Rubulotta G, Quadrelli EA (2019) Terpenes: a valuable family of compounds for the production of fine chemicals. In: Rubulotta G, Quadrelli EA (eds) Studies in surface science and catalysis, 1nd edn. Elsevier, Amsterdam, NL, pp 215–229

Saldanha AA, Vieira L, Ribeiro RIMA, Thomé RG, Santos HBD, Silva DB, Carollo CA, Oliveira FM, Lopes DO, Siqueira JM, Soares AC (2019) Chemical composition and evaluation of the anti-inflammatory and antinociceptive activities of *Duguetia furfuracea* essential oil: effect on edema leukocyte recruitment tumor necrosis factor alpha production iNOS expression and adenosinergic and opioidergic systems. J Ethnopharmacol 231:325–336. https://doi. org/10.1016/j.jep.2018.11.017

Sales VS, Monteiro ÁB, Delmondes GA, Nascimento EP, Figuêiredo FRSDN, Rodrigues CKS, Lacerda JFE, Fernandes CN, Barbosa MO, Brasil AX, Tintino SR, Gomez MCV, Coronel C,

Coutinho HDM, Costa JGM, Felipe CFB, Menezes IRA, Kerntopf MR (2018) Antiparasitic activity and essential oil chemical analysis of the *Piper tuberculatum* Jacq fruit. Iran J Pharm Res 17:268–275. https://doi.org/10.3390/molecules26195848

Sampaio MGV, Santos CRB, Vandesmet LCS, Santos BS, Santos IBS, Correia MTS, Martins ALB, Silva LCN, Menezes IRA, Gomez MCV, Silva MV (2021) Chemical composition antioxidant and antiprotozoal activity of *Eugenia gracillima* Kiaersk. leaves essential oil. Nat Prod Res 35:1914–1918. https://doi.org/10.3390/molecules26195848

Sanchez-Suarez J, Riveros I, Delgado G (2013) Evaluation of the leishmanicidal and cytotoxic potential of essential oils derived from ten Colombian plants. Iran J Parasitol 8:129–136

Santin MR, Santos AO, Nakamura CV, Dias Filho BP, Ferreira IC, Ueda-Nakamura T (2009) *In vitro* activity of the essential oil of *Cymbopogon citratus* and its major component (citral) on *Leishmania amazonensis*. Parasitol Res 105:1489–1496. https://doi.org/10.1007/s00436-009-1578-7

Santoro GF, Cardoso MG, Guimarães LG, Salgado AP, Menna-Barreto RF, Soares MJ (2007) Effect of oregano (*Origanum vulgare* L.) and thyme (*Thymus vulgaris* L.) essential oils on *Trypanosoma cruzi* (Protozoa: Kinetoplastida) growth and ultrastructure. Parasitol Res 100:783–790. https://doi.org/10.1007/s00436-006-0326-5

Santos AO, Ueda-Nakamura T, Dias Filho BP, Veiga Junior VF, Pinto AC, Nakamura CV (2008) Effect of Brazilian copaiba oils on *Leishmania amazonensis*. J Ethnopharmacol 120:204–208. https://doi.org/10.1016/j.jep.2008.08.007

Sen R, Ganguly S, Saha P, Chatterjee M (2010) Efficacy of artemisinin in experimental visceral leishmaniasis. Int J Antimicrob Agents 36:43–49. https://doi.org/10.1016/j.ijantimicag.2010.03.008

Sharifi-Rad M, Salehi B, Sharifi-Rad J, Setzer WN, Iriti M (2018) *Pulicaria vulgaris* Gaertn. essential oil: an alternative or complementary treatment for Leishmaniasis. Cell Mol Biol 64:18–21. https://doi.org/10.14715/cmb/2018.64.8.3

Sharma S, Barkauskaite S, Jaiswal AK, Jaiswal S (2021) Essential oils as additives in active food packaging. Food Chem 343:128403. https://doi.org/10.1016/j.foodchem.2020.128403

Silva ER, Castilho TM, Pioker FC, Silva CHTP, Floeter-Winter LM (2002) Genomic organisation and transcription characterisation of the gene encoding *Leishmania (Leishmania) amazonensis* arginase and its protein structure prediction. Int J Parasitol 32:727–737. https://doi.org/10.1016/S0020-7519(02)00002-4

Silva VD, Almeida-Souza F, Teles AM, Neto PA, Mondego-Oliveira R, Mendes Filho NE, Taniwaki NN, Abreu-Silva AL, Calabrese KS, Mouchrek Filho VE (2018) Chemical composition of *Ocimum canum* Sims. essential oil and the antimicrobial antiprotozoal and ultrastructural alterations it induces in *Leishmania amazonensis* promastigotes. Ind Crop Prod 119:201–208. https://doi.org/10.1016/j.indcrop.2018.04.005

Siqueira CAT, Oliani J, Sartoratto A, Queiroga CL, Moreno PRH, Reimão JQ, Tempone AG, Fischer DCH (2011) Chemical constituents of the volatile oil from leaves of *Annona coriacea* and in vitro antiprotozoal activity. Rev Bras Farmacogn 21:33–40. https://doi.org/10.1590/S0102-695X2011005000004

Siqueira CA, Serain AF, Pascoal AC, Andreazza NL, de Lourenço CC, Ruiz AL, de Carvalho JE, de Souza AC, Mesquita JT, Tempone AG, Salvador MJ (2015) Bioactivity and chemical composition of the essential oil from the leaves of *Guatteria australis* a.St.-Hil. Nat Prod Res 29:1966–1969. https://doi.org/10.1080/14786419.2015.1015017

Tasdemir D, Kaiser M, Demirci B, Demirci F, Baser KHC (2019) Antiprotozoal activity of Turkish origanum onites essential oil and its components. Molecules 24:1–16. https://doi.org/10.3390/molecules24234421

Teles AM, Rosa TDDS, Mouchrek AN, Abreu-Silva AL, Calabrese KDS, Almeida-Souza F (2019) *Cinnamomum zeylanicum* origanum vulgare and curcuma longa essential oils: chemical composition antimicrobial and antileishmanial activity. Evid Based Complement Altern Med 2019. https://doi.org/10.1155/2019/2421695

Ultee A, Bennik MHJ, Moezelaar R (2002) The phenolic hydroxyl group of carvacrol is essential for action against the food-borne pathogen Bacillus cereus. Appl Environ Microbiol 68:1561–1568. https://doi.org/10.1128/AEM.68.4.1561-1568.2002

Vandesmet LCS, Menezes SA, Portela BYM, Sampaio MGV, Santos CRB, Lermen VL, Gomez MCV, Silva MV, Menezes IRA, Correia MTS (2020) Leishmanicidal and trypanocidal potential of the essential oil of *Psidium myrsinites* DC. Nat Prod Res 19:1–5. https://doi.org/10.1080/14786419.2020.1844688

Vannier-Santos M, Castro S (2009) Electron microscopy in antiparasitic chemotherapy: a (close) view to a kill. Curr Drug Targets 10:246–260. https://doi.org/10.2174/138945009787581168

Vincendeau P, Gobert AP, Daulouède S, Moynet D, Mossalayi MD (2003) Arginases in parasitic diseases. Trends Parasitol 19:9–12. https://doi.org/10.1016/S1471-4922(02)00010-7

Wang Q, Gong J, Huang X, Yu H, Xue F (2009) In vitro evaluation of the activity of microencapsulated carvacrol against *Escherichia coli* with K88 pili. J Appl Microbiol 107:1781–1788. https://doi.org/10.1111/j.1365-2672.2009.04374.x

World Health Organization (WHO) (2021) Leishmaniasis. Available online: https://www.who.int/health-topics/leishmaniasis#tab=tab_1. Accessed on 31 Aug 2021

Yang L, Wen KS, Ruan X, Zhao YX, Wei F, Wang Q (2018) Response of plant secondary metabolites to environmental factors. Molecules 23:762. https://doi.org/10.3390/molecules23040762

Chapter 14
Anti-*Toxoplasma* Effect of Essential Oils Used as Food Ingredient

Sandra Alves de Araújo, Wendel F. F. de Moreira, Ailésio R. M. Filho, Tatiane A. da Penha-Silva, Fernando Almeida-Souza, and Ana Lucia Abreu-Silva

14.1 Introduction

Toxoplasma gondii is an obligate intracellular protozoan, belonging to the family Sarcocystidae (Apicomplexa) and is responsible for toxoplasmosis, a zoonotic disease that affects both human and animal health, and for that reason has great importance to the public health worldwide (Ouologuem and Roos 2014; Flegr et al. 2014; Arab-Mazar et al. 2017; Benitez et al. 2017). It is a heteroxenic parasite, requiring two hosts to complete its biological cycle (Fig. 14.1). The definitive host is usually felids and the intermediate the host is mainly birds and mammals, including humans. The parasite is transmitted by different routes: fecal-oral, blood transfusion, organ transplantation, and transplacental route (Álvarez-Garcia et al. 2021).

Feline animals are primarily responsible for parasite dissemination via the fecal-oral route. Infected animals shed oocysts along with their feces into the environment. Humans, in turn, become infected by eating raw or undercooked meat, or by consumption of water and other foods contaminated by the oocysts. The congenital toxoplasmosis, via transplacental route, is potentially the most serious form, since it

S. A. de Araújo
Universidade Estadual do Maranhão, São Luís, MA, Brazil

Programa de Pós-graduação em Biotecnologia/RENORBIO, Universidade Federal do Maranhão, São Luís, MA, Brazil

W. F. F. de Moreira · A. R. M. Filho · T. A. da Penha-Silva · A. L. Abreu-Silva
Universidade Estadual do Maranhão, São Luís, MA, Brazil

F. Almeida-Souza (✉)
Universidade Estadual do Maranhão, São Luís, MA, Brazil

Laboratório de Imunomodulação e Protozoologia, Instituto Oswaldo Cruz, Fiocruz, Rio de Janeiro, Brazil
e-mail: fernandosouza@professor.uema.br

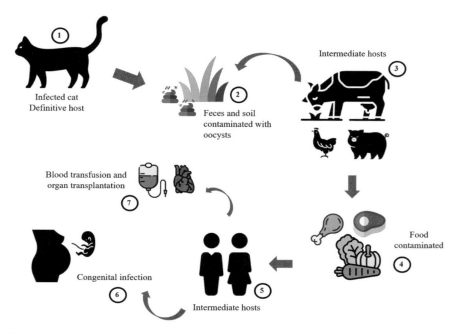

Fig. 14.1 Biological cycle and transmission routes of the *Toxoplasma gondii* protozoan. Infected cats, the definitive host, contaminate the soil with oocysts through their feces (1 and 2). Other animals, the intermediate hosts, become infected from oocysts present in the environment (3). Contaminated food such as meat and vegetables is consumed by humans (4 and 5). Congenital infection, blood transfusion and organ transplants are important routes of transmission (6 and 7)

is transmitted during pregnancy, resulting in severe sequelae for the fetus or even spontaneous abortions (Carmo et al. 2016; Minuzzi et al. 2020). Blood transfusion and organ transplantation are also considered routes of transmission, but are not so frequent (Hill and Dubey 2002). During its developmental stage, the protozoan has three infective forms: the tachyzoite, the bradyzoite, and the sporozoite (McAuley 2014).

The treatment of toxoplasmosis is done mainly through the combined use of drugs such as pyrimethamine, sulfadiazine and folinic acid, however, the disease is characterized by the absence of clinical symptoms in about 90% of cases, which makes the diagnosis, and consequently the treatment, difficult (Carmo et al. 2016; Dunay et al. 2018; Konstantinovic et al. 2019). The drugs currently available have several side effects, including megaloblastic anemia, leukopenia, and granulocytopenia, which justify the search for new and more effective treatment alternatives (Sanfelice et al. 2018; Giovati et al. 2018; Munera-López et al. 2019). In this sense, plants have been pointed out as potential alternatives in the development of new drugs.

Essential oils are complex and extremely volatile mixtures present in aromatic plants and are related to functions necessary for adaptation to the environment (Dhifi et al. 2016; Stashenko and Martinez 2019). Its commercial use is mainly concentrated in the food, cosmetics and perfume industries, but it has been explored in the development of alternative drugs due to the several pharmacological effects already described, among them, acaricide, antifungal and antiviral (Cruz et al. 2013; Gauch et al. 2014; Kim et al. 2017). Terpenes are the main constituents of essential oils and are thought to be responsible for their medicinal properties (Dhifi et al. 2016).

The use of essential oils in the treatment of toxoplasmosis, as well as the identification of active compounds, is still little explored. In this sense, it is necessary to search for the development of new drugs with fewer side effects, low costs and effectiveness for the treatment of this disease. The focus of this chapter is to explore the anti-toxoplasmosis potential and possible applications of essential oils against *T. gondii*, furthermore, explore the known relationships between the oil components and the biological activity of these blends, discuss the synergistic effects, the importance of encapsulation, and the possible mechanisms of action involved in the control of toxoplasmosis.

14.2   Essential Oils

Essential oils are secondary plant metabolite products. They are extremely volatile, hydrophobic in character and strong in odor, and are related to several important plant functions including attracting pollinators and plant defense (Dhifi et al. 2016). The essential oils can be extracted from flowers, leaves, bark, trunks, branches, roots, rhizomes, fruits or seeds and are stored in specialized structures called glandular trichomes, which when ruptured naturally release the essential oil (Aumeeruddy-Elalfi et al. 2018). The extraction methods to obtain essential oils are done by hydrodistillation and steam distillation techniques, using organic solvents, and by mechanical pressing (Stashenko and Martinez 2019; Wilkin et al. 2021).

Essential oils have a complex composition, with about 20 to 60 bioactive compounds in different concentrations, of which two to three compounds are called main compounds and are considered responsible for the aroma, application and biological properties (Bilia et al. 2014). The phytochemical analysis of the essential oils is performed by chromatographic and colorimetric techniques to identify the main classes of metabolites (Acquavia et al. 2021). Gas chromatography coupled to mass spectrometer (GC-MS) is an important tool in the evaluation of the complex composition of essential oils both in research laboratories and in industry (Stashenko and Martinez 2019). It is a technique widely used due to its accuracy and speed, simplicity of handling and because it is an efficient separation technique (Jalsenjak et al. 1987; Avato et al. 2005).

The phytochemical composition of these essential oils can vary and these variations are attributed to genetic and environmental factors, harvest time, extraction method, sensitivity of the active ingredient, as well as drying and storage method, which can imply in the oscillation of the constituents qualitative and/or quantitatively (Sanli and Karadogan 2017; Fokou et al. 2020). Thus, to obtain essential oils with constant composition, the plant material must be extracted under the same conditions, using the same plant organ, grown in similar soil, climate, and harvested at the same season (Hayet et al. 2017).

Essential oils are mainly constituted by terpenoids or isoprenoids, molecules essentially formed by isoprene (C_5H_8) (Mc Murry 2011). Monoterpenes consist of two isoprene units ($C_{10}H_{16}$), sesquiterpenes consist of three isoprene units ($C_{15}H_{24}$), diterpenes consist of four isoprene units ($C_{20}H_{32}$), triterpenes consist of six isoprene units ($C_{30}H_{48}$), and tetraterpenes have eight isoprene units. ($C_{40}H_{64}$) (Farkas and Mohácsi-Farkas 2014). The difference in the composition and quantity of these molecules in the essential oils is responsible for the differences in the effectiveness of biological activity and in the different uses of the essential oils.

14.3   Bioactivity of Essential Oils Against *Toxoplasma gondii*

Research on toxoplasmosis and the search for new chemotherapeutics address the importance of essential oils as a promising source to combat and/or control *T. gondii* (Table 14.1).

14.3.1   *In Vitro anti-*Toxoplasma *evaluation of essential oils*

Myristica fragrans (nutmeg) essential oil was investigated for its *in vitro* cytotoxicity on Vero cells and antiparasitic activity against *T. gondii* (Pillai et al. 2012). It was observed low cytotoxicity of the essential oil on normal cells with inhibitory concentration (IC_{50}) of 24.83 µg/mL, while in the anti-*T. gondii* assay, the essential oil showed activity with IC_{50} of 24.45 µg/mL, comparable to the positive control, clindamycin (IC_{50} = 16.57 µg/mL). The authors attribute the anti-*T. gondii* action of *M. fragrans* essential oil and the low cytotoxic effects on Vero cells to the terpenes present in the essential oil.

The antioxidant, cytotoxic and anti-*T. gondii* effect of *Psidium guajava* (guava) essential oil was observed by Lee et al. (2013). The antioxidant activity was evaluated through 2,2-diphenyl-1-picryl-hydrazyl-hydrate (DPPH) method, where essential oil was considered a moderate antioxidant (IC_{50} = 460.56 mg/mL) compared to ascorbic acid (IC_{50} = 18.41 mg/mL). Furthermore, the essential oil was not toxic to Vero cells (IC_{50} of 37.54 µg/mL), and the results demonstrated antiparasitic effect of *P. guajava* essential oil against *T. gondii* (IC_{50} = 3.94 µg/mL), when compared to the reference drug clindamycin (IC_{50} = 6.24 µg/mL). The anti-*T. gondii* activity of the essential oil was attributed to its antioxidant potential, suggested by free radical inhibition, as well as the presence of bioactive compounds. However, the authors highlighted the importance of studies on the isolation of these compounds in order to understand the possible mechanisms involved in its activity.

Khamesipour et al. (2021) reported the anti-*Toxoplasma* activity of *Dracocephalum kotschyi* essential oil. 1 µg of essential oil presented efficacy of

Table 14.1 Essential oils with anti-*Toxoplasma* activity, their respective active ingredients, assay type, and effective concentration and/or dose

Species	Main compounds	Assay	Concentration/ Dose	Reference
Thymus broussonetii	Carvacrol	*In vivo*	D = 20 µg	Dahbi et al. (2010)
Saturja khuzestanica	Carvacrol	*In vivo*	D = 0.2 and 0.3 mL/ kg	Mahmoudvand et al. (2017)
Origanum vulgare	Carvanol	*In vitro*	IC_{50} = 16.08 and 134.9 µg/mL	Yao et al. (2021b)
Zataria multiflora	Thymol; carvacrol; p-cymene	*In vivo*	D = 0.2 and 0.4 mL/ kg	Mahmoudvand et al. (2020)
Myrtus communis	α-pinene; 1,8-cineol; linalool	*In vivo*	D = 200 and 300 mg/ kg	Shaapan et al. (2021)
Pelargonium asperum	Linalool; geraniol	*In vitro*	IC_{50} = 1.4 mg/mL	Huang et al. (2021)
Bunium persicum	γ-Terpinene; cuminaldehyde	*In vivo*	D = 0.05 and 0.1 mL/ kg	Tavakoli-Kareshk et al. (2015)
Dracocephalum kotschyi	Copaene; methyl geranate; geranial; carvone	*In vitro; In vivo*	IC_{50} = 9.94 µg/mL; D = 200 mg/kg	Khamesipour et al. (2021)
Allium sativum	Diallyl disulfide; Diallyl trisulfide; allyl methyl trisulfide	*In vitro; In vivo*	IC_{50} = 66.9 µg/mL; D = 200, 400 and 600 mg/kg	Alnomasy (2021)
Cymbopogon nardus; Cymbopogon citratus	nd	*In vitro*	IC_{50} = 2.5 and 4.6 µg/ mL	Elazab et al. (2021)
Lavandula angustifólia	nd	*In vitro*	IC_{50} = 4.48 mg/mL	Yao et al. (2021a)
Myristica fragrans	nd	*In vitro*	IC_{50} = 24.45 µg/mL	Pillai et al. (2012)
Psidium guajava	nd	*In vitro*	IC_{50} = 3.94 µg/mL	Lee et al. (2013)

IC_{50} inhibitory concentration of 50% of the population; *D* dose; *nd* not determined

96.33, 90.66, and 86.66% on parasite viability after incubation at 30, 90 and 180 min, respectively. The effect of *D. kotschyi* essential oil against *T. gondii* was significant (IC_{50} = 9.94 µg/mL) compared to sulfadiazine (IC_{50} = 391.1 µg/mL) and pyrimethamine (IC_{50} = 84.2 µg/mL). The main substances identified in the essential oil were copaene (22.15%), methyl geranate (16.31%), geranial (13.78%) and carvone (11.34%). The authors suggested that the essential oil may also dose-dependent reduce ATP levels in *T. gondii,* interfering in the mitochondrial function of the parasite.

Elazab et al. (2021) verified the anti-*T. gondii* activity of different essential oils, including *Cymbopogon citratus* (lemon grass), *Origanum majorana* (marjoram), *Nasturtium officionale* (cress), *Salvia rosmarinus* (rosemary), *Cymbopogon nardus* (citronella), *Syzygium aromaticum* (clove) and *Ocimum basilicum* (basil). Lemon grass and citronella essential oils were more active against *T. gondii* with IC_{50} of 2.5 μg/mL and 4.6 μg/mL, respectively, when compared to the positive control, sulfadiazine (IC_{50} = 99.4 μg/mL). The activity of the oils against the parasite was associated with its chemical constituents.

The antiparasitic activity against *T. gondii* was also observed by Huang et al. (2021) using the commercial essential oils of *Eucalyptus globulus* (eucalyptus), *Cupressus sempervirens* (italian cypress), *Citrus aurantifolia* (lime), *Melaleuca alternifolia* (tea tree) and *Pelargonium asperum* (pelargonium). The latter exhibited better antiparasitic activity when compared to sulfadiazine, inhibiting the growth of *T. gondii* at IC_{50} = 1.4 mg/mL. Additionally, it was observed that the invasion rate in HFF cells by *T. gondii* in 60 min was reduced by 41.18% after treatment with 3.55 mg/mL of the oil compared to the untreated group (67.64%). The ultrastructural analysis of the parasites through scanning electron microscopy indicated that the parasites became smaller and contracted, and also presented a rough cellular membrane, showing to consider that the *P. asperum* essential oil acted directly on the integrity of the cell membrane of *T. gondii*. Its main constituents identified were linalool and geraniol.

Lavandula angustifolia (lavender) essential oil demonstrated antiparasitic activity on *T. gondii*, IC_{50} of 4.48 mg/mL (Yao et al. 2021a). The inhibition occurred in a dose-dependent manner at safe concentrations. Ultrastructural analysis of the parasites and the cell invasion assay were used to explain the mechanism of action. The invasion rates of *T. gondii* at 20, 40 and 60 min after treatment were 21.3%, 29.77% and 39.17%, respectively. While the untreated groups, at the same time intervals, the invasion was 38.50%, 51.51%, and 67.64%, respectively. The lavender essential oil was able to reduce the invasion of tachyzoites into host cells due to serious morphological changes in the tachyzoites.

The anti-*T. gondii* from *Origanum vulgare* (oregano) essential oil and its main compound, carvanol, were also verified by Yao et al. (2021b) with both displaying antiparasitic activity, IC_{50} of 134.9 and 43.93 μg/mL respectively. The inhibitory effect of cell invasion was also observed for carvanol (IC_{50} = 16.08 μg/mL) and *O. vulgare* essential oil (IC_{50} = 7.68 μg/mL). Furthermore, treatment with essential oil and carvanol altered the morphology of the tachyzoites, with distorted and wrinkled structures, compared to the untreated group. These results showed that oregano essential oil was able to inhibit the proliferation of *T. gondii*, reducing its invasion and acting directly on the mobility of tachyzoites.

Alnomasy (2021) evaluated the effect of *Allium sativum* (garlic) essential oil against the *T. gondii* (RH strain). Different concentrations of essential oils were analyzed (32.5, 75 and 150 μg/mL), however, the highest efficacy was observed at the concentration of 150 μg/mL where the mortality of tachyzoites was 100% after 2 h of incubation. Essential oil of *A. sativum* significantly decreased the mean number of intracellular parasites with IC_{50} of 66.9 μg/mL, lower than reference drug atovaquone, IC_{50} of 72.11 μg/mL. The main substances identified in the essential oil were diallyl disulfide (29.2%), diallyl trisulfide (28.6% and allyl methyl trisulfide (19.8%).

14.3.2   *In Vivo anti-*Toxoplasma *activity of essential oils*

The action of *Thymus broussonetii* (thyme) essential oil, rich in carvacrol (83.18%), on animals infected with *T. gondii* (PRU strain) were observed by Dahbi et al. (2010). For fifteen days, mice were treated with essential oil and then euthanized. The cysts in the brain were counted and the control group had an average of 121.5 cysts per encephalon, while the treated groups showed an absence of intracerebral cysts. The animals displayed no change in their behavior or health, demonstrating that the essential oil had no side effects.

Evaluating the efficacy of the *Bunium persicum* (black cumin) essential oil against acute toxoplasmosis, Tavakoli-Kareshk et al. (2015) intraperitoneally infected animals with *T. gondii*. After 24 h of infection, the animals were treated orally with black cumin essential oil for 5 days. The infected animals treated with essential oil had longer survival (8 to 9 days) when compared to the control group (5 days) and the mean number of parasites was lower for the treated animals when compared to the control group, demonstrating that essential oil was effective in treating acute toxoplasmosis. The main constituents of the black cumin essential oil were γ-terpinene (46.1%) and cuminaldehyde (15.5%), which were attributed the anti-*T. gondii* effect.

The prophylactic effects of *Satureja khuzestanica* (satureja) essential oil on mice infected with acute toxoplasmosis were observed by Mahmoudvand et al. (2017). The animals were treated orally, once a day for 15 days, with the oil and 24 h later were infected with *T. gondii* tachyzoites. The mortality rate of the infected mice was lower (7–8 days) after the administration of the essential oil when compared to the control group (5 days). Furthermore, the mean number of parasites significantly reduced in the treated groups compared to the control group. Thus, the *S. khuzestanica* essential oil exhibited prophylactic effects against acute toxoplasmosis in the animals. Among the compounds identified in the essential oil, the main compound was carvacrol (78,8%).

Mahmoudvand et al. (2020) evaluated the efficacy and safety of *Zataria multiflora* (satar) essential oil against acute toxoplasmosis in mice. Animals infected with *T. gondii* tachyzoites were treated and observed daily. *Z. multiflora* essential oil increased survival time in mice with acute toxoplasmosis. The mortality rate of animals in the control group was 100% on day 5, while mortality was observed from day 8 onward in the animals treated with essential oil. In addition, the mean number of tachyzoites in the treated animals were lower compared to the control group. The main compounds identified in the oil were thymol (45.42%), carvacrol (25.96%) and ρ-cymene (10.65%).

Dracocephalum kotschyi essential oil riched with copaene (22.15%), methyl geranate (16.31%), geranial (13.78%) and carvone (11.34%) showed anti-*Toxoplasma* activity in mice (Khamesipour et al. 2021). The infected animals treated orally with essential oil, sulfadiazine and pyrimethamine were monitored daily for 14 days. During this period, the mice treated with *D. kotschyi* essential oil died from day 8 and 12 after treatment. However, animals treated with sulfadiazine and pyrimethamine died from day 6 and 7, respectively. In this sense, *D. kotschyi* essential oil improved the survival rate of animals compared to the control groups. The authors further suggest that anti-*Toxoplasma* activity of the essential oil is due to the presence of the active molecules.

Shaapan et al. (2021) evaluated the prophylactic effects of *Myrtus communis* (myrtle) essential oil against chronic toxoplasmosis in mice induced by *T. gondii* (Tehran strain). The animals were administered orally with essential oil and also with atovaquone (positive control) for 21 days. On day 15, the mice were infected by intraperitoneal inoculation with 20–25 *T. gondii* tissue cysts. The mean number of *T. gondii* cysts as well as their mean diameter in the treated groups was significantly reduced in a dose-dependent manner compared to the control group. The main constituents identified in the essential oil were α-pinene (24.7%), 1,8-cineole (19.6%) and linalool (12.6%).

The garlic essential oil was also assayed in vivo against the RH strain in *T. gondii* murine infection. The animals pre-treated for 14 days with garlic essential oil at 600 µg/kg/day showed a decrease in mortality rate from the 6th day post-infection, three more days than untreated mice. The average number of tachyzoites was also reduced on the 5th day by 64.9%, 79.5% and 92.4%, for the treatment at 200, 400 and 600 µg/kg/day, respectively, compared to the atovaquone group (100 mg/kg), 86.7%. Garlic essential oil has prophylactic effects against *T. gondii*, increasing the survival rate of animals and reducing their parasite load. In addition, garlic essential oil acted on the immune system, stimulating the secretion of cytokines IFN-γ and IL-1β, and inhibiting liver damage in animals with toxoplasmosis (Alnomasy 2021).

Essential oils offer excellent potential activity against the protozoan *T. gondii*, however, the mechanisms of action are not yet well understood, although it is believed that the lipophilicity of the oils may be an important characteristic in the treatment of toxoplasmosis. Parasitic protozoa have a plasma membrane composed of lipid structures, mainly phospholipids and cholesterols, and some associated proteins (Vial et al. 2003). In this sense, essential oils can be easily solubilize fatty membranes and disintegrate important structures for parasites.

Huang et al. (2021) and Yao et al. (2021a) observed through ultrastructural analysis that *T. gondii* tachyzoites treated with essential oil of *Pelargonium X. asperum* (geranium) and lavender presented smaller, deformed, contracted and with rough cellular membrane. According to the authors, the essential oils affected the parasites cell membrane, damaging its permeability, inhibiting its movement and invasion of host cells. Furthermore, the components of the essential oils can interfere with the parasites metabolism, since they are able to disrupt ion channels, destroy the depolarization of the mitochondrial membrane, cause electrolyte leakage and make the mitochondria permeable, causing serious damage and consequently leading to the parasites death (Swamy et al. 2016).

Additionally, ultrastructural analysis was used to explain the mechanism of action of *Ocimum canum* (wild basil) essential oil on protozoa of the genus *Leishmania* (Silva et al. 2018). The essential oil induced autophagosome, lipid bodies, discontinuity of nuclear membrane and exocytic activity by the flagellar pocket of *Leishmania amazonensis*. Such actions can be explained by the hydrophobic nature of the EOs, allowing them to easily interact with fatty acids present in the dense cell membrane (Cox et al. 2000; Boyom et al. 2003; Burt 2004).

Immunomodulatory properties of the essential oils have also been pointed out as possible mechanisms of action on protozoa. Immunomodulators are molecules with the ability to act on the mechanisms of the host immune response (Anastasiou and Gerhard 2017; Sandner et al. 2020). In toxoplasmosis, Mahmoudvand et al. (2020) observed that *Z. multiflora* essential oil stimulated the immune response in infected animals, since previous studies reported the immunomodulatory activity of *Z. multiflora*, increasing the level of IFN-γ, stimulating phagocytosis, potentiating Th1 and humoral immune responses (Dupont et al. 2012; Mosleh et al. 2013; Boskabady et al. 2013). Although the immunostimulatory effects of essential oils are still unclear, there is evidence that essential oils have the potential to enhance some immune functions in organisms. Carrasco et al. (2009) verified the immunomodulatory activity of clove (*S. aromaticum*), ginger (*Zingiber officinale*) and sage (*Salvia officinalis*) essential oils in healthy and immunosuppressed animals. According to the authors, the essential oils were able to increase the number of leukocytes and the delayed-type hypersensitivity (DTH) response, and restore the cellular and humoral immune responses of the animals.

In leishmaniasis, Rodrigues et al. (2015) found that *Syzygium cumini* essential oil and its main component α-pinene increased phagocytic and lysosomal activities as well as nitric oxide production of peritoneal macrophages of mice infected with *L. amazonensis*, demonstrating that anti-*Leishmania* activity may be mediated by immunomodulation induced by the essential oils.

In addition to the essential oils, phytochemical isolates have been analyzed for their action on the toxoplasmosis parasite. Oliveira et al. (2016) verified the anti-*T. gondii* activity of estragole and thymol, compounds isolated from *Croton zehntneri* and *Lippia sidoides* (rosemary pepper), respectively. In the in vivo model using estragole, the subcutaneous and oral treatment reduced the number of animals death. While only subcutaneous treatment with thymol showed a similar effect. In addition, the compounds also exhibited modulation of the inflammatory response. The treatment with estragole induced high production of IL-12 and INF-γ, being fundamental in the generation of a potent Th1 response, which is efficient in the control of intracellular parasites. Thymol also showed antioxidant activity, conferring a protective role against protozoa, reducing oxidative stress generated during infection and providing less tissue damage.

In this chapter, other phytochemicals were identified and pointed out as the main responsible for the anti-*T. gondii* activity of essential oils, namely carvacrol, copaene, methyl geranate, geranial, α-pinene, 1,8-cineole, linalool, and carvone among others, as shown in Table 14.1. The complexity of chemical constituents present in essential oils may be directly involved in their biological activity. However, studies indicate that the different bioactivities of these oils may also be related to the possible synergistic effects among its phytochemicals (Wilkin et al. 2021).

Synergism occurs when the combined effect is greater than, or equal to, the sum of the individual effects (Davidson and Parish 1989; Dorman and Deans 2000; Burt

2004). The synergistic activity of essential oils on *T. gondii* is still unknown in the literature, however, the current treatment of toxoplasmosis is performed by the combined use of pyrimethamine and sulfadiazine, suggesting that these drugs act synergistically inhibiting enzymes important in the biosynthesis of pyrimidines essential for the survival and replication of the parasite (Teixeira et al. 2020).

Currently, synergism between essential oils and some commercial drugs is also observed in the literature. For example, in leishmaniasis, Jihene et al. (2020) observed that propolis essential oil combined with amphotericin B exhibited synergistic potential against promastigotes and amastigotes of *Leishmania major* and *Leishmania infantum*. Similar results were observed by Mubarak and Alnomani (2020). According to the authors, the combined use of *Eucalyptus camaldulensis* essential oil and meglumine antimoniate (Glucantime®) exerted a more efficient effect on the wound progression process at the base of the tail of animals with cutaneous leishmaniasis.

Synergism between essential oils, active compounds and commercial drugs has numerous therapeutic advantages including increased potency, decreased toxicity, reduced drug resistance, display fewer side effects and lower cost (Pourghanbari et al. 2016). Thus, essential oils and bioactive compounds may be a promising alternative in the treatment of toxoplasmosis when used alone and/or in combination with commercial drugs.

14.4 Optimizing the Use of Essential Oils

Essential oils are products characterized by high volatility, easily degraded by oxidation, heating, and exposure to light. Thus, nanotechnology through the encapsulation of oils is an important vehicle, since it reduces some problems associated with sensitivity and product solubility, in addition to increasing efficiency (Bedoya-Serna et al. 2018; Moradi and Barati 2019). Among the different nanoformulations, the most commonly used in the encapsulation of essential oils are emulsions (nanoemulsions and microemulsions), solid lipid nanoparticles (SBN), polymeric nanoparticles (PPN) and liposome (Asbahani et al. 2015; Echeverría and Alburquerque 2019; Ashaolu 2021).

In the literature, some studies have already reported the importance of the application of nanoformulations in increasing the efficiency of essential oils against different parasitic diseases. Baldissera et al. (2013) evaluated the *in vitro* activity of nanoemulsified *Schinus molle* essential oil against *Trypanosoma evansi* and observed that the nanoemulsion at 0.5% and 1% were able to reduce the number of live parasites by 81% and 100%, respectively, when compared to the non-emulsified essential oil (63% and 68%, respectively), demonstrating a greater potential of the nanoemulsion in causing parasite mortality.

Additionally, Shokri et al. (2017) also investigated the antiparasitic activity of nanoemulsified essential oils. The nanoemulsions of lavender and rosemary essential oils showed antileishmanial activity on promastigotes of *L. major* with $IC_{50} = 0.11$ µL/mL and $IC_{50} = 0.08$ µL/mL, respectively.

In the treatment of toxoplasmosis, there are no reports of the activity of nanoformulations using essential oils, however, Azami et al. (2018) observed that the nano-encapsulated drug atovaquone showed more efficient antiparasitic effects on *T. gondii* by in vitro and in vivo assays. Nanoemulsified atovaquone increased oral bioavailability, tissue distribution, survival time of the animals, reduced parasitemia and the number and size of brain cysts, highlighting the importance of nanoformulations.

Therefore, the use of nanostructures for the encapsulation of essential oils, bioactive molecules and/or drugs has numerous advantages, including stability for long periods of storage, protection against external factors, ease of handling, and also decreases marketing costs (Barradas and Silva 2020; Costa et al. 2021). However, it still requires a lot of development in order to achieve the optimization of the obtaining methods for industrial scale production.

14.5   Final Considerations

Essential oils have proven to be effective against various infectious diseases, including the *T. gondii* parasite. However, their biological activities depend basically on their phytochemical compounds. We have seen that essential oils have been playing an important role in folk medicine, as well as were subjective of few experimental studies against toxoplasmosis, with a majority evaluating essential oils that is also used as food ingredients. Understanding the mechanism of action and identifying the molecular targets involved could lead to the development of new therapeutic applications of essential oils and the treatment of various diseases, including toxoplasmosis. Thus, new pre-clinical and clinical studies must be carried out to elucidate and verify the real potential of essential oils as effective anti-*Toxoplasma* compounds.

Acknowledgements   This study was financed in part by the Coordenação de Aperfeiçoamento de Pessoal de Nível Superior – Brasil (CAPES) – Finance Code 001. Dr. Ana Lúcia Abreu-Silva are research fellow of National Council for Scientific and Technological Development – CNPq. Dr. Fernando Almeida-Souza is research fellow of CAPES.

References

Acquavia MA, Pascale R, Foti L, Carlucci G, Scrano L, Martelli G, Brienza M, Coviello D, Bianco G, Lelario F (2021) Analytical methods for extraction and identification of primary and secondary metabolites of apple (Malus domestica) fruits: a review. Separations 8:91. https://doi.org/10.3390/separations8070091
Alnomasy SF (2021) In vitro and in vivo anti-toxoplasma effects of Allium sativum essential oil against toxoplasma gondii RH strain. Infect Drug Resist 30:5057–5068. https://doi.org/10.2147/IDR.S337905

Álvarez-García G, Davidson R, Jokelainen P, Klevar S, Spano F, Seeber F (2021) Identification of Oocyst-driven toxoplasma gondii infections in humans and animals through stage-specific serology-current status and future perspectives. Microorganisms 13:2346. https://doi.org/10.3390/microorganisms9112346

Anastasiou C, Gerhard B (2017) Essential oils as immunomodulators: some examples. Open Chem 15:352–370. https://doi.org/10.1515/chem-2017-0037

Arab-Mazar Z, Kheirandish F, Rajaeian S (2017) Anti-toxoplasmosis activity of herbal medicines: narrative review. Herb Med J 1:43–49. https://doi.org/10.22087/hmj.v2i1.594

Asbahani EA, Miladi K, Badri W, Sala M, Addi A, Casabianca H, Mousadik AE, Hartmann D, Jilale A, Renaud FNR, Elaissari A (2015) Essential oils: from extraction to encapsulation. Int J Pharm 483:220–243. https://doi.org/10.1016/j.ijpharm.2014.12.069

Ashaolu TJ (2021) Nanoemulsions for health, food, and cosmetics: a review. Environ Chem Lett 19:3381–3395. https://doi.org/10.1007/s10311-021-01216-9

Aumeeruddy-Elalfi Z, Lall N, Fibrich B, Blom van Staden A, Hosenally M, Mahomoodally MF (2018) Selected essential oils inhibit key physiological enzymes and possess intracellular and extracellular antimelanogenic properties in vitro. J Food Drug Anal 26:17. https://doi.org/10.1016/j.jfda.2017.03.002

Avato P, Fortunato IM, Ruta C, D'Elia R (2005) Glandular hairs and essential oils in micropropagated plants of Salvia officinalis L. Plant Sci 169:29–36. https://doi.org/10.1016/j.plantsci.2005.02.004Get

Azami SJ, Teimouri A, Keshavarz H, Amani A, Esmaeili F, Hasanpour H, Elikaee S, Salehiniya H, Shojaee S (2018) Curcumin nanoemulsion as a novel chemical for the treatment of acute and chronic toxoplasmosis in mice. Int J Nanomedicine 13:7363–7374. https://doi.org/10.2147/IJN.S181896

Baldissera MD, Da Silva AS, Oliveira CB, Zimmermann CE, Vaucher RA, Santos RC, Rech VC, Tonin AA, Giongo JL, Mattos CB, Koester L, Santurio JM, Monteiro SG (2013) Trypanocidal activity of the essential oils in their conventional and nanoemulsion forms: in vitro tests. Exp Parasitol 134:356–361. https://doi.org/10.1016/j.exppara.2013.03.035

Barradas TN, Silva KG (2020) Nanoemulsions as optimized vehicles for essential oils. In: Sustainable agriculture reviews 44. Springer, Cham, pp 115–167. https://doi.org/10.1007/978-3-030-41842-7_4

Bedoya-Serna CM, Dacanal GC, Fernandes AM, Pinho SC (2018) Antifungal activity of nanoemulsions encapsulating oregano (Origanum vulgare) essential oil: in vitro study and application in Minas Padrão cheese. Braz J Microbiol 49:929–935. https://doi.org/10.1016/j.bjm.2018.05.004

Benitez ADN, Martins FDC, Mareze M, Santos NJR, Ferreira FP, Martins CM, Garcia JL, Mitsuka-Breganó R, Freire RL, Biondo AW, Navarro IT (2017) Spatial and simultaneous representative seroprevalence of anti-Toxoplasma gondii antibodies in owners and their domiciled dogs in a major city of southern Brazil. PLoS One 12:e0180906. https://doi.org/10.1371/journal.pone.0180906

Bilia AR, Guccione C, Isacchi B, Righeschi C, Firenzuoli F, Bergonzi MC (2014) Essential oils loaded in nanosystems: a developing strategy for a successful therapeutic approach. Evid-Based Compl Altern Med 651593:1–14. https://doi.org/10.1155/2014/651593

Boskabady MH, Mehrjardi SS, Rezaee A, Rafatpanah H, Jalali S (2013) The impact of Zataria multiflora Boiss extract on in vitro and in vivo Th1/Th2 cytokine (IFN-γ/IL4) balance. J Ethnopharmacol 150:1024–1031. https://doi.org/10.1016/j.jep.2013.10.003

Boyom FF, Ngouana V, Zollo PH, Menut C, Bessiere JM, Gut J, Rosenthal PJ (2003) Composition and anti-plasmodial activities of essential oils from some Cameroonian medicinal plants. Phytochemistry 64:1269–1275. https://doi.org/10.1016/j.phytochem.2003.08.004

Burt S (2004) Essential oils: their antibacterial properties and potential applications in foods—a review. Int J Food Microbiol 94:223–253. https://doi.org/10.1016/j.ijfoodmicro.2004.03.022

Carmo EL, Bogoevich-Morais RAP, Oliveira AS, Figueredo JE, Figueredo MC, Silva AV, Bichara CNC, Póvoa MM (2016) Soroepidemiologia da infecção pelo Toxoplasma gondii. Rev Pan-Amaz Saude 7:79–87. https://doi.org/10.5123/s2176-62232016000400010

Carrasco FR, Schmidt G, Romero AL, Sartoretto JL, Caparroz-Assef SM, Bersani-Amado CA, Cuman RK (2009) Immunomodulatory activity of Zingiber officinale Roscoe, Salvia officinalis L. and Syzygium aromaticum L. essential oils: evidence for humor- and cell-mediated responses. J Pharm Pharmacol 61:961–967. https://doi.org/10.1211/jpp/61.07.0017

Costa IN, Ribeiro M, Silva Franco P, da Silva RJ, de Araújo TE, Milián I, Luz LC, Guirelli PM, Nakazato G, Mineo JR, Mineo T, Barbosa BF, Ferro E (2021) Biogenic silver nanoparticles can control toxoplasma gondii infection in both human trophoblast cells and villous explants. Front Microbiol 11:623947. https://doi.org/10.3389/fmicb.2020.623947

Cox SD, Mann CM, Markham JL, Bell HC, Gustafson JE, Warmington JR, Wyllie SG (2000) The mode of antimicrobial action of essential oil of alternifola (tea tree oil). J Appl Microbiol 88:170–175

Cruz EM, Costa LM Jr, Pinto JA, Santos Dde A, de Araujo SA, Arrigoni-Blank Mde F, Bacci L, Alves PB, Cavalcanti SC, Blank AF (2013) Acaricidal activity of Lippia gracilis essential oil and its major constituents on the tick Rhipicephalus (Boophilus) microplus. Vet Parasitol 1:198–202. https://doi.org/10.1016/j.vetpar.2012.12.046

Dahbi A, Bellete B, Flori P, Hssaine A, Elhachimi Y, Raberin H, Chait A, Tran Manh Sung R, Hafid J (2010) The effect of essential oils from *Thymus broussonetii* Boiss on transmission of *Toxoplasma gondii* cysts in mice. Parasitol Res 107:55–58. https://doi.org/10.1007/s00436-010-1832-z

Davidson PM, Parish ME (1989) Methods for testing the efficacy of food antimicrobials. Food Technol 43:148–155

Dhifi W, Bellili S, Jazi S, Bahloul N, Mnif W (2016) Essential Oils' chemical characterization and investigation of some biological activities: a critical review. Medicines (Basel) 22:25. https://doi.org/10.3390/medicines3040025

Dorman HJD, Deans SG (2000) Antimicrobial agents from plants: antibacterial activity of plant volatile oils. J Appl Microbiol 88:308–316

Dunay IR, Gajurel K, Dhakal R, Liesenfeld O, Montoya JG (2018) Treatment of toxoplasmosis: historical perspective, animal models, and current clinical practice. Clin Microbiol Rev 12:e00057–e00017. https://doi.org/10.1128/CMR.00057-17

Dupont CD, Christian DA, Hunter CA (2012) Immune response and immunopathology during toxoplasmosis. Semin Immunopathol 34:793–813. https://doi.org/10.1007/s00281-012-0339-3

Echeverría J, Albuquerque R (2019) Nanoemulsions of essential oils: new tool for control of Vector-Borne diseases and in vitro effects on some parasitic agents. Medicines 6:1–12. https://doi.org/10.3390/medicines6020042

Elazab ST, Soliman AF, Nishikawa Y (2021) Effect of some plant extracts from Egyptian herbal plants against Toxoplasma gondii tachyzoites in vitro. J Vet Med Sci 14:100–107. https://doi.org/10.1292/jvms.20-0458

Farkas J, Mohácsi-Farkas C (2014) In: Motajermi Y (ed) Safety of foods and beverages: spices and seasonings. Encyclopedia of food safety. Volume 3: Foods, materials, technologies and risks, 1st edn. Elsevier, pp 324–330

Flegr J, Prandota J, Sovičková M, Israili ZH (2014) Toxoplasmosis--a global threat. Correlation of latent toxoplasmosis with specific disease burden in a set of 88 countries. PLoS One 24:e90203. https://doi.org/10.1371/journal.pone.0090203

Fokou JBH, Dongmo PMJ, Boyom FF (2020) Essential oil's chemical composition and pharmacological properties. In: Essential Oils Oils Nat, vol 13, p 13. https://doi.org/10.5772/intechopen.86573

Gauch LMR, Pedrosa SS, Esteves RA, Silveira-Gomes F, Gurgel ESC, Arruda AC, Marques-da-Silva SH (2014) Antifungal activity of Rosmarinus officinalis Linn. essential oil against Candida albicans, Candida dubliniensis, Candida parapsilosis and Candida krusei. Rev Pan-Amaz Saude 5:61–66. https://doi.org/10.5123/S2176-62232014000100007

Giovati L, Santinoli C, Mangia C, Vismarra A, Belletti S, D'Adda T, Fumarola C, Ciociola T, Bacci C, Magliani W, Polonelli L, Conti S, Kramer LH (2018) Novel activity of a synthetic Decapeptide against toxoplasma gondii Tachyzoites. Front Microbiol 20:753. https://doi.org/10.3389/fmicb.2018.00753

Hayet EK, Hocine L, Meriem EK (2017) Chemical composition and biological activities of the essential oils and the methanolic extracts of Bunium incrassatum and Bunium alpinum from Algeria. J Chil Chem Soc Concepción 62:3335–3341. https://doi.org/10.4067/S0717-97072017000100006

Hill D, Dubey JP (2002) Toxoplasma gondii: transmission, diagnosis and prevention. Clin Microbiol Infect 8:634–640. https://doi.org/10.1046/j.1469-0691.2002.00485.x

Huang SY, Yao N, He JK, Pan M, Hou ZF, Fan YM, Du A, Tao JP (2021) In vitro anti-parasitic activity of Pelargonium X. asperum essential oil against Toxoplasma gondii. Front Cell Dev Biol 9:616340. https://doi.org/10.3389/fcell.2021.616340

Jalsenjak V, Peljnajk S, Kustrak D (1987) Microcapsules of sage oil, essential oils content and antimicrobial activity. Pharmazie 42:419–420

Jihene A, Rym E, Ines KJ, Majdi H, Olfa T, Abderrabba M (2020) Antileishmanial potential of Propolis essential oil and its synergistic combination with Amphotericin B. Nat Prod Commun. https://doi.org/10.1177/1934578X19899566

Khamesipour F, Razavi SM, Hejazi SH, Ghanadian SM (2021) In vitro and in vivo anti-Toxoplasma activity of Dracocephalum kotschyi essential oil. Food Sci Nutr 9:522–531. https://doi.org/10.1002/fsn3.2021

Kim YW, You HJ, Lee S, Kim B, Kim DK, Choi JB, Kim JA, Lee HJ, Joo IS, Lee JS, Kang DH, Lee G, Ko GP, Lee SJ (2017) Inactivation of Norovirus by Lemongrass essential oil using a Norovirus surrogate system. J Food Prot 80:1293–1302. https://doi.org/10.4315/0362-028X.JFP-16-162

Konstantinovic N, Guegan H, Stäjner T, Belaz S, Robert-Gangneux F (2019) Treatment of toxoplasmosis: current options and future perspectives. Food Waterborne Parasitol 15:e00036. https://doi.org/10.1016/j.fawpar.2019.e00036

Lee WC, Mahmud R, Noordin R, Piaru SP, Perumal S, Ismail S (2013) Free radicals scavenging activity, cytotoxicity and anti-parasitic activity of essential oil of Psidium guajava L. Leaves against toxoplasma gondii. J Essent Oil-Bear Plants 16:32–38. https://doi.org/10.1080/0972060X.2013.764196

Mahmoudvand H, Beyranvand M, Nayebzadeh H, Fallahi S, Mirbadie SR, Kheirandish F, Kayedi MH (2017) Chemical composition and prophylactic effects of Saturja khuzestanica essential oil on acute toxoplasmosis in mice. Afr J Tradit Complement Altern Med 14:49–55. https://doi.org/10.21010/ajtcam.v14i5.7

Mahmoudvand H, Tavakoli Kareshk A, Nabi Moradi M, Monzote Fidalgo L, Mirbadie SR, Niazi M, Khatami M (2020) Efficacy and safety of Zataria multiflora Boiss essential oil against acute toxoplasmosis in mice. Iran J Parasitol 15:22–30

Mc Murry J (2011) 7° Ed. Química Orgânica - Combo. Cengage Learning, São Paulo, p 1344

McAuley JB (2014) Congenital toxoplasmosis. J Pediatr Infect Dis Soc 3(Suppl 1):S30–S35. https://doi.org/10.1093/jpids/piu077

Moradi S, Barati A (2019) Essential Oils Nanoemulsions: Preparation, Characterization and Study of Antibacterial Activity against Escherichia Coli. Int J Nanosci Nanotechnol 15:199–210

Mosleh N, Shomali T, Aghapour Kazemi H (2013) Effect of Zataria multiflora essential oil on immune responses and faecal virus shedding period in broilers immunized with live Newcastle disease vaccines. Iranian J Vet Res 14:220–25. https://doi.org/10.22099/IJVR.2013.1684

Minuzzi CE, Portella LP, Bräunig P, Sangioni LA, Ludwig A, Ramos LS, Pacheco L, Silva CR, Pacheco FC, Menegolla IA, Farinha LB, Kist PP, Breganó RM, Nino BSL, Cardoso Martins FD, Monica TC, Ferreira FP, Britto I, Signori A, Medici KC, Freire RL, Garcia JL, Navarro IT, Difante CM, Flores Vogel FS (2020) Isolation and molecular characterization of Toxoplasma gondii from placental tissues of pregnant women who received toxoplasmosis treatment during an outbreak in southern Brazil. PLoS One 30:e0228442. https://doi.org/10.1371/journal. pone.0228442

Mubarak SMH, Alnomani Y (2020) Synergistic effect of Eucalyptus camaldulensis essential oil with glucantime against cutaneous Leishmaniasis in vitro and in vivo. Meta Gene 25:100717. https://doi.org/10.1016/j.mgene.2020.100717

Munera-López J, Ganuza A, Bogado SS, Muñoz D, Ruiz DM, Sullivan WJ Jr, Vanagas L, Angel SO (2019) Evaluation of ATM Kinase Inhibitor KU-55933 as potential anti-toxoplasma gondii agent. Front Cell Infect Microbiol 13:9:26. https://doi.org/10.3389/fcimb.2019.00026

Oliveira CB, Meurer YS, Medeiros TL, Pohlit AM, Silva MV, Mineo TW, Andrade-Neto VF (2016) Anti-toxoplasma activity of Estragole and thymol in Murine Models of congenital and noncongenital toxoplasmosis. J Parasitol 102:369–376. https://doi.org/10.1645/15-848

Ouologuem DT, Roos DS (2014) Dynamics of the Toxoplasma gondii inner membrane complex. J Cell Sci 1:3320–3330. https://doi.org/10.1242/jcs.147736

Pillai S, Mahmud R, Lee WC, Perumal S (2012) Anti-parasitic activity of *Myristica fragrans* Houtt. essential oil against *Toxoplasma gondii* parasite. APCBEE Proc 2:92–96

Pourghanbari G, Nili H, Moattari A, Mohammadi A, Iraji A (2016) Antiviral activity of the oseltamivir and Melissa officinalis L. essential oil against avian influenza A virus (H9N2). Virus Dis 27:170–178. https://doi.org/10.1007/s13337-016-0321-0

Rodrigues KA, Amorim LV, Dias CN, Moraes DF, Carneiro SM, Carvalho FA (2015) Syzygium cumini (L.) Skeels essential oil and its major constituent α-pinene exhibit anti-Leishmania activity through immunomodulation in vitro. J Ethnopharmacol 3:32–40. https://doi.org/10.1016/j. jep.2014.11.024

Sandner G, Heckmann M, Weghuber J (2020) Immunomodulatory activities of selected essential oils. Biomol Ther 3:1139. https://doi.org/10.3390/biom10081139

Sanfelice RA, Bosqui LR, da Silva SS, Miranda-Sapla MM, Pana-Gio LA, Navarro IT, Conchon-Costa I, Pavanelli WR, Almeida RS, Costa IN (2018) Proliferation of Toxoplasma gondii (RH strain) is inhibited by the combination of pravastatin and simvastatin with low concentrations of conventional drugs used in toxoplasmosis. J Appl Biomed 16:29–33. https://doi.org/10.1016/J. jab.2017.10.009

Sanli A, Karadoğan T (2017) Geographical impact on essential oil composition of endemic Kundmannia anatolica hub.-mor. (Apiaceae). Afr J Tradit Complement Altern Med 23:131–137. https://doi.org/10.21010/ajtcam.v14i1.14

Shaapan RM, Al-Abodi HR, Alanazi AD, Abdel-Shafy S, Rashidipour M, Shater AF, Mahmoudvand H (2021) Myrtus communis essential oil; anti-parasitic effects and induction of the innate immune system in mice with toxoplasma gondii infection. Molecules 26:819. https://doi.org/10.3390/molecules26040819

Shokri A, Saeedi M, Fakhar M, Morteza-Semnani K, Keighobadi M, Hosseini Teshnizi S, Kelidari HR, Sadjadi S (2017) Antileishmanial activity of Lavandula angustifolia and Rosmarinus Officinalis essential oils and Nano-emulsions on Leishmania major (MRHO/IR/75/ER). Iran J Parasitol 12:622–631

Silva VD, Almeida-Souza F, Teles AM, Neto PA, Mondego-Oliveira R, Mendes Filho NE, Taniwaki NN, Abreu-Silva AL, Calabrese KS, Mouchrek-Filho VE (2018) Chemical composition of Ocimum canum Sims. essential oil and the antimicrobial, antiprotozoal and ultrastructural alterations it induces in Leishmania amazonensis promastigotes. Ind Crop Prod 119:201–208. https://doi.org/10.1016/j.indcrop.2018.04.005

Stashenko E, Martínez JR (2019) Study of essential oils obtained from tropical plants grown in Colombia, Essential oils - oils of nature, Hany A, El-Shemy. IntechOpen. https://doi. org/10.5772/intechopen.87199

Swamy MK, Akhtar MS, Sinniah UR (2016) Antimicrobial properties of plant essential oils against human pathogens and their mode of action: na updated review. Evid Based Complement Alternat Med 3012462. https://doi.org/10.1155/2016/3012462

Tavakoli-Kareshk A, Keyhani A, Mahmoudvand H, Tavakoli Oliaei R, Asadi A, Andishmand M, Azzizian H, Babaei Z, Zia-Ali N (2015) Efficacy of the Bunium persicum (Boiss) essential oil against acute toxoplasmosis in Mice Model. Iran J Parasitol 10:625–631

Teixeira SC, de Souza G, Borges BC, de Araújo TE, Rosini AM, Aguila FA, Ambrósio SR, Veneziani RCS, Bastos JK, Silva MJB, Martins CHG, de Freitas BB, Ferro EAV (2020) Copaifera spp. oleoresins impair Toxoplasma gondii infection in both human trophoblastic cells and human placental explants. Sci Rep 16:15158. https://doi.org/10.1038/s41598-020-72230-0

Vial HJ, Eldin P, Tielens AG, van Hellemond JJ (2003) Phospholipids in parasitic protozoa. Mol Biochem Parasitol 126:143–154. https://doi.org/10.1016/s0166-6851(02)00281-5

Wilkin PJ, Al-Yozbaki M, George A, Gupta GK, Wilson CM (2021) The undiscovered potential of essential oils for treating SARS-CoV-2 (COVID-19). Curr Pharm Des 27. https://doi.org/1 0.2174/1381612826666201015154611

Yao N, He JK, Pan M, Hou ZF, Xu JJ, Yang Y, Tao JP, Huang SY (2021a) In vitro evaluation of Lavandula angustifolia essential oil on anti-toxoplasma activity. Front Cell Infect Microbiol 11:755715. https://doi.org/10.3389/fcimb.2021.755715

Yao N, Xu Q, He JK, Pan M, Hou ZF, Liu D-D, Tao JP, Huang SY (2021b) Evaluation of Origanum vulgare essential oil and its active ingredients as potential drugs for the treatment of toxoplasmosis. Front Cell Infect Microbiol 11:793089. https://doi.org/10.3389/fcimb.2021.793089

Chapter 15
Essential Oil Antimalarial Activity

Jorddy Neves Cruz, Márcia Moraes Cascaes, Adriane Gomes Silva, Valdicley Vale, Mozaniel Santana de Oliveira, and Eloisa Helena de Aguiar Andrade

15.1 Introduction

Over the recent years, the incidence of malaria has declined. However, the number of new cases remains high, especially in Africa. In 2019 and 2020, 217 and 219 million cases of malaria were reported worldwide, respectively and there are estimates of an increase in the number of cases in several regions of the globe, especially in the Americas (Tse et al. 2019; Rasmussen and Ringwald 2021).

In Brazil, most people affected are composed of river-dwellers from the Amazon region or people who visit this location (Santos and Almeida 2018). The number of malaria cases in the country increased considerably from 1980 onwards with the disordered occupation of the Amazon, and due to the construction of hydroelectric dams, gold mines, and roads. In 1980, 169,871 cases were registered, a number three times greater than that observed in the previous decade since 52,469 cases were registered in 1970 (Souza et al. 2019). However, from the 1990s on, investments were made to fight malaria, and a gradual reduction in the number of reported cases was verified (MacDonald and Mordecai 2019; Monteiro et al. 2020).

J. N. Cruz (✉) · M. S. de Oliveira · E. H. de Aguiar Andrade
Adolpho Ducke Laboratory, Botany Coordination, Museu Paraense Emílio Goeldi, Belém, Pará, Brazil

M. M. Cascaes
Universidade Federal do Pará, Programa de Pós-Graduação em Química, Belém, Pará, Brazil

A. G. Silva
Faculty of Pharmacy, Centro Universitário, FIBRA, Belém, Pará, Brazil

V. Vale
Faculty of Pharmacy, Federal University of Pará, Belém, Pará, Brazil

© The Author(s), under exclusive license to Springer Nature Switzerland AG 2022
M. Santana de Oliveira (ed.), *Essential Oils*, https://doi.org/10.1007/978-3-030-99476-1_15

In 2014, also in Brazil, the number of malaria cases was approximately 146,000, and in the following year, this number reached almost 159,000, being reduced to 133,000 in 2016. However, in 2017, there were 217,928 cases (Recht et al. 2017; Daher et al. 2019). and in 2019, 193,534 incidents were registered (Lana et al. 2017). With this, the search for compounds with potential application for the control of malaria is still present, in this sense, the present work aims to carry out a literature survey on the potential control of malaria by essential oils.

15.2   Biological Cycle

The biological cycle of malaria (Fig. 15.1) was described only in 1900 by Ronald Ross, a British military doctor. It is a complex process that involves two hosts: the vertebrate host, in which asexual reproduction occurs (schizogonic cycle), and the vector, the female mosquito of genus *Anopheles* sp.., in which sexual reproduction happens (Sousa et al. 2019; Smith et al. 2020).

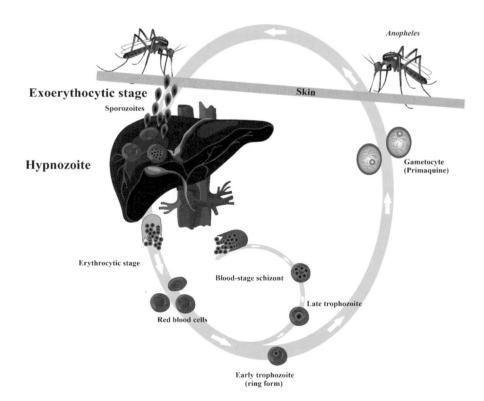

Fig. 15.1 Biologic cycle of malaria. (Adapted from Baer et al. 2007)

The cycle in vertebrates begins with the inoculation of parasite sporozoites into the host dermis during blood repast of the *Anopheles* females. Once in the blood-stream, they are directed to the liver and lodge in the hepatocytes, forming a protective layer called parasitophorous vacuole. Then they undergo transformation and multiplication by schizogony to become merozoites (extra-erythrocytic cycle) that are released into the bloodstream (Markus 2018; Charon et al. 2019).

In the extra-erythrocytic cycle, *P. vivax* and *P. ovale* have a particular characteristic: they convert into dormant forms known as hypnozoites, which can remain for months or years in hepatocytes and if not treated correctly, they can be released into the bloodstream causing a new episode of the disease (Fig. 15.1) (Nanvyat et al. 2018; van Biljon et al. 2018).

Once in the bloodstream, the merozoites go through a complex process, forming junctions in the membranes allowing the invasion of erythrocytes (Zhang et al. 2021). In the red blood cells, the merozoites undergo a series of morphological and structural transformations and become young trophozoites (rings). The young trophozoites undergo a maturation process, originating multinucleated schizonts, which rupture the cells, releasing merozoites that invade new cells and restart the cell cycle. Some schizonts are converted into sexed forms known as gametocytes, which circulate in the bloodstream of the host until ingested by the mosquito vector, initiating the biological cycle in the vector (sexual or sporogonic cycle) (Murray et al. 2017).

In the intestine of the phlebotomine sandfly, the gametocytes are converted into female (macrogamete) and male gametes (microgamete), forming the zygote. And, by meiosis, it becomes a new form known as the ookinete, which can move through the cells of the intestine (Vallone et al. 2018). In its basal membrane, it transforms again and forms the oocyst, where it remains until its final transformation into sporozoite, which migrates to the salivary glands of the mosquito and is inoculated in the blood repast, originating a new biological cycle (Li et al. 2019; Subudhi et al. 2020).

The clinical picture of the disease occurs when the schizonts break the erythrocytes and release, along with the merozoites, pro-inflammatory cytokines that activate the immune system of the person making them have a fever and chills, which can be followed by headache, myalgia, nausea, and vomiting (Warrell et al. 2017; Taffese et al. 2018). In some cases, infection by *P. falciparum* can evolve to severe malaria, which, besides the fever crisis, can be accompanied by hypoglycemia, respiratory problems with pulmonary edema, renal failure, metabolic acidosis, severe anemia, cerebral malaria, and consequent death (Ashley and Phyo 2018; Talapko et al. 2019). This severe condition is mainly caused by *P. falciparum*, but *P. knowlesi* and even *P. vivax* have also been reported in the literature.

The antimalarial chemotherapy started in the 17th century, when Jesuits observed that Peruvian Indians used plants of the genus *Cinchona* spp (Rubiaceae family) for the treatment of febrile diseases. Studies of this genus led to the isolation of the alkaloid quinine (Fig. 15.2a), which to this day is used to treat severe malaria (Krungkrai and Krungkrai 2016; Lee et al. 2018). Due to the toxicity of this drug, other synthetic antimalarials were implemented in the market, such as chloroquine

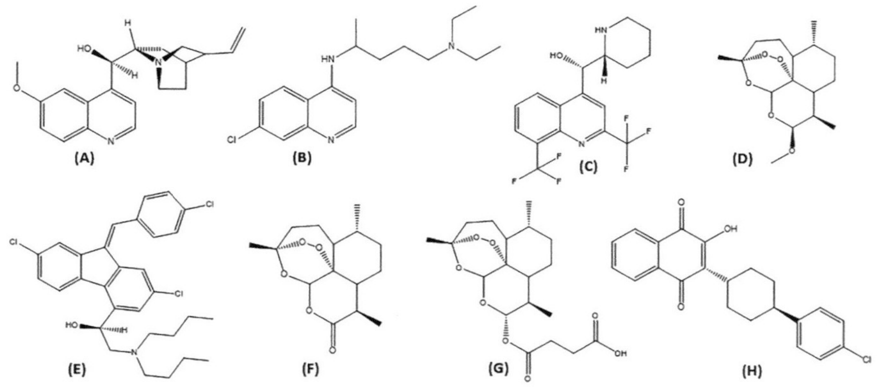

Fig. 15.2 Chemical structure of some antimalarials. (**a**) quinine; (**b**) chloroquine; (**c**) mefloquine; (**d**) artemether; (**e**) lumefantrine; (**f**) artemisinin; (**g**) artesunate; (**h**) atovaquone

(Fig. 15.2b) and mefloquine (Fig. 15.2c). Chloroquine has several advantages over quinine, including low cost and lower toxicity, which leads to greater acceptance of treatment (Walker and Cairns 2015; Gaillard et al. 2016).

Antimalarial drugs act mainly on the erythrocytic cycle of the parasite, thus preventing the conversion of young trophozoite forms into schizonts and consequently preventing the febrile crisis and the clinical picture of malaria. These drugs also break the parasite biological cycle (Mildenhall 2016; Vale et al. 2020; Costa et al. 2020). Among the targets of antimalarial drugs are inhibition of hemozoin polymerization, folate metabolism, and protein synthesis; radical generation; and alterations in mitochondrial electron transport (Ding and Li 2015; Goodman et al. 2017; Mounkoro et al. 2021).

In Brazil, the treatment of malaria caused by *P. vivax* and *P. ovale,* which present tissue forms, is performed with an association of chloroquine and primaquine. The objective of the treatment is to interrupt the blood cycle of the protozoan, thus avoiding blood schizogony, causing the death of tissue parasites (hypnozoites) and gametocytes (Frausin et al. 2015; Lima et al. 2015). In uncomplicated *P. falciparum* infection, treatment is performed with an association of artemether (Fig. 15.2d) and lumefantrine (Fig. 15.2e) (Vale et al. 2015; Abreu et al. 2019).

For many years, chloroquine was widely used in the treatment of malaria because it had low toxicity and was economically viable; however, in the late 1950s, strains resistant to this drug were discovered in Southeast Asia (Mwanza et al. 2016; Zhou et al. 2020). Then, an intensive research program for new antimalarial agents began, and more than 300,000 compounds were tested, with two compounds being active against *P. falciparum*-resistant strains: mefloquine, a quinoline methanol, and halofantrine, a phenanthrene methanol (Abdul and Al-Bari 2015). The most recently introduced drugs were artemisinin (Fig. 15.2f) and its derivatives artemether (Fig. 15.2d) and artesunate (Fig. 15.2g), and atovaquone (Fig. 15.2h). It should be noted that artemisinin was isolated from *Artemisia annua,* which has been used in

China since ancient times for the treatment of febrile diseases and death of hypnozoites and gametocytes of the parasite (Mbengue et al. 2015; Guo 2016; Yeo et al. 2017).

P. falciparum is the species with the greatest potential for drug insensitivity, and the first case of resistance was registered in 1910 with quinine (which was introduced in 1632) (Stallmach et al. 2015; Bhatt et al. 2015). Resistance has been observed in almost all drugs in an increasingly shorter time interval, such as chloroquine, synthesized in 1945 and that presented resistance in 1957; proguanil introduced in 1948 also presented drug insensitivity one year after its insertion in therapy; sulfadoxine-pyrimethamine inserted in 1967 presented drug tolerance in the same year, as well as atovaquone (inserted in 1996 and presented resistance in the same year) and mefloquine, which has been used since 1977 and presented resistance in 1982 (Snow 2015; Koopmans et al. 2015; Marciniak et al. 2016). More recently, even artemisinin, the most potent schizonticide currently used, has presented drug tolerance, registered in Cambodia (Landier et al. 2018; Maier et al. 2019; Boyle et al. 2019).

Computational methods have been applied in the design of new drugs against different diseases (Alves et al. 2020; Leão et al. 2020; Araújo et al. 2020) . One of the possibilities to fight drug resistance is to obtain analogues of active molecules, which should have equal or greater potency in terms of pharmacological effect, and if possible, less toxic (Carvalho et al. 2019; Oliveira et al. 2020; Silva et al. 2021). An example is the antimalarial drug atovaquone (Fig. 15.2h), a naphthoquinone analogue of lapachol, which proved to be more effective and less toxic when compared to other analogues (Duffy and Avery 2017; Greenshields et al. 2019). Its mechanism of action was attributed to competition for the catalytic site of ubiquinone in the parasite mitochondria, acting selectively in this organelle during the erythrocytic cycle. This causes a collapse in electron transport, which has a fundamental role for *Plasmodium*, because, through this mechanism, the supply of dihydroorotate dehydrogenase (DHODH) is interrupted, which acts in the biosynthesis of pyrimidines. Therefore, atovaquone acts to inhibit purine biosynthesis (Gao et al. 2018; Nonaka et al. 2018; Heller and Roepe 2018).

Based on the history of resistance, WHO (World Health Organization) advises the urgent search for new pharmaceutical alternatives, i.e., new drugs with antimalarial capacity (Kilonzi et al. 2019; Yobi et al. 2020). Plants appear as one of these alternatives because of the antimalarial drugs obtained from them, such as quinine and artemisinin. Even secondary metabolites have been used as prototypes for the synthesis of new drugs, such as chloroquine, which was planned from quinine. Artesunate and artemether were obtained from artemisinin, as well as atovaquone itself, which was produced from lapachol, a naphthoquinone obtained from the Ipe wood (lapacho or Brazilian walnut) (Nkoli Mandoko et al. 2018; Lum Ndamukong-Nyanga et al. 2021).

It is known that Brazil is the largest holder of the global flora, and presents several plants with antimalarial potential, such as plants of the genus *Aspidosperma*. In its chemical composition, alkaloids, such as ulein are present, which offers great potential against *Plasmodium* strains. These alkaloids probably act in the inhibition of heme formation (Spencer et al. 2018; Mahaman Moustapha et al. 2021) However, there is still a series of plant species that have not been studied yet and that may provide other chemical substances, presenting a different mechanism of action, such as species of the Iridaceae family, which produce naphthoquinones (Soleimani-Ahmadi et al. 2017; Vatandoost et al. 2018).

15.3   Antimalarial Activity of Essential Oils (EOs)

Malaria is still considered a major global health problem, affecting much of the world population, especially in developing countries. Thus, effective drug discovery is still one of the main efforts to control the spread of this disease (Pan et al. 2018). According to the World Health Organization (WHO), there were approximately 229 million malaria cases in 2019, with 409,000 deaths. And 94% of these cases were reported in the WHO African region. Children under 5 are the most vulnerable group affected by malaria, and in 2019, they accounted for about two-thirds of all malaria deaths worldwide (Sypniewska et al. 2017; Oo et al. 2019; Park et al. 2020).

There are five species of protozoa in the genus *Plasmodium* (*P. falciparum, P. vivax, P. ovale, P. malariae,* and *P. knowlesi*) that cause malaria in humans, but most cases are caused by *P. falciparum. P. vivax* is generally considered less dangerous than *P. falciparum,* although both can cause fatal complications in infected people (Pan et al. 2018). Some drugs, alone or in combination (chloroquine, primaquine, mefloquine, halofantrine, artemisinin, atovaquone, and others) have been used in chemotherapy of malaria (Nogueira and Lopes 2011). In addition, this disease is still the most destructive infection and it is getting worse due to the increased resistance of *P. falciparum* to most antimalarial drugs, which has been a challenge to the effectiveness of this chemotherapy (Kaur and Kaur 2017).

Historically, fixed and volatile compounds have influenced the development of new drugs (Ferreira et al. 2020; Neto et al. 2020; Santana de Oliveira et al. 2021; Castro et al. 2021). Natural products are the main sources of antimalarial drugs, such as low-molecular-weight compounds and essential oils (EOs), which present monoterpenes, sesquiterpenes, and phenylpropanoids (Mota et al. 2012; Bezerra et al. 2020a; Bezerra et al. 2020b). Some EOs exhibit anti-Plasmodium activity according to *in vitro* and *in vivo* studies, mainly on *P. falciparum* [85; 77]. For instance, the essential oil from the leaves of *Helichrysum gymnocephalum* (DC.) Humbert. (Asteraceae), composed mostly of 1,8-cineole (47.4%), bicyclosesquiphellandrene (5.6%), γ-curcumene (5.6%), α-amorphene (5.1%), and bicyclogermacrene (5%), was active against *P. falciparum* ($IC_{50} = 25 \pm 1$ mg.L^{-1}) (Afoulous et al. 2011).

The *in vitro* activity of *Origanum onites* L. EO was evaluated against *P. falciparum*. The main component of the oil was carvacrol (70.6%), followed by linalool (9.7%) and p-cymene (7%). This essential oil showed moderate anti-Plasmodium effect with IC_{50} value equal to 7.9 μg.mL^{-1} (Tasdemir et al. 2019). The effects of the leaf EOs of *Ocimum gratissimum* L. and *Cymbopogon citratus* (DC.) on the growth of *Plasmodium berghei* were investigated and both oils showed significant antimalarial activities. The main constituents of *O. gratissimum* EO were γ-terpinene (21.9%), β-felandrene (21.1%), limonene (11.4%), and thymol (11.2%), whereas

Table 15.1 Essential oils with proven antimalarial activity

Species	Botanical family	Plant part	Major substances (≥5%)	Malaria parasite	IC$_{50}$	Reference
Achillea filipendulina Lam.	Asteraceae	Aerial parts	Santolina alcohol (43.8%), 1,8-cineole (14.5%), and *cis*-chrysanthenyl acetate (12.5%)	*Plasmodium falciparum*	Chloroquine-sensitive (D6) (0.68 µg.mL^{-1}) and chloroquine-resistant (W2) (0.9 µg.mL^{-1})	Demirci et al. (2017)
Achillea magnifica Hiemerl ex Hub.-Mor.	Asteraceae	Aerial parts	Linalool (27.5%), spathulenol (5.8%), and terpinen-4-ol (5.5%)	*P. falciparum*	Chloroquine-sensitive (D6) (1.2 µg.mL^{-1}) and chloroquine-resistant (W2) (1.1 µg.mL^{-1})	Demirci et al. (2017)
Annona squamosa L.	Annonaceae	Leaves	(*E*)-Caryophyllene (27.4%), germacrene D (17.1%), and bicyclogermacrene (10.8%)	*P. falciparum*	14.7 ± 2.9 µg.mL^{-1}	Meira et al. (2015)
Annona vepretorum Mart.	Annonaceae	Leaves	Bicyclogermacrene (39.0%), spathulenol (14.0%), and α-phellandrene (11.5%)	*P. falciparum*	9.9 ± 0.7 µg.mL^{-1}	Meira et al. (2015)
Artemisia terrae-albae Krasch.	Asteraceae	Aerial parts	Camphor (35.4%), 1,8-cineol (20.4%), camphene (9.1%), and α-thujone (5.3%)	*P. falciparum*	Not informed	Suleimen et al. (2016)
Baccharis microdonta DC.	Asteraceae	Aerial parts	Spathulenol (24.74%) and kongol (22.22%)	*P. falciparum*	Chloroquine-sensitive (D6) (14.75 ± 3.80 µg.mL^{-1}) and chloroquine-resistant (W2) (23.93 ± 4.64 µg.mL^{-1})	Budel et al. (2018)
Baccharis pauciflosculosa DC.	Asteraceae	Aerial parts	β-pinene (18.33%) and limonene (18.77%)	*P. falciparum*	Chloroquine-sensitive (D6) (10.90 ± 0.98 µg.mL^{-1}) and chloroquine-resistant (W2) (14.20 ± 1.08 µg.mL^{-1})	Budel et al. (2018)

(continued)

Table 15.1 (continued)

Species	Botanical family	Plant part	Major substances (≥5%)	Malaria parasite	IC_{50}	Reference
Baccharis punctulata DC.	Asteraceae	Aerial parts	α-Bisabolol (23.63%)	*P. falciparum*	Chloroquine-sensitive (D6) (17.26 ± 0.83 µg.mL⁻¹) and chloroquine-resistant (W2) (19.73 ± 4.11 µg.mL⁻¹)	Budel et al. (2018)
Baccharis reticularioides Deble & A.S. Oliveira	Asteraceae	Aerial parts	α-Pinene (24.50%)	*P. falciparum*	Chloroquine-sensitive (D6) (20.32 ± 4.37 µg.mL⁻¹) and chloroquine-resistant (W2) (34.35 ± 10.15 µg.mL⁻¹)	Budel et al. (2018)
Baccharis sphenophylla Dusén ex Malme.	Asteraceae	Aerial parts	β-Pinene (15.24%), limonene (14.33%), and spathulenol (13.15%)	*P. falciparum*	Chloroquine-sensitive (D6) (27.58 ± 1.64 µg.mL⁻¹) and chloroquine-resistant (W2) (32.53 ± 16.5 µg.mL⁻¹)	Budel et al. (2018)
Croton zehntneri Pax & K. Hoffm.	Euphorbiaceae	Leaves	Estragole (76.80%), 1,8-cineole (7.0%), and eugenol (5.3%)	*P. falciparum*	15.20 µg.mL⁻¹	Mota et al. (2012)
Cymbopogon citratus (DC.)	Poaceae	Leaves	Geranial (32.8%), neral (29.0%), myrcene (16.2%), and β-pinene (10.5%)	*Plasmodium berghei*	Not informed	Tchoumbougnang et al. (2005)
Guatteria friesiana (W.A. Rodrigues) Erkens & Maas	Annonaceae	Leaves	β-Eudesmol (51.9%), γ-eudesmol (18.9%), and α-eudesmol (12.6%)	*P. falciparum*	0.5 µg.mL⁻¹	Meira et al. (2017)
Guatteria pogonopus Martius	Annonaceae	Leaves	Spathulenol (24.8%), γ-amorphene (14.7%), and germacrene D (11.8%)	*P. falciparum*	6.8 µg.mL⁻¹	Meira et al. (2017)
Helichrysum cymosum (L.) D. Don	Asteraceae	–	1,8-Cineole (20.4%), α-pinene (12.4%) (Z)-β-ocimene (9.5%), and limonene (7.2%)	*P. falciparum*	0.204 ± 0.05 µg.mL⁻¹	van Vuuren et al. (2006)

Helichrysum gymnocephalum	Asteraceae	Leaves	1,8-Cineole (47.4%), bicyclosesquiphellandrene (5.6%), γ-curcumene (5.6%), α-amorphene (5.1%), and bicyclogermacrene (5.0%)	*Plasmodium falciparum*	$25 \pm 1\ \mu g.mL^{-1}$	Afoulous et al. (2011)
Lippia sidoides Cham.	Verbenaceae	Leaves	Thymol (84.87%) and *p*-cymene (5.33%)	*P. falciparum*	$10.5\ \mu g.mL^{-1}$	Mota et al. (2012)
Ocimum gratissimum L.	Lamiaceae	Leaves	γ-Terpinene (21.9%), β- phellandrene (21.1%), limonene (11.4%), and thymol (11.2%)	*Plasmodium berghei*	Not informed	Tchoumbougnang et al. (2005)
Ocimum sanctum L.	Lamiaceae	Aerial parts	Eugenol (22.0%), β-elemene (19.2%), β-caryophyllene (19.1%), and Germacrene D (5.03%)	Not informed	$>10\ mg.mL^{-1}$	Hussain et al. (2017)
Origanum onites L.	Lamiaceae	Aerial parts	Carvacrol (70.6%), linalool (9.7%), and *p*-cymene (7%)	*P. falciparum*	$7.9\ \mu g.mL^{-1}$	Tasdemir et al. (2019)
Salvia lavandulifolia Vahl.	Lamiaceae	Aerial parts	β-Caryophyllene (11.87%), spathulenol (8.13%), neomenthol (7.75%), pulegone (6.97%), hexadecanoic acid (6.85%), and germacrene D (5.70%)	*P. falciparum*	Not informed	Ihsan et al. (2017)
Vanillosmopsis arborea (Gardner) Baker	Asteraceae	Stems	α-Bisabolol (80.43%)	*P. falciparum*	$7.00\ \mu g.mL^{-1}$	Mota et al. (2012)

the EO of *C. citratus* contained geranial (32.8%), neral (29.0%), myrcene (16.2%), and β-pinene (10.5%) (Tchoumbougnang et al. 2005).

The aromatic plants, whose utility and relative safety have been identified by ethnobotanical sources (do Nascimento et al. 2020; Cascaes et al. 2021), represent a valuable source for the discovery of antiparasitic and anti-plasmodic agents. Thus, in Table 15.1 some EOs with evaluated antimalarial activity are organized according to their major chemical constituents (>5%), botanical family, plant part studied, malaria-causing parasite, and IC_{50} values.

15.4   Final Considerations

Compounds of natural origin, whether volatile or fixed, are inspirational sources for the development of new drugs. Thus, the investigation of their potential biological activities *in vitro, in vivo,* and *in silico* is very important to enable more knowledge about their mechanism of interaction with molecular targets of therapeutic interest. In this chapter, it was possible to notice that the essential oils from several species had strong pharmacological action on different species of *Plasmodium*, the cause of malaria, at different stages of parasite development. Finally, we can suggest that these initial studies can serve as inspiration for further scientific investigations into the use of volatile compounds for the treatment of malaria and for the development of more chemically complex molecules.

Acknowledgements   The author Dr Mozaniel Santana de Oliveira, thanks PCI-MCTIC/MPEG, as well as CNPq for the scholarship process number: 302050/2021-3.

Declaration of Competing Interest   The authors declare no conflict of interest.

References

Abdul M, Al-Bari A (2015) Chloroquine analogues in drug discovery: new directions of uses, mechanisms of actions and toxic manifestations from malaria to multifarious diseases. J Antimicrob Chemother 70:1608–1621. https://doi.org/10.1093/JAC/DKV018

Abreu FVS de, Santos E dos, Mello ARL, Gomes LR, Alvarenga DAM de, Gomes MQ, Vargas WP, Bianco-Júnior C, Pina-Costa A de, Teixeira DS, Romano APM, Manso PP de A, Pelajo-Machado M, Brasil P, Daniel-Ribeiro CT, Brito CFA de, Ferreira-da-Cruz M de F, Lourenço-de-Oliveira R (2019) Howler monkeys are the reservoir of malarial parasites causing zoonotic infections in the Atlantic forest of Rio de Janeiro. PLoS Negl Trop Dis 13:e0007906. https://doi.org/10.1371/JOURNAL.PNTD.0007906

Afoulous S, Ferhout H, Raoelison EG, Valentin A, Moukarzel B, Couderc F, Bouajila J (2011) Helichrysum gymnocephalum essential oil: chemical composition and cytotoxic, antimalarial and antioxidant activities, attribution of the activity origin by correlations. Molecules 16:8273–8291. https://doi.org/10.3390/molecules16108273

Alves FS, Rodrigues Do Rego J de A, da Costa ML, Lobato Da Silva LF, da Costa RA, Cruz JN, Brasil DDSB (2020) Spectroscopic methods and in silico analyses using density functional theory to characterize and identify piperine alkaloid crystals isolated from pepper (Piper Nigrum L.). J Biomol Struct Dyn 38:2792–2799. https://doi.org/10.1080/07391102.2019.1639547

Araújo PHF, Ramos RS, da Cruz JN, Silva SG, Ferreira EFB, de Lima LR, Macêdo WJC, Espejo-Román JM, Campos JM, Santos CBR (2020) Identification of potential COX-2 inhibitors for the treatment of inflammatory diseases using molecular modeling approaches. Molecules 25:4183. https://doi.org/10.3390/molecules25184183

Ashley EA, Phyo AP (2018) Drugs in development for malaria. Drugs 78:9 78:861–879. https://doi.org/10.1007/S40265-018-0911-9

Baer K, Klotz C, Kappe SHI, Schnieder T, Frevert U (2007) Release of hepatic plasmodium yoelii Merozoites into the pulmonary microvasculature. PLoS Pathog 3:e171. https://doi.org/10.1371/journal.ppat.0030171

Bezerra FWF, do Nascimento Bezerra P, Cunha VMB, de Salazar MLAR, Barbosa JR, da Silva MP, de Oliveira MS, da Costa WA, Pinto RHH, da Cruz JN, de Carvalho Junior RN (2020a) Supercritical green solvent for Amazonian natural resources. In: Inamuddin, Asiri AM (eds) Advanced nanotechnology and application of supercritical fluids. Springer International Publishing, Cham, pp 15–31

Bezerra FWFWF, de Oliveira MSS, Bezerra PNN, Cunha VMBMB, Silva MPP, da Costa WAA, Pinto RHHHH, Cordeiro RMM, da Cruz JNN, Chaves Neto AMJMJ, Carvalho Junior RNN (2020b) Extraction of bioactive compounds. In: Green sustainable process for chemical and environmental engineering and science. Elsevier, pp 149–167

Bhatt S, Weiss DJ, Cameron E, Bisanzio D, Mappin B, Dalrymple U, Battle KE, Moyes CL, Henry A, Eckhoff PA, Wenger EA, Briët O, Penny MA, Smith TA, Bennett A, Yukich J, Eisele TP, Griffin JT, Fergus CA, Lynch M, Lindgren F, Cohen JM, Murray CLJ, Smith DL, Hay SI, Cibulskis RE, Gething PW (2015) The effect of malaria control on Plasmodium falciparum in Africa between 2000 and 2015. Nature 526:7572 526:207–211. https://doi.org/10.1038/nature15535

Boyle MJ, Chan JA, Handayuni I, Reiling L, Feng G, Hilton A, Kurtovic L, Oyong D, Piera KA, Barber BE, William T, Eisen DP, Minigo G, Langer C, Drew DR, de Labastida RF, Amante FH, Williams TN, Kinyanjui S, Marsh K, Doolan DL, Engwerda C, Fowkes FJI, Grigg MJ, Mueller I, McCarthy JS, Anstey NM, Beeson JG (2019) IgM in human immunity to Plasmodium falciparum malaria. Sci Adv 5:18. https://doi.org/10.1126/SCIADV.AAX4489/SUPPL_FILE/AAX4489_SM.PDF

Budel JM, Wang M, Raman V, Zhao J, Khan SI, Rehman JU, Techen N, Tekwani B, Monteiro LM, Heiden G, Takeda IJM, Farago PV, Khan IA (2018) Essential oils of five baccharis species: investigations on the chemical composition and biological activities. Molecules 23. https://doi.org/10.3390/molecules23102620

Carvalho RN de, Oliveira MS de MS de, Silva SG, Cruz JN da, Ortiz E, Costa WA da, Bezerra FWF, Cunha VMB, Cordeiro RM, Neto AM de JC, Andrade EH de A, Junior RN de C, Carvalho RN de, Oliveira MS de MS de (2019) Supercritical CO2 application in essential oil extraction. In: Inamuddin RM, Asiri AM (eds) Materials research foundations, 2nd edn. Materials Research Foundations, Millersville, pp 1–28

Cascaes MM, Dos O, Carneiro S, Diniz Do Nascimento L, Antônio Barbosa De Moraes Â, Santana De Oliveira M, Neves Cruz J, Skelding GM, Guilhon P, Helena De Aguiar Andrade E, Vico C, Cruz-Chamorro I (2021) Essential oils from Annonaceae species from Brazil: a systematic review of their Phytochemistry, and biological activities. Int J Mol Sci 22:12140. https://doi.org/10.3390/IJMS222212140

Castro ALG, Cruz JN, Sodré DF, Correa-Barbosa J, Azonsivo R, de Oliveira MS, de Sousa Siqueira JE, da Rocha Galucio NC, de Oliveira BM, Burbano RMR, do Rosário Marinho AM, Percário S, Dolabela MF, Vale VV (2021) Evaluation of the genotoxicity and mutagenicity of isoeleutherin and eleutherin isolated from Eleutherine plicata herb. using bioassays and in silico approaches. Arab J Chem 14:103084. https://doi.org/10.1016/j.arabjc.2021.103084

Charon J, Grigg MJ, Eden JS, Piera KA, Rana H, William T, Rose K, Davenport MP, Anstey NM, Holmes EC (2019) Novel RNA viruses associated with Plasmodium vivax in human malaria and Leucocytozoon parasites in avian disease. PLoS Pathog 15:e1008216. https://doi.org/10.1371/JOURNAL.PPAT.1008216

Costa EB, Silva RC, Espejo-Román JM, Neto MF de A, Cruz JN, Leite FHA, Silva CHTP, Pinheiro JC, Macêdo WJC, Santos CBR (2020) Chemometric methods in antimalarial drug design from 1,2,4,5-tetraoxanes analogues. SAR QSAR Environ Res 31:677–695. https://doi.org/10.1080/1062936X.2020.1803961

Daher A, Silva JCAL, Stevens A, Marchesini P, Fontes CJ, ter Kuile FO, Lalloo DG (2019) Evaluation of Plasmodium vivax malaria recurrence in Brazil 11 Medical and Health Sciences 1117 Public Health and Health Services. Malar J 18:1–10. https://doi.org/10.1186/S12936-019-2644-Y/TABLES/3

Demirci B, Başer KHC, Aytaç Z, Khan SI, Jacob MR, Tabanca N (2017) Comparative study of three Achillea essential oils from eastern part of Turkey and their biological activities. Rec Nat Prod 12:195–200. https://doi.org/10.25135/rnp.09.17.03.019

Ding H, Li D (2015) Identification of mitochondrial proteins of malaria parasite using analysis of variance. Amino Acids 47:329–333. https://doi.org/10.1007/s00726-014-1862-4

do Nascimento LD, de Moraes AAB, da Costa KS, Galúcio JMP, Taube PS, Costa CML, Cruz JN, Andrade EH de A, de Faria LJG (2020) Bioactive natural compounds and antioxidant activity of essential oils from spice plants: new findings and potential applications. Biomol Ther 10:1–37. https://doi.org/10.3390/biom10070988

Duffy S, Avery VM (2017) Plasmodium falciparum in vitro continuous culture conditions: a comparison of parasite susceptibility and tolerance to anti-malarial drugs throughout the asexual intra-erythrocytic life cycle. Int J Parasitol Drugs Drug Resist 7:295–302. https://doi.org/10.1016/J.IJPDDR.2017.07.001

Ferreira OO, Neves da Cruz J, de Jesus Pereira Franco C, Silva SG, da Costa WA, de Oliveira MS, de Aguiar Andrade EH (2020) First report on yield and chemical composition of essential oil extracted from myrcia eximia DC (Myrtaceae) from the Brazilian Amazon. Molecules 25:783. https://doi.org/10.3390/molecules25040783

Frausin G, Ari DFH, Lima RBS, Kinupp VF, Ming LC, Pohlit AM, Milliken W (2015) An ethnobotanical study of anti-malarial plants among indigenous people on the upper Negro River in the Brazilian Amazon. J Ethnopharmacol 174:238–252. https://doi.org/10.1016/J.JEP.2015.07.033

Gaillard T, Madamet M, Tsombeng FF, Dormoi J, Pradines B (2016) Antibiotics in malaria therapy: which antibiotics except tetracyclines and macrolides may be used against malaria? Malar J 15:1 15:1–10. https://doi.org/10.1186/S12936-016-1613-Y

Gao X, Liu X, Shan W, Liu Q, Wang C, Zheng J, Yao H, Tang R, Zheng J (2018) Anti-malarial atovaquone exhibits anti-tumor effects by inducing DNA damage in hepatocellular carcinoma. Am J Cancer Res 8:1697

Goodman CD, Buchanan HD, McFadden GI (2017) Is the Mitochondrion a good malaria drug target? Trends Parasitol 33:185–193. https://doi.org/10.1016/J.PT.2016.10.002

Greenshields AL, Fernando W, Hoskin DW (2019) The anti-malarial drug artesunate causes cell cycle arrest and apoptosis of triple-negative MDA-MB-468 and HER2-enriched SK-BR-3 breast cancer cells. Exp Mol Pathol 107:10–22. https://doi.org/10.1016/J.YEXMP.2019.01.006

Guo Z (2016) Artemisinin anti-malarial drugs in China. Acta Pharm Sin B 6:115–124. https://doi.org/10.1016/J.APSB.2016.01.008

Heller LE, Roepe PD (2018) Quantification of free Ferriprotoporphyrin IX Heme and Hemozoin for Artemisinin sensitive versus delayed clearance phenotype plasmodium falciparum malarial parasites. Biochemistry 57:6927–6934. https://doi.org/10.1021/ACS.BIOCHEM.8B00959

Hussain AI, Chatha SAS, Kamal GM, Ali MA, Hanif MA, Lazhari MI (2017) Chemical composition and biological activities of essential oil and extracts from Ocimum sanctum. Int J Food Proper 20:1569

Ihsan S, Zaki M, Khan S (2017) Chemical analysis and biological activities of Salvia lavandulifolia Vahl. Essent Oil. Chem Anal 7:71

Kaur R, Kaur H (2017) Plant derived antimalarial agents. J Med Plants Stud 5:346–363

Kilonzi M, Minzi O, Mutagonda R, Sasi P, Kamuhabwa A, Aklillu E (2019) Comparison of malaria treatment outcome of generic and innovator's anti-malarial drugs containing artemether-lumefantrine combination in the management of uncomplicated malaria amongst Tanzanian children. Malar J 18:1–8. https://doi.org/10.1186/S12936-019-2769-Z/FIGURES/2

Koopmans LC, van Wolfswinkel ME, Hesselink DA, Hoorn EJ, Koelewijn R, van Hellemond JJ, van Genderen PJJ (2015) Acute kidney injury in imported Plasmodium falciparum malaria. Malar J 14:1–7. https://doi.org/10.1186/S12936-015-1057-9/TABLES/3

Krungkrai SR, Krungkrai J (2016) Insights into the pyrimidine biosynthetic pathway of human malaria parasite Plasmodium falciparum as chemotherapeutic target. Asian Pac J Trop Med 9:525–534. https://doi.org/10.1016/J.APJTM.2016.04.012

Lana RM, Riback TIS, Lima TFM, da Silva-Nunes M, Cruz OG, Oliveira FGS, Moresco GG, Honório NA, Codeço CT (2017) Socioeconomic and demographic characterization of an endemic malaria region in Brazil by multiple correspondence analysis. Malar J 16:1–16. https://doi.org/10.1186/S12936-017-2045-Z/TABLES/4

Landier J, Parker DM, Thu AM, Lwin KM, Delmas G, Nosten FH, Andolina C, Aguas R, Ang SM, Aung EP, Baw NB, Be SA, B'Let S, Bluh H, Bonnington CA, Chaumeau V, Chirakiratinant M, Cho WC, Christensen P, Corbel V, Day NP, Dah SH, Delmas G, Dhorda M, Dondorp AM, Gaudart J, Gornsawun G, Haohankhunnatham W, Hla SK, Hsel SN, Htoo GN, Htoo SN, Imwong M, John S, Kajeechiwa L, Kereecharoen L, Kittiphanakun P, Kittitawee K, Konghahong K, Khin SD, Kyaw SW, Landier J, Ling C, Lwin KM, Lwin KSW, Ma NKY, Marie A, Maung C, Marta E, Minh MC, Miotto O, Moo PK, Moo KL, Moo M, Na NN, Nay M, Nosten FH, Nosten S, Nyo SN, Oh EKS, Oo PT, Oo TP, Parker DM, Paw ES, Phumiya C, Phyo AP, Pilaseng K, Proux S, Rakthinthong S, Ritwongsakul W, Salathibuphha K, Santirad A, Sawasdichai S, von Seidlein L, Shee PW, Shee PB, Tangseefa D, Thu AM, Thwin MM, Tun SW, Wanachaloemlep C, White LJ, White NJ, Wiladphaingern J, Win SN, Yee NL, Yuwapan D (2018) Effect of generalised access to early diagnosis and treatment and targeted mass drug administration on Plasmodium falciparum malaria in Eastern Myanmar: an observational study of a regional elimination programme. Lancet 391:1916–1926. https://doi.org/10.1016/S0140-6736(18)30792-X

Leão RP, Cruz JVJN, da Costa GV, Cruz JVJN, Ferreira EFB, Silva RC, de Lima LR, Borges RS, dos Santos GB, Santos CBR (2020) Identification of new rofecoxib-based cyclooxygenase-2 inhibitors: a bioinformatics approach. Pharmaceuticals 13:1–26. https://doi.org/10.3390/ph13090209

Lee RS, Waters AP, Brewer JM (2018) A cryptic cycle in haematopoietic niches promotes initiation of malaria transmission and evasion of chemotherapy. Nat Commun 9:1 9:1–9. https://doi.org/10.1038/s41467-018-04108-9

Li X, Kumar S, McDew-White M, Haile M, Cheeseman IH, Emrich S, Button-Simons K, Nosten F, Kappe SHI, Ferdig MT, Anderson TJC, Vaughan AM (2019) Genetic mapping of fitness determinants across the malaria parasite Plasmodium falciparum life cycle. PLoS Genet 15:e1008453. https://doi.org/10.1371/JOURNAL.PGEN.1008453

Lima RBS, Rocha Silva LF, Melo MRS, Costa JS, Picanço NS, Lima ES, Vasconcellos MC, Boleti APA, Santos JMP, Amorim RCN, Chaves FCM, Coutinho JP, Tadei WP, Krettli AU, Pohlit AM (2015) In vitro and in vivo anti-malarial activity of plants from the Brazilian Amazon. Malar J 14:1 14:1–14. https://doi.org/10.1186/S12936-015-0999-2

Lum Ndamukong-Nyanga J, Batandi N, Virginie H, Flore TC, Nadege FC (2021) The severity of malarial-anemia in pregnant women in Biyem-Assi, Yaounde. Arch Curr Res Int 57–68. https://doi.org/10.9734/ACRI/2021/V21I230233

MacDonald AJ, Mordecai EA (2019) Amazon deforestation drives malaria transmission, and malaria burden reduces forest clearing. Proc Natl Acad Sci U S A 116:22212–22218. https://doi.org/10.1073/PNAS.1905315116/-/DCSUPPLEMENTAL

Mahaman Moustapha L, Adamou R, Ibrahim ML, Abdoulaye Louis Padounou M, Diallo A, Courtin D, Testa J, Ndiaye JLA (2021) Evidence that seasonal malaria chemoprevention with SPAQ influences blood and pre-erythrocytic stage antibody responses of Plasmodium falciparum infections in Niger. Malar J 20:1–8. https://doi.org/10.1186/S12936-020-03550-9/TABLES/2

Maier AG, Matuschewski K, Zhang M, Rug M (2019) Plasmodium falciparum. Trends Parasitol 35:481–482. https://doi.org/10.1016/J.PT.2018.11.010

Marciniak S, Prowse TL, Herring DA, Klunk J, Kuch M, Duggan AT, Bondioli L, Holmes EC, Poinar HN (2016) Plasmodium falciparum malaria in 1st–2nd century CE southern Italy. Curr Biol 26:R1220–R1222. https://doi.org/10.1016/J.CUB.2016.10.016

Markus MB (2018) Biological concepts in recurrent Plasmodium vivax malaria. Parasitology 145:1765–1771. https://doi.org/10.1017/S003118201800032X

Mbengue A, Bhattacharjee S, Pandharkar T, Liu H, Estiu G, Stahelin RV, Rizk SS, Njimoh DL, Ryan Y, Chotivanich K, Nguon C, Ghorbal M, Lopez-Rubio JJ, Pfrender M, Emrich S, Mohandas N, Dondorp AM, Wiest O, Haldar K (2015) A molecular mechanism of artemisinin resistance in Plasmodium falciparum malaria. Nature 520:7549 520:683–687. https://doi.org/10.1038/nature14412

Meira CS, Guimarães ET, MacEdo TS, Da Silva TB, Menezes LRA, Costa EV, Soares MBP (2015) Chemical composition of essential oils from Annona vepretorum Mart. and Annona squamosa L. (Annonaceae) leaves and their antimalarial and trypanocidal activities. J Essent Oil Res 27:160–168. https://doi.org/10.1080/10412905.2014.982876

Meira CS, Menezes LRA, dos Santos TB, Macedo TS, Fontes JEN, Costa EV, Pinheiro MLB, da Silva TB, Teixeira Guimarães E, Soares MBP (2017) Chemical composition and antiparasitic activity of essential oils from leaves of Guatteria friesiana and Guatteria pogonopus (Annonaceae). J Essent Oil Res 29:156–162. https://doi.org/10.1080/10412905.2016.1210041

Mildenhall DC (2016) The role of forensic palynology in sourcing the origin of falsified antimalarial pharmaceuticals. Palynology 41:203–206. https://doi.org/10.1080/01916122.2016.1156587

Monteiro EF, Fernandez-Becerra C, Araujo M da S, Messias MR, Ozaki LS, Duarte AMR de C, Bueno MG, Catao-Dias JL, Chagas CRF, Mathias B da S, dos Santos MG, Santos SV, Holcman MM, de Souza JC, Kirchgatter K (2020) Naturally acquired humoral immunity against malaria parasites in non-human primates from the Brazilian Amazon, Cerrado and Atlantic Forest. Pathogens 9:525. https://doi.org/10.3390/PATHOGENS9070525

Mota ML, Lobo LTC, Galberto Da Costa JM, Costa LS, Rocha HAO, Rocha E Silva LF, Pohlit AM, De Andrade Neto VF (2012) In vitro and in vivo antimalarial activity of essential oils and chemical components from three medicinal plants found in Northeastern Brazil. Planta Med 78:658–664. https://doi.org/10.1055/s-0031-1298333

Mounkoro P, Michel T, Golinelli-Cohen MP, Blandin S, Davioud-Charvet E, Meunier B (2021) A role for the succinate dehydrogenase in the mode of action of the redox-active antimalarial drug, plasmodione. Free Radic Biol Med 162:533–541. https://doi.org/10.1016/J.FREERADBIOMED.2020.11.010

Murray L, Stewart LB, Tarr SJ, Ahouidi AD, Diakite M, Amambua-Ngwa A, Conway DJ (2017) Multiplication rate variation in the human malaria parasite Plasmodium falciparum. Sci Rep 7:1 7:1–8. https://doi.org/10.1038/s41598-017-06295-9

Mwanza S, Joshi S, Nambozi M, Chileshe J, Malunga P, Kabuya JBB, Hachizovu S, Manyando C, Mulenga M, Laufer M (2016) The return of chloroquine-susceptible Plasmodium falciparum malaria in Zambia. Malar J 15:1–6. https://doi.org/10.1186/S12936-016-1637-3/FIGURES/1

Nanvyat N, Mulambalah C, Barshep Y, Ajiji J, Dakul D, Tsingalia H (2018) Malaria transmission trends and its lagged association with climatic factors in the highlands of Plateau State, Nigeria. Trop Parasitol 8:18. https://doi.org/10.4103/TP.TP_35_17

Neto R de AMM, Santos CBRR, Henriques SVCC, Machado L de O, Cruz JN, da Silva CHT de PT de P, Federico LB, Oliveira EHC de C de, de Souza MPCC, da Silva PNBB, Taft CA, Ferreira IM, Gomes MRFF (2020) Novel chalcones derivatives with potential antineoplastic activity investigated by docking and molecular dynamics simulations. J Biomol Struct Dyn 1–13. https://doi.org/10.1080/07391102.2020.1839562

Nkoli Mandoko P, Sinou V, Moke Mbongi D, Ngoyi Mumba D, Kahunu Mesia G, Losimba Likwela J, Bi Shamamba Karhemere S, Muepu Tshilolo L, Tamfum Muyembe JJ, Parzy D

(2018) Access to artemisinin-based combination therapies and other anti-malarial drugs in Kinshasa. Med Mal Infect 48:269–277. https://doi.org/10.1016/J.MEDMAL.2018.02.003

Nogueira CR, Lopes LMX (2011) Antiplasmodial natural products. Molecules 16:2146–2190. https://doi.org/10.3390/molecules16032146

Nonaka M, Murata Y, Takano R, Han Y, Bin Kabir MH, Kato K (2018) Screening of a library of traditional Chinese medicines to identify anti-malarial compounds and extracts. Malar J 17:1–10. https://doi.org/10.1186/S12936-018-2392-4/FIGURES/6

Oliveira MS de, Cruz JN da, Costa WA da, Silva SG, Brito M da P, Menezes SAF de, Neto AM de JC, Andrade EH de A, Junior RN de C (2020) Chemical composition, antimicrobial properties of Siparuna guianensis essential oil and a molecular docking and dynamics molecular study of its major chemical constituent. Molecules 25:3852. https://doi.org/10.3390/molecules25173852

Oo WH, Gold L, Moore K, Agius PA, Fowkes FJI (2019) The impact of community-delivered models of malaria control and elimination: a systematic review. Malar J 18:1–16. https://doi.org/10.1186/S12936-019-2900-1/FIGURES/6

Pan WH, Xu XY, Shi N, Tsang SW, Zhang HJ (2018) Antimalarial activity of plant metabolites. Int J Mol Sci 19. https://doi.org/10.3390/ijms19051382

Park S, Nixon CE, Miller O, Choi NK, Kurtis JD, Friedman JF, Michelow IC (2020) Impact of malaria in pregnancy on risk of malaria in young children: systematic review and meta-analyses. J Infect Dis 222:538–550. https://doi.org/10.1093/INFDIS/JIAA139

Rasmussen C, Ringwald P (2021) Is there evidence of anti-malarial multidrug resistance in Burkina Faso? Malar J 20:1–5. https://doi.org/10.1186/S12936-021-03845-5/TABLES/1

Recht J, Siqueira AM, Monteiro WM, Herrera SM, Herrera S, Lacerda MVG (2017) Malaria in Brazil, Colombia, Peru and Venezuela: current challenges in malaria control and elimination. Malar J 16:1 16:1–18. https://doi.org/10.1186/S12936-017-1925-6

Santana de Oliveira M, Pereira da Silva VM, Cantão Freitas L, Gomes Silva S, Nevez Cruz J, de Aguiar Andrade EH (2021) Extraction yield, chemical composition, preliminary toxicity of Bignonia nocturna (Bignoniaceae) essential oil and in silico evaluation of the interaction. Chem Biodivers 18:cbdv.202000982. https://doi.org/10.1002/cbdv.202000982

Santos AS, Almeida AN (2018) The impact of deforestation on malaria infections in the Brazilian Amazon. Ecol Econ 154:247–256. https://doi.org/10.1016/J.ECOLECON.2018.08.005

Silva SG, de Oliveira MS, Cruz JN, da Costa WA, da Silva SHM, Barreto Maia AA, de Sousa RL, Carvalho Junior RN, de Aguiar Andrade EH (2021) Supercritical CO2 extraction to obtain Lippia thymoides Mart. & Schauer (Verbenaceae) essential oil rich in thymol and evaluation of its antimicrobial activity. J Supercrit Fluids 168:105064. https://doi.org/10.1016/j.supflu.2020.105064

Smith LM, Motta FC, Chopra G, Moch JK, Nerem RR, Cummins B, Roche KE, Kelliher CM, Leman AR, Harer J, Gedeon T, Waters NC, Haase SB (2020) An intrinsic oscillator drives the blood stage cycle of the malaria parasite Plasmodium falciparum. Science 368:754–759. https://doi.org/10.1126/SCIENCE.ABA4357/SUPPL_FILE/ABA4357_SMITH_SM.PDF

Snow RW (2015) Global malaria eradication and the importance of Plasmodium falciparum epidemiology in Africa. BMC Med 13:1–3. https://doi.org/10.1186/S12916-014-0254-7/FIGURES/1

Soleimani-Ahmadi M, Abtahi SM, Madani A, Paksa A, Abadi YS, Gorouhi MA, Sanei-Dehkordi A (2017) Phytochemical profile and mosquito Larvicidal activity of the essential oil from aerial parts of Satureja bachtiarica Bunge against malaria and lymphatic Filariasis Vectors. J Essent Oil Plant 20:328–336. https://doi.org/10.1080/0972060X.2017.1305919

Sousa A, Morales J, Vetter M, Aguilar-Alba M, García-Barrón L (2019) Influence of monthly temperatures on the intra-annual distribution of autochthonous malaria in Spain. Environ Sci Eng 2373–2377. https://doi.org/10.1007/978-3-030-51210-1_371

Souza PF, Xavier DR, Mutis MCS, da Mota JC, Peiter PC, de Matos VP, de Avelar Figueiredo Mafra Magalhães M, Barcellos C (2019) Spatial spread of malaria and economic frontier expansion in the Brazilian Amazon. PLoS One 14:e0217615. https://doi.org/10.1371/JOURNAL. PONE.0217615

Spencer LM, Peña-Quintero A, Canudas N, Bujosa I, Urdaneta N, Spencer LM, Peña-Quintero A, Canudas N, Bujosa I, Urdaneta N (2018) Antimalarial effect of two photo-excitable compounds in a murine model with Plasmodium berghei (Haemosporida: Plasmodiidae). Rev Biol Trop 66:880–891. https://doi.org/10.15517/RBT.V66I2.33420

Stallmach R, Kavishwar M, Withers-Martinez C, Hackett F, Collins CR, Howell SA, Yeoh S, Knuepfer E, Atid AJ, Holder AA, Blackman MJ (2015) Plasmodium falciparum SERA5 plays a non-enzymatic role in the malarial asexual blood-stage lifecycle. Mol Microbiol 96:368–387. https://doi.org/10.1111/MMI.12941/SUPPINFO

Subudhi AK, O'Donnell AJ, Ramaprasad A, Abkallo HM, Kaushik A, Ansari HR, Abdel-Haleem AM, Ben Rached F, Kaneko O, Culleton R, Reece SE, Pain A (2020) Malaria parasites regulate intra-erythrocytic development duration via serpentine receptor 10 to coordinate with host rhythms. Nat Commun 11:1 11:1–15. https://doi.org/10.1038/s41467-020-16593-y

Suleimen EM, Ibataev ZA, Iskakova ZB, Ishmuratova MY, Ross SA, Martins CHG (2016) Constituent composition and biological activity of essential oil from Artemisia terrae-albae. Chem Nat Compd 52:173–175. https://doi.org/10.1007/s10600-016-1584-9

Sypniewska P, Duda JF, Locatelli I, Althaus CR, Althaus F, Genton B (2017) Clinical and laboratory predictors of death in African children with features of severe malaria: a systematic review and meta-analysis. BMC Med 15:1–17. https://doi.org/10.1186/S12916-017-0906-5/TABLES/3

Taffese HS, Hemming-Schroeder E, Koepfli C, Tesfaye G, Lee MC, Kazura J, Yan GY, Zhou GF (2018) Malaria epidemiology and interventions in Ethiopia from 2001 to 2016. Infect Dis Poverty 7:1–9. https://doi.org/10.1186/S40249-018-0487-3/TABLES/2

Talapko J, Škrlec I, Alebić T, Jukić M, Včev A (2019) Malaria: the past and the present. Microorganisms 7:179 7:179. https://doi.org/10.3390/MICROORGANISMS7060179

Tasdemir D, Kaiser M, Demirci B, Demirci F, Hüsnü Can Baser K (2019) Antiprotozoal activity of Turkish Origanum onites essential oil and its components. Molecules 24:1–16. https://doi. org/10.3390/molecules24234421

Tchoumbougnang F, Amvam Zollo PH, Dagne E, Mekonnen Y (2005) In vivo antimalarial activity of essential oils from Cymbopogon citratus and Ocimum gratissimum on mice infected with Plasmodium berghei. Planta Med 71:20–23. https://doi.org/10.1055/s-2005-837745

Tse EG, Korsik M, Todd MH (2019) The past, present and future of anti-malarial medicines. Malar J 18:1 18:1–21. https://doi.org/10.1186/S12936-019-2724-Z

Vale VV, Vilhena TC, Trindade RCS, Ferreira MRC, Percário S, Soares LF, Pereira WLA, Brandão GC, Oliveira AB, Dolabela MF, de Vasconcelos F (2015) Anti-malarial activity and toxicity assessment of Himatanthus articulatus, a plant used to treat malaria in the Brazilian Amazon. Malar J 14:1–10. https://doi.org/10.1186/S12936-015-0643-1/FIGURES/4

Vale VV, Cruz JN, Viana GMR, Póvoa MM, Brasil D, do SB, Dolabela MF (2020) Naphthoquinones isolated from Eleutherine plicata herb: in vitro antimalarial activity and molecular modeling to investigate their binding modes. Med Chem Res 29:487–494. https://doi.org/10.1007/s00044-019-02498-z

Vallone A, D'Alessandro S, Brogi S, Brindisi M, Chemi G, Alfano G, Lamponi S, Lee SG, Jez JM, Koolen KJM, Dechering KJ, Saponara S, Fusi F, Gorelli B, Taramelli D, Parapini S, Caldelari R, Campiani G, Gemma S, Butini S (2018) Antimalarial agents against both sexual and asexual parasites stages: structure-activity relationships and biological studies of the Malaria Box compound 1-[5-(4-bromo-2-chlorophenyl)furan-2-yl]-N-[(piperidin-4-yl)methyl]methanamine (MMV019918) and analogues. Eur J Med Chem 150:698–718. https://doi.org/10.1016/J.EJMECH.2018.03.024

van Biljon R, Niemand J, van Wyk R, Clark K, Verlinden B, Abrie C, von Grüning H, Smidt W, Smit A, Reader J, Painter H, Llinás M, Doerig C, Birkholtz LM (2018) Inducing controlled cell cycle arrest and re-entry during asexual proliferation of Plasmodium falciparum malaria parasites. Sci Rep 8:1 8:1–14. https://doi.org/10.1038/s41598-018-34964-w

van Vuuren SF, Viljoen AM, van Zyl RL, van Heerden FR, Başer KHC (2006) The antimicrobial, antimalarial and toxicity profiles of helihumulone, leaf essential oil and extracts of Helichrysum cymosum (L.) D. Don subsp. cymosum. S Afr J Bot 72:287–290. https://doi.org/10.1016/j.sajb.2005.07.007

Vatandoost H, Rustaie A, Talaeian Z, Abai MR, Moradkhani F, Vazirian M, Hadjiakhoondi A, Shams-Ardekani MR, Khanavi M (2018) Larvicidal activity of Bunium persicum essential oil and extract against Malaria Vector, Anopheles stephensi. J Arthropod Borne Dis 12:85

Walker PGT, Cairns M (2015) Value of additional chemotherapy for malaria in pregnancy. Lancet Glob Health 3:e116–e117. https://doi.org/10.1016/S2214-109X(15)70081-1

Warrell DA, Gilles HM, Herbert M (2017) Clinical features of malaria:191–205. https://doi.org/10.1201/9780203756621-7

Yeo SJ, Liu DX, Kim HS, Park H (2017) Anti-malarial effect of novel chloroquine derivatives as agents for the treatment of malaria. Malar J 16:1–9. https://doi.org/10.1186/S12936-017-1725-Z/FIGURES/4

Yobi DM, Kayiba NK, Mvumbi DM, Boreux R, Kabututu PZ, Situakibanza HNT, Likwela JL, de Mol P, Okitolonda EW, Speybroeck N, Mvumbi GL, Hayette MP (2020) Molecular surveillance of anti-malarial drug resistance in Democratic Republic of Congo: high variability of chloroquinoresistance and lack of amodiaquinoresistance. Malar J 19:1–8. https://doi.org/10.1186/S12936-020-03192-X/TABLES/3

Zhang M, Wang C, Oberstaller J, Thomas P, Otto TD, Casandra D, Boyapalle S, Adapa SR, Xu S, Button-Simons K, Mayho M, Rayner JC, Ferdig MT, Jiang RHY, Adams JH (2021) The apicoplast link to fever-survival and artemisinin-resistance in the malaria parasite. Nat Commun 2021 12:1 12:1–15. https://doi.org/10.1038/s41467-021-24814-1

Zhou W, Wang H, Yang Y, Chen ZS, Zou C, Zhang J (2020) Chloroquine against malaria, cancers and viral diseases. Drug Discov Today 25:2012–2022. https://doi.org/10.1016/J.DRUDIS.2020.09.010

Part V
Essential Oil of Food Applications in Degenerative Diseases

Chapter 16
Neuroprotective Activity of the Essential Oils From Food Plants

Oliviu Voştinaru, Simona Codruţa Hegheş, and Lorena Filip

16.1 Introduction

In the last decades, the accelerated ageing of the global population has led to a constant increase of the number of patients suffering from age-related neurodegenerative diseases, which has sparked a significant number of studies aimed at finding an effective drug treatment (Gan et al. 2018). The neuroprotective effect of drugs helps maintaining the structural and functional integrity of nerve cells against acute injuries but also in chronic neurodegenerative disorders (Mallah et al. 2020). In the larger category of natural compounds capable of providing neuroprotection, essential oils play an key role, being able to mitigate neuronal lesions by a variety of general mechanisms like the reduction of inflammation and oxidative stress but also acting by specific, CNS-targeted mechanisms like the inhibition of glutamate-induced excitotoxicity, augmentation of cholinergic neurotransmission or inhibition of beta-amyloid aggregation (Ayaz et al. 2017).

Although there is currently no treatment capable of stopping the progression of neurodegenerative disease like Alzheimer or Parkinson, the dietary intake of natural compounds like essential oils could have the potential of reducing the symptoms of these devastating illnesses, improving the patients' quality of life. Due to their high lipophilicity and low molecular weight, essential oils can easily pass the

O. Voştinaru (✉)
Department of Pharmacology, Physiology and Physiopathology, Iuliu Hatieganu University of Medicine and Pharmacy, Cluj-Napoca, Romania

S. C. Hegheş
Department of Drug Analysis, Iuliu Hatieganu University of Medicine and Pharmacy, Cluj-Napoca, Romania

L. Filip
Department of Bromatology, Hygiene, Nutrition, Iuliu Hatieganu University of Medicine and Pharmacy, Cluj-Napoca, Romania

blood-brain barrier and can access a variety of molecular targets inside the central nervous system like NMDA receptors or synaptic acetylcholinesterase with potential protective effects, helping the preservation of long-term integrity and functions of neural structures (Ayaz et al. 2017; Lizarraga-Valderrama 2021).

16.2 Essential Oils From Food Plants: Chemical Composition

The idea that herbs and spices containing essential oils (EOs) could have an important medicinal value is as old as medicine itself, traditionally used spices being important sources of phytocompounds. A spice is the dried seed, fruit, root, or bark of a plant, primarily used for seasoning food. Most spices come from families such as *Lamiaceae* (basil, rosemary, mint, lavender, oregano), *Apiaceae* (dill, coriander, fennel), *Zingiberaceae* (cardamom, turmeric), *Fabaceae* (licorice), *Piperaceae* (pepper), *Myrtaceae* (cloves), *Myristicaceae* (nutmeg), *Lauraceae* (cinnamon) or *Rutaceae* (lemon, orange) (Sharma et al. 2021). The mentioned plants are a rich source of bioactive essential oils and have a wide range of therapeutic effects, including neuroprotective properties, acting by multiple molecular mechanisms (Price and Price 2007; Sharma et al. 2021) (Fig. 16.1).

Essential oils are complex mixtures containing mainly aromatic terpenes. They are characterized by the presence in quite high concentrations (20–70%) of two or three major components, along with others in smaller quantities or traces (Heghes

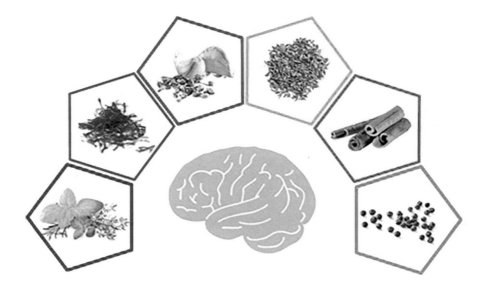

Fig. 16.1 Spices with neuroprotective properties

et al. 2019). Together, they are responsible for the diverse therapeutic effects of a plant, but it has also been shown that some of the pure isolated compounds can be responsible for a variety of effects in humans and other mammalian species (Ayaz et al. 2017; Lizarraga-Valderrama 2021). Thus, anxiolytic, antidepressant, sedative, and anticonvulsant effects have been already described for various compounds such as: α-pinene (Allenspach and Steuer 2021; Weston-Green et al. 2021), limonene (Eddin et al. 2021; Lorigooini et al. 2021), β-myrcene (Jansen et al. 2019; Surendran et al. 2021), b-caryophyllene (Machado et al. 2018), linalool (de Lucena et al. 2020; Airao et al. 2021; An et al. 2021; Caputo et al. 2021; Migheli et al. 2021; Weston-Green et al. 2021), eugenol (Irie 2012; Sun et al. 2020; Barot and Saxena 2021), citral (Quintans et al. 2011; Gonçalves et al. 2020; Charret et al. 2021), trans-anethole (Ryu et al. 2014; Memon et al. 2019), eugenol (Irie 2012; Ma et al. 2018), borneol, safranal (Sadeghnia et al. 2017; Zhao and Xi 2018; Li Puma et al. 2019), geraniol, citronellol (Qneibi et al. 2019), 2-phenylethanol (Ueno et al. 2019), carvone (Gonçalves et al. 2008; Dai et al. 2020), linalyl acetate (Malcolm and Tallian 2018; Wang and Heinbockel 2018), diallyl disulphide (Hazzaa et al. 2020), other properties like the neuroprotective effect being currently researched (Fig. 16.2).

The chemical composition of the essential oils from the most important food plants and spices is presented in Table 16.1.

16.3   Neuroprotective Effects of Essential Oils From Food Plants

16.3.1   Protection Against Glutamate-Induced Excitotoxicity

Although excitatory amino acids like glutamate or aspartate play important physiological roles in the central nervous system serving as neurotransmitters, their excessive level may lead to pathological consequences (Gan et al. 2018). Occasionally, the increased synaptic level of glutamate generates the "excitotoxicity phenomenon" where excessive stimulation of glutamate receptors favors the neural accumulation of large amounts of calcium ions with the development of intense mitochondrial activation and release of intracellular enzymes capable of inducing apoptosis and neural degeneration (Mattson 2008). In laboratory animals, the administration of monosodium glutamate in the early stages of life had a negative effect on behavioral and motor tests due to excitotoxic effects in the CNS (Gudiño-Cabrera et al. 2014). In humans, an excessive dietary intake of glutamate may lead to overweight and obesity but also toxic effects in various brain areas vital for cognitive functions (Garattini 2000). Over-activation of N-methyl-D-aspartate (NMDA) receptors of excitatory amino acids is directly involved in the pathogenesis of diseases like Alzheimer and schizophrenia, but also in acute ischemic stroke (Javitt 2004). Therefore, a protective effect of dietary essential oils against excitotoxic effects of glutamate, frequently added to various foods, could be an important tool for preserving the long-term integrity and functions of CNS (Fig. 16.3).

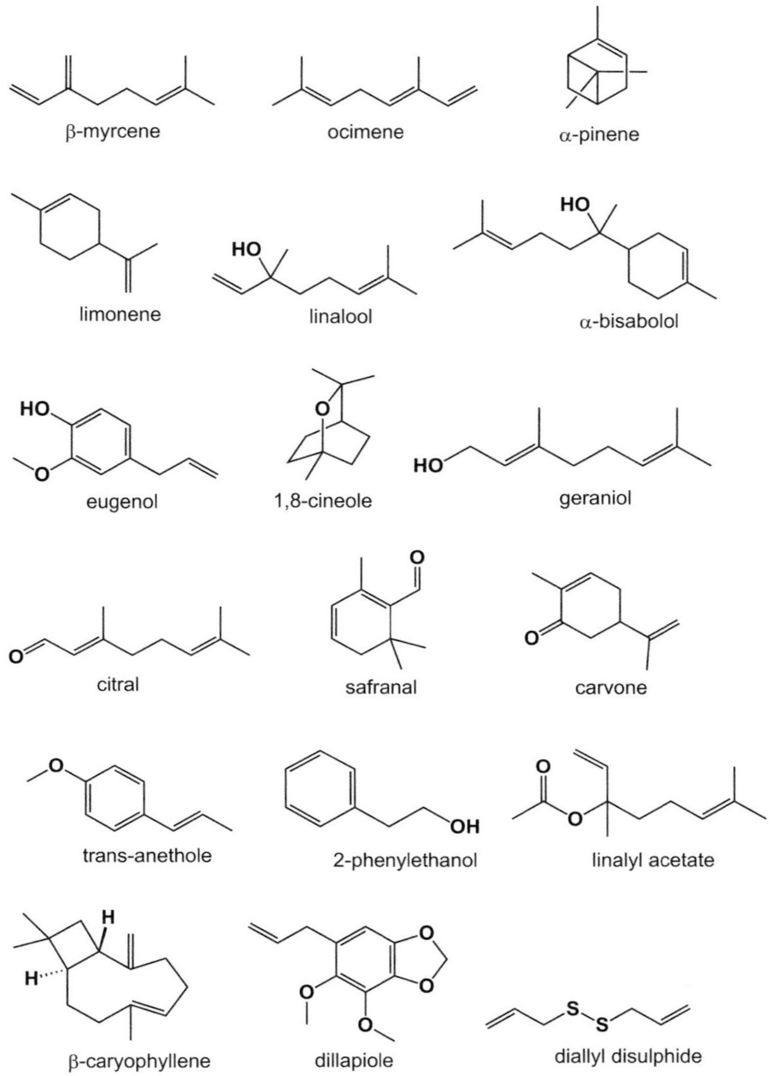

Fig. 16.2 Chemical structure of common active compounds in neuroprotective EOs

Several constituents of dietary essential oils have been tested in preclinical settings in a variety of experimental models, a particular molecule, linalool, proving significant capabilities of reducing excitotoxicity. The monoterpene linalool is a major constituent in the chemical composition of the essential oils from several plant species of dietary importance: *Citrus limon, Citrus aurantiifolia, Citrus sinensis (Rutaceae), Lavandula angustifolia* Mill.*, Rosmarinus officinalis* L.*, Mentha piperita* L.*, Mentha spicata* L. *Ocimum basilicum* L. *Organum majorana* L. (*Lamiaceae*) and *Crocus sativus* L. (*Iridaceae*).

Table 16.1 Chemical composition of EOs from representative food plants and spices

Plant species	Representative EOs compounds	References
Amaryllidaceae		
Allium sativum L. (Garlic)	diallyl disulphide, carvone, diallyl trisulfide, allyl tetrasulfide, and 1-allyl-3-(2-(allylthio) propyl) trisulfane	Satyal et al. (2017a), Hazzaa et al. (2020)
Anacardiaceae		
Pistacia lentiscus L. (Pistacia)	α-pinene, limonene, β-pinene, sabinene, β-myrcene, β-phellanderne, β-ocimene, terpinene-4-ol, trans-β-terpineol, longifolene	Dob et al. (2006), Negro et al. (2014)
Apiaceae		
Anethum graveolens L. (Dill)	limonene, α-phellandrene, β-phellandrene, carveol, eugenol, carvone, cis-dihydrocarvone, trans-dihydrocarvone, piperitone, apiole, dillapiole	Rădulescu et al. (2010), Sharopov et al. (2013), Ahl et al. (2015), Stanojević et al. (2016), Chahal et al. (2017)
Coriandrum sativum L. (Coriander)	γ-terpinene, limonene, α-pinene, β-myrcene, p-cymene, linalool, geraniol, terpinen-4-ol, camphor, geranyl acetate, linalyl acetate	Price and Price (2007), Orav et al. (2011), Mandal and Mandal (2015), Caputo et al. (2016)
Foeniculum vulgare L. (Fennel)	α-pinene, limonene, α-phellandrene, β-myrcene, β-phellandrene, γ-terpinene, cis-β-ocimene, α-terpinolene, p-cymene, fenchol, fenchone, methyl chavicol, cis-anethole, trans-anethole, 1,8-cineole	Miguel et al. (2010), Raal et al. (2012), Hammouda et al. (2014), Marín et al. (2016)
Pimpinella anisum L. (Aniseed)	γ-himachalene, anisol, cis-anethole, trans-anethole, methyl chavicol, anisaldehyde	Özcan and Chalchat (2006), Price and Price (2007), Orav et al. (2008), Saibi et al. (2012)
Fabaceae		
Glycyrriza glabra L. (Liquorice)	5-methyl-furfural, p-cymen, cumin aldehyde, carvone, piperitone, cinnamaldehyde, thymol, carvacrol, eugenol, methyl-eugenol	Quirós-Sauceda et al. (2016)
Iridaceae		
Crocus sativus L. (Saffron)	β-isophorone, linalool, α-isophorone, safranal, 6-oxoisophorone	Kosar et al. (2017)
Lamiaceae		
Lavandula angustifolia Mill. (Lavender)	cis-β-ocimene, trans-β-ocimene, limonene, β-caryophyllene, linalool, terpinen-4-ol, α-terpineol, borneol, lavandulol, 1,8-cineole, linalyl acetate, lavandulyl acetate	Tarakemeh et al. (2013), Caputo et al. (2016), Smigielski et al. (2018), Chen et al. (2020)
Mentha × piperita L. (Pepermint)	α-pinene, β-pinene, limonene, germacrene D, menthol, neomenthol, cis-carveol, terpinen-4-ol, menthone, isomenthone, neomenthone, pulegone, 1,8-cineole, menthyl acetate, menthofuran	Price and Price (2007), Şerban et al. (2012), Taherpour et al. (2017), Beigi et al. (2018), Ainane (2018)

(continued)

Table 16.1 (continued)

Plant species	Representative EOs compounds	References
Mentha spicata L. *(Spearmint)*	β-myrcene, limonene, β-caryophyllene, cis-carveol, menthol, α-terpineol, carvone, menthone, cis-dihydrocarvone, trans-dihydrocarvone, isomenthone, 1,8-cineole, dihydrocarvyl acetate, cis-carvyl acetate, trans-carvyl acetate, neoisodihydrocarveol acetate	Price and Price (2007), Snoussi et al. (2015), Ainane (2018), Mahboubi (2021)
Ocimum basilicum L. (Basil)	linalool, α-fenchyl alcohol, α-terpineol, β-elmene, eugenol, iso-eugenol, 1,8-cineole, methyl chavicol, methyl eugenol, methyl cinnamate	Joshi (2014), Pandey et al. (2014), Toncer et al. (2017)
Origanum majorana L. *(Majoram)*	sabinene, β-myrcene, α-terpinolene, α-pinene, cis-/trans-β-ocimenes, 3-carene, α-terpinene, γ-terpinene, β-caryophyllene, δ-cadinene, α-farnesene, germacrene D, benzene, p-cymene, terpinen-4-ol, thujan-4-ol, linalool, α-terpineol, carvacrol, geranyl acetate, linalyl acetate	Komaitis et al. (1992), Price and Price (2007), Brada et al. (2012), Rus et al. (2015), Bağci et al. (2017)
Rosmarinus officinalis L. *(Rosemary)*	α-pinene, borneol, verbenone, camphor, 1,8-cineole, bornyl acetate, camphene, limonene, α-terpineol, bornyl acetate	Price and Price (2007), Belkhodja et al. (2016), Satyal et al. (2017b), Verma et al. (2019)
Salvia officinalis L. *(Sage)*	α-pinene, β-pinene, camphene, β-myrcene, limonene, α-humulene, linalool, terpinen-4-ol, α-terpineol, borneol, viridiflorol, α-thujone, β-thujone, camphor, 1,8-cineole, bornyl acetate, linalyl acetate	Raal et al. (2007), Damyanova et al. (2016), El Euch et al. (2019)
Thymus officinalis L. *(Thyme)*	p-cymen, γ-terpinene, myrcene, thymol, carvacrol, pulegone	Porte and Godoy (2008), Satyal et al. (2016), Al-Asmari et al. (2017)
Lauraceae		
Cinnamomum verum *J.Presl* (True cinnamon)	α-pinene, β-myrcene, cinnamaldehyde, linalool, β-caryophyllene, eucalyptol, eugenol	Paranagama et al. (2001), Price and Price (2007), Alizadeh Behbahani et al. (2020)
Myristicaceae		
Myristica fragrans Houtt. (Nutmeg)	α-pinene, sabinene, β-pinene, α-terpinene, limonene, γ-terpinene, terpinolene, 4-terpineol, α-terpineol, safrole, isoeugenol, myristicin, elemicin	Muchtaridi et al. (2010), Kapoor et al. (2013)
Myrtaceae		
Syzygium aromaticum L. (Cloves)	eugenol, methyleugenol, pinene, β-caryophyllene, eugenyl acetate	Gaylor et al. (2014), Tahir et al. (2016), Kaur et al. (2019)
Piperaceae		
Piper nigrum L. (Black pepper)	β-caryophyllene, limonene, β-pinene, δ -3-carene, sabinene, α-pinene, camphene, eugenol, caryophyllene, terpinen-4-ol, eudesmol	Orav et al. (2004), Morshed et al. (2017)

(continued)

Table 16.1 (continued)

Plant species	Representative EOs compounds	References
Ranunculaceae		
Nigella sativa L. (Black cumin)	α-thujene, p-cymene, thymoquinone, γ-terpinene, α-thujene, carvacrol, β-pinene, limonene, thymol, β-caryophyllene, methyl linoleate, sabinene	Ghahramanloo et al. (2017), Kabir et al. (2020)
Rosaceae		
Rosa damascena Mill. (Rose)	2-phenylethanol, citronellol, geraniol, nerol, nonadecane, heneicosane, damascenone, β-ionone	Najem et al. (2011), Spasova Nunes and Graça Miguel (2017), Seify et al. (2018)
Rutaceae		
Citrus aurantiifolia (Christm.) Swingle (Lime)	α-pinene, limonene, γ-terpinene, β-pinene, β-myrcene, β -bisabolene, p-citral, linalool, α-terpineol, citronellal, neryl acetate	Spadaro et al. (2012), González-Mas et al. (2019), Lin et al. (2019)
Citrus limon L. (Lemon)	β-pinene, limonene, linalool, α-terpineol, linalyl acetate, geranyl acetate, nerolidol, neryl acetate, farnesol	Campêlo et al. (2011), Ben Hsouna et al. (2017), González-Mas et al. (2019)
Citrus x sinensis L. (Sweet orange)	limonene, β-myrcene, β-phellandrene, linalool, carvone	Njoroge et al. (2009), González-Mas et al. (2019)
Schisandraceae		
Illicium verum Hook.f. (Star anise)	trans-anethole, limonene, chavicol, anisaldehyde	Price and Price (2007), Wang et al. (2011)
Zingiberaceae		
Zingiberis officinale Roscoe (Ginger)	α-pinene, camphene, β-phellandrene, zingiberene, sesquiphellandrene, ar-curcumene, limonene, farnesene, β-bisabolene, neral, geranial	Mahboubi (2019), Al-Dhahli et al. (2020), Oforma et al. (2020)

A study on rat cerebral cortex proved that (±)-linalool blocked the binding of L-[3H]-glutamate to NMDA receptors, confirming glutamate receptor antagonism (Elisabetsky et al. 1995). Two additional neurochemical studies performed in mice showed that (±)-linalool, blocked in a non-competitive manner the binding of [3H]-MK801 to glutamate receptors, with an IC_{50} of 2.97 mM, indicating NMDA antagonist effects, and also inhibited K-stimulated release of glutamate in cortical synaptosomes (Silva Brum et al. 2001a; Silva Brum et al. 2001b). Another study also investigated neuroprotective effects of essential oils and their terpene components using SH-SY5Y cells, finding that linalool and linalyl acetate were capable of binding to NMDA receptor with Ki of 2.3 and 0.54 μL/mL, respectively, although they have interacted also with another neuropharmacological target, the serotonin transporter SERT (López et al. 2017).

Other terpenes present in the chemical composition of essential oils from foods were also studied for their neuroprotective effect against excitotoxicity. In a new

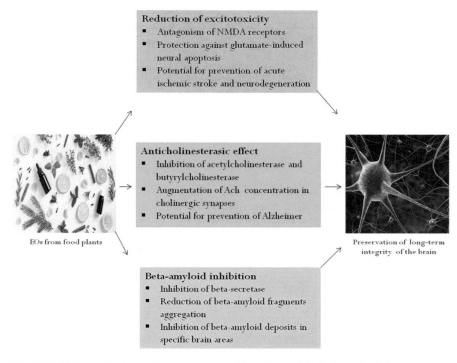

Fig. 16.3 Main mechanisms of neuroprotective effect of essential oils from food plants

study, a series of terpenes (linalool, terpinen-4-ol, cis-hexen-1-ol, 1-octen-3-ol, nerol, citronellol, geraniol, and α-terpineol), present in traditional Japanese liquors, were tested for their neuroprotective effects in a model using *Xenopus* oocytes. The compounds proved inhibitory activity of GluN1/GluN2A and GluN1/GluN2B subtype NMDA receptors at concentrations of 2.5 mM. Additionally, the intraperitoneal administration of linalool (89 mg/kg) and citronellol (16 mg/kg) improved the overall performances of treated mice in the elevated plus maze test, suggesting a positive effect in the management of brain disorders like Alzheimer or depression (Yamada et al. 2015).

An *in vivo* study in Wistar rats, showed that the oral treatment with a powder from *Allium sativum* containing carvone and diallyl disulphide (200 mg/kg for 7 days) prevented the manifestations of monosodium glutamate-induced neurotoxicity by the reduction of caspase-3 and calretinin expression with the preservation of normal brain architecture (Hazzaa et al. 2020).

16.3.2   Anticholinesterase Effect

In Alzheimer's disease, a reduction of the functionality of central cholinergic neurotransmission is a key pathogenetic mechanism (Bondi et al. 2017). Acetylcholinesterase enzyme (AchE) is an ester hydrolase responsible for the breakdown of acetylcholine (Ach) in the cholinergic synapses therefore an inhibition of the enzyme generates an accumulation of Ach in the synaptic cleft which may gradually lead to a reduction of Alzheimer's symptoms. Another related enzyme, butyrylcholinesterase (BchE) is not specific for Ach, degrading other substrates like butyrylcholine. Anticholinesterase drugs which effectively inhibit acetylcholinesterase enzyme are a major pharmacological class, three molecules which pass the blood-brain barrier being already authorized for the treatment of Alzheimer's disease worldwide (donepezil, rivastigmine, galantamine). A series of preclinical studies have investigated the anticholinesterase potential of natural compounds as better tolerated alternatives to the existing drugs, several essential oils from food plants proving promising results. The majority of the studies used the Ellman assay, a colorimetric in vitro experimental model based on the reaction between thiocholine with the sulfhydryl group of a chromogen like Ellman reagent (Rashed et al. 2021). Only one study used an *in vivo* technique on genetically modified mice with immunohistochemical evaluation of enzyme expression (Liu et al. 2020).

A study on *Citrus* species (*C. aurantifolia, C. aurantium* and *C. bergamia*) investigated their capacity of inhibiting cholinesterases. Significant effects were observed for C. *aurantifolia* and *C. aurantium* essential oils which inhibited AChE with IC_{50} values of 139.3 and 147.5 µg/mL, respectively. An inferior effect was observed against BChE with IC_{50} values ranging from 235.5 to 266.6 µg/mL for *C. aurantifolia* and *C. aurantium*, respectively (Tundis et al. 2012). An anterior study identified limonene as an important monoterpene responsible for the anticholinesterasic effect, with IC_{50} values of 225.9 and 456.2 µg/mL against AChE and BChE, although other terpenes like linalool or γ-terpinene could contribute to the effect (Menichini et al. 2009). Two essential oils from the peels and seeds of sweet orange, *Citrus sinensis* L. Osbeck, another plant from *Citrus* genus, was tested for a possible anticholinesterasic effect, the results of the study showing a dose-dependent inhibition of acetylcholinesterase with IC_{50} of 2.64 µg/mL for peel EO and 3.54 µg/mL for seed EO (Ademosun et al. 2016).

A recently published study investigated the anticholinesterase effect of the essential oil from true cinnamon (*Cynnamomum verum*), showing an inhibitory effect of 99% with AChE and BuChE inhibitory activities of 278.72 and 330.72 µg galanthamine equivalents GALAEs/g sample (Sihoglu Tepe and Ozaslan 2020).

Another experimental study used genetically engineered mice (APP/PS1) to evaluate the effect of lemon essential oil on cholinergic transmission in the hippocampus. The results showed a level of Ach increased with 31% compared to wild type mice after 30 days of treatment with lemon essential oil. Also, the expression of acetylcholinesterase determined by immunohistochemical techniques was decreased in the studied tissue, the experimental data confirming a significant reduction of AchE-positive cells in specific area of the hippocampus, after the treatment with the essential oil (Liu et al. 2020).

The essential oil from star aniseed *(Ilicium verum* Hook f.) and its main chemical component anethole showed anticholinesterase properties, inhibiting primarily acetylcholinesterase with IC_{50} of 36.00 µg/mL and 39.89 µg/mL, respectively (Bhadra et al. 2011).

Mentha spicata L. is another aromatic plant used for food flavoring with a significant content of essential oil. A study investigating chemical composition and the effects of the essential oil from spearmint showed a carvone chemotype but also proved a significant inhibition of acetylcholinesterase (IC_{50} of 23.1 µg/mL) and butyrylcholinesterase (IC_{50} of 35.0 µg/mL) (Ali-Shtayeh et al. 2019).

16.3.3   Inhibition of β-Amyloid Plaque Formation

Beta-amyloid is directly involved in the formation of senile plaques, an important hallmark of Alzheimer's disease. Generated by the sequential action of a family of secretase enzymes on the amyloid precursor protein (APP), beta-amyloid (subtypes Aβ1-40 and Aβ1-42) can form dimers and trimers that can accumulate in the neurons and self-assemble into senile plaques, with toxic consequences (Bondi et al. 2017). Therefore, the inhibition of beta-amyloid represents an important strategy to counteract one of the main pathological mechanisms of Alzheimer disease, leading to the recent authorization by the FDA of aducanumab, a monoclonal antibody directed against beta-amyloid fragments. Other drugs, either inhibitors of secretase enzymes or inhibitors of beta-amyloid aggregation are in various phases of preclinical investigations. In this context, several studies were focused on the potential inhibitory effect of essential oils on beta-amyloid plaque formation.

A complex study used transgenic APP/PS1 mice which spontaneously develop high levels of beta-amyloid which can form plaques in cortical areas. The treatment of the animals with lemon essential oil for 30 days induced a clear reduction of amyloid deposits and a 26% reduction of cortical areas affected by amyloid plaques (Liu et al. 2020).

The essential oil from ginger (*Zingiber officinale*) was tested in vivo using aluminum chloride for induction of Alzheimer disease. The oral administration of the essential oil for 12 weeks produced a reduction of the formation of amyloid plaques, demonstrated by histopathological examination (Fathy et al. 2015; Talebi et al. 2021; Schepici et al. 2021).

Linalool, present in the chemical composition of coriander and lavender essential oils was tested in vitro using PC12 cell cultures against toxicity induced by Aβ1-42. The results showed that linalool but also the essential oils were able to prevent nuclear morphological abnormalities in the treated cells, by inhibiting the activation of caspase-3, an enzyme induced by the beta-amyloid fragment (Caputo et al. 2021).

16.3.4 Protection Against Other Neurodegenerative Disorders

Oxidative stress is involved in the apparition of other neurodegenerative disorders like Parkinson disease, characterized by a severe loss of dopaminergic neurons in the substantia nigra. Free radicals formed excessively in the central nervous system during dopamine abnormal metabolism may attack proteins, lipid structures or DNA, inflicting considerable damage to neuronal structures controlling movement, with a possible development of Parkinson disease. Numerous essential oils from thyme, clove, cinnamon, basil, coriander, cumin were found to have significant antioxidant potential (Tomaino et al. 2005). Therefore, the antioxidant effect of essential oils has the potential of slowing down the onset of Parkinson disease and was tested in several *in vitro* and *in vivo* experimental models.

A study investigated *in vitro* on PC12 cell cultures the protective effects of essential oils from true cinnamon and cassia cinnamon (*Cinnamomum verum* and *Cinnamomum cassia*) against toxic effects of 6-OH-dopamine. Pretreatment with essential oils at 20 µg/mL increased the viability of the cells and decreased the content of reactive oxygen species, demonstrating a clear potential of preventing neural lesions (Ramazani et al. 2020).

An *in vivo* study evaluated the protective effects of eugenol, the main component of the essential oil from *Szygium aromaticum* L. in a mouse model of Parkinson induced by 6-OH-dopamine. The results showed that eugenol prevented the reduction of dopamine in striatal region of the brain by antioxidant mechanisms (increase of SOD activity and reduced glutathione concentration (Kabuto et al. 2007).

Myrcene, a terpene present in the essential oils of many food plants and spices was administered in doses of 200 mg/kg to C57Bl/J6 mice, in an ischemia/reperfusion injury model, protecting the brain against oxidative lesions by augmenting glutathione peroxidase and superoxide dismutase (Surendran et al. 2021).

References

Ademosun AO, Oboh G, Olupona AJ, Oyeleye SI, Adewuni TM, Nwanna EE (2016) Comparative study of chemical composition, in vitro inhibition of Cholinergic and Monoaminergic enzymes, and antioxidant potentials of essential oil from peels and seeds of sweet Orange (Citrus Sinensis [L.] Osbeck) fruits. J Food Biochem 40:53–60. https://doi.org/10.1111/JFBC.12187

Ahl HAHS, Sarhan AM, Dahab A, Dahab MA (2015) Essential oils of Anethum graveolens L.: chemical composition and their antimicrobial activities at vegetative , flowering and fruiting stages of development. Int J Plant Sci Ecol 1:98–102

Ainane A (2018) Chemical study by GC-MS of the essential oils of certain mints grown in the region of Settat (Morocco): Mentha Piperita, Mentha Pulegium and Mentha Spicata. Drug Des Intellect Prop Int J 1:124–127. https://doi.org/10.32474/DDIPIJ.2018.01.000120

Airao V, Buch P, Sharma T, Vaishnav D, Parmar S (2021) Linalool protects hippocampal neurons and improves functional outcomes following experimental ischemia/reperfusion in rats. J Appl Biol Biotechnol 9:174–181. https://doi.org/10.7324/JABB.2021.9218

Al-Asmari AK, Athar MT, Al-Faraidy AA, Almuhaiza MS (2017) Chemical composition of essential oil of Thymus vulgaris collected from Saudi Arabian market. Asian Pac J Trop Biomed 7:147–150. https://doi.org/10.1016/J.APJTB.2016.11.023

Al-Dhahli AS, Al-Hassani FA, Mohammed Alarjani K, Mohamed Yehia H, Al Lawati WM, Najmul Hejaz Azmi S, Alam Khan S (2020) Essential oil from the rhizomes of the Saudi and Chinese Zingiber officinale cultivars: comparison of chemical composition, antibacterial and molecular docking studies. J King Saud Univ - Sci 32:3343–3350. https://doi.org/10.1016/J.JKSUS.2020.09.020

Ali-Shtayeh MS, Jamous RM, Abu-Zaitoun SY, Khasati AI, Kalbouneh SR (2019) Biological properties and bioactive components of Mentha spicata L. essential oil: focus on potential benefits in the treatment of obesity, Alzheimer's disease, Dermatophytosis, and drug-resistant infections. Evid Based Compl Altern Med:834265. https://doi.org/10.1155/2019/3834265

Alizadeh Behbahani B, Falah F, Lavi Arab F, Vasiee M, Tabatabaee Yazdi F (2020) Chemical composition and antioxidant, antimicrobial, and antiproliferative activities of Cinnamomum zeylanicum Bark essential oil. Evid Based Compl Altern Med 2020. https://doi.org/10.1155/2020/5190603

Allenspach M, Steuer C (2021) α-Pinene: A never-ending story. Phytochemistry 190:112857. https://doi.org/10.1016/J.PHYTOCHEM.2021.112857

An Q, Ren J-N, Li X, Fan G, Qu S-S, Song Y, Li Y, Pan S-Y (2021) Recent updates on bioactive properties of linalool. Food Funct 12:10370–10389. https://doi.org/10.1039/D1FO02120F

Ayaz M, Sadiq A, Junaid M, Ullah F, Subhan F, Ahmed J (2017) Neuroprotective and anti-aging potentials of essential oils from aromatic and medicinal plants. Front Aging Neurosci 9. https://doi.org/10.3389/FNAGI.2017.00168

Bağci Y, Kan Y, Doğu S, Çelik SA, Bağci Y (2017) The essential oil compositions of Origanum majorana L. cultivated in Konya and collected from Mersin-Turkey. Indian J Pharm Educ Res 51:463–9. https://doi.org/10.5530/ijper.51.3s.68

Barot J, Saxena B (2021) Therapeutic effects of eugenol in a rat model of traumatic brain injury: a behavioral, biochemical, and histological study. J Tradit Complement Med 11:318–327. https://doi.org/10.1016/J.JTCME.2021.01.003

Beigi M, Torki-Harchegani M, Pirbalouti AG (2018) Quantity and chemical composition of essential oil of peppermint (Mentha × piperita L.) leaves under different drying methods. Int J Food Prop 21:267–276. https://doi.org/10.1080/10942912.2018.1453839

Belkhodja H, Meddah B, Touil AT, Şekeroğlu N, Sonnet P (2016) Chemical composition and properties of essential oil of Rosmarinus Officinalis and Populus Alba. World. J Pharm Pharm Sci 5041:108–119. https://doi.org/10.20959/wjpps20169-7558

Ben Hsouna A, Ben Halima N, Smaoui S, Hamdi N (2017) Citrus lemon essential oil: chemical composition, antioxidant and antimicrobial activities with its preservative effect against Listeria monocytogenes inoculated in minced beef meat. Lipids Health Dis 16:1–11. https://doi.org/10.1186/S12944-017-0487-5/FIGURES/3

Bhadra S, Mukherjee PK, Kumar NS, Bandyopadhyay A (2011) Anticholinesterase activity of standardized extract of Illicium verum Hook. f. fruits. Fitoterapia 82:342–346. https://doi.org/10.1016/J.FITOTE.2010.11.003

Bondi MW, Edmonds EC, Salmon DP (2017) Alzheimer's disease: past, present, and future. J Int Neuropsychol Soc 23:818–831. https://doi.org/10.1017/S135561771700100X

Brada M, Saadi A, Wathelet JP, Lognay G (2012) The essential oils of Origanum majorana l. and Origanum floribundum munby in Algeria. J Essent Oil-Bearing Plants 15:497–502. https://doi.org/10.1080/0972060X.2012.10644078

Campêlo LML, de Lima SG, Feitosa CM, de Freitas RMD (2011) Evaluation of central nervous system effects of Citrus limon essential oil in mice. Rev Bras Farmacogn 21:668–673. https://doi.org/10.1590/S0102-695X2011005000086

Caputo L, Piccialli I, Ciccone R, de Caprariis P, Massa A, De Feo V, Pannaccione A (2021) Lavender and coriander essential oils and their main component linalool exert a protective effect against amyloid-β neurotoxicity. Phyther Res 35:486–493. https://doi.org/10.1002/PTR.6827

Caputo L, Souza LF, Alloisio S, Cornara L, De Feo V (2016) *Coriandrum sativum* and *Lavandula angustifolia* essential oils: chemical composition and activity on central nervous system. Int J Mol Sci 17:1999:1-1999:12. https://doi.org/10.3390/ijms17121999

Chahal KK, Kumar A, Bhardwaj U, Kaur R (2017) Chemistry and biological activities of Anethum graveolens L. (dill) essential oil : a review. J Pharmacogn Phytochem 6:295–306

Charret TS, Pereira MTM, Pascoal VDB, Lopes-Cendes I, Pascoal ACRF (2021) Citral effects on the expression profile of brain-derived neurotrophic factor and inflammatory cytokines in Status Epilepticus-induced rats using the Lithium-Pilocarpine Model. J Med Food 24:916–924. https://doi.org/10.1089/JMF.2020.0073

Chen X, Zhang L, Qian C, Du Z, Xu P, Xiang Z (2020) Chemical compositions of essential oil extracted from Lavandula angustifolia and its prevention of TPA-induced inflammation. Microchem J 153:104458. https://doi.org/10.1016/J.MICROC.2019.104458

Dai M, Wu L, Yu K, Xu R, Wei Y, Chinnatambi A, Alahmadi TA, Zhou M (2020) D-carvone inhibit cerebral ischemia/reperfusion induced inflammatory response TLR4/NLRP3 signaling pathway. Biomed Pharmacother 132:110879. https://doi.org/10.1016/j.biopha.2020.110870

Damyanova S, Mollova S, Stoyanova A, Gubenia O (2016) Chemical composition of Salvia officinalis l. essential oil from Bulgaria. Ukr Food Technol 5:695–700. https://doi.org/10.24263/2304

de Lucena JD, Gadelha-Filho CVJ, da Costa RO, de Araújo DP, Lima FAV, Neves KRT, de Barros Viana GS (2020) L-linalool exerts a neuroprotective action on hemiparkinsonian rats. Naunyn Schmiedeberg's Arch Pharmacol 393:1077–1088. https://doi.org/10.1007/S00210-019-01793-1

Dob T, Dahmane D, Chelghoum C (2006) Chemical composition of the essential oils of Pistacia lentiscus L. from Algeria. J Essential Oil Res 18(3):335–338. https://doi.org/10.1080/1041290 5.2006.9699105

Eddin LB, Jha Kumar N, Nagoor MM, Kumar Kesari K, Beiram R, Ojha S (2021) Neuroprotective potential of Limonene and Limonene containing natural products. Molecules 26:4535. https://doi.org/10.3390/molecules26154535

El Euch SK, Hassine DB, Cazaux S, Bouzouita N, Bouajila J (2019) Salvia officinalis essential oil: chemical analysis and evaluation of anti-enzymatic and antioxidant bioactivities. South African J Bot 120:253–260. https://doi.org/10.1016/J.SAJB.2018.07.010

Elisabetsky E, Marschner J, Onofre Souza D (1995) Effects of Linalool on glutamatergic system in the rat cerebral cortex. Neurochem Res 20:461–465. https://doi.org/10.1007/BF00973103

Fathy MM, Eid HH, Hussein MA, Ahmed HH, Hussein AA (2015) The role of Zingiber officinale in the treatment of Alzheimer's disease: in-vitro and in-vivo evidences. Res J Pharm Biol Chem Sci 6:735–749

Gan L, Cookson MR, Petrucelli L, La Spada AR (2018) Converging pathways in neurodegeneration, from genetics to mechanisms. Nat Neurosci 21:1300–1309. https://doi.org/10.1038/S41593-018-0237-7

Garattini S (2000) Glutamic acid, twenty years later. J Nutr 130:901S–909S. https://doi.org/10.1093/JN/130.4.901S

Gaylor R, Gaylor R, Michel J, Thierry D, Panja R, Fanja F, Pascal D (2014) Bud, leaf and stem essential oil composition of Syzygium aromaticum from Madagascar, Indonesia and Zanzibar. Int J Basic Appl Sci 3:224–233. https://doi.org/10.14419/ijbas.v3i3.2473

Ghahramanloo KH, Kamalidehghan B, Akbari Javar H, Teguh Widodo R, Majidzadeh K, Noordin MI (2017) Comparative analysis of essential oil composition of Iranian and Indian Nigella sativa L. extracted using supercritical fluid extraction and solvent extraction. Drug Des Devel Ther 11:2221–2226. https://doi.org/10.2147/DDDT.S87251

Gonçalves JCR, de Sousa Oliveira F, Benedito RB, de Sousa DP, de Almeida RN, de Araujo DAM (2008) Antinociceptive activity of carvone: evidence of association with association with decreased peripheral nerve excitability. Biol Pharm Bull 31(5):1017–1020. https://doi.org/10.1248/bpb.31.1017

Gonçalves ECD, Assis PM, Junqueira LA, Cola M, Santos ARS, Raposo NRB, Dutra RC (2020) Citral inhibits the inflammatory response and Hyperalgesia in Mice: the role of TLR4, TLR2/Dectin-1, and CB2 Cannabinoid Receptor/ATP-sensitive K+ channel pathways. ACS

Appl Mater Interfaces. https://doi.org/10.1021/ACS.JNATPROD.9B01134/SUPPL_FILE/ NP9B01134_SI_001.PDF

González-Mas MC, Rambla JL, López-Gresa MP, Amparo Blázquez M, Granell A (2019) Volatile compounds in citrus essential oils: a comprehensive review. Front Plant Sci 10:12. https://doi. org/10.3389/FPLS.2019.00012/BIBTEX

Gudiño-Cabrera G, Ureña-Guerrero ME, Rivera-Cervantes MC, Feria-Velasco AI, Beas-Zárate C (2014) Excitotoxicity triggered by neonatal monosodium glutamate treatment and blood-brain barrier function. Arch Med Res 45:653–659. https://doi.org/10.1016/J.ARCMED.2014.11.014

Hammouda F, Saleh M, Abdel-Azim N, Shams K, Ismail S, Shahat A, Saleh I (2014) Evaluation of the essential oil of Foeniculum Vulgare Mill (Fennel) fruits extracted by three different extraction methods by GC/MS. Afr J Tradit Compl Altern Med 11:277–279. https://doi. org/10.4314/ajtcam.v11i2.8

Hazzaa SM, Abdelaziz SAM, Eldaim MAA, Abdel-Daim MM, Elgarawany GE (2020) Neuroprotective potential of Allium sativum against Monosodium Glutamate-induced excito- toxicity: impact on short-term memory, gliosis, and oxidative stress. Nutrients 12. https://doi. org/10.3390/NU12041028

Heghes SC, Vostinaru O, Rus LM, Mogosan C, Iuga CA, Filip L (2019) Antispasmodic effect of essential oils and their constituents: a review. Molecules 24. https://doi.org/10.3390/ molecules24091675_rfseq1

Irie Y (2012) Effects of Eugenol on the central nervous system: its possible application to treat- ment of Alzheimers disease, depression, and Parkinsons disease. Curr Bioact Compd 2:57–66. https://doi.org/10.2174/1573407210602010057

Jansen C, Shimoda L, Kawakami J, Ang L, Bacani A, Baker J, Badowski C, Speck M, Stokes A, Small-Howard A, Turner H (2019) Channels Myrcene and terpene regulation of TRPV1. Channels 13:344–366. https://doi.org/10.1080/19336950.2019.1654347

Javitt DC (2004) Glutamate as a therapeutic target in psychiatric disorders. Mol Psychiatry 9:984–997. https://doi.org/10.1038/sj.mp.4001551

Joshi RK (2014) Chemical composition and antimicrobial activity of the essential oil of Ocimum basilicum L. (sweet basil) from Western Ghats of North West Karnataka, India. Anc Sci Life 33:151. https://doi.org/10.4103/0257-7941.144618

Kabir Y, Akasaka-Hashimoto Y, Kubota K, Komai M (2020) Volatile compounds of black cumin (Nigella sativa L.) seeds cultivated in Bangladesh and India. Heliyon 6. https://doi. org/10.1016/J.HELIYON.2020.E05343

Kabuto H, Tada M, Kohno M (2007) Eugenol [2-methoxy-4-(2-propenyl)phenol] prevents 6-hydroxydopamine-induced dopamine depression and lipid peroxidation inductivity in mouse striatum. Biol Pharm Bull 30:423–427. https://doi.org/10.1248/BPB.30.423

Kapoor IPS, Singh B, Singh G, De Heluani CS, De Lampasona MP, Catalan CAN (2013) Chemical composition and antioxidant activity of essential oil and Oleoresins of Nutmeg (Myristica fragrans Houtt.) fruits. Int J Food Prop 16:1059–1070. https://doi.org/10.1080/1094291 2.2011.576357

Kaur K, Kaushal S, Rani R (2019) Chemical composition, antioxidant and antifungal potential of Clove (Syzygium aromaticum) essential oil, its major compound and its derivatives. J Essent Oil Bear Plants 22:1195–1217. https://doi.org/10.1080/0972060X.2019.1688689

Komaitis ME, Ifanti-Papatragianni N, Melissari-Panagiotou E (1992) Composition of the essential oil of marjoram (Origanum majorana L.). Food Chem 45:117–118. https://doi. org/10.1016/0308-8146(92)90020-3

Kosar M, Demirci B, Goger F, Kara I, Baser KHC (2017) Volatile composition, antioxidant activ- ity, and antioxidant components in saffron cultivated in Turkey. Int J Food Prop 20:S746–S754. https://doi.org/10.1080/10942912.2017.1311341

Li Puma S, Landini L, Macedo SJ, Seravalli V, Marone IM, Coppi E, Patacchini R, Geppetti P, Materazzi S, Nassini R, De Logu F (2019) TRPA1 mediates the antinociceptive properties of the constituent of Crocus sativus L., safranal. J Cell Mol Med 23:1976–1986. https://doi. org/10.1111/JCMM.14099

Lin LY, Chuang CH, Chen HC, Yang KM (2019) Lime (Citrus aurantifolia (Christm.) Swingle) essential oils: volatile compounds, antioxidant capacity, and hypolipidemic effect. Foods 8. https://doi.org/10.3390/FOODS8090398

Liu B, Kou J, Li F, Huo D, Xu J, Zhou X, Meng D, Ghulam M, Artyom B, Gao X, Ma N, Han D (2020) Lemon essential oil ameliorates age-associated cognitive dysfunction via modulating hippocampal synaptic density and inhibiting acetylcholinesterase. Aging (Albany NY) 12:8622–8639. https://doi.org/10.18632/AGING.103179

Lizarraga-Valderrama LR (2021) Effects of essential oils on central nervous system: focus on mental health. Phyther Res 35:657–679. https://doi.org/10.1002/ptr.6854

López V, Nielsen B, Solas M, Ramírez MJ, Jäger AK (2017) Exploring pharmacological mechanisms of Lavender (Lavandula angustifolia) essential oil on central nervous system targets. Front Pharmacol 8. https://doi.org/10.3389/FPHAR.2017.00280

Lorigooini Z, Boroujeni SN, Sayyadi-Shahraki M, Rahimi-Madiseh M, Bijad E, Amini-Khoei H (2021) Limonene through attenuation of Neuroinflammation and nitrite level exerts antidepressant-like effect on Mouse Model of maternal separation stress. Behav Neurol 2021. https://doi.org/10.1155/2021/8817309

Ma L, Mu Y, Zhang Z, Sun Q (2018) Eugenol promotes functional recovery and alleviates inflammation, oxidative stress, and neural apoptosis in a rat model of spinal cord injury. Restor Neurol Neurosci 36(5):659–668. https://doi.org/10.3233/RNN-180826

Machado KDC, Islam MT, Ali ES, Rouf R, Uddin SJ, Dev S, Shilpi JA, Shill MC, Reza HM, Das AK, Shaw S, Mubarak MS, Mishra SK, Melo-Cavalcante AADC (2018) A systematic review on the neuroprotective perspectives of betacaryophyllene. Phytother Res 32(12):2376–2388. https://doi.org/10.1002/ptr.6199

Mahboubi M (2021) Mentha spicata L. essential oil, phytochemistry and its effectiveness in flatulence. J Tradit Complement Med 11:75–81. https://doi.org/10.1016/J.JTCME.2017.08.011

Mahboubi M (2019) Zingiber officinale Rosc. essential oil, a review on its composition and bioactivity. Clin Phytosci 51 5:1–12. https://doi.org/10.1186/S40816-018-0097-4

Malcolm BJ, Tallian K (2018) Essential oils of lavender in anxiety disorders: ready for prime time? Ment Health Clin 7(4):147–155. https://doi.org/10.9740/mhc.2017.07.147

Mallah K, Couch C, Borucki DM, Toutonji A, Alshareef M, Tomlinson S (2020) Anti-inflammatory and neuroprotective agents in clinical trials for CNS disease and injury: where do we go from here? Front Immunol 11:2021. https://doi.org/10.3389/FIMMU.2020.02021/BIBTEX

Mandal S, Mandal M (2015) Coriander (Coriandrum sativum L.) essential oil: chemistry and biological activity. Asian Pac J Trop Biomed 5:421–428. https://doi.org/10.1016/J.APJTB.2015.04.001

Marín I, Sayas-Barberá E, Viuda-Martos M, Navarro C, Sendra E (2016) Chemical composition, antioxidant and antimicrobial activity of essential oils from organic Fennel, Parsley, and Lavender from Spain. Foods 5:18:1–18:10. https://doi.org/10.3390/foods5010018

Mattson MP (2008) Glutamate and neurotrophic factors in neuronal plasticity and disease. Ann N Y Acad Sci 1144:97–112. https://doi.org/10.1196/ANNALS.1418.005

Memon T, Yarishkin O, Reilly CA, Krizaj D, Olivera BM, Teichert RW (2019) Trans-anethole of fennel oil is a selective and nonelectrophilic agonist of the TRPA1 Ion channel. Mol Pharmacol 95:433–441. https://doi.org/10.1124/MOL.118.114561/-/DC1

Menichini F, Tundis R, Loizzo MR, Bonesi M, Marrelli M, Statti GA, Menichini F, Conforti F (2009) Acetylcholinesterase and butyrylcholinesterase inhibition of ethanolic extract and monoterpenes from Pimpinella anisoides V Brig. (Apiaceae). Fitoterapia 80:297–300. https://doi.org/10.1016/J.FITOTE.2009.03.008

Migheli R, Lostia G, Galleri G, Rocchitta G, Serra PA, Bassareo V, Acquas E, Peana AT (2021) Neuroprotective effect of (R)-(−)-linalool on oxidative stress in PC12 cells. Phytomedicine Plus 1:100073. https://doi.org/10.1016/J.PHYPLU.2021.100073

Miguel MG, Cruz C, Faleiro L, Simões MTF, Figueiredo AC, Barroso JG, Pedro LG (2010) Foeniculum vulgare essential oils: chemical composition, antioxidant and antimicrobial activities. Nat Prod Commun 5:319–328

Morshed S, Hossain MD, Ahmad M, Junayed M (2017) Physicochemical characteristics of essential oil of black pepper (Piper nigrum) cultivated in Chittagong. J Food Qual Hazards Control 4:66–69

Muchtaridi SA, Apriyantono A, Mustarichie R (2010) Identification of compounds in the essential oil of Nutmeg seeds (Myristica fragrans Houtt.) that inhibit locomotor activity in mice. Int J Mol Sci 11:4771. https://doi.org/10.3390/IJMS11114771

Najem W, El Beyrouthy M, Hanna Wakim L, Neema C, Ouaini N, Wafaa Najem B (2011) Essential oil composition of Rosa damascena Mill. from different localities in Lebanon. Acta Bot Gall 158:365–373. https://doi.org/10.1080/12538078.2011.10516279

Negro C, De Belis L, Miceli A (2014) Chemical composition and antioxidant activity of Pistacia lentiscus essential oil from southern Italy (Apulia). J Essential Oil Res 27(1):23–29. https://doi.org/10.1080/10412905.2014.973614

Njoroge SM, Phi NTL, Sawamura M (2009) Chemical composition of Peel essential oils of sweet oranges (Citrus sinensis) from Uganda and Rwanda. J Essent Oil Bear Plants 12:26–33. https://doi.org/10.1080/0972060X.2009.10643687

Oforma CC, Udourioh GA, Ojinnaka CM (2020) Characterization of essential oils and fatty acids composition of stored ginger (Zingiber officinale Roscoe). J Appl Sci Environ Manag 23:2231–2238. https://doi.org/10.4314/jasem.v23i12

Orav A, Arak E, Raal A (2011) Essential oil composition of Coriandrum sativum L. fruits from different countries. J Essent Oil Bear Plants 14:118–123. https://doi.org/10.1080/0972060X.2011.10643910

Orav A, Raal A, Elma A (2008) Essential oil composition of Pimpinella anisum L. fruits from various European countries. Nat Prod Res 22:227–232. https://doi.org/10.1080/14786410701424667

Orav A, Stulova I, Kailas T, Müürisepp M (2004) Effect of storage on the essential oil composition of Piper nigrum L. fruits of different ripening states. J Agric Food Chem 52:2582–2586. https://doi.org/10.1021/JF030635S

Özcan MM, Chalchat JC (2006) Chemical composition and antifungal effect of anise (Pimpinella anisum L.) fruit oil at ripening stage. Ann Microbiol 56:353–358. https://doi.org/10.1007/BF03175031

Pandey AK, Singh P, Tripathi NN (2014) Chemistry and bioactivities of essential oils of some Ocimum species: an overview. Asian Pac J Trop Biomed 4:682–694. https://doi.org/10.12980/APJTB.4.2014C77

Paranagama PA, Wimalasena S, Jayatilake GS, Jayawardena AL, Senanayake UM, Mubarak AM (2001) A comparison of essential oil constituents of bark, leaf, root and fruit of cinnamon (Cinnamomum zeylanicum Blum) grown in Sri Lanka. J Natl Sci Found Sri Lanka 29:147–153. https://doi.org/10.4038/JNSFSR.V29I3-4.2613

Porte A, Godoy RLO (2008) Chemical composition of Thymus vulgaris L. (thyme) essential oil from the Rio de Janeiro State (Brazil). J Serbian Chem Soc 73:307–310. https://doi.org/10.2298/JSC0803307P

Price S, Price L (2007) Essential oils for general use in health-care settings. In: Price S, Price L (eds) Aromatherapy for health professionals, 3rd edn. Churchill Livingstone Elsevier, London, pp 385–477

Qneibi M, Jaradat N, Emwas N (2019) Effect of geraniol and citronellol essential oils on the biophysical gating properties of AMPA receptors. Appl Sci 9(21):4693. https://doi.org/10.3390/app9214693

Quintans LJ, Guimarães AG, de Santana MT, Araújo BES, Moreira FV, Bonjardim LR, Araújo AAS, Siqueira JS, Ângelo AR, Botelho MA, Almeida JRGS, Santos MRV (2011) Citral reduces nociceptive and inflammatory response in rodents. Rev Bras Farmacogn 21:497–502. https://doi.org/10.1590/S0102-695X2011005000065

Quirós-Sauceda AE, Ovando-Martínez M, Velderrain-Rodríguez GR, González-Aguilar GA, Ayala-Zavala JF (2016) Licorice (Glycyrrhiza glabra Linn.) oils. Essent Oils Food Preserv Flavor Saf:523–530. https://doi.org/10.1016/B978-0-12-416641-7.00060-2

Raal A, Orav A, Arak E (2012) Essential oil composition of Foeniculum vulgare Mill. fruits from pharmacies in different countries. Nat Prod Res 26:1173–1178. https://doi.org/10.108 0/14786419.2010.535154

Raal A, Orav A, Arak E (2007) Composition of the essential oil of Salvia officinalis L. from various European countries. Nat Prod Res 21:406–411. https://doi.org/10.1080/14786410500528478

Rădulescu V, Popescu ML, Ilieş DC (2010) Chemical composition of the volatile oil from different plant parts of *Anethum graveolens* L. (Umbelliferae) cultivated in Romania. Farmacia 58:594–600. https://doi.org/10.1021/ja909321d

Ramazani E, YazdFazeli M, Emami SA, Mohtashami L, Javadi B, Asili J, Tayarani-Najaran Z (2020) Protective effects of Cinnamomum verum, Cinnamomum cassia and cinnamaldehyde against 6-OHDA-induced apoptosis in PC12 cells. Mol Biol Rep 474(47):2437–2445. https://doi.org/10.1007/S11033-020-05284-Y

Rashed AA, Rahman AZA, Rathi DNG (2021) Essential oils as a potential neuroprotective remedy for age-related neurodegenerative diseases: a review. Molecules 26:1–61. https://doi.org/10.3390/molecules26041107

Rus C, Pop G, Alexa E, Sumalan RM, Copolovici DM (2015) Antifungal activity and chemical composition of *Origanum Majorana* L. essential oil. Res J Agric Sci 47:179–185

Ryu S, Seol GH, Park H, Choi IY (2014) Trans-anethole protects cortical neuronal cells against oxygen-glucose deprivation/reoxygenation. Neurol Sci 35:1541–1547. https://doi.org/10.1007/S10072-014-1791-8

Sadeghnia HR, Shaterzadeh H, Forouzanfar F, Hosseinzadeh H (2017) Neuroprotective effect of safranal, an active ingredient of Crocus sativus , in a rat model of transient cerebral ischemia. Folia Neuropathol 55:206–213. https://doi.org/10.5114/FN.2017.70485

Saibi S, Belhadj M, Benyoussef E-H (2012) Essential oil composition of Pimpinella anisum from Algeria. Anal Chem Lett 2:401–404. https://doi.org/10.1080/22297928.2012.10662624

Satyal P, Craft JD, Dosoky NS, Setzer WN (2017a) The chemical compositions of the volatile oils of garlic (Allium sativum) and wild garlic (Allium vineale). Foods 6:1–10. https://doi.org/10.3390/FOODS6080063

Satyal P, Jones TH, Lopez EM, McFeeters RL, Ali NAA, Mansi I, Al-Kaf AG, Setzer WN (2017b) Chemotypic characterization and biological activity of Rosmarinus officinalis. Foods 6:20:1-20:15. https://doi.org/10.3390/foods6030020

Satyal P, Murray BL, McFeeters RL, Setzer WN (2016) Essential oil characterization of Thymus vulgaris from various geographical locations. Foods 5:1–12. https://doi.org/10.3390/foods5040070

Schepici G, Contestabile V, Valeri A, Mazzon E (2021) Ginger, a possible candidate for the treatment of dementias? Molecules 26. https://doi.org/10.3390/MOLECULES26185700

Seify Z, Yadegari M, Pirbalouti AG (2018) Essential oil composition of Rosa damascena Mill. Produced with different storage temperatures and durations. Hortic Sci Technol 36:552–559. https://doi.org/10.12972/KJHST.20180055

Şerban ES, Socaci SA, Tofana M, Maier SC, Bojita MT (2012) Advantages of "Headspace" technique for GC/MS analysis of essential oils. Farmacia 60(2):249–256.

Sharma N, Tan MA, An SSA (2021) Mechanistic aspects of apiaceae family spices in ameliorating Alzheimer's disease. Antioxidants 10:1–18. https://doi.org/10.3390/antiox10101571

Sharopov FS, Wink M, Gulmurodov IS, Isupov SJ, Zhang H, Setzer WN (2013) Composition and bioactivity of the essential oil of Anethum graveolens L. from Tajikistan. Int J Med Aromat Plants 3:125–130

Sihoglu Tepe A, Ozaslan M (2020) Anti-Alzheimer, anti-diabetic, skin-whitening, and antioxidant activities of the essential oil of Cinnamomum zeylanicum. Ind Crop Prod 145. https://doi.org/10.1016/J.INDCROP.2019.112069

Silva Brum LF, Elisabetsky E, Souza D (2001a) Effects of linalool on [(3)H]MK801 and [(3)H] muscimol binding in mouse cortical membranes. Phyther Res 15:422–425. https://doi.org/10.1002/PTR.973

388 O. Voştinaru et al.

Silva Brum LF, Emanuelli T, Souza DO, Elisabetsky E (2001b) Effects of linalool on glutamate release and uptake in mouse cortical synaptosomes. Neurochem Res 26:191–194. https://doi.org/10.1023/A:1010904214482

Smigielski K, Prusinowska R, Stobiecka A, Kunicka-Styczyñska A, Gruska R (2018) Biological properties and chemical composition of essential oils from flowers and aerial parts of Lavender (*Lavandula angustifolia*). J Essent Oil-Bearing Plants 21:1303–1314. https://doi.org/10.1080/0972060X.2018.1503068

Snoussi M, Noumi E, Trabelsi N, Flamini G, Papetti A, De Feo V (2015) *Mentha spicata* essential oil: chemical composition, antioxidant and antibacterial activities against planktonic and biofilm cultures of vibrio spp. strains. Molecules 20:14402–14424. https://doi.org/10.3390/molecules200814402

Spadaro F, Costa R, Circosta C, Occhiuto F (2012) Volatile composition and biological activity of key lime *Citrus aurantifolia* essential oil. Nat Prod Commun 7:1523–1526

Spasova Nunes H, Graça Miguel M (2017) Rosa damascena essential oils: a brief review about chemical composition and biological properties. Trends Phytochem Res 1:111–128

Stanojević LP, Stanković MZ, Cvetković DJ, Danilović BR, Stanojević JS (2016) Dill (*Anethum graveolens* L.) seeds essential oil as a potential natural antioxidant and antimicrobial agent. Biol Nyssana 7:31–39. https://doi.org/10.5281/zenodo.159101

Sun X, Wang D, Zhang T, Lu X, Duan F, Ju L, Zhuang X, Jiang X (2020) Eugenol attenuates cerebral ischemia-reperfusion injury by enhancing autophagy via AMPK-mTOR-P70S6K pathway. Front Pharmacol 11:84. https://doi.org/10.3389/FPHAR.2020.00084/BIBTEX

Surendran S, Qassadi F, Surendran G, Lilley D, Heinrich M (2021) Myrcene—what are the potential health benefits of this Flavouring and Aroma agent? Front Nutr 8:400. https://doi.org/10.3389/FNUT.2021.699666/BIBTEX

Taherpour AA, Khaef S, Yari A, Nikeafshar S, Fathi M, Ghambari S (2017) Chemical composition analysis of the essential oil of Mentha piperita L. from Kermanshah, Iran by hydrodistillation and HS/SPME methods. J Anal Sci Technol 8:1–6. https://doi.org/10.1186/S40543-017-0122-0/FIGURES/1

Tahir HU, Sarfraz RA, Ashraf A, Adil S (2016) Chemical composition and antidiabetic activity of essential oils obtained from two spices (Syzygium aromaticum and Cuminum cyminum). Int J Food Prop 19:2156–2164. https://doi.org/10.1080/10942912.2015.1110166

Talebi M, İlgün S, Ebrahimi V, Talebi M, Farkhondeh T, Ebrahimi H, Samarghandian S (2021) Zingiber officinale ameliorates Alzheimer's disease and cognitive impairments: lessons from preclinical studies. Biomed Pharmacother 133:111088. https://doi.org/10.1016/J.BIOPHA.2020.111088

Tarakemeh A, Rowshan V, Najafian S (2013) Essential oil content and composition of *Lavandula Angustifolia* Mill. as affected by drying method and extraction time. Anal Chem Lett 2:244–249. https://doi.org/10.1080/22297928.2012.10648276

Tomaino A, Cimino F, Zimbalatti V, Venuti V, Sulfaro V, De Pasquale A, Saija A (2005) Influence of heating on antioxidant activity and the chemical composition of some spice essential oils. Food Chem 89:549–554. https://doi.org/10.1016/J.FOODCHEM.2004.03.011

Toncer O, Karaman S, Diraz E, Tansi S (2017) Essential oil composition of Ocimum basilicum L. at different phenological stages in semi-arid environmental conditions. Fresenius Environ Bull 26:5441–5446

Tundis R, Loizzo MR, Bonesi M, Menichini F, Mastellone V, Colica C, Menichini F (2012) Comparative study on the antioxidant capacity and cholinesterase inhibitory activity of Citrus aurantifolia Swingle, C. aurantium L., and C. bergamia Risso and Poit. Peel Essential Oils. J Food Sci 77:H40–H46

Ueno H, Shimada A, Suemitsu S, Murakami S, Kitamura N, Wani K, Matsumoto Y, Okamoto M, Ishihara T (2019) Anti-depressive-like effect of 2-phenylethanol inhalation in mice. Biomed Pharmacother 111:1499–1506. https://doi.org/10.1016/J.BIOPHA.2018.10.073

Verma RS, Padalia RC, Chauhan A, Upadhyay RK, Singh VR (2019) Productivity and essential oil composition of rosemary (Rosmarinus officinalis L.) harvested at different growth stages

under the subtropical region of north India. J Essent Oil Res 32:144–149. https://doi.org/10.108
0/10412905.2019.1684391

Wang Z-J, Heinbockel T (2018) Essential oils and their constituents targeting the GABA-ergic system and sodium channels as treatment of neurological diseases. Molecules 23(5):1061. https://
doi.org/10.3390/molecules23051061

Wang G-W, Hu W-T, Huang B-K, Qin L-P (2011) Illicium verum: a review on its botany, traditional
use, chemistry and pharmacology. J Ethnopharmacol 136:10–20. https://doi.org/10.1016/J.
JEP.2011.04.051

Weston-Green K, Clunas H, Jimenez Naranjo C (2021) A review of the potential use of Pinene
and Linalool as Terpene-based medicines for brain health: discovering novel therapeutics
in the Flavours and Fragrances of Cannabis. Front Psych 12:1309. https://doi.org/10.3389/
FPSYT.2021.583211/BIBTEX

Yamada Y, Masuda S, Yamamoto S, Izu H (2015) Effects of compounds in Sake and shochu on
neuronal receptors. Aroma Res 16:66–73

Zhao Y, Xi G (2018) Safranal-promoted differentiation and survival of dopaminergic neurons in
an animal model of Parkinson's disease. Pharm Biol 56:450. https://doi.org/10.1080/1388020
9.2018.1501705

Chapter 17
An Overview of Essential Oil Anticancer Activity

Marcelli Geisse de Oliveira Prata da Silva,
Ingryd Nayara de Farias Ramos, Chrystiaine Helena Campos de Matos,
Mozaniel Santana de Oliveira, André Salim Khayat, Jorddy Neves Cruz,
and Eloisa Helena de Aguiar Andrade

17.1   Introduction

The human body constitution is made up of trillions of cells in different stages, which normally proliferate and divide according to the body's metabolic needs (Todd et al. 2017). There are well-regulated control mechanisms associated with signals that lead cells to stop growing and perform a programmed cell death. However, if these mechanisms present errors, cells may start to display abnormal characteristics such as disordered growth what can lead to tumor formation (Kulesz-Martin et al. 2018).

The main classes of genes affected in the carcinogenesis process are the oncogenes; positive growth and proliferation regulators; tumor suppressor genes; that prevent cell growth and division and repair genes, which work by repairing DNA damage (Klaunig 2020). Oncogenes and repair genes normally undergo hyperactivation changes what leads to an increase in their expression, while tumor suppressor genes undergo function loss, having their expression decreased or interrupted (Abel and DiGiovanni 2011). These changes in genes can be translated into increased production of proteins that induce cell division, interruption in the expression of cell cycle arrest proteins, or even expression of structurally abnormal proteins that have altered functioning (Weinstein et al. 2013).

Neoplastic cells have specific properties that alter their functions and induce their transformation, a process called carcinogenesis. Carcinogenesis follows three

M. G. de Oliveira Prata da Silva · I. N. de Farias Ramos
Center for Research in Oncology/Institute of Biological Sciences, Federal University of Pará, Belém, PA, Brazil

C. H. C. de Matos
Federal Institute of Education, Science and Technology of Goiás, Rio Verde, GO, Brazil

M. S. de Oliveira · A. Salim Khayat · J. N. Cruz · E. H. de Aguiar Andrade (✉)
Adolpho Ducke Laboratory, Botany Coordination, Museu Paraense Emilio Goeldi, Belém, PA, Brazil

very characteristic stages: initiation, promotion, and progression. Initiation is characterized by damage (mutations) on the genetic material of cells, which promotes the first changes in key genes for cell maintenance and growth (Malarkey et al. 2018). These mutated cells acquire selective advantages over normal cells within their microenvironment, then proliferate in greater quantities and begin to form the tumor mass, proceeding to the promotion stage. In tumor promotion, the expansion of clones of altered cells takes place, which produces a larger population of cells at risk of developing new genetic alterations and undergoing transformation (Patterson et al. 2018). Finally, the progression consists of the stage in which the modification of pre-neoplastic cells occurs, which start to express the malignant phenotype and with more invasive characteristics over time, which may reach metastasis, when tumor cells migrate and invade distant tissues (Malarkey et al. 2018).

Tumor progression is an irreversible step as it results from progressive genomic instability and uncontrolled cell/tissue growth. Even with the evidence of this ordered cascade of events, both molecular and phenotypic tumor formation seems to be even more complex, having as a determining factor the set of these alterations and not the order or stage in which they occur (Weinstein et al. 2013).

Given this high complexity, cancer is still one of the diseases with the highest incidence and mortality rates in the world, appearing as a public health problem. According to data from the International Agency for Research on Cancer (IARC), around 19.3 million new cases of cancer and approximately 10.0 million deaths from cancer took place in 2020 (Sung et al. 2021). It is estimated for the year 2040 an increase of 47% of new cases of cancer, compared to 2020, reaching 28.4 million episodes. This scenario is largely due to the increased prevalence of risk factors such as smoking, unhealthy diet, sedentary lifestyle, in addition to the aging of the world population (Siegel et al. 2021).

Even with all the advances in research and clinical management, cancer still presents itself as a challenge, as there are still flaws in therapeutic schemes and high rates of recurrence that can be associated with the late diagnosis in a large number of patients (Zugazagoitia et al. 2016).

17.2 Screening and Diagnosis

17.2.1 Background

Cancer is a group of heterogeneous diseases that can affect almost any part of the body and has many anatomical and molecular subtypes, each requiring specific diagnostic and treatment strategies. Comprehensive cancer control consists of essential components – (i) prevention, (ii) early diagnosis and screening, (iii) treatment, (iv) palliative care and (v) survival care (Moses et al. 2018). Thus, the high incidence and mortality rates could be reduced with an early diagnosis and effective treatment. The early detection of cancer helps in choosing the best therapeutic strategy and, consequently, in a better prognosis for the patient, placing this clinical approach as essential in controlling the disease (Hawkes 2019).

There are two main components in the conduct of early cancer detection, including previous/early diagnosis and screening. The basic difference between them is the appearance of signs and symptoms of the disease, where early diagnosis aims to identify cancer as early as possible in patients who already have symptoms, while in screening healthy individuals from a target population are tested for traits of the disease or for the possibility of developing it before any symptoms (Markopoulos et al. 2017; Goodall et al. 2017).

When cancer is detected early, it is more likely to respond effectively to treatment, leading to longer survival and lower morbidity. This greatly benefits affected patients since delays and barriers in the accurate identification of the disease are reduced. The early diagnosis of symptomatic cancers is relevant for most types of cancer, providing care as early as possible and, therefore, is an important public health strategy in all environments (Hamilton et al. 2016).

The conducts of early diagnosis have two main elements that underlie their actions, which are: (i) increased awareness about the early signs of cancer among health professionals, as well as among the general population; and (ii) possibility of more adequate access and availability of these services (Wardle et al. 2015). Cancer control is a complex activity that is only successful when the health system has capacity in all these essential domains and when investments are effectively prioritized (Hamilton et al. 2016).

Although implementing early diagnosis often improves outcomes, not all cancers are equally favored. Those that preferentially benefit from early diagnosis procedures are the types of cancer that are possible to identify considering signs and symptoms, as well as those in which early treatment is essential for a better clinical outcome(Cancer et al. 2001a). Therefore, efficiently achieving the role of early diagnosis allows the selection and implementation of programs that provide to a target population the benefits of finding cancer as soon as possible and, consequently, better results and adequate use of resources (Smith et al. 2019).

Differently from the above, there is still a form of screening through the triage of individuals with findings suggestive of cancer or with precursor lesions, before presenting characteristic symptoms. It is usually performed through tests, exams, diagnostic imaging, or other procedures that can be applied quickly and are easily accessible (Hamilton et al. 2016).

During screening, an entire target population, apparently healthy, is evaluated for unrecognized cancer or precancerous lesions, but most individuals tested will not have the disease, this is what differentiates screening from early diagnosis (Sabik and Adunlin 2017).

Screening is considered a process, as it does not stop at the simple performance of a particular test, exam, or procedure. It, however, comprises a complex and coordinated information system, which ranges from inviting the target population to participate, performing the screening test, supervising the results with the application of additional tests to establish a diagnosis (or not), in cases where abnormalities are detected, until the access to effective treatment, if necessary (Cancer et al. 2001b).

There is a classification of screening tests that distinguishes them according to the type of intervention in the patient, among them are (Wardle et al. 2015):

- Physical examination and family history – examine general signs of health, including checking for signs of illness (cancer), such as granules or other unusual signs, and collects information about family health history and past illnesses.
- Laboratory tests – a set of procedures that examine samples of different types, such as blood, urine, tissues, in order to look for markers that may be altered, indicating the possible presence of cancer.
- Imaging exams – specialized techniques that take images of the interior of the body, allowing the identification of abnormalities in areas of specific organs, as well as the monitoring of existing conditions.
- Genetic tests – laboratory tests that aim to determine changes in genes or chromosomes, through detailed research on the DNA molecule. Such changes can show whether the patient has, or is at increased risk for developing a disease such as cancer.

Cancer screening programs are mainly based on age and associated risk factors displayed by the target population. Their real effectiveness must be demonstrated before starting the tests, since a certain diversity of resources is needed to confirm diagnoses, for the treatment and monitoring of those with abnormal results (Markopoulos et al. 2017; Goodall et al. 2017). Therefore, screening is a much more complex and costly process when compared to early diagnosis, and it proves to be effective only for some types of cancer. This scenario demonstrates that effective screening tests are those that detect cancer early, reduce the chances of death from the disease, and have more potential benefits than harm; the most well-succeeded examples of cancer screening methods are mammography screening to detect breast cancer, Papanicolaou smear test for cervical cancer, colonoscopy, sigmoidoscopy and fecal occult blood test to identify colorectal cancer, PSA test for prostate cancer, transvaginal ultrasound for localization of ovarian and cervical cancer, among others (Wang 2017). All of these screening tests are essential tools in current cancer management and most often have a beneficial outcome for patients and overall disease indexes (Tikkinen et al. 2018).

Early diagnosis and screening technologies differ in their assessment, where diagnostic studies primarily involve estimating the accuracy of a test to determine which patients with suspected cancer abnormalities have the disease or not. While in the screening processes, individuals with no suspects at all or even symptoms are tested to investigate potential risks of developing cancer (Wardle et al. 2015). However, both mechanisms must be employed in clinical practice and an evidence-based assessment must be performed before choosing between a cancer screening or early diagnosis program (Sabik and Adunlin 2017).

17.2.2 Machine Learning for Cancer Diagnosis

Accurate diagnosis is essential for the treatment and monitoring of cancer progression, in recent years many advances have been made in research and clinical practice with several methods, such as screening and new strategies for early diagnosis and treatment of cancer (Wong and Yip 2018). With the arrival of these new technologies, large amounts of data, in different methodologies have been collected and are available to the academic community, but their precise interpretation is not always achieved. This occurs because the accurate prediction of the results related to a particular disease is a very challenging task since many diagnostic findings are not unique markers and they cannot unambiguously certify the presence or absence of cancer, for example (Bertsimas and Wiberg 2020).

In this context, machine learning (ML) methods have become a popular tool for medical researchers. These techniques can discover and identify patterns and correlations between diagnostic findings from complex data sets, being able to effectively predict future outcomes for a given type of cancer (Kourou et al. 2015).

Machine learning is a subset of artificial intelligence, where neural network-based algorithms are developed to allow the machine to learn and solve problems similar to the human brain. Through different programming techniques, machine learning algorithms can trigger large collections of complex data and extract useful information from them. In this way, it is possible to improve previous iterations by learning from the data that is provided (Cruz and Wishart 2006).

Machine learning also has the potential to greatly modify oncology studies and its introduction into a cancer diagnosis is being made possible thanks to the digitization of patient data, including the use of electronic medical records. Technological advances in machine learning pave the way for independent tools in disease diagnosis, exploring big data sets to detect human diseases at a very early stage, especially cancer (Iqbal et al. 2021).

Clinical approaches are currently guided by medical guidelines and accumulated experience. Machine learning methods add a greater level of precision to this process since algorithms can generate individualized predictions by systemizing data from large databases allowing a personalized medicine approach that takes into account the unique characteristics of the patient (Adamson and Welch 2019). It is evident from this context that the integration of heterogeneous data, combined with the application of different techniques to select and classify patterns can yield promising tools for cancer diagnosis (Iqbal et al. 2021).

Despite being a good promise for better diagnosis and screening of cancer, there are still several challenges in the adoption of machine learning in clinical practice, since the methods depend on the availability of large-scale structured data. In this regard, a significant problem arises from the fact that the capture of data between departments and healthcare systems is highly variable, which creates significant challenges in creating cohesive datasets. The very implementation of machine learning in healthcare workflows presents substantial obstacles. Machine learning models must gain the trust of physicians through faster and more effective interpretation, as well as through collaboration between researchers and experts, and especially the prospective validation of methods (Hirasawa et al. 2018). Based on the analysis of their results, it is evident that the integration of multidimensional heterogeneous data, combined with the application of different techniques to select and classify patterns can lead to promising tools for inference in the cancer domain.

The diagnosis, screening, and treatment of cancer have been advancing fast over time and invested research efforts. Certainly, the implementation of new techniques, models, and practices must be conducted in order to obtain significant improvements in the management of the disease (Kwekkeboom et al. 2020).

17.3   Cancer Treatments

The growing knowledge of the molecular biology of cancer enables new advances for its treatment, however, the most effective therapy has not been established yet, which is why combined therapeutic approaches and sequence of treatments have been studied for decades (Fares et al. 2020).

The main goals of treatment are the cure and prolongation and improvement in quality of life. About 33% of existing cancers are curable, including breast cancer, cervical cancer, cancer of the oral cavity (mouth), and cancer of the colon and rectum (bowel) when they are early detected and treated according to the best clinical practices (Abed et al. 2020). Several techniques are used in the treatment of cancer to obtain a better prognosis for the patient. Among them, we can mention surgery, chemotherapy and/or radiotherapy, immunotherapy, and targeted (hormonal) therapy. Actually, surgery is still the main chance of cure in some types of cancer, followed by chemotherapy and radiotherapy (Kourou et al. 2015).

17.3.1   Biomarkers

A biomarker is a measurable characteristic that can be used as an indicator of normal biological and pathogenic processes and pharmacological responses to a specific therapeutic intervention. Biomarkers can be determined from the analysis of genetic material and proteins from various biological materials, for instance, from body fluids easily obtained, such as plasma, serum, and urine, as well as from tissues, which require more invasive techniques to be obtained (Costa-Pinheiro et al. 2015; Zhang et al. 2016).

Specifically, tumor biomarkers are biological molecules that suggest the presence of cancer in a patient and/or characterize diagnosed tumors. Also, they can be produced by the tumor itself or by the body in response to the tumor (Shaw et al. 2015).

These biomarkers can be subcategorized into diagnostic biomarkers (which determine the presence of a type of cancer), prognostic biomarkers (which generate information on the effects of the tumor on the clinical picture), and predictive biomarkers (which help in identifying the most appropriate treatment for the patient, considering their genetic peculiarities). Prognostic biomarkers can be useful in selecting patients for a particular treatment, even if it is not possible to predict the response to this specific approach (Italiano 2011; Dhama et al. 2019; Ben-Hamo et al. 2020).

Regarding cancer treatment, prognostic biomarkers are the most widely used because they can be applied in clinical trials to stratify patients into randomized groups when testing new treatments, as well as to estimate prognosis. For instance, in melanomas, somatic DNA changes in single genes, such as mutations in BRAF V600E, may indicate a better response to BRAF/MEK inhibitors. Additionally, Ben-Hamo et al. (2020) demonstrated that CREB and NFAT pathways are highly significant biomarkers for predicting response to BRAF/MEK inhibitors (Jin et al. 2015; Hu and Dignam 2019).

A study with the MET inhibitor (AMG 337) in 13 patients with gastroesophageal adenocarcinoma concluded that AMG 337 is a promising chemotherapeutic product for this neoplasm since 8 patients presented partial or nearly complete responses when treated with this compound. Similarly, the FGFR gene is currently a target of interest for the treatment of gastric cancer, as well as dovitinib and AZD4547 (FGFR inhibitors) (Deng et al. 2012; Xie et al. 2013; Hughes et al. 2016).

Breast cancer patients, who present ER (E-cadherin and estrogen receptor) and PR (progesterone receptor) positive, have a favorable prognosis with risk of mortality lower (83% five-year survival) than those with ER and PR negative (double negative). Additionally, the presence of the predictive biomarker HER2 suggested that trastuzumab may be an effective treatment for this type of tumor (Weinstein et al. 2013).

Ben-Hamo et al. (2020) demonstrated that lung cancer cell lines, which have high AIF pathway activity, show better responses to treatment with microtubule inhibitors. This may culminate in the release of Bax proteins (pro-apoptotic proteins) and disturbance of the balance of apoptotic cells (Siegel et al. 2021).

In prostate cancer patients, the androgen receptor (AR) is one of the most important oncogenic drivers of the disease (a mutation in the androgen receptor called AR-V7), which sets up greater resistance to enzalutamide and abiraterone. In contrast, the presence of AR-V7 does not seem to impair the taxane response. In addition to this mechanism, the AR amplification or point mutation may also confer resistance to next-generation anti-RA therapies (Sweeney et al. 2015).

Thus, the use of biomarkers in cancer still presents several challenges, since it is a multifactorial and highly mutable disease and the utilization of a single biomarker has limited detection. Additionally, no biomarker has been established as an "ideal" cancer screening tool that can meet diagnostic, prognostic, and predictive requirements simultaneously (Wu and Qi 2019).

Thus, the validation of new cancer biomarkers that have clinical relevance and applicability is essential.

17.3.2 Hormone Therapy

Another type of treatment that has been taking the lead in the fight against cancer is hormonal therapy, which consists of a treatment that slows or stops the growth of neoplastic cells by blocking hormone production or hormone receptors. This type of treatment is also known as hormonal treatment or endocrine therapy (Klaunig 2020).

The main types of cancer in which hormone therapy is used are breast and prostate cancers, since most breast cancers have estrogen (ER) or progesterone (PR) receptors, or both, which means that they need these hormones to grow and spread. On the other hand, prostate cancer needs testosterone and other male sex hormones, such as dihydrotestosterone (DHT), to grow and spread (Majumder et al. 2017).

Among the hormonal therapies directed to breast cancer we can highlight aromatase inhibitors (anastrozole Arimidex®, letrozole Femara®, and exemestane Aromasin®); tamoxifen (Nolvadex®); raloxifene (Evista®); and toremifene (Fareston®), which are known therapies for patients presenting ER-positive tumors. Additionally, patients who present HER2 and ER-positive have effective responses to treatment with tamoxifen (ER antagonist) and trastuzumab (a monoclonal antibody that interferes with HER2 receptors) (Majeed et al. 2014).

On the other hand, ER-positive and HER2-negative patients benefit from therapy with Fulvestrant (Faslodex), which binds to estrogen receptors, completely preventing the hormone from binding to the receptors (Smith et al. 2019).

For prostate cancer, hormonal therapy can complement the treatment, especially in processes of tumor metastasis. This type of therapy includes luteinizing hormone-releasing hormone (LHRH) agonists, also called LHRH analogues or GnRH agonists, such as Leuprolide (Lupron®, Eligard®), Goserelin (Zoladex®), Triptorelin (Trelstar®) or Histrelin (Vantas®). These drugs prevent the testicles from receiving messages from the body to produce testosterone by blocking intracellular signals. Also, they can be injected or placed as a small implant under the skin (Bolton and Lynch 2018; Shim et al. 2019).

Another class of hormones used in clinical practice are the LHRH antagonists. This class of drugs, also called gonadotropin-releasing hormone (GnRH) antagonists, prevent the testicles from producing testosterone, culminating in a reduction in testosterone levels more rapidly. FDA has approved Degarelix (Firmagon), administered by monthly injection, to treat advanced prostate cancer (Kawahara and Miyamoto 2014).

In order to induce a better response to treatment and reduction of side effects, scientific research has included the therapy of other anti-androgen isolates (androgen receptor inhibitors and/or androgen synthesis inhibitors), which prevents the body from producing testosterone, as a means of preventing this hormone from driving the growth of prostate cancer. Older RA inhibitors include bicalutamide (Casodex), flutamide (available as a generic drug), and nilutamide (Nilandron), which are taken as pills. Newer RA inhibitors include apalutamide (Erleada), darolutamide (Nubeqa), and enzalutamide (Xtandi). On the other hand, androgen synthesis inhibitors include abiraterone acetate (Zytiga) and ketoconazole (Nizoral) (David Crawford and Schally 2020).

Reinforcing this idea, a meta-analysis study demonstrated that the combined use of a non-steroidal antiandrogen at the time of ADT initiation could lead to a 3% increase in 5-year survival (O Dalesio et al. 2000).

17.3.3   Imunnotherapy

Immunotherapy was one of the main scientific advances in the treatment of cancer, because this type of therapy stimulates the organism itself to identify the cancerous cells and attack them employing drugs that modify the immune response. While traditional chemotherapy directly attacks cancerous cells, in immunotherapy, the immune system itself is stimulated to perform this action (Waldman et al. 2020).

Initially, the immune system, under normal conditions, uses checkpoints, which work as extracellular signals that activate or deactivate the immune response mediated by T and B cells. Basically, this process starts with the release of chemoattractants by the cells of the tissue that was injured by an aggressive agent, and then there is the recruitment of phagocytic/natural killer cells (NK) and antigen-presenting cells (APC). APCs will then go to the lymph nodes to stimulate the maturation and release of T and B lymphocytes, generating a more specific immune response (Male et al. 2014).

However, how does the immune system recognize what is foreign and what is proper to the organism? Through the recognition of a protein released by normal cells (PDL1), which binds to the membrane receptor present on the immune system cells, and signals the lymphocytes that that cell is proper to the organism. This regulatory mechanism keeps immune responses within a desirable physiological level and protects the host from autoimmunity (Farkona et al. 2016).

Thus, cancer cells sometimes use these checkpoints to avoid being attacked by the immune system, possessing the ability to evade programmed cell death. In order to overcome this ability, scientific research has been directed to the development of immunotherapeutic drugs that target these checkpoints and that have additional immunostimulatory strategies (Velcheti and Schalper 2016).

Regarding the immunological therapies currently available, we can categorize them into (1) drugs that target tumor immune evasion by blocking negative regulatory signals; and (2) agents that directly stimulate immunogenic pathways (e.g., co-stimulatory receptor agonists). Additional immunostimulatory strategies include antigen presentation enhancers (e.g., vaccines), the use of exogenous recombinant cytokines, oncolytic viruses, and cellular therapies using native or modified antigen-competent immune cells (dos Reis and Machado 2020).

Drugs that aim at blocking negative regulatory signals are anti-CTLA-4 (ipilimumab) and anti-PD1/PDL1(Pembrolizumab; Nivolumab; Atezolizumab; Avelumab; Durvalumab). Both CTLA-4 and PD1/PDL1 act as negative regulators of the immune response; therefore, blocking the activity of these proteins will culminate in activation of the immune response. Several studies have already identified positive results of immunotherapy in different types of cancers; however, it is worth noting that only 20–40% of patients respond to this type of treatment (Ribas and Wolchok n.d.).

An example of an agent that directly stimulates immunogenic pathways is the glucocorticoid-induced tumor necrosis factor receptor (GITR) activated, which is a member of the tumor necrosis factor receptor superfamily and is expressed on the surface of various types of immune cells. Its activation culminates in increased proliferation of defense cells, also generating an increase in host immune response. The use of these GITR agonist antibodies in combination with PD-1 blockade has been shown to be synergistic in murine models (Hirasawa et al. 2018).

Additionally, antigen presentation is the first step to generating an immune response. Thus, cancer vaccines have been extensively studied as an additional immunostimulatory strategy. However, there are major challenges to overcome in order to achieve an optimal vaccine, as the antitumor responses are restricted to the peptide target and, therefore, may not be sufficient for a clinically meaningful tumor response. Solid tumors such as melanoma and lung cancer have high somatic mutation rates and may result in multiple mutant neoantigens. Personalized vaccines targeting specific mutant neoepitopes detected in a specific tumor and combined with the patient HLA are under investigation (Adamson and Welch 2019).

Finally, the pathways of immunotherapy still face many challenges, as cancers present differently in different patients, and tumors have high somatic mutation rates, which can result in multiple mutated neoantigens. Thus, it becomes necessary to implement further research to achieve effective cancer treatments (Hegde and Chen 2020).

17.3.4 Chemotherapy

Chemotherapy is a type of treatment that uses drugs to prevent tumor development and progression by inducing cell death by different intracellular mechanisms (Lemjabbar-Alaoui et al. 2015). Such therapy can be designated as neoadjuvant chemotherapy when it is performed before surgery to reduce tumor size, or adjuvant chemotherapy when it is performed after surgery or radiotherapy to destroy remaining tumor cells (Withrow et al. 2013).

FDA has already approved several chemotherapeutic drugs that are used in clinical practice. For instance, for gastric cancer, 6-cycle regimens of epirubicin, cisplatin, and infusional fluorouracil (ECF) are used, whereas for lung cancer, 4 cycles of cisplatin-based regimens are used, such as the association of pemetrexed + cisplatin (Gould et al. 2013; Laxague and Schlottmann 2021).

Cisplatin is a well-known chemotherapeutic agent. This drug intercalates with DNA purines, interfering with DNA repair mechanisms and inducing programmed cell death. This drug is also used in the treatment of several types of neoplasms including lung, bladder, head and neck, lung, ovarian, and testicular cancers (Dasari and Bernard Tchounwou 2014).

However, the lack of specificity causes evident side effects to the patient because this treatment not only attacks the tumor cells but also causes damage to normal cells. Among the most common side effects, we can mention fatigue, mouth sores, nausea, and hair loss ([CSL STYLE ERROR: reference with no printed form.]).

Thus, the search for therapeutic conducts that have a major action on tumor cells and little toxicity on normal cells becomes necessary.

17.3.5 Radiation Therapy

Radiation therapy or radiotherapy uses ionizing radiation to damage the DNA of tumor cells inducing programmed cell death by apoptosis (Mavragani et al. 2019).

Since the early twentieth century, radiation therapy has been widely used to treat different tumor types. Although it is known to destroy healthy tissue in its attempt to kill cancer cells, there are technological advances that have allowed for precise therapies (Baskar et al. 2012).

Given this perspective, randomized studies have observed that partial breast irradiation reduces toxicity compared to total breast irradiation. They also demonstrated the important role of regional nodal irradiation in patients with severe disease (Shah et al. 2020).

Radiation therapy can be performed in two different ways: external radiotherapy, also called targeted radionuclide therapy, which emits ionizing radiation at the site to be treated; and brachytherapy, which uses a radioactive source in or near the area to be treated. It is worth noting that radiation generated by the source affects only areas very close to the site that will be treated, thus protecting healthy tissues (Sgouros et al. 2020).

Radiotherapy is an important part of cancer treatment and can contribute to the improvement of the patient quality of life since this therapeutic approach is used in combination with other types of therapy to reduce tumor size (Sampath 2016).

Although ionizing radiation remains one of the most effective tools in cancer therapy, continuous advances in radiotherapy are still necessary, which will culminate in the continuous improvement of treatments (Han et al. 2017).

17.3.6 Surgery

There are different surgical procedures in the treatment of cancer, which depend on the tumor location, size, and the amount of tissue to be removed, being used mainly in solid tumors (Arruebo et al. 2011).

Surgery can be classified into: palliative surgery (when it is performed to relieve the side effects caused by the tumor, but without the prospect of a cure), total removal surgery (when it is possible to remove the tumor and healthy margins), and debulking (when it is possible to remove only part of the tumor, but not all of it) (Broomfield et al. 2005).

For instance, in colorectal cancer, surgery is the main treatment for this type of tumor. It can be performed through local excision or polypectomy, in which the tumor is removed by colonoscopy. In local excision, the procedure is a bit more extensive and can remove superficial tumors and a small amount of tissue near the colon wall. In more advanced stages, colectomy is applied, which removes part or the whole colon and the nearby lymph nodes (Schmoll et al. 2012).

Despite the great benefits of oncologic surgery, the use of this type of treatment in isolation still has several limitations, mainly because it leaves behind remaining cells of the tumor, which can culminate in recurrence or even metastasis. Thus, this therapeutic option is commonly associated with radiotherapy and chemotherapy, in an attempt to eradicate cancer (Tavare et al. 2012).

17.3.7   Stem Cell Transplant

Hematopoietic stem cells (HSCs) have exacerbated proliferative capacity, with the potential to differentiate into various tissue cells. Therapies that are based on the use of stem cells have gained substantial attention in the treatment of different diseases, including cancer (Eaves 2015).

Stem cell therapy cannot fight or attack tumor cells, but rather help the organism of the patient to produce stem cells especially after very intense therapeutic regimens, such as radiation and chemotherapy. For instance, a pilot study demonstrated that high doses of chemotherapy followed by autologous transplantation of purified peripheral blood HSCs provided higher survival rates among patients with metastatic breast cancer than unmanipulated autologous peripheral blood transplantation (median overall survival = 60 versus 28 months) (Müller et al. 2012).

In hematological cancers such as myeloma, lymphomas, and leukemias, stem cell transplantation can be an effective therapeutic modality. In such cases, the infusion of donated cells (graft) differentiate, generating white blood cells that compose the immune system and begin to attack the mutated cells. This process is called graft-versus-tumor (Ratajczak 2019).

Specifically, in leukemias, patients who are candidates for stem cell transplantation receive high doses of chemotherapy and radiotherapy in order to destroy the bone marrow cells, and then they receive an infusion of new cells (Clarke 2019).

There are three types of transplantation: autologous, allogeneic, and syngeneic. In autologous transplantation, the patient receives back their own stem cells. However, this type of transplant is not a standard treatment for leukemia patients because their marrow contains abnormal cells. In allogeneic transplantation, the patient receives stem cells from a donor. The best results are obtained when the cells of the donor are compatible with those of the patient. In syngeneic transplantation, the patient receives stem cells from an identical twin (Sackett et al. 2016).

Despite the great advances in this field of research, there are still many obstacles in converting basic research into clinical applications. Moreover, another significant obstacle to stem cell-based therapies is clinical safety (Herberts et al. 2011).

17.4 Preliminary Studies on the Use of Essential Oils (EOs) Against Cancer Cells

In recent years, research into the anticancer properties of essential oils has received more attention worldwide. The chemical compounds that constitute essential oils include monoterpenes, sesquiterpenes, oxygenated monoterpenes, and phenolic and oxygenated sesquiterpenes (Ferreira et al. 2020; Silva et al. 2021; Santana de Oliveira et al. 2021). These substances exhibit various properties, such as antioxidant, antimutagenic, antiproliferative, enhancement of immune function, enzyme induction, increased detoxification, modulation of multidrug resistance, and synergistic mechanism of volatile constituents (da Costa et al. 2019; do Nascimento et al. 2020; Cascaes et al. 2021). In addition, several studies have demonstrated *in vitro* and *in vivo* antitumor activity of many essential oils obtained from plants. For this reason, EOs play an important role in pharmaceutical, agricultural, and food sciences (Sobral et al. 2014; de Oliveira et al. 2019, 2020).

EOs can act on multiple targets for cancer prevention, as can be seen in Fig. 17.1. Among the strategies that have already been identified are cell cycle arrest, apoptosis, and DNA repair. Also, essential oils can reduce proliferation, metastasis, and MDR in cancer cells, which makes them potential candidates for adjuvant therapeutic agents against cancer (Gautam et al. 2014).

Thus, the mechanism of action of the compounds that present antitumoral activity is of interest to improve both drug application and therapeutic techniques. Some compounds from Salvia, for example, inhibit proliferation, angiogenesis, and metastasis, or may reverse multidrug resistance of cancer cells, while other

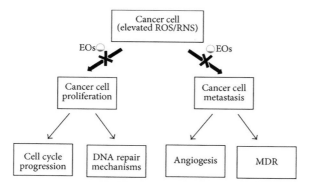

Fig. 17.1 Route of action of essential oils against cancer cells. (Gautam et al. 2014)

substances disrupt the cell cycle and induce apoptosis of tumor cells or enhance immunological activities (Hao et al. 2017; Silva et al. 2019).

Among EO compounds, we can highlight the derivatives of ρ-cymene, which have already demonstrated remarkable anticancer activity in several preclinical investigations. These studies aimed at evaluating their feasibility and therapeutic efficacy when combined with other chemotherapeutic agents (Bezerra et al. 2020a; Hassan et al. 2020).

Several chemical compounds synthesized by plants have already demonstrated great potential for the treatment of various types of cancer. The importance of pointing out their efficacy not only in cell culture systems but also in animal models shows that these agents may indeed be a promising source of chemotherapy and/or immunotherapy in the near future [96]. Thus, clinical studies of these natural products are necessary for the development of new drugs with applications in cancer therapies (Carvalho et al. 2015).

17.4.1 Antioxidant Properties of OEs

Compounds that present antioxidant activity are widely used in food and pharmaceutical sciences against pathological processes, which increase intracellular levels of free radicals caused by oxidative stress (Gulcin 2020; Bezerra et al. 2020b). Essential oils from secondary metabolites possess various biological functions, including antioxidant properties. Among the producing species, EOs from the leaves of *E. citriodora*, *E. urophyla,* and *E. deglupta* can be highlighted (Insuan et al. 2021). In addition, oils extracted from other plant parts may also present potential antioxidant activity, such as the fruits of *Elaeagnus umbellata* Thunb (Nazir et al. 2021).

The antioxidant action of some EOs comes from phenolic compounds, which increase the inhibitory force and the rate of hydrogen transfer to free radicals. One example is the essential oil of *Cedrus atlantica* Manetti, extracted from wood tar and sawdust. Experiments conducted with oils extracted from *C. atlantica* at different locations showed increased free radical scavenging activity with increasing essential oil concentration. The high antioxidant activity of tar oils may be related to the presence of methyl-1,4-cyclohexadiene, whose CH reaction mechanism is a process in which hydrogen is transported as a proton to the DPPH radical with its accompanying electron from the π-diene system. The oxygenated monoterpene 6-camphenol and the monoterpene cis-sabinene-hydrate also influence the antioxidant activity of this essential oil (de Carvalho et al. 2019; Jaouadi et al. 2021).

The antioxidant activity can also be attributed to the presence of monoterpenes, such as β-pinene, γ-terpinene, and limonene. *Peucedanum dhana* essential oil shows antioxidant activity that may be correlated to the properties of such volatile compounds (Khruengsai et al. 2021). In the case of *Thymus vulgaris* and *Mentha spicata* oils, their radical scavenging capacity decreased as the concentrations of the main constituents decreased as well during the 25-week postharvest experimental period (Ćavar Zeljković et al. 2021).

Furthermore, the great antioxidant effect of essential oils on biological and non-biological oxidants may be linked to the strong synergism between oxygenated monoterpenes and phenolic monoterpenes present in the complex mixture. Such interaction is present in *Oliveria decumbens* oil since compounds such as thymol and carvacrol exhibit strong antioxidant capacity *in vitro* (Jamali et al. 2021).

17.4.2   Antiproliferative Activity

Antiproliferative assays performed with *Aloysia polystachya* essential oil against three human cancer cell lines (colon H T-29; prostate PC-3; and breast MCF-7) showed IC_{50} of 5.85, 6.74, and 9.53 μg-mL^{-1}, and selectivity indices of 4.75, 4.12, and 2.92 for HT-29, PC-3, and MCF-7, respectively. The chemical composition of this oil included R-carvone, R-limonene, and dihydrocarvone as the major constituents, whose structures are shown in Fig. 17.2 (Moller et al. 2021). Such results may represent, in the near future, natural-based treatments.

A study with *Artemisia gmelinii* essential oil showed strong antiproliferative activity against different human cancer cell lines in tissues of different origins. Also, the oil successfully inhibited the migration of cancer cells, showing antimetastatic activity (Qadir et al. 2020). Essential oils of *O. basillicum* and *P. undulatapertence*, when evaluated for their antiproliferative activity against MCF7, HT29, and HCT116 cell lines showed strong results. However, the activity of *P. undulata* oil

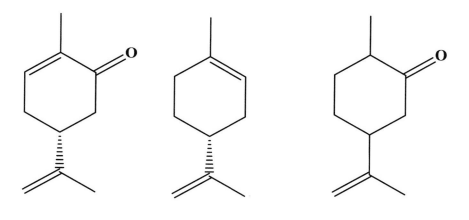

R-Carvone   *R*-Limonene   Dihdrocarvone

Fig. 17.2 Structure of the major compounds of *Aloysia polystachya* essential oil. (Moller et al. 2021)

(9.6–18.6 µg.mL^{-1}) was lower compared to that of *O. basilicum* oil (2.8–3.3 µg. mL^{-1}) (Mohammed et al. 2020).

The oil extracted from *Parrotiopsis jacquemontiana,* on the other hand, was evaluated against HCCLM3 and MDA-MB 231 cells and successfully inhibited their proliferation, migration, and invasion. This experiment with annexin V showed maximum percentage of apoptosis against cancer cells within 72 hours (Ali et al. 2021). The leaf essential oil of *Abies pindrow* showed activity against three cancer cell lines: human breast carcinoma cell line (MCF-7), human breast epithelial tumor cell line (T47D), and human lung adenocarcinoma epithelial cell line (A549) (Zubaid-ul-khazir et al. 2021). Moreover, the essential oil of *Oliveria decumbens* was shown to possess appropriate selectivity between cancer and normal cells and could control the proliferation of A549 cancer cell lines. Thus, results like these have provided new insight into the use of essential oils and their main constituents in the development of new antioxidant and anticancer drugs (Jamali et al. 2021).

17.4.3   Antimutagenic Activity

Some essential oils have also exhibited antiproliferative effects, such as the oil from the leaves of *C. citratus,* which showed activity against prostate carcinoma cells and glioblastoma cell lines. In addition, it showed a reduction in the initial development of proliferative/pre-neoplastic lesions in the mammary gland, colon, and bladder of mice via apoptotic activity (Karami et al. 2021; da Silva Júnior et al. 2021). Another species, *Origanum majorana,* has also been studied for its properties against tumors. Several *in vitro* investigations based on cell culture tests showed that *O. majorana* essential oil and extracts exhibited antiproliferative effects against different cancer cell lines (Bouyahya et al. 2021; Castro et al. 2021).

Some constituents of *Teucrium polium* EO also exhibited cytotoxic properties and antiproliferative effects on different cell lines. One of the constituents of the essential oil, α-pinene, demonstrated antiproliferative and cytotoxic effects on neuroblastoma N2a cells. Moreover, all isomers of eudesmol (α-eudesmol, β-eudesmol, and γ-eudesmol), have already shown cytotoxic activities in different cancer cell lines (Hashem-Dabaghian et al. 2020).

Still in time, it has already been reported that the essential oils of *C. sinensis* and *C. latifolia* can act by several antimutagenic mechanisms, being able to reduce damage to alkylated DNA through a reduction in the expression of base substitution mutations. In addition, the oils of both species also show antimutagenic activity (Toscano-Garibay et al. 2017).

Although there are consistent data evaluating the great potential of essential oils regarding their anticancer properties, more clinical trials are necessary so that new essential oil-based procedures and drugs can be designed, which can aid and/or change the approaches currently used for cancer treatment.

17.5 Conclusions

Cancer is a complex disease triggered by the interaction of multiple factors. From the development of carcinogenesis, the cells of the body that are affected carry out proliferation, changes in their microenvironment and are able to invade other tissues. In this chapter, general aspects of cancer were discussed, as well as multiple ways to intervene clinically against this disease. We also report on early studies of essential oils against some types of cancer. The preliminary studies that have been reported need further scientific investigations so that their real impact on immunotherapy is elucidated.

References

Abed J, Maalouf N, Manson AL, Earl AM, Parhi L, Emgård JEM, Klutstein M, Tayeb S, Almogy G, Atlan KA, Chaushu S, Israeli E, Mandelboim O, Garrett WS, Bachrach G (2020) Colon cancer-associated fusobacterium nucleatum may originate from the oral cavity and reach colon tumors via the circulatory system. Front Cell Infect Microbiol 10. https://doi.org/10.3389/fcimb.2020.00400

Abel EL, DiGiovanni J (2011) Multistage Carcinogenesis. Curr Cancer Res 6:27–51. https://doi.org/10.1007/978-1-61737-995-6_2

Adamson AS, Welch HG (2019) Machine learning and the cancer-diagnosis problem — no gold standard. N Engl J Med 381:2285–2287. https://doi.org/10.1056/NEJMP1907407

Ali S, Khan MR, Khan A, Khan R (2021) In vitro anticancer activity of extracted oil from Parrotiopsis jacquemontiana (Decne) Rehder. Phytomedicine:91. https://doi.org/10.1016/j.phymed.2021.153697

Arruebo M, Vilaboa N, Sáez-Gutierrez B, Lambea J, Tres A, Valladares M, González-Fernández Á (2011) Assessment of the evolution of cancer treatment therapies. Cancers 3:3279–3330

Baskar R, Lee KA, Yeo R, Yeoh KW (2012) Cancer and radiation therapy: current advances and future directions. Int J Med Sci 9:193–199

Ben-Hamo R, Jacob Berger A, Gavert N, Miller M, Pines G, Oren R, Pikarsky E, Benes CH, Neuman T, Zwang Y, Efroni S, Getz G, Straussman R (2020) Predicting and affecting response to cancer therapy based on pathway-level biomarkers. Nat Commun:11. https://doi.org/10.1038/s41467-020-17090-y

Bertsimas D, Wiberg H (2020) Machine learning in oncology: methods, applications, and challenges. 101200/CCI2000072 885–894. https://doi.org/10.1200/CCI.20.00072

Bezerra FWF, do Nascimento Bezerra P, VMB C, de LAR SM, Barbosa JR, da Silva MP, de Oliveira MS, da Costa WA, RHH P, da Cruz JN, de Carvalho Junior RN (2020a) Supercritical green solvent for Amazonian natural resources. In: Inamuddin, Asiri AM (eds) Advanced nanotechnology and applicationef of supercritical fluids. Springer, Cham, pp 15–31

Bezerra FWFWF, de Oliveira MSS, Bezerra PNN, Cunha VMBMB, Silva MPP, da Costa WAA, Pinto RHHHH, Cordeiro RMM, da Cruz JNN, Chaves Neto AMJMJ, Carvalho Junior RNN (2020b) Extraction of bioactive compounds. In: Green sustainable process for chemical and environmental engineering and science. Elsevier, pp 149–167

Bolton EM, Lynch T (2018) Are all gonadotrophin-releasing hormone agonists equivalent for the treatment of prostate cancer? A systematic review. BJU Int 122:371–383

Bouyahya A, Chamkhi I, Benali T, Guaouguaou FE, Balahbib A, El Omari N, Taha D, Belmehdi O, Ghokhan Z, El Menyiy N (2021) Traditional use, phytochemistry, toxicology, and phar-

macology of Origanum majorana L. J Ethnopharmacol 265. https://doi.org/10.1016/j.
jep.2020.113318

Broomfield S, Currie A, van der Most RG, Brown M, van Bruggen I, Robinson BWS, Lake
RA (2005) Partial, but not complete, tumor-debulking surgery promotes protective antitumor memory when combined with chemotherapy and adjuvant immunotherapy. Cancer Res
65:7580–7584. https://doi.org/10.1158/0008-5472.CAN-05-0328

Cancer I of M (US) and NRC (US) C on the ED of B, Patlak M, Nass SJ, Henderson IC, Lashof
JC (2001a) Theory and principles of cancer screening and diagnosis

Cancer I of M (US) and NRC (US) C on the ED of B, Patlak M, Nass SJ, Henderson IC, Lashof
JC (2001b) Theory and principles of cancer screening and diagnosis

Carvalho AA, Andrade LN, De Sousa ÉBV, De Sousa DP (2015) Antitumor phenylpropanoids
found in essential oils. Biomed Res Int:2015. https://doi.org/10.1155/2015/392674

de Carvalho RN, de MS de Oliveira MS, Silva SG, da Cruz JN, Ortiz E, da Costa WA, Bezerra
FWF, Cunha VMB, Cordeiro RM, de JC Neto AM, de A Andrade EH, de C Junior RN (2019)
Supercritical CO2 application in essential oil extraction. In: Inamuddin RM, Asiri AM (eds)
Materials research foundations, 2nd edn, Millersville PA, pp 1–28

Cascaes MM, Dos O, Carneiro S, Diniz Do Nascimento L, Antônio Barbosa De Moraes Â, Santana
De Oliveira M, Neves Cruz J, Skelding GM, Guilhon P, Helena De Aguiar Andrade E, Vico
C, Cruz-Chamorro I (2021) Essential oils from annonaceae species from Brazil: a systematic
review of their phytochemistry, and biological activities. Int J Mol Sci 22:12140. https://doi.
org/10.3390/IJMS222212140

Castro ALG, Cruz JN, Sodré DF, Correa-Barbosa J, Azonsivo R, de Oliveira MS, de Sousa Siqueira
JE, da Rocha Galucio NC, de Oliveira BM, Burbano RMR, do Rosário Marinho AM, Percário
S, Dolabela MF, Vale VV (2021) Evaluation of the genotoxicity and mutagenicity of isoeleutherin and eleutherin isolated from Eleutherine plicata herb. using bioassays and in silico
approaches. Arab J Chem 14:103084. https://doi.org/10.1016/j.arabjc.2021.103084

Ćavar Zeljković S, Smékalová K, Kaffková K, Štefelová N (2021) Influence of post-harvesting
period on quality of thyme and spearmint essential oils. J Appl Res Med Aromat Plants:25.
https://doi.org/10.1016/j.jarmap.2021.100335

Clarke MF (2019) Clinical and therapeutic implications of cancer stem cells. N Engl J Med
380:2237–2245. https://doi.org/10.1056/nejmra1804280

Costa-Pinheiro P, Montezuma D, Henrique R, Jerónimo C (2015) Diagnostic and prognostic epi-
genetic biomarkers in cancer. Epigenomics 7:1003–1015

Cruz JA, Wishart DS (2006) Applications of machine learning in cancer prediction and prognosis.
Cancer Informat 2:59–78

da Costa KS, Galúcio JM, da Costa CHS, Santana AR, dos Santos CV, do Nascimento LD, Lima
AH LE, Neves Cruz J, Alves CN, Lameira J (2019) Exploring the potentiality of natural products from essential oils as inhibitors of odorant-binding proteins: a structure- and ligand-based
virtual screening approach to find novel mosquito repellents. ACS Omega 4:22475–22486.
https://doi.org/10.1021/acsomega.9b03157

da Silva Júnior OS, de MS Franco C, de Moraes AAB, Cruz JN, da Costa KS, do Nascimento
LD, de A Andrade EH (2021) In silico analyses of toxicity of the major constituents of
essential oils from two Ipomoea L. species. Toxicon 195:111–118. https://doi.org/10.1016/j.
toxicon.2021.02.015

Dasari S, Bernard Tchounwou P (2014) Cisplatin in cancer therapy: molecular mechanisms of
action. Eur J Pharmacol 740:364–378

David Crawford E, Schally AV (2020) The role of FSH and LH in prostate cancer and cardiometa-
bolic comorbidities

de Oliveira MS, da Cruz JN, Gomes Silva S, da Costa WA, de Sousa SHB, Bezerra FWF, Teixeira
E, da Silva NJN, de Aguiar Andrade EH, de Jesus Chaves Neto AM, de Carvalho RN (2019)
Phytochemical profile, antioxidant activity, inhibition of acetylcholinesterase and interaction
mechanism of the major components of the Piper divaricatum essential oil obtained by supercritical CO2. J Supercrit Fluids 145:74–84. https://doi.org/10.1016/j.supflu.2018.12.003

Deng N, Goh LK, Wang H, Das K, Tao J, Tan IB, Zhang S, Lee M, Wu J, Lim KH, Lei Z, Goh G, Lim QY, Tan ALK, Poh DYS, Riahi S, Bell S, Shi MM, Linnartz R, Zhu F, Yeoh KG, Toh HC, Yong WP, Cheong HC, Rha SY, Boussioutas A, Grabsch H, Rozen S, Tan P (2012) A comprehensive survey of genomic alterations in gastric cancer reveals systematic patterns of molecular exclusivity and co-occurrence among distinct therapeutic targets. Gut 61:673–684. https://doi.org/10.1136/gutjnl-2011-301839

Dhama K, Latheef SK, Dadar M, Samad HA, Munjal A, Khandia R, Karthik K, Tiwari R, Yatoo MI, Bhatt P, Chakraborty S, Singh KP, Iqbal HMN, Chaicumpa W, Joshi SK (2019) Biomarkers in stress related diseases/disorders: diagnostic, prognostic, and therapeutic values. Frontiers in Molecular Biosciences 6

do Nascimento LD, de Moraes AAB, da Costa KS, Galúcio JMP, Taube PS, Costa CML, Cruz JN, de A Andrade EH, de Faria LJG (2020) Bioactive natural compounds and antioxidant activity of essential oils from spice plants: new findings and potential applications. Biomol Ther 10:1–37. https://doi.org/10.3390/biom10070988

Eaves CJ (2015) Review series hematopoietic stem cells: concepts, definitions, and the new reality. https://doi.org/10.1182/blood-2014

Fares J, Fares MY, Khachfe HH, Salhab HA, Fares Y (2020) Molecular principles of metastasis: a hallmark of cancer revisited. Signal Transduct Target Ther:5

Farkona S, Diamandis EP, Blasutig IM (2016) Cancer immunotherapy: the beginning of the end of cancer? BMC Med 14

Ferreira OO, Neves da Cruz J, de Jesus Pereira Franco C, Silva SG, da Costa WA, de Oliveira MS, de Aguiar Andrade EH (2020) First report on yield and chemical composition of essential oil extracted from myrcia eximia DC (Myrtaceae) from the Brazilian Amazon. Molecules 25:783. https://doi.org/10.3390/molecules25040783

Gautam N, Mantha AK, Mittal S (2014, 2014) Essential oils and their constituents as anticancer agents: a mechanistic view. Biomed Res Int. https://doi.org/10.1155/2014/154106

Goodall J, Mateo J, Yuan W, Mossop H, Porta N, Miranda S, Perez-Lopez R, Dolling D, Robinson DR, Sandhu S, Fowler G, Ebbs B, Flohr P, Seed G, Rodrigues DN, Boysen G, Bertan C, Atkin M, Clarke M, Crespo M, Figueiredo I, Riisnaes R, Sumanasuriya S, Rescigno P, Zafeiriou Z, Sharp A, Tunariu N, Bianchini D, Gillman A, Lord CJ, Hall E, Chinnaiyan AM, Carreira S, de Bono JS (2017) Circulating cell-free DNA to guide prostate cancer treatment with PARP inhibition. Cancer Discov 7:1006–1017. https://doi.org/10.1158/2159-8290.CD-17-0261

Gould MK, Donington J, Lynch WR, Mazzone PJ, Midthun DE, Naidich DP, Wiener RS (2013) Evaluation of individuals with pulmonary nodules: when is it lung cancer? Chest 143:e93S–e120S. https://doi.org/10.1378/chest.12-2351

Gulcin İ (2020) Antioxidants and antioxidant methods: an updated overview

Hamilton W, Walter FM, Rubin G, Neal RD (2016) Improving early diagnosis of symptomatic cancer. Nat Publ Group. https://doi.org/10.1038/nrclinonc.2016.109

Han X, Zhou Y, Liu W (2017) Precision cardio-oncology: understanding the cardiotoxicity of cancer therapy. NPJ Precis Oncologia 1:31. https://doi.org/10.1038/s41698-017-0034-x

Hao D-C, Ge G-B, Xiao P-G (2017) Anticancer drug targets of salvia phytometabolites: chemistry, biology and omics. Curr Drug Targets 19:1–20. https://doi.org/10.2174/1389450117666161207141020

Hashem-Dabaghian F, Shojaii A, Asgarpanah J, Entezari M (2020) Anti-mutagenicity and apoptotic effects of Teucrium polium L. essential oil in HT29 cell line. Jundishapur J Nat Pharm Products:15. https://doi.org/10.5812/jjnpp.79559

Hassan SMH, Ray P, Hossain R, Islam MT, Salehi B, Martins N, Sharifi-Rad J, Amarowicz R (2020) p-Cymene metallo-derivatives: an overview on anticancer activity. Cell Mol Biol 66:28. https://doi.org/10.14715/cmb/2020.66.4.5

Hawkes N (2019) Cancer survival data emphasise importance of early diagnosis. BMJ 364:l408. https://doi.org/10.1136/BMJ.L408

Hegde PS, Chen DS (2020) Top 10 challenges in cancer immunotherapy. Immunity 52:17–35

Herberts CA, Kwa MSG, Hermsen HPH (2011) Risk factors in the development of stem cell therapy. J Transl Med 9

Hirasawa T, Aoyama K, Tanimoto T, Ishihara S, Shichijo S, Ozawa T, Ohnishi T, Fujishiro M, Matsuo K, Fujisaki J, Tada T (2018) Application of artificial intelligence using a convolutional neural network for detecting gastric cancer in endoscopic images. Gastric Cancer 21:653–660. https://doi.org/10.1007/S10120-018-0793-2/FIGURES/4

Hu C, Dignam JJ (2019) Biomarker-driven oncology clinical trials: key design elements, types, features, and practical considerations https://doi.org/10.1200/PO.19

Hughes PE, Rex K, Caenepeel S, Yang Y, Zhang Y, Broome MA, Kha HT, Burgess TL, Amore B, Kaplan-Lefko PJ, Moriguchi J, Werner J, Damore MA, Baker D, Choquette DM, Harmange JC, Radinsky R, Kendall R, Dussault I, Coxon A (2016) In vitro and in vivo activity of AMG 337, a potent and selective MET kinase inhibitor, in MET-dependent cancer models. Mol Cancer Ther 15:1568–1579. https://doi.org/10.1158/1535-7163.MCT-15-0871

Insuan O, Thongchuai B, Chaiwongsa R, Khamchun S, Insuan W (2021) Antioxidant and anti-inflammatory properties of essential oils from three Eucalyptus species. Chiang Mai Univ J Nat Sci 20:1–15. https://doi.org/10.12982/CMUJNS.2021.091

Iqbal MJ, Javed Z, Sadia H, Qureshi IA, Irshad A, Ahmed R, Malik K, Raza S, Abbas A, Pezzani R, Sharifi-Rad J (2021) Clinical applications of artificial intelligence and machine learning in cancer diagnosis: looking into the future. Cancer Cell Int 21:1–11. https://doi.org/10.1186/S12935-021-01981-1

Italiano A (2011) Prognostic or predictive? It's time to get back to definitions! J Clin Oncol 29:4718–4718. https://doi.org/10.1200/JCO.2011.38.3729

Jamali T, Kavoosi G, Jamali Y, Mortezazadeh S, Ardestani SK (2021) In-vitro, in-vivo, and in-silico assessment of radical scavenging and cytotoxic activities of Oliveria decumbens essential oil and its main components. Sci Rep 11:1–19. https://doi.org/10.1038/s41598-021-93535-8

Jaouadi I, Cherrad S, Bouyahya A, Koursaoui L, Satrani B, Ghanmi M, Chaouch A (2021) Chemical variability and antioxidant activity of Cedrus atlantica Manetti essential oils isolated from wood tar and sawdust. Arab J Chem 14:103441. https://doi.org/10.1016/j.arabjc.2021.103441

Jin Z, Jiang W, Wang L (2015) Biomarkers for gastric cancer: progression in early diagnosis and prognosis (review). Oncol Lett 9:1502–1508. https://doi.org/10.3892/ol.2015.2959

Karami S, Yargholi A, Lamardi SNS, Soleymani S, Shirbeigi L, Rahimi R (2021) A review of ethnopharmacology, phytochemistry and pharmacology of cymbopogon species. Res J Pharm 8:83–112. https://doi.org/10.22127/rjp.2021.275223.1682

Kawahara T, Miyamoto H (2014) Send orders for reprints to reprints@benthamscience.net androgen receptor antagonists in the treatment of prostate cancer

Khruengsai S, Sripahco T, Rujanapun N, Charoensup R, Pripdeevech P (2021) Chemical composition and biological activity of Peucedanum dhana A. Ham essential oil. Sci Rep 11:1–11. https://doi.org/10.1038/s41598-021-98717-y

Klaunig JE (2020) Carcinogenesis. An introduction to interdisciplinary toxicology, pp 97–110. https://doi.org/10.1016/B978-0-12-813602-7.00008-9

Kourou K, Exarchos TP, Exarchos KP, Karamouzis MV, Fotiadis DI (2015) Machine learning applications in cancer prognosis and prediction. Comput Struct Biotechnol J 13:8–17. https://doi.org/10.1016/J.CSBJ.2014.11.005

Kulesz-Martin M, Ouyang X, Barling A, Gallegos JR, Liu Y, Medler T (2018) Multistage carcinogenesis: cell and animal models. Comprehensive toxicology, vol 7–15, 3rd edn, pp 11–35. https://doi.org/10.1016/B978-0-12-801238-3.02218-2

Kwekkeboom KL, Wieben A, Stevens J, Tostrud L, Montgomery K (2020) Guideline-recommended symptom management strategies that cross over two or more cancer symptoms. Oncol Nurs Forum 47:498–511. https://doi.org/10.1188/20.ONF.498-511

Laxague F, Schlottmann F (2021) Esophagogastric junction adenocarcinoma: preoperative chemoradiation or perioperative chemotherapy? World J Clin Oncol 12:557–564. https://doi.org/10.5306/wjco.v12.i7.557

Lemjabbar-Alaoui H, Hassan OU, Yang Y-W, Buchanan P (2015) Lung cancer: biology and treatment options. Biochim Biophys Acta 1856:189–210. https://doi.org/10.1016/j.bbcan.2015.08.002

Majeed W, Aslam B, Javed I, Khaliq T, Muhammad F, Ali A, Raza A (2014) Breast cancer: major risk factors and recent developments in treatment. Asian Pac J Cancer Prev 15:3353–3358. https://doi.org/10.7314/APJCP.2014.15.8.3353

Majumder A, Singh M, Tyagi SC (2017) Post-menopausal breast cancer: from estrogen to androgen receptor. Oncotarget 8:102739–102758. https://doi.org/10.18632/oncotarget.22156

Malarkey DE, Hoenerhoff MJ, Maronpot RR (2018) Carcinogenesis: manifestation and mechanisms. Fundamentals of toxicologic pathology, 3rd edn, pp 83–104. https://doi.org/10.1016/B978-0-12-809841-7.00006-X

Male D, Brostoff J, Roth DB, Roitt IM (2014) Imunologia, 8o edição. Elsevier Brasil

Markopoulos GS, Roupakia E, Tokamani M, Chavdoula E, Hatziapostolou M, Polytarchou C, Marcu KB, Papavassiliou AG, Sandaltzopoulos R, Kolettas E (2017) A step-by-step microRNA guide to cancer development and metastasis. Cell Oncol 40:303–339. https://doi.org/10.1007/s13402-017-0341-9

Mavragani IV, Nikitaki Z, Kalospyros SA, Georgakilas AG (2019) Ionizing radiation and complex DNA damage: from prediction to detection challenges and biological significance. Cancers 11

Mohammed ABA, Yagi S, Tzanova T, Schohn H, Abdelgadir H, Stefanucci A, Mollica A, Mahomoodally MF, Adlan TA, Zengin G (2020) Chemical profile, antiproliferative, antioxidant and enzyme inhibition activities of Ocimum basilicum L. and Pulicaria undulata (L.) C.A. Mey. grown in Sudan. S Afr J Bot 132:403–409. https://doi.org/10.1016/j.sajb.2020.06.006

Moller AC, Parra C, Said B, Werner E, Flores S, Villena J, Russo A, Caro N, Montenegro I, Madrid A (2021) Antioxidant and anti-proliferative activity of essential oil and main components from leaves of aloysia polystachya harvested in central Chile. Molecules 26:1–10. https://doi.org/10.3390/molecules26010131

Moses C, Garcia-Bloj B, Harvey AR, Blancafort P (2018) Hallmarks of cancer: the CRISPR generation. Eur J Cancer 93:10–18. https://doi.org/10.1016/J.EJCA.2018.01.002

Müller AMS, Kohrt HEK, Cha S, Laport G, Klein J, Guardino AE, Johnston LJ, Stockerl-Goldstein KE, Hanania E, Juttner C, Blume KG, Negrin RS, Weissman IL, Shizuru JA (2012) Long-term outcome of patients with metastatic breast cancer treated with high-dose chemotherapy and transplantation of purified autologous hematopoietic stem cells. Biol Blood Marrow Transplant 18:125–133. https://doi.org/10.1016/j.bbmt.2011.07.009

Nazir N, Zahoor M, Uddin F, Nisar M (2021) Chemical composition, in vitro antioxidant, anticholinesterase, and antidiabetic potential of essential oil of Elaeagnus umbellata Thunb. BMC Complement Med Ther 21:1–13. https://doi.org/10.1186/s12906-021-03228-y

Dalesio O, van Tinteren H, Clarke M, Peto R, Schroder FH et al (2000) Maximum androgen blockade in advanced prostate cancer: anoverview of the randomised trials. Lancet 355:1491–1498

de Oliveira MS, da Cruz JN, da Costa WA, Silva SG, da Paz Brito M, de Menezes SAF, de Jesus Chaves Neto AM, de Aguiar Andrade EH, de Carvalho Junior RN (2020) Chemical composition, antimicrobial properties of Siparuna guianensis essential oil and a molecular docking and dynamics molecular study of its major chemical constituent. Molecules 25:3852. https://doi.org/10.3390/molecules25173852

Patterson AD, Gonzalez FJ, Perdew GH, Peters JM (2018) Molecular regulation of carcinogenesis: friend and foe. Toxicol Sci 165:277–283. https://doi.org/10.1093/toxsci/kfy185

Qadir M, Maurya AK, Waza AA, Agnihotri VK, Shah WA (2020) Chemical composition, antioxidant and cytotoxic activity of Artemisia gmelinii essential oil growing wild in Kashmir valley. Nat Prod Res 34:3289–3294. https://doi.org/10.1080/14786419.2018.1557178

Ratajczak MZ (2019) Stem cells – therapeutic applications. Springer, Cham

dos Reis AP, Machado JAN (2020) Immunotherapy and immune checkpoint inhibitors in cancer. Arquivos de Asma, Alergia e Imunologia 4. https://doi.org/10.5935/2526-5393.20200005

Ribas A, Wolchok JD (n.d.) Cancer immunotherapy using checkpoint blockade

Sabik LM, Adunlin G (2017) The ACA and cancer screening and diagnosis. Cancer J (Sudbury, MA) 23:151. https://doi.org/10.1097/PPO.0000000000000261

Sackett SD, Brown ME, Tremmel DM, Ellis T, Burlingham WJ, Odorico JS (2016) Modulation of human allogeneic and syngeneic pluripotent stem cells and immunological implications for transplantation. Transplant Rev 30:61–70. https://doi.org/10.1016/j.trre.2016.02.001

Sampath S (2016) Treatment: radiation therapy. In: Cancer treatment and research. Kluwer Academic Publishers, pp 105–118

Santana de Oliveira M, Pereira da Silva VM, Cantão Freitas L, Gomes Silva S, Nevez Cruz J, de Aguiar Andrade EH (2021) Extraction yield, chemical composition, preliminary toxicity of Bignonia nocturna (Bignoniaceae) essential oil and in silico evaluation of the interaction. Chem Biodivers 18:cbdv.202000982. https://doi.org/10.1002/cbdv.202000982

Schmoll HJ, van Cutsem E, Stein A, Valentini V, Glimelius B, Haustermans K, Nordlinger B, van de Velde CJ, Balmana J, Regula J, Nagtegaal ID, Beets-Tan RG, Arnold D, Ciardiello F, Hoff P, Kerr D, Köhne CH, Labianca R, Price T, Scheithauer W, Sobrero A, Tabernero J, Aderka D, Barroso S, Bodoky G, Douillard JY, el Ghazaly H, Gallardo J, Garin A, Glynne-jones R, Jordan K, Meshcheryakov A, Papamichail D, Pfeiffer P, Souglakos I, Turhal S, Cervantes A (2012) Esmo consensus guidelines for management of patients with colon and rectal cancer. A personalized approach to clinical decision making. Ann Oncol 23:2479–2516. https://doi.org/10.1093/annonc/mds236

Sgouros G, Bodei L, McDevitt MR, Nedrow JR (2020) Radiopharmaceutical therapy in cancer: clinical advances and challenges. Nat Rev Drug Discov 19:589–608

Shah C, Bauer-Nilsen K, McNulty RH, Vicini F (2020) Novel radiation therapy approaches for breast cancer treatment. Semin Oncol 47:209–216

Shaw A, Bradley MD, Elyan S, Kurian KM (2015) Tumour biomarkers: diagnostic, prognostic, and predictive. BMJ (Online) 351

Shim M, Bang WJ, Oh CY, Lee YS, Cho JS (2019) Effectiveness of three different luteinizing hormone-releasing hormone agonists in the chemical castration of patients with prostate cancer: Goserelin versus triptorelin versus leuprolide. Investig Clin Urol 60:244–250. https://doi.org/10.4111/icu.2019.60.4.244

Siegel RL, Miller KD, Fuchs HE, Jemal A (2021) Cancer statistics, 2021. CA Cancer J Clin 71:7–33. https://doi.org/10.3322/CAAC.21654

Silva SG, da Costa RA, de Oliveira MS, da Cruz JN, Figueiredo PLB, do Socorro Barros Brasil D, Nascimento LD, de Jesus Chaves Neto AM, de Carvalho Junior RN, de Aguiar Andrade EH (2019) Chemical profile of lippia thymoides, evaluation of the acetylcholinesterase inhibitory activity of its essential oil, and molecular docking and molecular dynamics simulations. PLoS One 14:e0213393. https://doi.org/10.1371/journal.pone.0213393

Silva SG, de Oliveira MS, Cruz JN, da Costa WA, da Silva SHM, Barreto Maia AA, de Sousa RL, Carvalho Junior RN, de Aguiar Andrade EH (2021) Supercritical CO_2 extraction to obtain Lippia thymoides Mart. & Schauer (Verbenaceae) essential oil rich in thymol and evaluation of its antimicrobial activity. J Supercrit Fluids 168:105064. https://doi.org/10.1016/j.supflu.2020.105064

Smith RA, Andrews KS, Brooks D, Fedewa SA, Manassaram-baptiste D, Saslow D, Wender RC (2019) Cancer Screening in the United States, 2019 : A Review of Current American Cancer Society Guidelines and Current Issues in Cancer Screening, pp 184–210. https://doi.org/10.3322/caac.21557

Sobral MV, Xavier AL, Lima TC, De Sousa DP (2014) Antitumor activity of monoterpenes found in essential oils. Sci World J:2014. https://doi.org/10.1155/2014/953451

Sung H, Ferlay J, Siegel RL, Laversanne M, Soerjomataram I, Jemal A, Bray F (2021) Global Cancer Statistics 2020: GLOBOCAN Estimates of Incidence and Mortality Worldwide for 36 Cancers in 185 Countries. CA Cancer J Clin 71:209–249. https://doi.org/10.3322/CAAC.21660

Sweeney CJ, Chen Y-H, Carducci M, Liu G, Jarrard DF, Eisenberger M, Wong Y-N, Hahn N, Kohli M, Cooney MM, Dreicer R, Vogelzang NJ, Picus J, Shevrin D, Hussain M, Garcia JA, DiPaola

RS (2015) Chemohormonal therapy in metastatic hormone-sensitive prostate cancer. N Engl J Med 373:737–746. https://doi.org/10.1056/nejmoa1503747

Tavare AN, Perry NJS, Benzonana LL, Takata M, Ma D (2012) Cancer recurrence after surgery: direct and indirect effects of anesthetic agents. Int J Cancer 130:1237–1250

Tikkinen KAO, Dahm P, Lytvyn L, Heen AF, Vernooij RWM, Siemieniuk RAC, Wheeler R, Vaughan B, Fobuzi AC, Blanker MH, Junod N, Sommer J, Stirnemann J, Yoshimura M, Auer R, MacDonald H, Guyatt G, Vandvik PO, Agoritsas T (2018) Prostate cancer screening with prostate-specific antigen (PSA) test: a clinical practice guideline. BMJ 362. https://doi.org/10.1136/BMJ.K3581

Todd A, Groundwater PW, Gill JH (2017) Cellular and molecular basis of cancer. In: Anticancer therapeutics. Wiley, Chichester, UK, pp 39–80

Toscano-Garibay JD, Arriaga-Alba M, Sánchez-Navarrete J, Mendoza-García M, Flores-Estrada JJ, Moreno-Eutimio MA, Espinosa-Aguirre JJ, González-Ávila M, Ruiz-Pérez NJ (2017) Antimutagenic and antioxidant activity of the essential oils of Citrus sinensis and Citrus latifolia. Sci Rep 7:1–9. https://doi.org/10.1038/s41598-017-11818-5

Velcheti V, Schalper K (2016) Basic overview of current immunotherapy approaches in cancer

Waldman AD, Fritz JM, Lenardo MJ (2020) A guide to cancer immunotherapy: from T cell basic science to clinical practice. Nat Rev Immunol 20:651–668

Wang L (2017) Early diagnosis of breast cancer. Sensors 17:1572. https://doi.org/10.3390/S17071572

Wardle J, Robb K, Vernon S, Waller J (2015) Screening for prevention and early diagnosis of cancer. Am Psychol 70:119. https://doi.org/10.1037/A0037357

Weinstein JN, Collisson EA, Mills GB, Shaw KRM, Ozenberger BA, Ellrott K, Shmulevich I, Sander C, Stuart JM (2013) The cancer genome atlas pan-cancer analysis project. Nature Genetics 45:1113–1120. https://doi.org/10.1038/ng.2764

Withrow SJ, Page R, Vall DM (2013) Clinical oncology, vol 5. Elsevier

Wong D, Yip S (2018) Machine learning classifies cancer. Nature 555:446–447. https://doi.org/10.1038/d41586-018-02881-7

Wu L, Qi X (2019) Cancer biomarker detection: recent achievements and challenges. In: AIChE Annual Meeting, Conference Proceedings. American Institute of Chemical Engineers

Xie L, Su X, Zhang L, Yin X, Tang L, Zhang X, Xu Y, Gao Z, Liu K, Zhou M, Gao B, Shen D, Zhang L, Ji A, Gavine PR, Zhang J, Kilgour E, Zhang X, Ji Q (2013) FGFR2 gene amplification in gastric cancer predicts sensitivity to the selective FGFR inhibitor AZD4547. Clin Cancer Res 19:2572–2583. https://doi.org/10.1158/1078-0432.CCR-12-3898

Zhang J, Chen K, Fan ZH (2016) Circulating tumor cell isolation and analysis. In: Advances in clinical chemistry. Academic Press Inc, pp 1–31

Zubaid-ul-khazir YGN, Wani H, Shah SA, Zargar MI, Rather MA, Banday JA (2021) Gas chromatographic-mass spectrometric analysis, antibacterial, antioxidant and antiproliferative activities of the needle essential oil of Abies pindrow growing wild in Kashmir. India Microb Pathog 158. https://doi.org/10.1016/j.micpath.2021.105013

Zugazagoitia J, Guedes C, Ponce S, Ferrer I, Molina-Pinelo S, Paz-Ares L (2016) Current challenges in cancer treatment. Clin Ther 38. https://doi.org/10.1016/j.clinthera.2016.03.026

Part VI
Essential Oil and In Silico Study

Chapter 18
Molecular Modeling Approaches to Investigate Essential Oils (Volatile Compounds) Interacting with Molecular Targets

Suraj Narayan Mali, Srushti Tambe, Amit P. Pratap, and Jorddy Neves Cruz

18.1 Introduction to Molecular Modeling

The term molecular modeling comprises two words, "molecular' and 'modeling'. The term 'molecular' itself denotes the fact that molecules are involved, wherein, the second term 'modeling' indicates the process of representing various molecular structures numerically and correlating or expressing them so as to correlate with their biological activity or to model or mimic the behaviour of molecules (Verma et al. 2010). This has been done with the help of various quantum and classical physics equations (Vanommeslaeghe et al. 2014).

Since last decade, a new drug designing approach called CADD (Computer-aided drug design) has emerged as crucial technique for the drug discovery processes including identifying potential hits and development of a potential lead (Abdolmaleki et al. 2017) . Some of key examples are dorzolamide (carbonic anhydrase inhibitor); captopril (the angiotensin-converting enzyme); ritonavir, and indinavir (anti- human immunodeficiency virus (HIV), etc. It is proven that CADD approach utilizes more target-based searches as compared with traditional approach of finding hits (Pinto et al. 2019). Thus, this technique is not only

S. N. Mali · S. Tambe
Department of Pharmaceutical Sciences and Technology, Institute of Chemical Technology, Main Campus at Mumbai, (Deemed University; Nathalal Parekh Marg, Mumbai, Maharashtra, India

A. P. Pratap
Department of Oils, Oleochemicals and Surfactants Technology, Institute of Chemical Technology, Main Campus at Mumbai, Deemed University, Nathalal Parekh Marg, Mumbai, Maharashtra, India

J. N. Cruz (✉)
Adolpho Ducke Laboratory, Botany Coordination, Museu Paraense Emílio Goeldi, Belém, PA, Brazil

417

capable of explaining various molecular basis involved for pharmacological activities but also useful to predict plausible bioactivities of various synthesized derivatives. (Vucicevic et al. 2019).

It is also important to note that molecular modeling techniques look at biological processes at the molecular level while trying to understand the root cause of underlying disease conditions (Sun and Scott 2010).

Usually, this technique has been classified into two categories as (1) direct drug designing (the fact that 3Dimensional structure of the receptor is known) and (2) indirect drug designing (where, 3D structure of the receptor is not known and based on active and in-active ligand sets, a hypothetical receptor site would be assumed) (Santos et al. 2020). It is well evident that such techniques have a common feature depicting the atomistic level description of whole system (Leelananda and Lindert 2016). This involves two fundamental approaches (1) a molecular mechanics approach and (2) a quantum chemistry approach. Molecular modeling techniques have wide range of applications such as their use in drug discovery, computational biology, materials science, and in drug designing. The pharmaceutical field has been largely benefited from this technique. Considering the recent pandemic of COVID-19, such techniques would play important role in identifying possible hits against such virus within short span of time (Wang et al. 2017; Prajapat et al. 2020; Gurung et al. 2021).

18.2 Molecular Modeling Methods

18.2.1 Molecular Descriptors

Molecular descriptors are usually physicochemical properties. Such properties would contribute towards biological activity of molecule (Redžepović and Furtula 2021). This was also defined by Todeschini and Consonni as: "The molecular descriptor is the final result of a logic and mathematical procedure which transforms chemical information encoded within a symbolic representation of a molecule into a useful number or the result of some standardized experiment."(Alves et al. 2020; Pinzi et al. 2021) Although many physicochemical properties have been studied by medicinal chemists, only three of them are highly important and those are (1) hydrophobic (e.g., partition coefficient (P)), (2) steric and (3) electronic properties (e.g., Hammett substitution constant) or descriptors (Grisoni et al. 2018; Costa et al. 2020).

18.2.2   SAR and QSAR

In general, biological properties of compounds are dependent on their chemical structure. Furthermore, it is believed that structurally similar molecules would show similar properties (Huang et al. 2021b). Thus, the understanding of such relationships has given rise to a concept called structure–activity relationship (SAR). The structure activity relationships (SAR) are basically a qualitative expression. However, same relationship when established in a mathematical form by utilizing a set of molecular properties or descriptors along with their corresponding bioactivities would give rise to **Q**uantitative **S**tructure–**A**ctivity **R**elationship models (QSAR models) (Idakwo et al. 2019; Almeida et al. 2021). QSAR models are regression based or classification-based models. QSAR regression models relate two variables; (X) 'prediction variable' (physico-chemical properties or theoretical molecular descriptors) to the potency of the response variable (Y). Statistically robust and validated QSAR models can be also be used for predicting biological activity of newer chemical structures (Halder et al. 2018).

Quantitative structure–activity relationship models (QSAR models) can be expressed in the form of a mathematical model:

$$\text{Biological Activity} = f(\text{physiochemical properties and/or structural properties}) + \text{error}$$

In order to quantify the activity of a set of molecules, one need to usually have Half maximal inhibitory concentration (IC_{50}) or inhibition constant (Ki) measures. QSAR models, unlike various pharmacophoric models can be useful to see how particular features to drug molecule can have positive or negative effects upon introductions (Zhong et al. 2018). The selection of a proper set of molecular descriptors governs successful QSAR model development. Furthermore, its ability to predict biological activity has also been taken into consideration while deciding best QSAR model among various developed QSAR models. Various statistical measures would be applied to decide best QSAR model (Gupta et al. 2018). For the development of a good predictive QSAR model, one need to have enough biological activity data (training data), otherwise QSARs cannot perform well. MLR (multivariable linear regression) and Machine learning approaches (neural networks (NN) and support vector machine (SVM)) methods can be also used for building successful QSAR models. MLR methods can only be used when there is linear relationship between descriptors and activity (Achary 2020; Hadrup et al. 2021). Principal component analysis (PCA) technique would simplify the complexity of selecting molecular descriptors and building QSAR models by removing descriptors that are not independent. Various statistical validations were reported by various researchers (Sharma and Bhatia 2020). Although, good QSAR models have better predictivities still they should be used cautiously and applied only to the particular set of compounds with varied structural features on similar scaffold (Fukuchi et al. 2019).

18.2.3 Molecular Docking

The study of how two molecular structures would fit into each other, usually drug molecule and receptor or enzyme or proteins is called as 'molecular docking'. In a simpler way, it is a technique used to see or predict binding interactions of small molecules with target forming a complex that may indicate inhibition or enhancement of biological activity (Saikia and Bordoloi 2018; Pinzi and Rastelli 2019). Such behaviour of ligands (small molecules) can be established with molecular docking simulations by predicting affinity between the small molecules and proteins (Ramos et al. 2020). Based on such behaviours, docking can be classified into three types viz., (1) protein-ligand docking; (2) protein–nucleic acid docking; and (3) protein–protein docking (Torres et al. 2019; Mohammad et al. 2021). The protein-ligand docking is comparatively simple than protein-protein docking. As proteins are flexible in nature, their conformational space is so wide and thus making protein-protein docking more complex. Docking simulations are based on varieties of search algorithms like e.g., genetic algorithms (GAs), distance geometry methods, MC methods, fragment-based methods, Tabu searches, etc. (Li et al. 2019; Castro et al. 2021). Docking methodology typically includes three main steps as depicted below:

1. Retrieving X-ray co-crystallized structure from the protein data bank (PDB), and identifying active site. (Protein Preparation)
2. Ligand Preparation (Drawing of chemical structures and converting into 3D form, generating least energy conformers, etc.)
3. Docking of ligand into active site via Grid generation or site mapping.

Several docking engines have been reported over last decades which include *Glide, GOLD, AutoDock, iGEMDOCK, DOCK,* etc. Identifying correct binding site, re-docking validation and setting up of input files for docking are crucial steps in the molecular docking to get suitable acceptable results (Pagadala et al. 2017; Liu et al. 2018b).

18.2.4 Molecular Dynamics Simulations

Molecular dynamics simulation (MDs) is extensively used molecular modeling tool for understanding protein motions and conformational space (Van Der Spoel et al. 2005; Neves Cruz et al. 2020). There are many famous and widely used MD simulation software packages available such as GROMACS, AMBER, NAMD, Desmond, etc. One must note that for it has typical timescale ranges from nanoseconds to microseconds (Salomon-Ferrer et al. 2013; Lima et al. 2020). Basically, MD simulation is computer-based method to analyse physical movements of atoms. MD simulation typically finds its application in material science, chemical science, and in biophysics (Moradi et al. 2019). Apart from several MD simulation success stories, the application of MD simulation is still limited due to two main challenges: (1) the force field used and (2) high computational demand. For example, if someone wants to run a 1 microsecond simulation for a smaller system of 25,000 atoms using 24 processors, it will still take several months to complete the same (Liu et al. 2018a). Moreover, force fields are also approximations of the quantum-mechanical reality. The MD simulation is poorly suitable for systems, where quantum effects are important (Venable et al. 2019).

18.2.5 Binding Free Energy Calculations

In order to estimate binding affinity of the binding affinity of target–ligand complexes, binding free energy calculations are used. Binding affinity calculations can be used to understand the effects of target mutations. Moreover, the drug potency can be correlated directly with binding affinities (Gohlke and Case 2004; Cournia et al. 2017; Leão et al. 2020; Neto et al. 2020).

$$\Delta G_{bind} = \Delta G_{complex} - \left(\Delta G_{protein} + \Delta G_{ligand} \right)$$

Where,

ΔG_{bind} = the free energy of binding,

$\Delta G_{complex}$ = the free energy of the protein–ligand complex,

$\Delta G_{protein}$ and ΔG_{ligand} = the free energies of the protein and ligand, respectively.

Rigorous approaches are considered as most accuratHe approaches to calculate binding free energies. The FEP (free energy perturbation) methods and thermodynamic integration (TI) methods are the two important rigorous binding free energy approaches. The FEP methods were introduced by Zwanzig in the 1950s. Such method uses molecular dynamics and Monte Carlo simulations. Another method called BEDAM (binding energy distribution analysis method) is also used to calculate binding free energy calculations. It is well understood that the free energy is overall sum of all local energy minima (Wang et al. 2019; Kuhn et al. 2020).

18.2.6 In-silico ADMEtox Properties

After obtaining hit molecules, lead optimization would be carried out. During the lead optimization, various parameters should be taken into consideration like drug safety, pharmacokinetic properties and ADME profiles (absorption, distribution, metabolism, and excretion/elimination) (Bueno 2020; Araújo et al. 2020). Thus, carrying out ADME analysis is a crucial step. It is important to note that affinity changes with atom modifications. Considering drug absorption, permeability and solubility are two most important factors for the enhancement of drug potency. Henceforth, in-silico ADME analysis is important for predicting solubility and membrane permeability (Farouk and Shamma 2019; dos Santos et al. 2020). The experimental measurement of solubility is quite tedious, while in-silico solubility calculations are faster. One of published review on computational approaches explains various approaches to predict drug solubility. Human intestinal absorption is important while considering bioavailability of drug. Thus, the Lipinski's 'Rule of 5' (there should not be more than 5 H-bond donors, Log P is over 5, more than 10 H-bond acceptors, and the molecular weight is over 500) would be taken into consideration (Li 2001; Alqahtani 2017). The calculation of the Lipinski's 'Rule of 5' via computational methods would help medicinal chemists to design drug molecule with high bioavailability. QikProp, admetSAR, FAF-Drugs2, etc. are some of widely used ADMET calculation programs. For generating ADME models and calculations, 'VolSurf' package can be utilized. Qikprop, a program by Schrodinger is able to calculate large number of physically significant physicochemical properties, toxicity indicating descriptors for small molecules (Huang et al. 2021a). Even though many experimental verifications are required to assess the pharmacokinetic properties and toxicity of molecules, in-silico ADMET analysis offers several benefits by reducing the actual costs. The assessment of ADME properties is a key step in drug screening. However, one must take into consideration of several limitations of computational methodologies and thus, would use such techniques with caution (Stouch et al. 2003; Durán-Iturbide et al. 2020).

18.3 Investigation of the Mechanism of Action of Volatile Compounds

18.3.1 Background

Medicinal plants have been used to treat human diseases since antiquity as the world's greatest biochemical and pharmacological living reservoirs. Natural products originating from plants are an important option in the quest for therapeutic agents because they contain a diverse range of bioactive chemical components (Fowler 2006; de Carvalho et al. 2019). Phytochemicals have biological pre-validation concerning drug-like properties: their basic scaffolds can be seen as natural structures in drug discovery because they have interacted with diverse enzymes and proteins during their biosynthesis (Bezerra et al. 2020a; Barbosa et al. 2021). They thereby fall into the biologically relevant chemical region, which is predetermined for interaction with drug targets. Computational chemistry, in conjunction with bioinformatics, has aided in the development of new drugs with various biological activities (Kellenberger et al. 2011; Maier 2015).

Natural products are, unfortunately, disadvantaged since their isolation is difficult and time-consuming, and because of their high structural complexity and relatively large molecular weight their total synthesis is not as favorable for large-scale manufacture (de Oliveira et al. 2020). In addition, these traits can transmit poor absorption, distribution, metabolism, discharge, and toxicity profiles (ADMET) (Hazzaa et al. 2020). Molecular docking is a computer-based technology that predicts the positioning (orientation and configuration) of the ligand (drug or molecule of therapeutic interest) at a target site of interaction and helps comprehend the biological activity of volatile compounds. Thus, for therapeutic compounds, molecular docking serves as a predictive model that can help with *in vivo* pharmacological activity evaluations (Meng et al. 2011; Bezerra et al. 2020b). Plants that produce volatile compounds are classified into more than 17,500 species of plants from many angiosperm families, e.g., *Rutaceae, Alliaceae, Lamiaceae, Apiaceae, Poaceae, Asteraceae*, and *Myrtaceae* (de Paulo et al. 2020). They are well-known for their ability to produce commercial and therapeutic volatile compounds. Volatile compounds are complex chemicals with a strong odor that are produced as secondary metabolites by aromatic plants (Michel et al. 2020). Methyl-d-erythritol-4-phosphate (MEP), mevalonic acid, and malonic acid pathways are responsible for the synthesis of volatile oils in the cytoplasm and plastids of plant cells. They are found as liquid droplets in the roots, stems, fruits, flowers, bark and leaves of the plants, and are generated and preserved in secretory cavities, glands, and resin conduits which are some of the complex secretory structures (Arsenijevic et al. 2021). Volatile oils are exceedingly complex combinations of predominantly terpenoids phenylpropanoids, and terpenes, while comprising two or three major components at a level of 20–70% (Ferreira et al. 2020). The other components are aromatic and aliphatic constituents, all characterized by low molecular weight and are present in trace amounts. They may also comprise several other compounds such as sulfur derivatives fatty, oxides, and fatty acids. These primary components, in general, determine the biological features of volatile oils. Terpenes are divided into two categories based on their structural and functional features (Aremu and Van Staden 2013). They are the most common molecules, accounting for 90% of volatile oils and allowing for a wide range of configurations. They are made up of isoprene, which is a compound made up of multiple 5-carbon-base (C5) units. Monoterpenes ($C_{10}H_{16}$) and sesquiterpenes ($C_{15}H_{24}$) are the most common terpenes, but diterpenes ($C_{20}H_{32}$), triterpenes ($C_{30}H_{40}$), and other longer chains occur as well (Maltarollo et al. 2015). Examples of terpene compounds include limonene, pinene, p-cymene, sabinene, and terpinene. The aromatic compounds are found in lesser proportions than the terpenes. Figure 18.1 represents the chemical structures of few volatile components. The design of target metabolites, as well as the mechanism of action of pharmacologically active compounds, can be determined through molecular docking studies (Ma et al. 2011b).

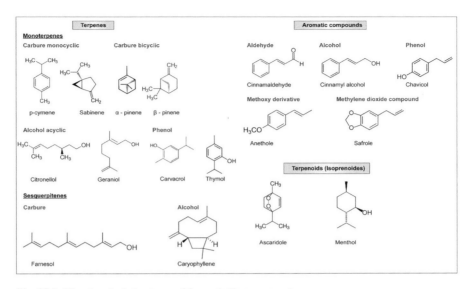

Fig. 18.1 The chemical structures of few volatile compounds

18.3.2   Molecular Modeling of Volatile Compounds with Antimicrobial Activity

Volatile compounds are secondary metabolites that are vital for plant defence because they often possess antibacterial capabilities (De Oliveira et al. 2019; Do Nascimento et al. 2020). De la Croix used volatile oil vapours to test the bactericidal activities of secondary metabolites for the first time in 1881. Since then, volatile oils and their components have been found to exhibit antibacterial effects across a wide range of bacteria. Volatile oils contain complex combinations of up to 45 distinct ingredients, making it difficult to identify the most active antibacterial molecules. The antibacterial effects of most volatile compounds are due to the disruption of bacterial membranes (Ooms 2012). Damage to membrane proteins (such as enzymes), motive proton force depletion, cell content leakage (leakage of cellular ions, Na^+, H^+, and K^+), and cytoplasm coagulation all seem to be common side effects. After treatment with volatile oils, disruption of plasma membrane integrity leads to efflux of DNA, RNA, and proteins, which has been identified as a key antimicrobial mode of action (Diao et al. 2014). Reduced membrane potentials, disruption of proton pumps, and ATP depletion are all linked to volatile compounds' antimicrobial properties as well (Carson et al. 2002). Nonetheless, inhibition of efflux pumps, which are responsible for antibiotic resistance, has been considered as a specific target for volatile compounds (Costa et al. 2019). This change in cell arrangement could trigger a cascade effect, affecting other cell organelles as well. These effects are almost certainly the outcome of the volatile compound's initial mode of bacterial membrane instability. Because of the effective hydroxyl group in chemical structures of volatile compounds, phenolic content in them exhibits greater specificity for the inhibition of microbial growth that contributes in disruption of plasma membrane structure and hence disorganization of membrane permeability, particularly, by altering the activity of the enzymes involved in Krebs's cycle. However, the terpenoids in volatile oils have a significant impact on plasma membrane fatty acids, resulting in changes in membrane dynamicity, permeability, and cytoplasmic constituent leakage (Bouyahya et al. 2017; Antunes et al. 2021). The lipophilic characteristic of volatile oils is closely linked to their antibacterial activity. The major target of volatile oils and bioactive components is the cell wall and plasma membrane, which leads to interactions with cellular polysaccharides, fatty acids, and phospholipids (Burt 2004). Changes in antibacterial action between gram-positive bacteria and gram-negative are explained by differences in cell wall construction, with gram-positive strains being far more sensitive to volatile compounds. In various bacterial species, volatile compounds suppress cell-to-cell transmission and biofilm development (Calo et al. 2015). Moreover, an efficient breakdown in the sensory transmission is triggered by the impact of volatile compounds on biofilm formation inhibitions in bacterial species. The mechanism of quorum sensing modulation via volatile compounds involves complicated interactions of the compounds with bacterial cell wall receptors, which lowers signal molecule reception and impairs cell-to-cell signal transmission (Camele et al. 2019). The antibacterial activity of volatile oils is mainly attributed to the low proportion of terpenoids and phenolic compounds present in them, thereby exhibiting antibacterial activity in their pure form. The primary components of volatile oils from plants in the *Lamiaceae* family, carvacrol and thymol, have the most well-researched antibacterial action. 1,8-cineole, α-pinene, citral, perillaldehyde, eugenol, terpinen-4-ol, and geraniol are some of the other constituents with antibacterial activity (Singh et al. 2009). The anti-bacterial mechanism of action of volatile compounds is shown in Fig. 18.2.

Several volatile oils are currently being investigated as a potential treatment for viral infections. Clove and oregano volatile oils have potent antiviral properties against a variety of non-enveloped RNA and DNA viruses, including poliovirus, coxsackievirus B1, and adenovirus type 3 (Allahverdiyev et al. 2004) . Antiviral activity of some sesquiterpenes, triterpenes, and phenylpropanes has been confirmed against various herpesviruses and rhinoviruses (Hayashi et al. 1996). Volatile oils are thought to mask viral components or influence the viral envelope that is required for adsorption or entrance into host cells, according to most studies (Niedermeyer et al. 2005). They inhibit the virus replication by hindering cellular DNA polymerase and alter the phenylpropanoid pathways. Monoterpenes, in particular, increase the fluidity and permeability of the cytoplasmic membrane and disrupt the order of membrane-embedded proteins. Virion envelopes are found to be more sensitive to volatile oils than host-cell membranes (Benencia and Courrèges 1999). Because volatile oils are lipophilic, their antiviral activity is thought to disrupt or interfere with viral membrane proteins involved in host cell attachment. The schematic representation of the anti-viral mechanism of volatile compounds is shown in Fig. 18.3 (Schuhmacher et al. 2003).

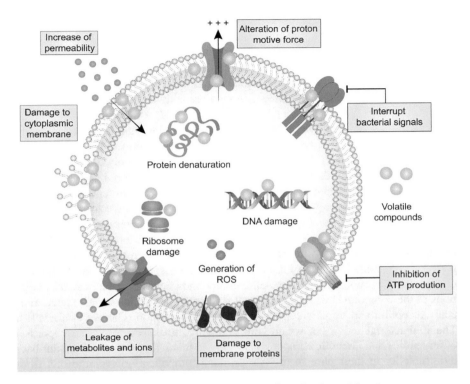

Fig. 18.2 The mechanism of action of volatile compounds against bacterial pathogens

Volatile oils have also been shown to have marked antifungal properties. Different species of fungus, including dermatophytes fungi, moulds, phytopathogenic fungi, and yeasts, have been reported to exhibit anti-fungal properties. The antifungal activity of volatile oils is governed by the existence of many active ingredients such as monoterpenes, sesquiterpenes, phenols, aldehyde, and ketones, all of which interact to produce synergistic, additive, and complementary effects (Soković et al. 2010). The majority of hypotheses about volatile compounds' antifungal effect have been postulated because of their hydrophobic character, which affects ergosterol synthesis in fungi's plasma membrane. Ergosterol is a sterol found only in the fungal plasma membrane, where it is responsible for maintaining membrane fluidity, viability, and integrity, as well as assisting in the biogenesis of certain membrane-bound enzymes (Hyldgaard et al. 2012).

The direct disruption of the plasma membrane is another important mechanism of anti-fungal action. When volatile compounds destabilize the plasma membrane, critical cellular ions like K^+, Ca^{2+}, and Mg^{2+} leak out. Volatile compounds have a significant impact on plasma membrane fluidity and permeability, causing damage to the structures of the membrane proteins. Furthermore, the cellular organelles such as the Golgi body, mitochondria, ribosome, and the endoplasmic reticulum are

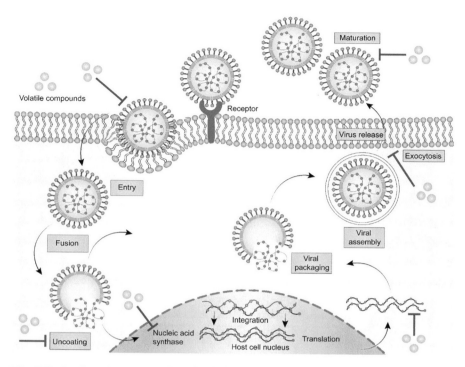

Fig. 18.3 A schematic representation of anti-viral mechanism of volatile compounds

also able to interact with the volatile compounds, resulting in reduced membrane potential (Ma et al. 2011a). This leads to proton pump disintegration, and eventually inhibition of the ATP generating enzyme, H^+-ATPase, which helps to develop electrochemical gradients and maintain cell pH across the membrane. The normal growth and reproduction of fungal cells is also hampered by the volatile compounds due to damage to nuclear contents (Diniz et al. 2021). The mechanism of action of volatile compounds against fungi is shown in Fig. 18.4.

Nowadays, many researchers have carried out molecular docking of essential oil components to find out the possible mechanism of action for their observed antimicrobial activities (Sun et al. 2009). Depending on type of antimicrobial analysis, one can choose rightly protein database id (pdb id) for molecular docking analysis. The selection of appropriate pdb id is a crucial step while carrying out molecular docking and is based on the resolution of crystal structure of protein or enzyme. One should select the pdb id of the target with the lesser resolution based on previous literature analysis. Recently, Melaku et al., 2021 carried out a molecular docking analysis of essential oil components of plant *Ocimum cufodontii* ((Lanza) A.J. Paton) (Aliye et al. 2021). Their results suggested that essential oil components of this plant have strong interactions with bacterial DNA gyrase. The docking analysis was carried out with the help of AutoDock Vina (Chen et al. 2017). Further, elaboration of the use of molecular docking analysis has been summarized in Table 18.1.

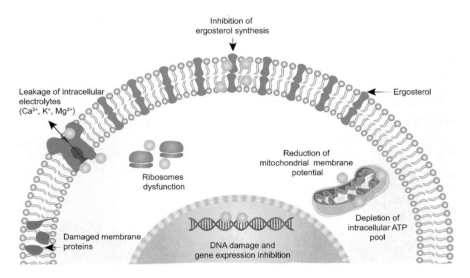

Fig. 18.4 A schematic representation of anti-fungal mechanism of volatile compounds

Table 18.1 Compounds present in essential oils used in molecular modeling

Plant name	Component used	Type of microorganism	Molecular modeling technique used	Ref.
Mentha species (Lamiaceae)	Carvone (55.71%), limonene (18.83%), trans-carveol (3.54%), cis-carveol (2.72%), beta-bourbonene (1.94%), and caryophyllene oxide (1.59%)	*Candida albicans* and *Candida parapsilosis; Salmonella enterica serotype Typhimurium (ATCC 14028), Escherichia coli (ATCC 25922), Pseudomonas aeruginosa (ATCC 27853), Shigella flexneri serotype 2b (ATCC 12022), Staphylococcus aureus (ATCC 25923)*	Molecular docking	Jianu et al. (2021)
Siparuna guianensis	Trans-β-Elemenone (11.78%) and Atractylone (18.65%), followed by δ-Elemene (5.38%), β-Elemene (3.13%), β- Yerangene (4.14%), γ-Elemene (7.04%), Germacrene D (7.61%), Curzerene (7.1%), and Germacrone (5.26%)	*Streptococcus mutans* (ATCC 3440), *Enterococcus faecalis* (ATCC 4083), *Escherichia coli* (ATCC 25922), and *Candida albicans* (ATCC 10231)	Molecular docking (Molegro Virtual Docker 6); Molecular Dynamics (MD) Simulation; and Free Energy Calculations	de Oliveira et al. (2020)

Table 18.1 (continued)

Plant name	Component used	Type of microorganism	Molecular modeling technique used	Ref.
Eryngium campestre	Essential Oils	*Staphylococcus aureus* (ATCC 6538), *S. epidermidis* (ATCC 12228), *Streptococcus pyogenes* (ATCC 19615), *Enterococcus faecalis* (ATCC 19433); *Escherichia coli* (ATCC 8739), *Pseudomonas aeruginosa* (ATCC 9027), *Proteus mirabilis* (ATCC 12453), *Klebsiella pneumoniae* (ATCC 10031)	Molecular docking (Molegro Virtual Docker 6)	Matejić et al. (2018)

18.3.3   Molecular Modeling of Volatile Compounds with Anticancer Activity

Cancer has recently emerged as one of the most pressing public health issues, as well as the second leading cause of death after heart disease (da Silva Júnior et al. 2021). Cancer is defined by uncontrolled cell proliferation that results in tumor formation. It develops as a result of somatic mutations in upstream cell signalling pathways or genetic abnormalities in any gene that encodes cell cycle proteins. Many standard therapeutic approaches have been unsuccessful against many malignant cancers due to cancer cell metastasis, recurrence, heterogeneity, and resistance to chemotherapy and radiotherapy (Siegel et al. 2016; de Oliveira et al. 2021). Another explanation for therapy failure has been linked to cancer cells' ability to evade immune responses. Natural products have recently become more popular as a therapy option for various types of cancers. The majority of volatile oils were first discovered and utilized to treat inflammatory and oxidative disorders. These volatile compounds demonstrate anticancer properties owing to the relationship between the production of ROS (reactive oxygen species) and the onset of inflammation and oxidation, both of which are known to cause cancer in humans (Sun 2015; Cascaes et al. 2021b). It is difficult to pinpoint a single mode of action for volatile compounds because of their highly varied compositions. A chemical may, in fact, affect one form of the tumor but not on others. Murata et al., for example, discovered that 1,8-cineole/eucalyptol causes apoptosis in human colon cancer cells (Jackson and Loeb 2001). This chemical, on the other hand, does not influence the survival of prostate cancer and glioblastoma cells. Furthermore, depending on the concentration of active chemicals, multiple processes, such as an effect on the cell cycle, cell proliferation, and/or death, may be observed (Murata et al. 2013; Silva et al. 2021).

Apoptosis is one of volatile oil's cancer-prevention methods which can be triggered by effects on genetic material, multiple signalling pathways, and other cellular events such as intracellular protein alterations by volatile compounds. In cancer cells, the cleavage of poly (ADP-ribose) polymerase-1 (PARP) by volatile oil components is an indication of both alteration of the DNA repair process and apoptosis (Cardile et al. 2009). The aberrant cells also undergo apoptosis as a result of elevated ROS levels. Cell death as a result of volatile oils treatment in cancer cells is characterized by reduced levels of cellular antioxidants like glutathione as well as increased production of ROS in the presence of the volatile oils (Santana de Oliveira et al. 2021). Increased ROS production damages DNA, which often leads the cancer cells towards cell death. This activity is particularly detrimental to cancer cells, whilst it does not affect normal cells (Itani et al. 2008). One of the unique aspects of volatile compounds is that, while they are cytotoxic to cancer cells, they promote normal cell proliferation. Downregulation of repair genes (DNA polymerases α, δ, and ε) volatile compounds may prove to be a viable approach for preventing DNA damage. The protein kinase B, often known as Akt, which regulates p53, is another target for volatile oils (Kelley et al. 2001). It has been demonstrated that upregulation of p21, which occurs from the deactivation of mdm2 as a result of the dephosphorylation of the Akt protein, causes the cell cycle to be interrupted in lung carcinoma cells. The G1-S phase transition was suppressed by increasing the binding of p21 to cyclins (Legault et al. 2003). A transcription factor (TF) called Nuclear factor, often known as NF- κB, is triggered in cancer cells. As a result, it is a promising target for developing anticancer therapeutics. Another TF called AP-1 (Activator protein-1) is involved in a variety of cell activities including differentiation, proliferation, transformation, and apoptosis. MAPK proteins, which are likewise impacted by volatile oils therapy in cancer cells, govern its activity. Furthermore, various MAPKs, such as p38 kinase, ERK, and JNK are the key signalling molecules in the MAPK pathway that are implicated in cancer cell apoptosis (Jaafari et al. 2007).

Volatile compounds are highly potent anticancer agents because they target several cell cycles phases in cancer cells. Volatile compounds such as thymol, carvacrol, and geraniol have shown to inhibit different phases of cell cycle (Frank et al. 2009). Monoterpenes exert their effects through modulating the expression of cell cycle regulators. Volatile oils have also shown to possess antimetastatic and antiangiogenic properties. They have shown to suppress tumor growth and metastasis (Mitoshi et al. 2012). The major sign of antiangiogenic behavior demonstrated by the volatile compounds is the suppression of vascular endothelial growth factor (VEGF), which is vital in the process of angiogenesis. In cell line models, certain volatile compounds function as inducers of several detoxifying enzymes (catalase, CAT; superoxide dismutase, SOD; glutathione reductase, GR; and glutathione peroxidase, GPx) preventing induced damage and even cancer (Suhail et al. 2011). A marked increase in these antioxidant enzymes after the treatment with volatile oils has been demonstrated as a chemo preventive activity (Seal et al. 2012). The cancer cell cycle can be seen in Fig. 18.5.

Natural essential oils are beneficial to human health. They are important to prevent as well as to treat varieties of cancers. A large number of essential oil

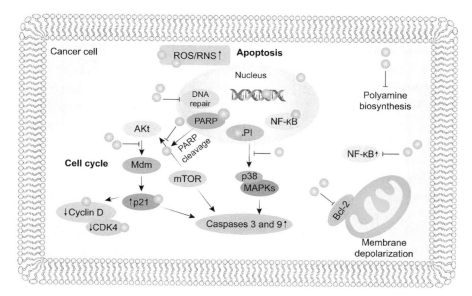

Fig. 18.5 Cancer cell cycle

components from varieties of aromatic herbs and dietary plants have been reported (Kim et al. 2000; Manjamalai and Grace 2012). These include oxygenated monoterpenes, oxygenated sesquiterpenes, phenolics, monoterpenes, sesquiterpenes, etc. (Chidambara Murthy et al. 2012). It is also known that various mechanisms such as antimutagenic, antiproliferative, enzyme induction, detoxification, modulation of drug resistance, antioxidant, etc. would be responsible for the chemoprotection properties of volatile oils (Cha et al. 2009). There are a large number of literatures reports available depicting the anticancer activity of volatile oils or essential oil components against various cancer types using molecular modeling techniques (Jaafari et al. 2012). Below are few examples showing implications of molecular modeling to predict the anticancer mechanism of volatile or essential oils from plants, Table 18.2.

18.3.4 Molecular Modeling of Volatile Compounds Against Neglected Diseases

A disease of poverty (DoP) is defined by the WHO (World Health Organization) Special Programme for Research and Training in Tropical Diseases (WHO-TDR) as a disease that mostly affects the poor in developing nations and is split into two classe. The "big three" DoPs are included in the first class: malaria, HIV/AIDS, and tuberculosis (Cascaes et al. 2021a). The community has paid close attention to these diseases and has invested much in their eradication. Around 70% of pharmaceutical

Table 18.2 Compounds present in essential oils used in potential cancer treatment and mechanism of action

Plant name	Component used	Type of cancer cell line	Molecular modeling technique used	Ref.
Ocimum viride Willd. (family: Lamiaceae)	Thymol (~50%) and γ-terpinene (~18%)	DU-145 (prostate), HEP-2 (liver), IMR-32 (neuroblastoma), HT-29 (colon), 502,713 (colon) and SW-620 (colon)	Molecular docking	Bhagat et al. (2020)
Ocimum basilicum (sweet basil) (family: Lamiaceae)	Essential Oil components	HeLa and FemX	Molecular docking	Zarlaha et al. (2014)
Mentha longifolia, M. spicata, and Origanum. majorana	Essential Oil components (Carvone (35.14%), limonene (27.11%), germacrene D (4.73%), β-caryophyllene (3.02%), γ-muurolene (2.75%), and α-bourbonene (2.27%))	Antioxidant and Anticancer	Molecular docking	Farouk et al. (2021)

development is devoted to these disorders. The other is a group of tropical diseases that are often overlooked, called Neglected Tropical Diseases (NTD) (Lenk et al. 2018). There are 17 NTDs, and they affect groups that have minimal visibility and political power. They create discrimination and stigma, as well as having a significant impact on morbidity and mortality; these diseases are mostly ignored by researchers, yet they can be prevented, controlled, and, in many cases, eliminated with the right solutions (Chen et al. 2017).

Leprosy, commonly known as Hansen's disease, is one of the neglected diseases which is caused by *Mycobacterium leprae*, an intracellular parasitic mycobacterium that causes skin lesions and nerve damage (Fotakis et al. 2020). Various plant-derived antileprotic agents have been found to be extremely effective in the management of leprosy. *Centella asiatica,* commonly known as Gotu kola or kodavan is a well-known and reputed herbal medicinal plant that constitutes saponin-containing triterpene acids along with sugar esters such as madecassic acid, asiatic acid, and asiaticosides (asiaticoside A, asiaticoside B, and asiaticoside) (Sharma et al. 2020). Asiaticosides have shown to accelerate wound healing and alleviate the symptoms of leprosy. Other volatile oils exhibiting antileprotic activity are Chaulmoogra oil (chaulmoogric acid and hydnocarpic acid), *Abutilon indicum* (β-sitosterol and α-amyrin), *Azadirachta indica* (azadirachtin), *Hemidesmus indicus* (hemidesmins and hemidesmosides A-C), Butea monosperma (Butin), etc. (Balasubramani et al. 2018).

Malaria kills one to three million people globally each year, the most portion involving pregnant women and children, but it remains a low priority for public health. Resistance to chloroquine, the first-line antimalarial treatment, has reached 90% in many parts of Africa, and resistance to sulfadoxine pyrimethamine is also on the rise (Vatandoost et al. 2018). Below are few examples showing the usefulness of molecular docking to predict the mechanism of volatile or essential oils from plants against two neglected diseases; malaria and dengue, the information is summarized in the Table 18.3.

Trypanosomiases are parasitic protozoan trypanosome illnesses caused by Trypanosoma genus parasites. The Chagas disease, Human African trypanosomiasis, and leishmaniases are all classified as neglected tropical illnesses by the WHO. There are roughly 20 Trypanosoma species, but only two species, *Trypanosoma brucei* (*T. brucei*) and *Trypanosoma cruzi* (*T. cruzi*) are the species that mainly infect humans. *T. cruzi* is the parasite that causes American trypanosomiasis, generally known as Chagas disease, which is found all over America. Triatominae insects, also known as "kissing bugs," spread it (de Morais et al. 2020). The parasite multiplies in the bloodstream and can spread to other organs such as the liver, spleen, and heart, where it can cause serious damage. African trypanosomiasis, sometimes known as sleeping sickness, is caused by *T. brucei*, which is most typically seen in equatorial Africa. If left untreated, both forms of trypanosomes infect the brain, causing mental degeneration, coma, and death. Several volatile oils from various species have found to be biologically active against trypanosomiasis (Bottieau and Clerinx 2019). Some volatile oils activity may be linked to the lipophilic properties of their constituents. Lipophilic substances can pass the cell membrane and interact with several proteins, inactivating enzymes and influencing cellular activity once within the cells (Yang and Hinner 2015). Depolarization of the mitochondrial membrane is linked to alterations in calcium channels and the production of ROS, both of which can lead to cell death via apoptosis and necrosis. Cell death through necrosis is characterized by a discontinuous plasma membrane, which indicates that the parasite has lost its integrity (Yoon et al. 2000). There are also changes to the mitochondria, ROS production, ATP depletion, and cytoplasm vacuolization in this kind of cell death. The essential oils of *Melaleuca alternifolia*, *Xylopia frutescens*, *Xylopia laevigata*, *Cymbopogon citratus*, exert this type of

Table 18.3 Molecular docking in neglected diseases

Plant name	Components of Oil Detected	neglected disease	Molecular modeling technique used	Ref.
Artemisia vulgaris	α-humulene (0.72%), βcaryophyllene (0.81%)	Dengue Fever	Molecular docking	Balasubramani et al. (2018)
Neem (Azadirachta indica)	Bitter principles of neem oil	Malaria	Molecular docking	Ghosh et al. (2021)
Eucalyptus globulus and Syzygium aromaticum	1,8-Cineol (78.20%), 2-methoxy-3-(2-propenyl) (77.04%)	Malaria	Molecular docking	Sheikh et al. (2021)

action (Giorgio et al. 2018). Loss of mitochondrial membrane potential, cytoplasmic blebbing, nuclear chromatin condensation, cell volume reduction, and DNA fragmentation are among the changes that occur during apoptosis. Such characteristics were also observed from the volatile oils of *Cinnamomum verum*, *Lippia dulcis*, *Achyrocline satureioides* (Menna-Barreto et al. 2005).

18.4   Conclusion and Future Perspectives

This chapter emphasizes the relevance of volatile oils investigations, particularly those involving pharmacology and bioinformatics/computational tools, which are now complementing and facilitating the identification of new compounds by steering and orienting studies toward specific molecular targets. The diversity of volatile compounds that make up volatile oils are becoming increasingly well characterized. Similarly, the range of biological activity of volatile oils and their constituents is beginning to be known and comprehended. Computational methods contribute to the selection of chemical structures with the highest probability of biological activity and the rationalization of natural volatile compounds. Moreover, these methods aid in the identification of chemical and structural descriptors thus providing insight into the active molecules' modes of action, and all of this information can be used to build novel structures that can be synthesized as small molecules. The discovery of new leads may thus provide an interesting platform for this research avenue in the future. Nonetheless, there is a broad scope for utilizing volatile oils not only as antimicrobial and anticancer agents but also in the treatment of neglected diseases in an array of settings, providing those critical issues such as effective delivery systems and potential toxicity the environment is addressed. Furthermore, pre-clinical studies are needed to ensure the security of the use of these compounds in humans. Likewise, administration strategies should be studied to enhance the effect of such compounds.

References

Abdolmaleki A, Ghasemi J, Ghasemi F (2017) Computer aided drug design for multi-target drug design: SAR/QSAR, molecular docking and pharmacophore methods. Curr Drug Targets 18:556–575. https://doi.org/10.2174/1389450117666160101120822

Achary PGR (2020) Applications of Quantitative Structure-Activity Relationships (QSAR) based virtual screening in drug design: a review. Mini Rev Med Chem 20:1375–1388. https://doi.org/10.2174/1389557520666200429102334

Aliye M, Dekebo A, Tesso H et al (2021) Molecular docking analysis and evaluation of the antibacterial and antioxidant activities of the constituents of Ocimum cufodontii. Sci Rep 11:10101. https://doi.org/10.1038/s41598-021-89557-x

Allahverdiyev A, Duran N, Ozguven M, Koltas S (2004) Antiviral activity of the volatile oils of Melissa officinalis L. against Herpes simplex virus type-2. Phytomedicine 11:657–661. https://doi.org/10.1016/j.phymed.2003.07.014

Almeida VM, Dias ÊR, Souza BC et al (2021) Methoxylated flavonols from Vellozia dasypus Seub ethyl acetate active myeloperoxidase extract: in vitro and in silico assays. J Biomol Struct Dyn:1–10. https://doi.org/10.1080/07391102.2021.1900916

Alqahtani S (2017) In silico ADME-Tox modeling: progress and prospects. Expert Opin Drug Metab Toxicol 13:1147–1158. https://doi.org/10.1080/17425255.2017.1389897

Alves FS, de Arimatéia Rodrigues Do Rego J, Da Costa ML et al (2020) Spectroscopic methods and in silico analyses using density functional theory to characterize and identify piperine alkaloid crystals isolated from pepper (Piper Nigrum L.). J Biomol Struct Dyn 38:2792–2799. https://doi.org/10.1080/07391102.2019.1639547

Antunes SS, Won-Held Rabelo V, Romeiro NC (2021) Natural products from Brazilian biodiversity identified as potential inhibitors of PknA and PknB of M. tuberculosis using molecular modeling tools. Comput Biol Med 136(104694). https://doi.org/10.1016/J.COMPBIOMED.2021.104694

Araújo PHF, Ramos RS, da Cruz JN et al (2020) Identification of potential COX-2 inhibitors for the treatment of inflammatory diseases using molecular modeling approaches. Molecules 25:4183. https://doi.org/10.3390/molecules25184183

Aremu AO, Van Staden J (2013) The genus Tulbaghia (Alliaceae)—a review of its ethnobotany, pharmacology, phytochemistry and conservation needs. J Ethnopharmacol 149:387–400. https://doi.org/10.1016/J.JEP.2013.06.046

Arsenijevic D, Stojanovic B, Milovanovic J et al (2021) Hepatoprotective effect of mixture of dipropyl polysulfides in concanavalin A-induced hepatitis. Nutrients 13:1022. https://doi.org/10.3390/NU13031022

Balasubramani S, Sabapathi G, Moola AK et al (2018) Evaluation of the leaf essential oil from Artemisia vulgaris and its larvicidal and repellent activity against dengue fever vector Aedes aegypti—an experimental and molecular docking investigation. ACS Omega 3:15657–15665. https://doi.org/10.1021/acsomega.8b01597

Barbosa SM, do Couto Abreu N, de Oliveira MS et al (2021) Effects of light intensity on the anatomical structure, secretory structures, histochemistry and essential oil composition of Aeollanthus suaveolens Mart. ex Spreng. (Lamiaceae). Biochem Syst Ecol 95(104224). https://doi.org/10.1016/j.bse.2021.104224

Benencia F, Courrèges MC (1999) Antiviral activity of sandalwood oil against Herpes simplex viruses-1 and -2. Phytomedicine 6:119–123. https://doi.org/10.1016/S0944-7113(99)80046-4

Bezerra FWF, do Nascimento Bezerra P, VMB C et al (2020a) Supercritical green solvent for Amazonian natural resources. In: Nanotechnology in the Life Sciences. Springer Science and Business Media B.V., pp 15–31

Bezerra FWFWF, De Oliveira MSS, Bezerra PNN et al (2020b) Extraction of bioactive compounds. In: Green sustainable process for chemical and environmental engineering and science. Elsevier, pp 149–167

Bhagat M, Sangral M, Kumar A et al (2020) Chemical, biological and in silico assessment of Ocimum viride essential oil. Heliyon 6:e04209. https://doi.org/10.1016/j.heliyon.2020.e04209

Bottieau E, Clerinx J (2019) Human African trypanosomiasis: progress and stagnation. Infect Dis Clin 33:61–77. https://doi.org/10.1016/j.idc.2018.10.003

Bouyahya A, Dakka N, Et-Touys A et al (2017) Medicinal plant products targeting quorum sensing for combating bacterial infections. Asian Pac J Trop Med 10:729–743. https://doi.org/10.1016/j.apjtm.2017.07.021

Bueno J (2020) ADMETox: bringing nanotechnology closer to Lipinski's rule of five. Nanotechnol Life Sci 61–74. https://doi.org/10.1007/978-3-030-43855-5_5

Burt S (2004) Essential oils: their antibacterial properties and potential applications in foods - a review. Int J Food Microbiol 94:223–253. https://doi.org/10.1016/j.ijfoodmicro.2004.03.022

Calo JR, Crandall PG, O'Bryan CA, Ricke SC (2015) Essential oils as antimicrobials in food systems – a review. Food Control 54:111–119. https://doi.org/10.1016/j.foodcont.2014.12.040

Camele I, Elshafie HS, Caputo L, De Feo V (2019) Anti-quorum sensing and antimicrobial effect of mediterranean plant essential oils against phytopathogenic bacteria. Front Microbiol 10:2619

Cardile V, Russo A, Formisano C et al (2009) Essential oils of Salvia bracteata and Salvia rubifolia from Lebanon: chemical composition, antimicrobial activity and inhibitory effect on human melanoma cells. J Ethnopharmacol 126:265–272. https://doi.org/10.1016/j.jep.2009.08.034

de Carvalho RN, de Oliveira MS, Silva SG et al (2019) Supercritical CO2 application in essential oil extraction. In: Inamuddin RM, Asiri AM (eds) Materials research foundations, 2nd edn, Millersville PA, pp 1–28

Cascaes MM, Dos O, Carneiro S et al (2021a) Essential oils from Annonaceae species from Brazil: a systematic review of their phytochemistry, and biological activities. Int J Mol Sci 22:12140. https://doi.org/10.3390/IJMS222212140

Cascaes MM, Silva SG, Cruz JN et al (2021b) First report on the Annona exsucca DC. Essential oil and in silico identification of potential biological targets of its major compounds. Nat Prod Res. https://doi.org/10.1080/14786419.2021.1893724

Castro ALG, Cruz JN, Sodré DF et al (2021) Evaluation of the genotoxicity and mutagenicity of isoeleutherin and eleutherin isolated from Eleutherine plicata herb. Using bioassays and in silico approaches. Arab J Chem 14(103084). https://doi.org/10.1016/j.arabjc.2021.103084

Cha J-D, Moon S-E, Kim H-Y et al (2009) Essential oil of Artemisia Capillaris induces apoptosis in KB cells via mitochondrial stress and caspase activation mediated by MAPK-stimulated signaling pathway. J Food Sci 74:T75–T81. https://doi.org/10.1111/j.1750-3841.2009.01355.x

Chen Y, De Bruyn KC, Kirchmair J (2017) Data resources for the computer-guided discovery of bioactive natural products. J Chem Inf Model 57:2099–2111. https://doi.org/10.1021/ACS.JCIM.7B00341

Chidambara Murthy KN, Jayaprakasha GK, Patil BS (2012) D-limonene rich volatile oil from blood oranges inhibits angiogenesis, metastasis and cell death in human colon cancer cells. Life Sci 91:429–439. https://doi.org/10.1016/j.lfs.2012.08.016

Costa EB, Silva RC, Espejo-Román JM et al (2020) Chemometric methods in antimalarial drug design from 1,2,4,5-tetraoxanes analogues. SAR QSAR Environ Res 31:677–695. https://doi.org/10.1080/1062936X.2020.1803961

Costa RA, Cruz JN, Nascimento FCA et al (2019) Studies of NMR, molecular docking, and molecular dynamics simulation of new promising inhibitors of cruzaine from the parasite Trypanosoma cruzi. Med Chem Res 28:246–259. https://doi.org/10.1007/s00044-018-2280-z

Cournia Z, Allen B, Sherman W (2017) Relative binding free energy calculations in drug discovery: recent advances and practical considerations. J Chem Inf Model 57:2911–2937. https://doi.org/10.1021/ACS.JCIM.7B00564

da Silva Júnior OS, de JP Franco C, de Moraes AAB et al (2021) In silico analyses of toxicity of the major constituents of essential oils from two Ipomoea L. species. Toxicon 195:111–118. https://doi.org/10.1016/j.toxicon.2021.02.015

de Oliveira MS, Cruz JN, Ferreira OO et al (2021) Chemical composition of volatile compounds in apis mellifera propolis from the northeast region of Pará State, Brazil. Molecules 26:3462. https://doi.org/10.3390/molecules26113462

De Oliveira MS, Da Cruz JN, Mitre GP et al (2019) Antimicrobial, cytotoxic activity of the Syzygium aromaticum essential oil, molecular docking and dynamics molecular studies of its major chemical constituent. J Comput Theor Nanosci 16:355–364. https://doi.org/10.1166/jctn.2019.8108

de Paulo Farias D, Neri-Numa IA, de Araújo FF, Pastore GM (2020) A critical review of some fruit trees from the Myrtaceae family as promising sources for food applications with functional claims. Food Chem 306:125630. https://doi.org/10.1016/J.FOODCHEM.2019.125630

Diao WR, Hu QP, Zhang H, Xu JG (2014) Chemical composition, antibacterial activity and mechanism of action of essential oil from seeds of fennel (Foeniculum vulgare Mill.). Food Control 35:109–116. https://doi.org/10.1016/j.foodcont.2013.06.056

Diniz LRL, Perez-Castillo Y, Elshabrawy HA et al (2021) Bioactive terpenes and their derivatives as potential SARS-CoV-2 proteases inhibitors from molecular modeling studies. Biomol 11:74. https://doi.org/10.3390/BIOM11010074

Do Nascimento LD, de Moraes AAB, da Costa KS et al (2020) Bioactive natural compounds and antioxidant activity of essential oils from spice plants: new findings and potential applications. Biomol Ther 10:1–37. https://doi.org/10.3390/biom10070988

dos Santos KLB, Cruz JN, Silva LB et al (2020) Identification of novel chemical entities for adenosine receptor type 2a using molecular modeling approaches. Molecules 25:1245. https://doi.org/10.3390/molecules25051245

Durán-Iturbide NA, Díaz-Eufracio BI, Medina-Franco JL (2020) In silico ADME/Tox profiling of natural products: a focus on BIOFACQUIM. ACS Omega 5:16076–16084. https://doi.org/10.1021/ACSOMEGA.0C01581/SUPPL_FILE/AO0C01581_SI_001.PDF

Carson CF, Mee BJ, Riley TV (2002) Mechanism of action of Melaleuca alternifolia (tea tree) oil on Staphylococcus aureus determined by time-kill, lysis, leakage, and salt tolerance assays and electron microscopy. Antimicrob Agents Chemother 46:1914–1920. https://doi.org/10.1128/AAC.46.6.1914-1920.2002

Farouk A, Mohsen M, Ali H et al (2021) Antioxidant activity and molecular docking study of volatile constituents from different aromatic Lamiaceous plants cultivated in Madinah Monawara. Saudi Arabia Molecules 26. https://doi.org/10.3390/molecules26144145

Farouk F, Shamma R (2019) Chemical structure modifications and nano-technology applications for improving ADME-Tox properties, a review. Arch Pharm (Weinheim) 352:1800213. https://doi.org/10.1002/ARDP.201800213

Ferreira OO, Neves da Cruz J, de Jesus Pereira Franco C et al (2020) First report on yield and chemical composition of essential oil extracted from myrcia eximia DC (Myrtaceae) from the Brazilian Amazon. Molecules 25:783. https://doi.org/10.3390/molecules25040783

Fotakis AK, Vågene ÅJ, Denham SD et al (2020) Multi-omic detection of mycobacterium leprae in archaeological human dental calculus. Philos Trans R Soc B 375:20190584. https://doi.org/10.1098/RSTB.2019.0584

Fowler MW (2006) Plants, medicines and man. J Sci Food Agric 86:1797–1804. https://doi.org/10.1002/jsfa.2598

Frank MB, Yang Q, Osban J et al (2009) Frankincense oil derived from Boswellia carteri induces tumor cell specific cytotoxicity. BMC Complement Altern Med 9:6. https://doi.org/10.1186/1472-6882-9-6

Fukuchi J, Kitazawa A, Hirabayashi K, Honma M (2019) A practice of expert review by read-across using QSAR Toolbox. Mutagenesis 34:49–54. https://doi.org/10.1093/MUTAGE/GEY046

Ghosh S, Mali SN, Bhowmick DN, Pratap AP (2021) Neem oil as natural pesticide: pseudo ternary diagram and computational study. J Indian Chem Soc 98:100088. https://doi.org/10.1016/j.jics.2021.100088

Giorgio V, Guo L, Bassot C et al (2018) Calcium and regulation of the mitochondrial permeability transition. Cell Calcium 70:56–63. https://doi.org/10.1016/j.ceca.2017.05.004

Gohlke H, Case DA (2004) Converging free energy estimates: MM-PB(GB)SA studies on the protein-protein complex Ras-Raf. J Comput Chem 25:238–250. https://doi.org/10.1002/jcc.10379

Grisoni F, Consonni V, Todeschini R (2018) Impact of molecular descriptors on computational models. Methods Mol Biol 1825:171–209. https://doi.org/10.1007/978-1-4939-8639-2_5

Gupta A, Müller AT, Huisman BJH et al (2018) Generative recurrent networks for De novo drug design. Mol Inform 37:1700111. https://doi.org/10.1002/MINF.201700111

Gurung AB, Ali MA, Lee J et al (2021) An updated review of computer-aided drug design and its application to COVID-19. Biomed Res Int 2021. https://doi.org/10.1155/2021/8853056

Hadrup N, Frederiksen M, Wedebye EB et al (2021) Asthma-inducing potential of 28 substances in spray cleaning products—assessed by quantitative structure activity relationship (QSAR) testing and literature review. J Appl Toxicol. https://doi.org/10.1002/JAT.4215

Halder AK, Moura AS, Cordeiro MNDS (2018) QSAR modelling: a therapeutic patent review 2010-present. Expert Opin Ther Pat 28:467. https://doi.org/10.1080/13543776.2018.1475560

Hayashi K, Hayashi T, Ujita K, Takaishi Y (1996) Characterization of antiviral activity of a ses-quiterpene, triptofordin C-2. J Antimicrob Chemother 37:759–768. https://doi.org/10.1093/jac/37.4.759

Hazzaa SM, Abdelaziz SAM, Eldaim MAA et al (2020) Neuroprotective Potential of Allium sati-vum against Monosodium Glutamate-Induced Excitotoxicity: Impact on Short-Term Memory, Gliosis, and Oxidative Stress. Nutrients 12:1028. https://doi.org/10.3390/NU12041028

Huang DZ, Baber JC, Bahmanyar SS (2021a) The challenges of generalizability in artificial intelligence for ADME/Tox endpoint and activity prediction. Expert Opin Drug Discov 16:1045–1056. https://doi.org/10.1080/17460441.2021.1901685

Huang T, Sun G, Zhao L et al (2021b) Quantitative Structure-Activity Relationship (QSAR) Studies on the Toxic Effects of Nitroaromatic Compounds (NACs): A Systematic Review. Int J Mol Sci 22:8557. https://doi.org/10.3390/IJMS22168557

Hyldgaard M, Mygind T, Meyer RL (2012) Essential oils in food preservation: mode of action, synergies, and interactions with food matrix components. Front Microbiol 3:12

Idakwo G, Luttrell J IV, Chen M et al (2019) A review of feature reduction methods for QSAR-based toxicity prediction. Adv Comput Chem Phys 30:119–139. https://doi.org/10.1007/978-3-030-16443-0_7

Itani WS, El-Banna SH, Hassan SB et al (2008) Anti colon cancer components from Lebanese sage (Salvia libanotica) essential oil: mechanistic basis. Cancer Biol Ther 7:1765–1773. https://doi.org/10.4161/cbt.7.11.6740

Jaafari A, Mouse HA, Rakib EM et al (2007) Chemical composition and antitumor activity of dif-ferent wild varieties of Moroccan thyme. Rev Bras Farmacogn 17:477–491

Jaafari A, Tilaoui M, Mouse HA et al (2012) Comparative study of the antitumor effect of natural monoterpenes: relationship to cell cycle analysis. Rev Bras Farmacogn 22:534–540

Jackson AL, Loeb LA (2001) The contribution of endogenous sources of DNA damage to the mul-tiple mutations in cancer. Mutat Res Mol Mech Mutagen 477:7–21. https://doi.org/10.1016/S0027-5107(01)00091-4

Jianu C, Stoin D, Cocan I et al (2021) In silico and in vitro evaluation of the antimicrobial and anti-oxidant potential of Mentha × smithiana R. GRAHAM essential oil from Western Romania. Foods 10:10.3390/foods10040815

Kellenberger E, Hofmann A, Quinn RJ (2011) Similar interactions of natural products with bio-synthetic enzymes and therapeutic targets could explain why nature produces such a large pro-portion of existing drugs. Nat Prod Rep 28:1483–1492. https://doi.org/10.1039/C1NP00026H

Kelley MR, Cheng L, Foster R et al (2001) Elevated and altered expression of the multifunc-tional DNA Base excision repair and redox enzyme Ape1/ref-1 in prostate cancer. Clin Cancer Res 7:824

Kim DW, Sovak MA, Zanieski G et al (2000) Activation of NF-κB/Rel occurs early during neo-plastic transformation of mammary cells. Carcinogenesis 21:871–879. https://doi.org/10.1093/carcin/21.5.871

Kuhn M, Firth-Clark S, Tosco P et al (2020) Assessment of binding affinity via alchemical free-energy calculations. J Chem Inf Model 60:3120–3130. https://doi.org/10.1021/ACS.JCIM.0C00165/SUPPL_FILE/CI0C00165_SI_002.ZIP

Leão RP, Cruz JVJN, da Costa GV et al (2020) Identification of new rofecoxib-based cyclooxygen-ase-2 inhibitors: a bioinformatics approach. Pharmaceuticals 13:1–26. https://doi.org/10.3390/ph13090209

Leelananda SP, Lindert S (2016) Computational methods in drug discovery. Beilstein J Org Chem 12:2694–2718

Legault J, Dahl W, Debiton E et al (2003) Antitumor activity of balsam fir oil: production of reactive oxygen species induced by α-humulene as possible mechanism of action. Planta Med 69:402–407. https://doi.org/10.1055/s-2003-39695

Lenk EJ, Redekop WK, Luyendijk M et al (2018) Socioeconomic benefit to individuals of achiev-ing 2020 targets for four neglected tropical diseases controlled/eliminated by innovative and intensified disease management: human African trypanosomiasis, leprosy, visceral leishmaniasis, Chagas disease. PLoS Negl Trop Dis 12:e0006250. https://doi.org/10.1371/journal. pntd.0006250

Li AP (2001) Screening for human ADME/Tox drug properties in drug discovery. Drug Discov Today 6:357–366. https://doi.org/10.1016/S1359-6446(01)01712-3

Li J, Fu A, Zhang L (2019) An overview of scoring functions used for Protein–Ligand Interactions in Molecular Docking. Interdiscip Sci Comput Life Sci 112(11):320–328. https://doi. org/10.1007/S12539-019-00327-W

Lima A d M, Siqueira AS, MLS M et al (2020) In silico improvement of the cyanobacterial lectin microvirin and mannose interaction. J Biomol Struct Dyn. https://doi.org/10.1080/0739110 2.2020.1821782

Liu X, Shi D, Zhou S et al (2018a) Molecular dynamics simulations and novel drug discovery. Expert Opin Drug Discov 13:23–37

Liu Z, Liu Y, Zeng G et al (2018b) Application of molecular docking for the degradation of organic pollutants in the environmental remediation: a review. Chemosphere 203:139–150. https://doi. org/10.1016/J.CHEMOSPHERE.2018.03.179

Ma DL, Chan DSH, Lee P et al (2011a) Molecular modeling of drug–DNA interactions: virtual screening to structure-based design. Biochimie 93:1252–1266. https://doi.org/10.1016/J. BIOCHI.2011.04.002

Ma DL, Chan DSH, Leung CH (2011b) Molecular docking for virtual screening of natural product databases. Chem Sci 2:1656–1665. https://doi.org/10.1039/C1SC00152C

Maier ME (2015) Design and synthesis of analogues of natural products. Org Biomol Chem 13:5302–5343

Maltarollo VG, Gertrudes JC, Oliveira PR, Honorio KM (2015) Applying machine learning techniques for ADME-Tox prediction: a review. Expert Opin Drug Metab Toxicol 11:259–271. https://doi.org/10.1517/17425255.2015.980814

Manjamalai A, Grace VMB (2012) Antioxidant activity of essential oils from Wedelia chinensis (Osbeck) in vitro and in vivo lung cancer bearing C57BL/6 mice. Asian Pac J Cancer Prev 13:3065–3071

Matejić JS, Stojanović-Radić ZZ, Ristić MS et al (2018) Chemical characterization, in vitro biological activity of essential oils and extracts of three Eryngium L. species and molecular docking of selected major compounds. J Food Sci Technol 55:2910–2925. https://doi.org/10.1007/ s13197-018-3209-8

Meng XY, Zhang H-X, Mezei M, Cui M (2011) Molecular docking: a powerful approach for structure-based drug discovery. Curr Comput Aided Drug Des 7(2):146–157

Menna-Barreto RFS, Henriques-Pons A, Pinto AV et al (2005) Effect of a β-lapachone-derived naphthoimidazole on Trypanosoma cruzi: identification of target organelles. J Antimicrob Chemother 56:1034–1041. https://doi.org/10.1093/jac/dki403

Michel J, Abd Rani NZ, Husain K (2020) A review on the potential use of medicinal plants from Asteraceae and Lamiaceae plant family in cardiovascular diseases. Front Pharmacol 11:852. https://doi.org/10.3389/FPHAR.2020.00852/BIBTEX

Mitoshi M, Kuriyama I, Nakayama H et al (2012) Effects of essential oils from herbal plants and citrus fruits on DNA polymerase inhibitory, cancer cell growth inhibitory, Antiallergic, and antioxidant activities. J Agric Food Chem 60:11343–11350. https://doi.org/10.1021/jf303377f

Mohammad T, Mathur Y, Hassan MI (2021) InstaDock: a single-click graphical user interface for molecular docking-based virtual high-throughput screening. Brief Bioinform 22:1–8. https:// doi.org/10.1093/BIB/BBAA279

Moradi S, Nowroozi A, Shahlaei M (2019) Shedding light on the structural properties of lipid bilayers using molecular dynamics simulation: a review study. RSC Adv 9:4644–4658. https:// doi.org/10.1039/C8RA08441F

de Morais MC, de Souza JV, da Silva, Maia, Bezerra Filho C et al (2020) Trypanocidal essential oils: a review. Molecules 25. https://doi.org/10.3390/molecules25194568

Murata S, Shiragami R, Kosugi C et al (2013) Antitumor effect of 1, 8-cineole against colon cancer. Oncol Rep 30:2647–2652

Neto R d AM, Santos CBRR, Henriques SVCC et al (2020) Novel chalcones derivatives with potential antineoplastic activity investigated by docking and molecular dynamics simulations. J Biomol Struct Dyn:1–13. https://doi.org/10.1080/07391102.2020.1839562

Neves Cruz J, Santana de Oliveira M, Gomes Silva S et al (2020) Insight into the interaction mechanism of nicotine, NNK, and NNN with cytochrome P450 2A13 based on molecular dynamics simulation. J Chem Inf Model 60:766–776. https://doi.org/10.1021/acs.jcim.9b00741

Niedermeyer THJ, Lindequist U, Mentel R et al (2005) Antiviral Terpenoid Constituents of Ganoderma pfeifferi. J Nat Prod 68:1728–1731. https://doi.org/10.1021/np0501886

de Oliveira MS, da Cruz JN, da Costa WA et al (2020) Chemical composition, antimicrobial properties of Siparuna guianensis essential oil and a molecular docking and dynamics molecular study of its major chemical constituent. Molecules 25:3852. https://doi.org/10.3390/molecules25173852

Ooms F (2012) Molecular modeling and computer aided drug design. Examples of their applications in medicinal chemistry. Curr Med Chem 7:141–158. https://doi.org/10.2174/0929867003375317

Pagadala NS, Syed K, Tuszynski J (2017) Software for molecular docking: a review. Biophys Rev 9:91–102. https://doi.org/10.1007/s12551-016-0247-1

Pinto V d S, JSC A, Silva RC et al (2019) In silico study to identify new antituberculosis molecules from natural sources by hierarchical virtual screening and molecular dynamics simulations. Pharmaceuticals 12:36. https://doi.org/10.3390/ph12010036

Pinzi L, Rastelli G (2019) Molecular Docking: Shifting Paradigms in Drug Discovery. Int J Mol Sci 20:4331. https://doi.org/10.3390/IJMS20184331

Pinzi L, Tinivella A, Rastelli G (2021) Chemoinformatics analyses of Tau Ligands reveal key molecular requirements for the identification of potential drug candidates against tauopathies. Mol 26:5039. https://doi.org/10.3390/MOLECULES26165039

Prajapat M, Sarma P, Shekhar N et al (2020) Drug targets for corona virus: a systematic review. Indian J Pharmacol 52:56. https://doi.org/10.4103/IJP.IJP_115_20

Ramos RS, Macêdo WJC, Costa JS et al (2020) Potential inhibitors of the enzyme acetylcholinesterase and juvenile hormone with insecticidal activity: study of the binding mode via docking and molecular dynamics simulations. J Biomol Struct Dyn 38:4687–4709. https://doi.org/10.1080/07391102.2019.1688192

Redžepović I, Furtula B (2021) Comparative study on structural sensitivity of eigenvalue–based molecular descriptors. J Math Chem 59:476–487. https://doi.org/10.1007/S10910-020-01202-6/FIGURES/6

Saikia S, Bordoloi M (2018) Molecular docking: challenges, advances and its use in drug discovery perspective. Curr Drug Targets 20:501–521. https://doi.org/10.2174/1389450119666181022153016

Salomon-Ferrer R, Case DA, Walker RC (2013) An overview of the Amber biomolecular simulation package. Wiley Interdiscip Rev Comput Mol Sci 3:198–210. https://doi.org/10.1002/wcms.1121

Santana de Oliveira M, Pereira da Silva VM, Cantão Freitas L et al (2021) Extraction yield, chemical composition, preliminary toxicity of Bignonia nocturna (Bignoniaceae) essential oil and in silico evaluation of the interaction. Chem Biodivers, cbdv.202000982 18. https://doi.org/10.1002/cbdv.202000982

Santos CBR, Santos KLB, Cruz JN et al (2020) Molecular modeling approaches of selective adenosine receptor type 2A agonists as potential anti-inflammatory drugs. J Biomol Struct Dyn. https://doi.org/10.1080/07391102.2020.1761878

Schuhmacher A, Reichling J, Schnitzler P (2003) Virucidal effect of peppermint oil on the enveloped viruses herpes simplex virus type 1 and type 2 in vitro. Phytomedicine 10:504–510. https://doi.org/10.1078/094471103322331467

Seal S, Chatterjee P, Bhattacharya S et al (2012) Vapor of volatile oils from Litsea cubeba seed induces apoptosis and causes cell cycle arrest in lung cancer cells. PLoS One 7:e47014. https://doi.org/10.1371/journal.pone.0047014

Sharma R, Singh P, McCoy RC et al (2020) Isolation of mycobacterium lepromatosis and development of molecular diagnostic assays to distinguish mycobacterium leprae and M. lepromatosis. Clin Infect Dis 71:e262–e269. https://doi.org/10.1093/CID/CIZ1121

Sharma S, Bhatia V (2020) Recent trends in QSAR in modelling of drug-protein and protein-protein interactions. Comb Chem High Throughput Screen 24:1031–1041. https://doi.org/10.2174/1386207323666201209093537

Sheikh Z, Amani A, Basseri HR et al (2021) Repellent efficacy of Eucalyptus globulus and Syzygium aromaticum essential oils against malaria vector, anopheles ste-phensi (Diptera: Culicidae). Iran J Public Health 50:10.18502/ijph.v50i8.6813

Siegel RL, Miller KD, Jemal A (2016) Cancer statistics, 2016. CA Cancer J Clin 66:7–30. https://doi.org/10.3322/caac.21332

Silva SG, de Oliveira MS, Cruz JN et al (2021) Supercritical CO2 extraction to obtain Lippia thymoides Mart. & Schauer (Verbenaceae) essential oil rich in thymol and evaluation of its antimicrobial activity. J Supercrit Fluids 168(105064). https://doi.org/10.1016/j.supflu.2020.105064

Singh N, Dueñas-González A, Lyko F, Medina-Franco JL (2009) Molecular modeling and molecular dynamics studies of hydralazine with human DNA methyltransferase 1. ChemMedChem 4:792–799. https://doi.org/10.1002/CMDC.200900017

Soković M, Glamočlija J, Marin PD et al (2010) Antibacterial effects of the essential oils of commonly consumed medicinal herbs using an in vitro model. Molecules 15:7532–7546. https://doi.org/10.3390/molecules15117532

Stouch TR, Kenyon JR, Johnson SR et al (2003) In silico ADME/Tox: why models fail. J Comput Mol Des 172(17):83–92. https://doi.org/10.1023/A:1025358319677

Suhail MM, Wu W, Cao A et al (2011) Boswellia sacra essential oil induces tumor cell-specific apoptosis and suppresses tumor aggressiveness in cultured human breast cancer cells. BMC Complement Altern Med 11:129. https://doi.org/10.1186/1472-6882-11-129

Sun H, Scott DO (2010) Structure-based drug metabolism predictions for drug design. Chem Biol Drug Des 75:3–17. https://doi.org/10.1111/J.1747-0285.2009.00899.X

Sun Y (2015) Translational horizons in the tumor microenvironment: harnessing breakthroughs and targeting cures. Med Res Rev 35. https://doi.org/10.1002/med.21338

Sun Y, Xun K, Wang Y, Chen X (2009) A systematic review of the anticancer properties of berberine, a natural product from Chinese herbs. Anti-Cancer Drugs 20:757–769. https://doi.org/10.1097/CAD.0B013E328330D95B

Torres PHM, Sodero ACR, Jofily P, Silva-Jr FP (2019) Key topics in molecular docking for drug design. Int J Mol Sci (20):4574. https://doi.org/10.3390/IJMS20184574

Van Der Spoel D, Lindahl E, Hess B et al (2005) GROMACS: fast, flexible, and free. J Comput Chem 26:1701–1718

Vanommeslaeghe K, Guvench O, MacKerell AD (2014) Molecular mechanics. Curr Pharm Des 20:3281–3292. https://doi.org/10.2174/13816128113199990600

Vatandoost H, Rustaie A, Talaeian Z et al (2018) Larvicidal activity of Bunium persicum essential oil and extract against malaria vector, Anopheles stephensi. J Arthropod Borne Dis 12:85

Venable RM, Krämer A, Pastor RW (2019) Molecular dynamics simulations of membrane permeability. Chem Rev 119:5954–5997. https://doi.org/10.1021/ACS.CHEMREV.8B00486

Verma J, Khedkar VM, Coutinho EC (2010) 3D-QSAR in drug design – a review. Curr Top Med Chem 10:95–115. https://doi.org/10.2174/156802610790232260

Vucicevic J, Nikolic K, Mitchell JBO (2019) Rational drug design of antineoplastic agents using 3D-QSAR, cheminformatic, and virtual screening approaches. Curr Med Chem 26:3874–3889. https://doi.org/10.2174/0929867324666170712115411

Wang B-C, Wang L-J, Jiang B et al (2017) Application of fluorine in drug design during 2010–2015 years: a mini-review. Mini Rev Med Chem 17(8):683–692

Wang E, Sun H, Wang J et al (2019) End-point binding free energy calculation with MM/PBSA and MM/GBSA: strategies and applications in drug design. Chem Rev 119:9478–9508. https://doi.org/10.1021/acs.chemrev.9b00055

Yang NJ, Hinner MJ (2015) Getting across the cell membrane: an overview for small molecules, peptides, and proteins. Site-specific protein labeling. Methods Mol Biol (Clifton, N.J.) 1266:29–53

Yoon J, Ben-Ami HC, Hong YS et al (2000) Novel mechanism of massive photoreceptor degeneration caused by mutations in the trp gene of drosophila. J Neurosci 20:649. https://doi.org/10.1523/JNEUROSCI.20-02-00649.2000

Yu W, Mackerell AD (2017) Computer-aided drug design methods. Methods Mol Biol 1520:85–106. https://doi.org/10.1007/978-1-4939-6634-9_5

Zarlaha A, Kourkoumelis N, Stanojkovic TP, Kovala-Demertzi D (2014) Cytotoxic activity of essential oil and extracts of ocimum basilicum against human carcinoma cells. Molecular docking study of isoeugenol as a potent cox and lox inhibitor. Dig J Nanomater Bios 9

Zhong F, Xing J, Li X et al (2018) Artificial intelligence in drug design. Sci China Life Sci 61(10):1191–1204. https://doi.org/10.1007/S11427-018-9342-2

Index

Printed in the United States
by Baker & Taylor Publisher Services